Conversion Factors

Physical Quantity	Symbol	Conversion Factor
Pressure	p	$6894.8 \text{ N/m}^2 = 1 \text{ lbf/in}^2$
		$101.325 \text{ kN/m}^2 = 1 \text{ atm}$
Specific entropy	s	$4.187 \text{ kJ/kg.K} = 1 \text{ Btu/lbm-R}$
Specific heat	c_p, c_v	$4.187 \text{ kJ/kg} \cdot \text{K} = 1 \text{ Btu/lbm-R}$
Specific volume	v	$0.06243 \text{ m}^3/\text{kg} = 1 \text{ ft}^3/\text{lbm}$
Specific energy	q, w, h, u, e	$2.3261 \text{ kJ/kg} = 1 \text{ Btu/lbm}$
Temperature	T	$(9/5)T(\text{K}) = T(^\circ\text{R})$
		$[T(^\circ\text{C})](9/5) + 32 = T(^\circ\text{F})$
		$[T(\text{K}) - 273.15](9/5) + 32 = T(^\circ\text{F})$
Thermal conductivity	λ	$1.731 \text{ W/m.K} = 1 \text{ Btu/hr-ft-}^\circ\text{F}$
Thermal diffusivity	α	$0.0929 \text{ m}^2/\text{s} = 1 \text{ ft}^2/\text{sec}$
		$2.581 \times 10^{-5} \text{ m}^2/\text{s} = 1 \text{ ft}^2/\text{hr}$
Velocity	v	$0.3048 \text{ m/s} = 1 \text{ ft/sec}$
		$0.44703 \text{ m/s} = 1 \text{ mph}$
Viscosity, dynamic	μ	$1.488 \text{ N} \cdot \text{s/m}^2 = 1 \text{ lbm/ft-sec}$
		$0.001 \text{ N} \cdot \text{s/m}^2 = 1 \text{ centipoise}$
Viscosity, kinematic	ν	$0.0929 \text{ m}^2/\text{s} = 1 \text{ ft}^2/\text{sec}$
		$2.581 \times 10^{-5} \text{ m}^2/\text{s} = 1 \text{ ft}^2/\text{hr}$
Volume	V	$0.02832 \text{ m}^3 = 1 \text{ ft}^3$
		$1.6387 \times 10^{-5} \text{ m}^3 = 1 \text{ in.}^3$
		$0.003785 \text{ m}^3 = 1 \text{ gal (U.S. liq.)}$

Engineering Thermodynamics with Applications

The Harper & Row Series in Mechanical Engineering

Norman H. Beachley and Howard Harrison, *Introduction to Dynamic System Analysis*

M. David Burghardt, *Engineering Thermodynamics With Applications,* 2d Edition

Ira Cochin, *Analysis and Design of Dynamic Systems*

Howard L. Harrison and John G. Bollinger, *Introduction to Automatic Controls,* 2d Edition

Burgess H. Jennings, *Environmental Engineering: Analysis and Practice*

Burgess H. Jennings, *The Thermal Environment: Conditioning and Control*

Frank Kreith, *Principles of Heat Transfer,* 3d Edition

Frank Kreith and William Z. Black, *Basic Heat Transfer*

Edward F. Obert, *Internal Combustion Engines and Air Pollution*

Reuben Olson, *Essentials of Engineering Fluid Mechanics,* 4th Edition

James Todd and Herbert Ellis, *Applied Heat Transfer*

Robert K. Vierck, *Vibration Analysis,* 2d Edition

Kenneth Wark and Cecil F. Warner, *Air Pollution: Its Origin and Control,* 2d Edition

Engineering Thermodynamics with Applications

SECOND EDITION

M. David Burghardt
U. S. Merchant Marine Academy
Kings Point, New York

HARPER & ROW, PUBLISHERS, New York
Cambridge, Philadelphia, San Francisco,
London, Mexico City, São Paulo, Sydney

1817

Sponsoring Editor: Cliff Robichaud
Project Editor: Eleanor Castellano
Production Manager: Marion A. Palen
Compositor: Science Typographers
Printer and Binder: The Murray Printing Co.
Art Studio: J & R Art Services, Inc.

Engineering Thermodynamics with Applications, Second Edition

Library of Congress Cataloging in Publication Data

Burghardt, M. David.
  Engineering thermodynamics with applications.

  (The Harper & Row series in mechanical engineering)   Bibliography: p.
  Includes index.
  1. Thermodynamics. I. Title. II. Series.
TJ265.B87 1982   621.402'1   81-13435
ISBN 0-06-041042-6   AACR2

To Linda

Contents

PREFACE   xiii

1   INTRODUCTION   1

1.1   Elementary Steam Power Plant   *1*
1.2   Combustion Engines   *2*
1.3   Direct Energy Conversion   *4*
1.4   Geothermal Power Plant   *6*
1.5   Solar Energy   *6*
1.6   Founders of Thermodynamics   *7*

2   DEFINITIONS AND UNITS   9

2.1   Macroscopic and Microscopic Analysis   *9*
2.2   Substances   *10*
2.3   Systems—Fixed Mass and Fixed Space   *10*
2.4   Properties, Intensive and Extensive   *12*
2.5   Phases of a Substance   *13*
2.6   Processes and Cycles   *13*
2.7   Units of Force and Mass   *14*
2.8   Specific Volume   *17*
2.9   Pressure   *18*
2.10   Equality of Temperature   *23*
2.11   Zeroth Law of Thermodynamics   *23*
2.12   Temperature Scales   *24*

3   CONSERVATION OF MASS AND ENERGY   28

3.1   Conservation of Mass   *28*
3.2   Energy Forms   *31*

3.3   First Corollary of the First Law   *43*
3.4   Energy as a Property   *45*
3.5   Second Corollary of the First Law   *46*
3.6   System Boundaries   *53*

4   PROPERTIES OF PURE SUBSTANCES   59

4.1   Pure Substance   *59*
4.2   Liquid–Vapor Equilibrium   *60*
4.3   Saturated Properties   *60*
4.4   Critical Properties   *61*
4.5   Solid–Liquid–Vapor Equilibrium   *62*
4.6   Quality   *64*
4.7   Three-Dimensional Surface   *64*
4.8   Tables of Thermodynamic Properties   *66*

5   IDEAL GAS AND SPECIFIC HEAT   77

5.1   Ideal-Gas Equations of State   *77*
5.2   Actual-Gas Equations of State   *80*
5.3   Boyle's Law   *84*
5.4   Charles' Law   *84*
5.5   Specific Heat   *87*
5.6   Gas Tables   *92*

**6   PROCESSES USING TABLES AND
      GAS EQUATIONS OF STATE   97**

6.1   Equilibrium and Nonequilibrium Processes   *97*
6.2   Closed Systems   *98*
6.3   Open Systems   *107*
6.4   Polytropic Process—Ideal Gas   *111*
6.5   Three-Process Cycles   *116*
6.6   Transient Flow   *118*

**7   THE SECOND LAW OF THERMODYNAMICS
      AND THE CARNOT CYCLE   128**

7.1   The Second Law of Thermodynamics   *128*
7.2   Energy Level   *128*
7.3   Second Law for a Cycle   *129*
7.4   Carnot Cycle   *131*
7.5   Carnot Engine   *131*
7.6   Mean Effective Pressure   *135*
7.7   Reversed Carnot Engine   *137*
7.8   First Corollary of the Second Law   *139*

7.9   Second Corollary of the Second Law   *140*
7.10 Thermodynamic Temperature Scale   *140*

8   ENTROPY   146

8.1   Clausius Inequality   *146*
8.2   Derivation of Entropy   *148*
8.3   Third Law of Thermodynamics   *151*
8.4   Equilibrium State   *151*
8.5   Entropy Change of a Closed System   *152*
8.6   Calculation of Entropy Change for Ideal Gases   *153*
8.7   Relative Pressure and Relative Specific Volume   *155*
8.8   Entropy of a Pure Substance   *156*
8.9   Carnot Cycle Using $T-S$ Coordinates   *158*
8.10 Heat and Work as Areas   *159*
8.11 The Second Law for Open Systems   *160*
8.12 Further Considerations   *162*

9   AVAILABLE ENERGY AND AVAILABILITY   166

9.1   Available Energy for Systems with Heat Transfer   *166*
9.2   Open Systems, Steady Flow   *172*
9.3   Further Considerations of Available Energy—Availability   *176*
9.4   Second-Law Efficiency   *179*

10   THERMODYNAMIC RELATIONSHIPS   188

10.1 Interpreting Differentials and Partial Derivatives   *188*
10.2 An Important Relationship   *191*
10.3 Application of Mathematical Methods to
     Thermodynamic Relations   *193*
10.4 Maxwell's Relations   *194*
10.5 Specific Heats, Enthalpy, and Internal Energy   *195*
10.6 Clapeyron Equation   *199*
10.7 Important Physical Coefficients   *201*
10.8 Real-Gas Behavior   *204*

11   VAPOR POWER CYCLES   209

11.1 Carnot Vapor Cycle   *209*
11.2 The Rankine Cycle   *210*
11.3 Rankine Cycle Components   *212*
11.4 Efficiencies   *217*
11.5 Regenerative Cycles   *220*
11.6 Reheat Cycles   *228*
11.7 Reheat–Regenerative Cycle   *229*

11.8   Supercritical and Binary Vapor Cycles   *232*
11.9   Steam-Turbine Reheat Factor and Condition Curve   *236*
11.10 Geothermal Energy   *238*

12   REFRIGERATION SYSTEMS   247

12.1   Reversed Carnot Cycle   *247*
12.2   Refrigerant Considerations   *248*
12.3   Vapor-Compression Cycle   *250*
12.4   Multistage Vapor-Compression Systems   *254*
12.5   Absorption Refrigeration Systems   *259*
12.6   Heat Pump   *267*
12.7   Low Temperature and Liquefaction   *269*

13   MIXTURES: GAS–GAS AND GAS–VAPOR   276

13.1   Ideal-Gas Mixtures   *276*
13.2   Gas–Vapor Mixtures   *282*
13.3   Psychrometer   *289*
13.4   Psychrometric Chart   *290*
13.5   Air-Conditioning Processes   *292*
13.6   Cooling Towers   *296*

14   REACTIVE SYSTEMS   303

14.1   Hydrocarbon Fuels   *303*
14.2   Combustion Process   *304*
14.3   Theoretical Air   *305*
14.4   Air–Fuel Ratio   *307*
14.5   Products of Combustion   *309*
14.6   Enthalpy of Formation   *312*
14.7   First-Law Analysis for Steady-State Reacting Systems   *314*
14.8   Adiabatic Flame Temperature   *318*
14.9   Enthalpy of Combustion, Heating Value   *320*
14.10 Second-Law Analysis   *322*
14.11 Chemical Equilibrium and Dissociation   *327*
14.12 Steam Generator Efficiency   *334*
14.13 Fuel Cells   *335*

15   GAS COMPRESSORS   344

15.1   Compressors Without Clearance   *344*
15.2   Reciprocating Compressors with Clearance   *347*
15.3   Volumetric Efficiency   *350*
15.4   Multistage Compression   *353*
15.5   Compressor Performance Factors   *356*
15.6   Rotative Compressors   *357*

16 INTERNAL-COMBUSTION ENGINES 366

16.1 Air-Standard Cycles *366*
16.2 Open-Cycle Analysis *382*
16.3 Actual Diesel and Otto Cycles *386*
16.4 Cycle Comparisons *389*
16.5 Engine Performance Analysis *389*
16.6 Wankel Engine *391*
16.7 Engine Efficiencies *392*
16.8 Power Measurement *394*

17 GAS TURBINES 403

17.1 Fundamental Gas-Turbine Cycle *403*
17.2 Cycle Analysis *404*
17.3 Efficiencies *407*
17.4 Open-Cycle Analysis *410*
17.5 Combustion Efficiency *413*
17.6 Regeneration *413*
17.7 Reheating, Intercooling *419*
17.8 Combined Cycle *423*
17.9 Aircraft Gas Turbines *426*

18 FLUID FLOW AND NOZZLES 438

18.1 Conservation of Mass *438*
18.2 Conservation of Momentum *439*
18.3 Acoustic Velocity *442*
18.4 Stagnation Properties *444*
18.5 Mach Number *446*
18.6 First-Law Analysis *446*
18.7 Nozzles *447*
18.8 Supersaturation *453*
18.9 Shock Waves *457*
18.10 Diffuser *458*
18.11 Flow Measurement *460*
18.12 Wind Power *461*

19 HEAT TRANSFER AND HEAT EXCHANGERS 469

19.1 Modes of Heat Transfer *469*
19.2 Laws of Heat Transfer *470*
19.3 Combined Modes of Heat Transfer *477*
19.4 Conduction Through A Composite Wall *479*
19.5 Conduction in Cylindrical Coordinates *480*
19.6 Critical Insulation Thickness *483*
19.7 Heat Exchangers *484*

REFERENCES 501

LIST OF SYMBOLS 503

APPENDIX TABLES 505

A.1   Gas Constants and Specific Heats at Low Pressures *507*
A.2   Properties of Air at Low Pressures *507*
A.3   Products—400 Percent Theoretical Air—at Low Pressures *510*
A.4   Products—200 Percent Theoretical Air—at Low Pressures *512*
A.5   Saturated Steam Temperature Table *514*
A.6   Saturated Steam Pressure Table *517*
A.7   Superheated Steam Vapor Table *520*
A.8   Compressed Liquid Table *528*
A.9   Saturated Ammonia Table *530*
A.10 Superheated Ammonia Table *532*
A.11 Saturated Freon-12 Table *535*
A.12 Superheated Freon-12 Table *537*
A.13 Properties of Selected Materials at 20°C *541*
A.14 Physical Properties of Selected Fluids *542*
B.1   Mollier (Enthalpy-Entropy) Diagram for Steam *544*
B.2   Temperature-Entropy Diagram for Steam *546*
B.3   Ammonia-Water Equilibrium Chart *547*
B.4   Psychrometric Chart *548*
C.1   Enthalpies of Formation, Gibbs Function of Formation, and Absolute Entropy at 25°C and 1 atm Pressure *549*
C.2   Enthalpy of Formation at 25°C, Ideal-Gas Enthalpy, and Absolute Entropy at 0.1 MPa Pressure *550*
C.3   Enthalpy of Combustion (Heating Value) of Various Compounds *559*
C.4   Natural Logarithm of Equilibrium Constant K *560*

ANSWERS TO SELECTED PROBLEMS 561

INDEX 565

Preface

The purpose of this text is to help undergraduate engineering students learn thermodynamics—the theoretical foundations of the science and their application to realistic situations. A second aim is to offer practicing engineers a wide array of applications that will be helpful to them in their work. The first section of the book is devoted to developing the student's ability to undertake thermodynamic analyses; the second section is devoted to the application of these skills.

Since students are able to use their new knowledge effectively only if they fully understand it, they must first master thermodynamics before they can reach the goal of undertaking systems analysis. To this end, a large number of examples are included in the text. It is my hope that when students study a theory, see its application in examples, and perform similar analyses, they will be able to understand and apply the principles of thermodynamics. The book demonstrates that the same four thermodynamic postulates are repeatedly applicable regardless of the system to which they are applied.

The text provides a solid theoretical background, because the emphasis and discussion of the fundamental principles are coordinated with their application. The juxtaposition of what is happening theoretically to what is happening physically is emphasized strongly. Because engineering students leave college as potential professional engineers, they must be provided with the necessary tools of analysis. This book provides these tools.

This edition of the text uses SI units only. The number of problems has been significantly increased, and the applications areas have been expanded to include such topics as geothermal energy, wind power, and fuel cells. The topic coverage of available energy and availability has been increased, and the concept of second-law efficiencies for processes and cycles has been introduced. In addition, the first half of the book has been restructured—reflecting current thermodynamic practice.

For course use, the text is suitable for either lecture or combined lecture-and-laboratory presentation. The material forms the basis of three courses in thermodynamics. The last two courses include a laboratory coordinated with the lecture sequence.

The book represents an amalgam of numerous ideas, influences, and data from many sources; yet, the ultimate responsibility for the material is my own. It is my hope that the text will form a useful part of an educational program in engineering and a useful reference for the practicing engineer.

M. David Burghardt

Engineering Thermodynamics with Applications

Chapter 1
Introduction

The study of thermodynamics covers many areas of engineering, from power plant analysis to the analysis of fuel cells. What follows in this chapter is an indication of some, by no means all, of the types of situations and systems that may be analyzed thermodynamically. The strength of thermodynamics lies in one's ability to analyze with the use of a few tenets—four to be exact—a wide range of systems. It may also be of interest to the student to know of some of the great men and their discoveries that advanced the science of thermodynamics.

1.1 ELEMENTARY STEAM POWER PLANT

One of the first systems to look at is the steam power plant, an energy device basic to modern life. It is necessary for the generation of electric power as well as in transportation systems, ships, for example. (Many other power systems are also used and these, too, yield to thermodynamic probing.) Figure 1.1 depicts a simple steam power plant. Just as in the oil-fired boiler in the basements of many homes, fuel is burned to generate heat. This heat is used to boil water under pressure in the steam generator (boiler). The steam leaves the generator and passes through superheater tubes, where more heat is added to the steam; it then passes through the turbine, where it increases in volume, decreases in pressure, and performs work, which is used to generate electric power or drive a ship. The steam is then condensed, liquefied, and pumped back to the steam generator.

This system seems simple enough and not too great a challenge for an engineer, but life and circumstances are not that kind or simple. Several additions have been made to the plant, and we will consider these in greater detail later. Water heaters, for instance, preheat the water

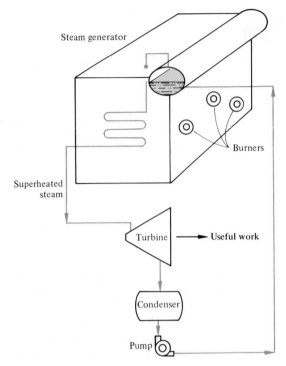

Figure 1.1 Simple steam power plant.

before it enters the steam generator; also, some of the steam is taken from the turbine, reheated, and returned to the turbine; and before the air reaches the fuel-oil burners, it is reheated, which improves the combustion process. These are some of the considerations that an engineer using thermodynamic analysis must include in analyzing a steam power plant.

1.2 COMBUSTION ENGINES

Another standard power plant that we use almost every day is the gasoline engine—the automotive engine (Figure 1.2). Innovations have been made, such as the Wankel engine (Figure 1.3), but even this follows the same principles. The engine may be viewed as a small power plant: Fuel is burned and the energy from the burning fuel is transferred to the pistons, which, through gears, turn the wheels, thus moving the automobile.

There are many problems associated with the engine: the combustion process in the piston and containing the energy of the burning flame, to mention two. The thermodynamic analysis seeks to determine ahead of time how much work we may expect from an engine and through

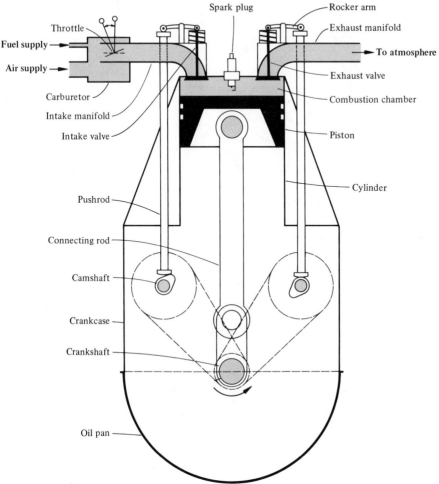

Figure 1.2   Automotive engine.

experiments how efficiently the engine is performing. This is very important if the pollution from exhausts is to be minimized.

The gas turbine (Figure 1.4) is another automotive power source,
more commonly found in jet planes. There is an upsurge in the development of gas-turbine plants, in both electric power generation and ship
propulsion. Air is compressed and energy added to it by burning fuel in a
combustion chamber; this mixture—the products of combustion, air, and
burned fuel—expands through a turbine, doing work, which drives the
electric generator or the ship. The analysis is similar to that of most
power plants, and all these analyses have a common purpose, which is to
consider how efficiently the chemical energy of the fuel is converted into
mechanical energy. The processes of converting the energy are different
but the principle of energy conversion remains the same.

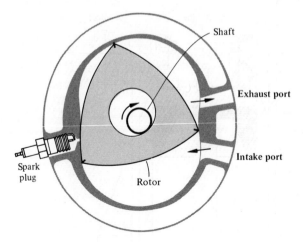

Figure 1.3   Wankel engine.

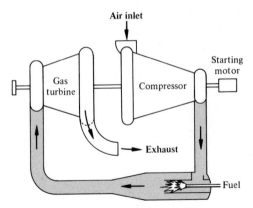

Figure 1.4   A gas-turbine unit.

1.3 DIRECT ENERGY CONVERSION

There are energy converters that do not rely on an intermediate device to produce the desired energy form, in this case work or electric power. These devices are called direct energy conversion devices and two will be mentioned. The one that may be most familiar is the fuel cell, in which chemical energy is converted directly into electric energy. In Figure 1.5, a simplified fuel cell using hydrogen and oxygen is shown. The hydrogen is oxidized at the cathode, giving up two electrons, and the oxygen is reduced at the anode by picking up the two electrons. Connected to the two poles is the load.

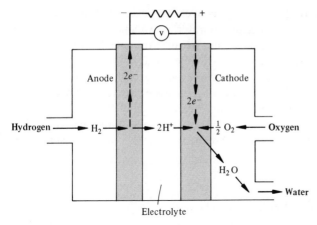

Figure 1.5   *The hydrogen–oxygen fuel cell.*

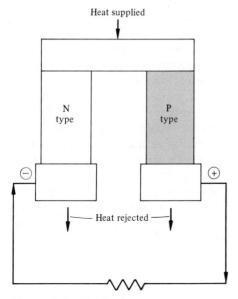

Figure 1.6   *Thermoelectric generator.*

The half-cell reactions are

$$H_2(g) \rightarrow 2H^+ + 2e^- \qquad V = 0.0 \text{ volts}$$

$$\tfrac{1}{2}O_2(g) + 2H^+ + 2e^- \rightarrow H_2O(1) \qquad V = 1.23 \text{ volts}$$

Thus the voltage of the cell is 1.23 volts, and if the load resistance is known, the current flowing through the load may be determined by using Ohm's law.

A second type of direct energy conversion device that may not be so well known is the thermoelectric generator, in which heat is supplied to the junction of two dissimilar metals. Because the metals are dissimilar, there is an electron flow, caused by the electrical potential difference of the two metals at the same temperature. If the leads of the dissimilar metals are joined by a load (Figure 1.6), then the circuit is connected and current will flow. If a battery is connected in place of the load, a current is passed through a junction of dissimilar metals and heat is either liberated or absorbed. If it is absorbed, then the junction acts as a refrigerator.

1.4 GEOTHERMAL POWER PLANT

A variation of the steam power plant is the geothermal power plant. In this instance the energy to produce steam is supplied by high-temperature brine below the earth's surface. Because no fuel is consumed in producing the steam, this method is environmentally beneficial in two ways: petroleum reserves are not used, preserving a natural resource, and there is no air pollution due to stack gas, as occurs in nonnuclear power plants. There are, however, many problems associated with the utilization of geothermal energy, one of which is using the high-temperature, highly corrosive brine. The brine is used to heat freshwater, producing steam, in a heat exchanger. The brine is very corrosive, which means a short lifetime for the heat exchanger. Also, the brine must be returned to the earth; otherwise the resulting void would weaken the earth's structure in the area. This assumes greater importance when we realize that these brine sources are located near faults in the earth's surface. There is another type of geothermal power plant in which steam generated by the flashing brine is used directly in the turbine; however, there are problems associated with this method such as corrosion of turbine blades and nozzles and disposing of the hot brine.

1.5 SOLAR ENERGY

Another energy form, one that will assume increasing importance, is solar energy. It is free for the using, but means must be developed to use it economically. The difficulty is economic. Solar energy can be used to heat water, which can be stored and later used for washing and heating. Solar energy can be used to power fuel cells or provide the high-temperature heat source for direct energy converters. Parabolic collectors have been used for cooking, the food being placed at the focal point of the collector. The basic problem associated with all attempts to use solar energy is that the concentration, or intensity of radiation, is low. Hence collecting enough to heat a house requires a large surface area and a large storage capacity, making solar-energy heating more expensive than conventional methods. Research is actively under way to reduce the storage costs and design an economical system for the supplemental heating of homes. Thermodynamics concerns itself with all energy-related matters, and an understanding of it is essential in developing new ways to deal with the energy problems that now affect the entire world.

1.6 FOUNDERS OF THERMODYNAMICS

The question may still lie in the student's mind as to how the subject of thermodynamics came into being. Who were its founders and what did they do? What follows is by no means an exhaustive or even extensive list of the contributors to the field of thermodynamics. It is, rather, a list of those individuals whose lives were vital to the founding of the subject and interesting as well. Who wants boring forefathers?

One thermodynamicist, Antoine Laurent Lavoisier (1743–1794), was the founder of modern chemistry, no mean feat in itself, and was particularly interested in the compounding of gunpowder. One of his assistants was the founding Du Pont of the American company of that name. It was Lavoisier's unfortunate fate that he lived during the French Revolution. Before the revolution he was a recognized scientist and contributed to the analysis of the combustion process. Also, he led in developing the concept of elements; he proposed the name of "oxygen" and described its function in the combustion process. He was born into a prosperous middle-class family and pursued an education in the sciences. He was active in politics before the revolution, seeking greater social reform. Even during the revolution and its aftermath he pursued various scientific topics, including the development of the metric system. However, his history as a tax collector of sorts did not endear him to the Jacobins, and he lost his head. A rather unfitting end to such a valuable person.

Another of the foremost contributors is Nicolas Leonard Sadi Carnot (1796–1832), whom you will never forget. His observations led to the naming of a theoretical thermodynamic cycle after him. He lived his youth in Napoleon's France, and his father, one of Napoleon's military organizers, was able to provide Sadi with a sound technical education. Sadi had the good sense to avoid political machinations; he kept his head and lived to retire from the army at the age of 24. He moved to Paris and life began, thermodynamically speaking. It was then that he wrote his famous *Reflections on the Motive Power of Fire* in which he tried for the first time to go beyond an actual engine and find the basic principles governing the conversion of heat to work. This, as we will see in a later chapter, led to the second law of thermodynamics.

A German professor by the name of Rudolf J. Clausius (1822–1888) used some of Carnot's reflections and spent the mid-1860s developing ideas about matter and publishing papers dealing with an abstract quantity called "entropy." He was endeavoring to formulate the mathematical representation of physical quantities, and we shall discover more about this in the chapter developing the concept of entropy.

Lord Kelvin (1824–1907), a nineteenth-century British physicist developed the thermodynamic temperature scale, and an absolute temperature scale was named after him. From his experiments with the concepts of work and heat, he deduced the doctrine of available energy, one of the fundamental doctrines of energy analysis. His work was not limited to thermodynamics; he also helped to develop the foundations of electrical engineering. He was knighted for his efforts in laying the first transatlantic cable.

James Prescott Joule (1818–1889) was another scientist of the mid-nineteenth century who ran careful and thoughtful experiments that established the equivalence of heat and work. This was of considerable importance because the scientific theory of the day supposed heat to be caused by some mysterious "caloric." Joule's work routed this theory, and he developed the conversion equivalents between units of mechanical work and thermal energy (heat).

The last thermodynamics pioneer we will discuss is J. Willard Gibbs (1839–1903), a comparatively unknown American who had a tremendous effect on the establishment of thermodynamics, statistical mechanics, chemistry, and mathematics. One reason he remains without fame is that he was not given to boisterousness; in fact, he was modest in the extreme. He studied and taught at Yale University and worked for nine years without pay. Eventually his worth was recognized, at least in academic circles, and he was salaried. The great surge in the chemical industry, of which thermodynamics is a cornerstone, rests to a great extent on Gibbs' work. Gibbs' writings are another indication of the man: There is not a word more than necessary, everything reduced to its essence.

Chapter 2
Definitions and Units

"Thermodynamics" is derived from the Greek words *therme*, meaning "heat," and *dynamis*, meaning "strength," particularly applied to motion. Literally, then, "thermodynamics" would mean "heat strength," implying such things as the heat liberated by the burning of wood, coal, or oil. Actually, if the word "energy" is substituted for "heat," one can come to grips with the meaning and scope of thermodynamics. It is the science that deals with energy transformations; the conversion of heat into work, or chemical energy into electrical energy, for example. Both of these are energy transformations, and thermodynamics is the science that provides the tools to analyze them.

There are four laws of thermodynamics: the zeroth, first, second, and third. These laws are the basis of all thermodynamic analysis, and throughout this text the attempt is made to show how all problems relate to these basic postulates.

2.1 MACROSCOPIC AND MICROSCOPIC ANALYSIS

This text deals with macroscopic, as opposed to microscopic or statistical, thermodynamics. In microscopic thermodynamics we must look at every molecule and analyze collective molecular action by statistical methods. In macroscopic thermodynamics we concern ourselves with the overall effect of the individual molecular interactions. The macroscopic level is the level on which we live. We measure distance in meters, time in seconds. These measurements are very large compared with the measurement of events on the molecular level; because they are very large, they are called "macroscopic." It is convenient to use this method of analysis because we have been familiar with the measurements since childhood. Temperature is a macroscopic effect, and more will be said of it later in the chapter. The microscopic point of view will be used only to explain some phenomena that cannot be understood by classical (macroscopic) means.

2.2 SUBSTANCES

What follows will be illustrations of the jargon of thermodynamics. One must be able to solve problems, and to do that the parts of the problem must be enumerated. The first consideration is that there must be something performing the energy transformation. This something is called a *substance*. This substance in the trusty family automobile engine is usually the mixture of gasoline and air. In a stream turbine the substance is probably the steam. A lot of "iffy" words are used in thermodynamics because the science covers many areas of energy transfer, and to specify that a substance must be a fluid is in general, too restrictive.

The substance may be further divided into subcategories. The scientific penchant for classification is alive in thermodynamicists. The substance is a pure substance if it is homogeneous in nature—that is, if it does not undergo chemical reactions and is not a mechanical mixture of different species. What does this mean? The substance, to be pure, must be homogeneous—that is, the same chemical and physical composition at one point as at another point. It must not undergo chemical reaction. What would happen if it did? There would be at least two substances present, the reactant and the product, and hence it would not meet the criteria of homogeneity. A mechanical mixture would also violate the homogeneity requirement. Oxygen by itself is a pure substance; but when mixed with another pure substance, say, nitrogen, the mixture is not a pure substance. Air therefore is a mixture, not a pure substance. It may seem that at normal temperatures and pressures the mixture of oxygen and nitrogen is the same composition everywhere. This is true. Oxygen and nitrogen condense at different temperatures, however, so it is possible to have conditions under which air is not homogeneous throughout and hence is a mixture.

2.3 SYSTEMS—FIXED MASS AND FIXED SPACE

A substance does not exist alone; it must be contained. This brings us to the concept of a system. In thermodynamics a system is defined as any collection of matter or space of fixed identity, and the concept is one of the most important in thermodynamics. The picking of the system is up to the individual and requires some skill. Later we will see some problems associated with determining system boundaries.

What is the term "boundary"? When a system is defined, let us say the fluid in a cylinder (Figure 2.1), what separates the fluid from the

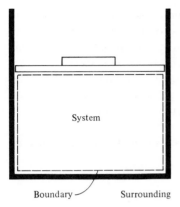

Boundary   Surrounding

Figure 2.1   Piston–cylinder, illustrating system, boundary, and surrounding.

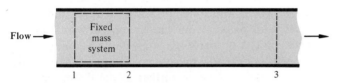

Figure 2.2   Flow through a pipe.

cylinder wall and the piston and everything external to the piston–cylinder? It is the system boundary. Everything not in the system is called the "surroundings." Note that the piston can be raised or lowered, but the system, matter of fixed identity, is constant. Let us backtrack a moment: A system is defined as a collection of matter or space of fixed identity. You ask, "How do I know whether to choose matter or space as a system?" Consider a long pipe (Figure 2.2) filled with a flowing fluid. Pass imaginary planes through the pipe at points 1 and 2 at the same instant and consider the fluid between them as the system. This mass of fluid is our system; each molecule is identified. Now consider the problem of finding out what happens to the fluid as it flows from 1 to 3. The molecules of our system become mixed with molecules from the surroundings, so to find out what happens, we must account for the movements of each molecule. This certainly is not feasible. What if the system were defined differently for this problem? Consider our system to be a collection of space and, for this problem, consider it to be all the space in the pipe between the planes at points 1 and 3. Our system is fixed space; it is not moving, even though fluid is moving through the space. To find out what happens to the fluid, all we need to do is note the condition of the fluid at the system boundaries at planes 1 and 3, and any changes are what has happened to the fluid in the pipe.

What conclusions can we draw from this? In the case of the piston–cylinder, where there was no flow of matter, the system was considered to be a matter of fixed identity. No problem in analysis occurred even when the volume changed. This is called a "closed system," a system closed to matter flow. When there is matter flow, then the system is considered to be a volume of fixed identity, a control volume. This system is open to the flow of matter, and is called an "open system." It may seem as if the definition of a system has been greatly belabored, but a clear understanding of the two concepts eliminates a great deal of difficulty in performing analysis.

2.4 PROPERTIES, INTENSIVE AND EXTENSIVE

In the previous discussion of properties we talked about the condition of a substance. "Condition" is not the accepted thermodynamic term; the correct term is "state." The state of a substance completely describes how the substance exists. It has temperature, pressure, density, and other macroscopic properties, and by knowing these properties we can determine the state of the fluid. A property is a characteristic quality of the entire system and depends not on how the system changes state but only on the final system state. Before looking further at the definition of a property, let us subdivide properties into two classes, intensive properties and extensive properties. An extensive property is one that depends on the size or extent of the system, such as mass and volume. An intensive property is independent of the size of the system, for example, temperature and pressure. Note that an extensive property per unit mass, such as specific volume (the volume divided by the mass), is an intensive property.

Getting back to the general definition of a property, we see that a property must be characteristic of the system. Consider a column of water. The temperature can be the same everywhere, and hence it is a property of this system. Is the pressure? The pressure cannot be uniform throughout the system, it is greater on the bottom than on the top, so the pressure is not a property of this system. The other part of the definition means that if a system goes from one state to another state, a property must depend only on the state, not on the changes the system underwent from one state to the other. For example, let us consider the system in Figure 2.1. Let it have a temperature, T_1; a pressure, p_1, and a specific volume, v_1. Add more weight to the piston and the system will reach a new state characterized by a temperature, T_2, and a pressure, p_2, and a specific volume, v_2. Consider, now, that in going from 1 to 2, weights were added and subtracted, finally resulting in the same state 2. The final temperature, pressure, and specific volume must be the same as they were when just a single weight was added to achieve state 2, since these properties depend only on the system state.

Properties are defined only when they are uniform throughout a closed system; the system is then in equilibrium with respect to that property. For instance, the water in the pipe is in thermal equilibrium because temperature is uniform, but the water is not in mechanical equilibrium because the pressure is not uniform. The air in Figure 2.1 is in both thermal and mechanical equilibrium. There is an equilibrium associated with each intensive property, but we are primarily concerned with mechanical and thermal equilibrium. Chemical equilibrium must be considered in some instances, as delineated in Chapter 14.

From the definition of properties we know that all the states we can potentially analyze must be equilibrium states, or there would be no properties defined. When a closed system is in thermodynamic equilibrium, then the properties are uniform throughout the system and may be defined. Thus thermodynamics is a misnomer, for we can define properties, the tools of analysis, only when they are at equilibrium. We shall discover, though, that this does not present a very great hardship in problem solving.

2.5 PHASES OF A SUBSTANCE

Certain groups of states of a substance may be called phases of that substance. Water has solid, liquid, and vapor phases, and any pure substance may exist in any combination of the phases. There are specific terms to characterize phase transitions. "Melting" occurs when a solid turns to a liquid; freezing or solidifying occurs when a liquid turns to a solid. Vaporization occurs when a liquid turns to a gas (vapor); condensation occurs when a gas (vapor) turns to a liquid. Sublimation occurs when a solid turns to a gas.

2.6 PROCESSES AND CYCLES

What is used to denote a change in the system state? A *process*. A process is simply a change in the system state. Just as there is an infinite number of ways to go between two points (Figure 2.3), so is there an infinite number of ways for a system to change from state 1 to state 2. The path describes the infinite number of system states that occur when a system undergoes a particular process from state 1 to state 2. A thermodynamic cycle is a collection of two or more processes for which the initial and final states are the same. The cycle in Figure 2.3 illustrates a closed system that undergoes process A and process B. The system is returned to state 1 from which it started. As an example of a simple cycle, consider an isothermal, or constant-temperature, process from state 1 to state 2. By denoting that the process is isothermal, the path the process will follow is defined. Let path *A* be the isothermal path as denoted on Figure 2.3. Since a cycle is two or more processes returning to the initial state,

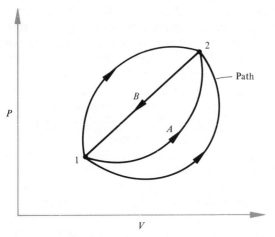

Figure 2.3   *Graphical description of system path.*

let the return process be isothermal also. The system will proceed from state 2 to state 1 along path *A*, but in the opposite direction. Is this the same path? No, because by definition a path is a succession of system states, and the succession from 1 to 2 differs from the succession from 2 to 1.

2.7 UNITS OF FORCE AND MASS

In this section the units of force and mass will be considered in the International System of Units, SI. The basic units of measure are mass, length, and time. All other units are derived from these. Mass is in kilograms (kg), length in meters (m), and time in seconds (s).

Confusion exists between mass and force. Since mass is a fundamental unit of measure, let us determine force. From Newton's second law of motion we know that the force on a body is given by

$$F = \frac{d}{dt}(mv) \qquad (2.1)$$

where v is the velocity of the body and *m* is the mass of the body. This may not look familiar, so take the derivative

$$F = v\frac{dm}{dt} + m\frac{dv}{dt}$$

Most often the mass is constant, and the change of velocity with respect to time is acceleration *a*, so

$$F = ma \qquad (2.2)$$

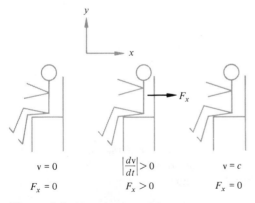

Figure 2.4   Visualization of force.

We know that if we are sitting in a chair (Figure 2.4), there is no force acting on the back of the chair. If the chair is accelerated, as happens in a moving car, there will be a force developed, given by Equation (2.2). When the acceleration ceases, the velocity is constant, and the force becomes zero. The units of force are said to be newtons (N). Perform a unit balance on Equation (2.2),

$$N = (kg)(m/s^2)$$

This does not balance, but if we multiply Equation (2.2) by a constant and demand that the units of the constant be such that a balance does occur, there is no inconsistency,

$$N = \frac{C(N)}{(kg)(m/s^2)}(kg)(m/s^2) \tag{2.3}$$

The units on the constant must be $(N \cdot s^2)/(kg \cdot m)$. All that remains is to determine the constant C. The standard unit 1 N is the force resulting from the action of one unit acceleration (1 m/s^2) on one kilogram. We substitute into Equation (2.2) and solve for the constant C.

$$C = 1 \ (N \cdot s^2)/ (kg \cdot m) \tag{2.4}$$

Since Equation (2.2) is valid for all cases, if C is determined for a given case, it must be true for all other cases. So goes the logic in determining the value of C. Since C has a value of unity, it is neglected from the force equation, which remains

$$F = ma \tag{2.5}$$

where the mass is in kilograms, the force in newtons and the acceleration in m/s^2.

Example 2.1

As an illustration, consider the horizontal force on the chair in Figure 2.4 if the person has a mass of 70 kg and is uniformly accelerated at 2 m/s².

$$F_x = ma = (70 \text{ kg})(2 \text{ m/s}^2)$$
$$F_x = 140 \text{ N}$$

Thus far we have not mentioned weight. Take the person in the previous example, and let him/her be standing on the ground. The person experiences a force on his/her body equal and opposite to the force exerted on the earth. This force is the person's weight. The person is acted upon by the local gravitational field g, if the gravitational field is located so that g has a value of 9.8 m/s², then the force is

$$F = (70 \text{ kg})(9.8 \text{ m/s}^2) = 686.0 \text{ N}$$

If the person were standing on a tall mountain where the gravitational field g had a value of 9.7 m/s², then the force on the ground, the weight, would be

$$F = (70 \text{ kg})(9.7 \text{ m/s}^2) = 679.0 \text{ N}$$

Thus in space the weightless condition is caused by the fact that $g = 0$; hence the force or weight of the body mass is zero. Since weight is a force, the units associated with it are newtons. In colloquial usage, however, such as in supermarkets, food is purchased by its weight in kilograms, no newtons. This is derived from a metric system that uses the concept of a kilogram force; however, this and other specialized systems will not be considered. The SI units of energy will be brought up in the discussion of energy problems in the next chapter.

There are conventions to be followed in using the SI units, which may be unfamiliar to many of us. The meter is the basic length unit, but at times it is either too large or too small to be convenient. In these cases we may use the prefixes listed in Table 2.1. This is not a complete listing, but will cover those necessary for thermodynamics. Thus 1000 meters (m) is 1 kilometer (km),

$$1000 \text{ m} = 1 \text{ km}$$

Punctuation marks are not used with SI units. Thus 1000 m. is incorrect; furthermore, a single space must be left between the 1000 and the m. The definition of a newton illustrates how a combination of symbols may be expressed:

$$N = \text{kg} \cdot \text{m/s}^2 \quad \text{or} \quad \text{kg} \cdot \text{m} \cdot \text{s}^{-2}$$

Table 2.1

Multiplier	Prefix	Symbol
10^9	giga	G
10^6	mega	M
10^3	kilo	k
10^2	hecto	h
10^1	deca	da
10^{-1}	deci	d
10^{-2}	centi	c
10^{-3}	milli	m
10^{-6}	micro	μ
10^{-9}	nano	n

Table 2.2

Physical Quantity	SI Unit	SI Abbreviation
length	meter	m
mass	kilogram	kg
time	second	s
electric current	ampere	A
thermodynamic temperature	kelvin	K
luminous intensity	candela	cd
amount of mass	mole	mol

Note that for a *product* of two or more unit symbols, a raised dot is used between the symbols.

For numerals of more than four digits, the digits are written in groups of three with a space between consecutive groups. Commas are not used. If there is a decimal point, groupings should be made on both sides of the decimal point. Thus

$$80,000 \text{ (incorrect)} = 80\,000 \text{ (correct)}$$

$$356142.61984 \text{ (incorrect)} = 356\,142.619\,84 \text{ (correct)}$$

Start at the decimal point and group left and right. Sequences of four numbers are not usually divided into groups:

$$8,000 \text{ (incorrect)} = 8000 \text{ (correct)}$$

$$4\,913.1593 \text{ (incorrect)} = 4913.1593 \text{ (correct)}$$

Table 2.2 lists the basic SI units with their names and symbols.

2.8 SPECIFIC VOLUME

There are several properties that are familiar to us but must be rigorously defined so we know exactly what is meant. *Specific volume* is the volume of a substance divided by its mass. But is there a point at which this is no longer true? Would specific volume have a meaning if we were to select a single molecule as the substance we want to measure? No. First, specific volume is a macroscopic phenomenon; second, for a property to be macroscopic, a continuum must exist. That is to say, macroscopic properties must vary continuously from region to region without a discontinuity. We say region instead of point because a point is too small (it is infinitesimal) to contain enough particles of substance (does it contain any?); thus the smallest space a macroscopic substance can occupy and still have macroscopic properties is a region with a characteristic volume, $\delta V'$. If the specific volume is designated as v, and δV is a small substance volume with mass δm, then

$$v = \lim_{\delta V \to \delta V'} \frac{\delta V}{\delta m} \qquad (2.6)$$

Equation (2.6) is precise, but it could prove to be a bit tedious to work with. If the system substance of which the specific volume is desired is of finite size and homogeneous, then the specific volume at one part of the system is the same as at another part of the system, and for a system volume V and mass m, the specific volume is

$$v = \frac{V}{m} \; m^3/kg \qquad (2.7)$$

If a system is considered to be a mixture of ice and water at the same temperature and pressure, can the specific volume of the system be found? No. First, the specific volume by Equation (2.7) would depend on whether ice or water was picked as the region; and, second, the mass of the system exists in two phases, or states, so the system is not homogeneous.

There is also a property known as *density*, (ρ) which may be defined as the mass divided by the volume, or the reciprocal of the specific volume; that is,

$$\rho = \frac{1}{v} \; kg/m^3 \qquad (2.8)$$

2.9 PRESSURE

The second property to be rigorously defined is *pressure*. If we consider a small cubical volume, $\delta V'$, of a substance, there will be forces acting normally in the x, y, and z directions as well as forces acting at angles to these directions. In solids these pressures are known as the stress of the solid. In fluids at equilibrium the nonnormal forces are zero, and the pressure that is measured is the average of pressures, forces per unit area, acting in the three directions. If the average normal force is δF_n and the area that is characteristic of $\delta V'$ is $\delta A'$, then the pressure p is

$$p = \lim_{\delta A \to \delta A'} \frac{\delta F_n}{\delta A} \tag{2.9}$$

As in the case of specific volume, if the substance in the system is homogeneous and at equilibrium, then the pressure acting on a finite area, A, with normal force, F, is

$$p = \frac{F}{A} \, \mathrm{N/m^2} \tag{2.10}$$

The unit of pressure is defined as 1 unit force per unit area. This is called 1 pascal. Thus,

$$1 \, \mathrm{Pa} = 1 \, \mathrm{N/m^2} \tag{2.11}$$

where Pa is the abbreviation for pascal.

The pressure that is measured most often is the difference between the pressure of the surroundings and that of the system. Consider, for instance, the Bourdon tube pressure gage, shown schematically in Figure 2.5. The pressure of the fluid acts against the inside tube surface, causing the tube to move. This movement is relayed through linkages to a pointer, which, when a calibrated dial is attached, indicates the difference

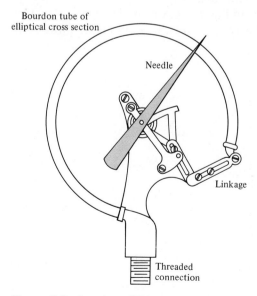

Figure 2.5   *Bourdon tube pressure gage.*

in pressure between the system fluid inside the tube and the surrounding fluid outside the tube. This pressure is called "gage pressure" (psig). The absolute pressure of the system is the sum of the system gage pressure and the surrounding absolute pressure:

$$p_{abs} = p_{gage} + p_{surr} \tag{2.12}$$

Most often the pressure gage is located in a room, and the pressure acting on it is atmospheric pressure. The air in the atmosphere has mass that is acted upon by gravity, and this force per unit area has an average value of 101 325 Pa at sea level. Because this is a very large number to work with, we make use of the prefixes in Table 2.1; thus

101 325 Pa = 101.325 kPa (kilopascals)

= 0.101 325 MPa (megapascals)

Very often atmospheric pressure is assumed to be 100 kPa for simplicity.

What happens when the gage pressure is zero, or less than atmospheric pressure? If the gage pressure is zero, then the absolute system pressure is equal to that of the surroundings. Should the gage pressure be less than the surrounding pressure, it is considered negative, in that the reading is subtracted from the surrounding absolute pressure to obtain the system absolute pressure. A pressure less than atmospheric pressure is called a "vacuum pressure" or vacuum.

Let us consider the sketch in Figure 2.6, where gage A reads 200 kPa (gage), gage B reads 120 kPa (gage), and the atmospheric pressure is

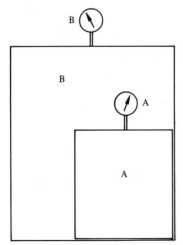

Figure 2.6 *Two containers, one within the other, at different pressures.*

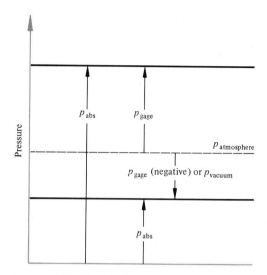

Figure 2.7 Graphical representation of pressure.

101.3 kPa. Find the absolute pressure of the air in box B and box A.

$$p_{B_{abs}} = p_{B_{gage}} + p_{surr}$$

$$p_{B_{abs}} = 120 \text{ kPa (gage)} + 101.3 \text{ kPa} = 221.3 \text{ kPa}$$

$$p_{A_{abs}} = p_{A_{gage}} + p_{surr}$$

$$p_{A_{abs}} = 200 \text{ kPa (gage)} + 221.3 \text{ kPa} = 421.3 \text{ kPa}$$

Figure 2.7 depicts the various pressures in a graphical form. The negative pressure, or vacuum, if often measured by a manometer, which has readings of millimeters of mercury or whatever other fluid the manometer may contain. This pressure may also be called kPa (vacuum).

Consider Figure 2.8, in which a manometer is used to measure the system pressure in the container. The system has a pressure p, the fluid in the manometer has a density ρ, and the surroundings are atmospheric with pressure p_{atm}. The difference in pressure between the system and the atmosphere is able to support the fluid in the manometer with density ρ a distance L. This may be expressed by

$$p - p_{atm} = \Delta p = (\rho \text{ kg/m}^3)(L \text{ m})(g \text{ m/s}^2) = \text{N/m}^2$$

$$\Delta p = \rho L g \text{ Pa}$$

Note that the units in the previous problem balance; the checking of units is a necessary step in all engineering practice, and is especially necessary in thermodynamics.

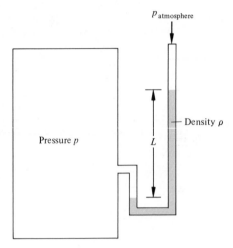

Figure 2.8   Diagram relating pressure and density in a manometer.

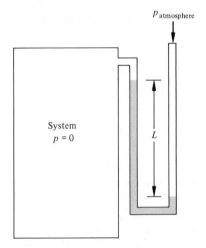

Figure 2.9   Schematic diagram illustrating concept of inches of mercury vacuum.

There are several ways of measuring pressure, and it is necessary to know the equivalence between them. For instance,

101.325 kPa = 1 atmosphere (atm) = 760 mm Hg

One atmosphere of pressure exerts a force of 101.325 kPa, or supports a column of mercury 760 mm high. A pressure less than atmospheric could not support so high a column of mercury, and zero pressure absolute could not support the column at all. This is said to be a perfect vacuum. Looking at Figure 2.9, we see that when the system has zero pressure, the height L to which the atmospheric pressure can push the mercury is 760 mm. The system pressure would be

0 kPa = 760 mm Hg (vacuum) = 101.325 kPa (vacuum)

2.10 EQUALITY OF TEMPERATURE

Temperature is not possible to define, but equality of temperature is possible to define. Consider two blocks of material, say, iron; if these two blocks are brought together and there is no change in any observable property, then the two blocks are said to be in thermal equilibrium and their temperature is the same. But if one block has a higher temperature than the other, it will also have a different pressure, or stress, a different electrical resistivity, and a different density. When the blocks are brought together, heat will flow from the hotter to the colder body, causing a change in these three properties; when thermal equilibrium is reached, no more property changes occur, and the temperatures are equal. Note that temperature equality is measured by changes in other properties. This will have greater significance when we discuss thermometers.

2.11 ZEROTH LAW OF THERMODYNAMICS

The zeroth law of thermodynamics states that when two bodies are in thermal equilibrium with a third body, they are in thermal equilibrium with each other and hence are at the same temperature. Fig. 2.10 illustrates this law. In this case the third body is a thermometer. As a result of the zeroth law, a third body, the thermometer, may be used for relating temperatures of two bodies without bringing them in contact

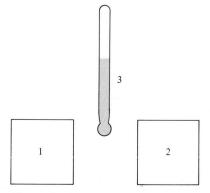

Figure 2.10   Diagram illustrating the equality of temperature, the zeroth law of thermodynamics.

with each other, since they are all related to the thermometer. It is the thermometer's scale that permits such a comparison.

2.12 TEMPERATURE SCALES

All sorts of devices, from hands to glass bubbles, have been used for measuring temperature, but the most common thermometers are sealed glass capillary tubes filled with colored alcohol or mercury. Note that the thermometer relies on change in another property, density, to indicate a temperature. There have been many researchers in the field of thermometry, but for the most part they have based the temperature scales on two points: the triple point of water (where ice, liquid, and vapor coexist), and the boiling point of water at 1 atmosphere. (See Figure 2.11.)

One of the common temperature scales, the Fahrenheit scale, has a slightly different beginning. Gabriel Daniel Fahrenheit was interested in thermometry and visited an astronomer friend, Romer, who devised a temperature scale of 60 degrees. There are 60 seconds to a minute, 60 minutes in an hour, so why not 60 degrees for a temperature scale? The lower limit was an ice, salt, and water mixture that corresponded to 0 degrees Fahrenheit. Anyway, Fahrenheit thought the scale too rough, and increased the number of divisions to 240; why not 360 divisions remains a mystery. He originally set the human body temperature to be 90, but later decided it should be 96; the scale's temperature having been fixed by body temperature, and an ice-salt-water mixture, the triple point became 32 degrees and water boiled at 212 degrees. To eliminate these inconvenient numbers, a man by the name of Celsius devised a scale that

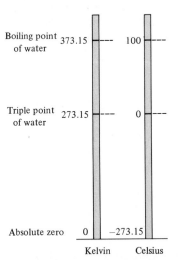

Figure 2.11   *Graphical illustration of two temperature scales.*

started at 100 (triple point) and went to 0 (boiling point). A friend suggested he reverse them, which resulted in the Celsius temperature scale with 0 at the triple point and 100 at the boiling point.

A Frenchman by the name of Guillaume Amontons first proposed that there was an absolute zero temperature. He was rather ignored, but in 1851 Lord Kelvin developed the idea further and devised an absolute temperature scale, the Kelvin scale, which has the same degree division as the Celsius scale. The relationship between the scales is

$$T_K = T_{°C} + 273.15$$

where T_K is temperature in degrees kelvins (K) and $T_{°C}$ is temperature in degrees Celsius (°C). The symbol T denotes temperature throughout this text.

PROBLEMS

1. Referring to Figure 2.6, the atmospheric pressure is 100 kPa and the pressure gages A and B read 210 kPa (gage). Determine the reading of gages A and B in (a) kPa; (b) mm Hg absolute.

2. A manometer is used to measure pressure as indicated in Figure 2.8 of the text. The fluid in the column is mercury with a density of 13.6 times that of water. If atmospheric pressure is 95 kPa and the height of the column is 1.5 m, determine the pressure of the system.

3. During takeoff in a space ship, an 80-kg astronaut is subjected to an acceleration equal to five times the pull of the earth's standard gravity. If the takeoff is vertical, what force does he exert on the seat?

4. A system has a mass of 20 kg. Determine the force necessary to accelerate it 10 m · s^{-2} (a) horizontally, assuming no friction; (b) vertically, where $g = 9.6$ m · s^{-2}.

5. A pump discharges into a 3-m-per-side cubical tank. The flow rate is 300 liters per minute and the fluid has a density 1.2 times that of water (density of water $= 1000.0$ kg/m^3). Determine (a) the flow rate in kilograms per second; (b) the time it takes to fill the tank.

6. Someone proposes a new absolute temperature scale in which the boiling and freezing points of water at atmospheric pressure are 500°X and 100°X, respectively. Develop a relation to convert this scale to degrees Celsius.

7. A vertical column of water will be supported to what height by standard atmospheric pressure?

8. A tank contains a mixture of 20 kg of nitrogen and 20 kg of carbon monoxide. The total tank volume is 20 m^3. Determine the density and specific volume of the mixture.

9. An automobile has a 1200-kg mass and is accelerated to 7 m/s^2. Determine the force required to perform this acceleration.

10. A hiker is carrying a barometer that measures 101.3 kPa at the base of the mountain. The barometer reads 85.0 kPa at the top of the mountain. The average air density is 1.21 kg/m³. Determine the height of the mountain.

11. Convert 225 kPa into (a) atmospheres; (b) millimeters of mercury absolute.

12. A skin diver wants to determine the pressure exerted by the water on her body after a descent of 35 m to a sunken ship. The specific gravity of seawater is 1.02 times that of pure water (1000 kg/m³). Determine the pressure.

13. A new temperature scale is desired with freezing of water at 0°X and boiling occurring at 1000°X. Derive a conversion between degrees Celsius and degrees X. What is absolute zero in degrees X?

14. A 50-kg frictionless piston fits inside a 20-cm diameter vertical pipe and is pulled upward to a height of 6.1 meters. The pipe's lower end is immersed in water and atmospheric pressure is 100 kPa. The gravitational acceleration is 9.45 m/s². Determine (a) the force required to hold the piston at the 6.1 m mark; (b) the pressure of the water on the underside of the piston.

15. A spring scale is used to measure force and to determine the mass of a sample of moon rocks on the moon's surface. The springs were calibrated for the earth's gravitational acceleration of 9.8 m/s². The scale reads 4.5 kg and the moon's gravitational attraction is 1.8 m/s². Determine the sample mass. What would the reading be on a beam balance scale?

16. Determine the pressure at points A and B if the density of mercury is 13 590.0 kg/m³ and that of water is 1000.0 kg/m³.

17. The following information is known:

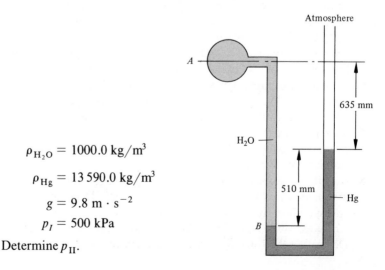

$$\rho_{H_2O} = 1000.0 \text{ kg/m}^3$$

$$\rho_{Hg} = 13\,590.0 \text{ kg/m}^3$$

$$g = 9.8 \text{ m} \cdot \text{s}^{-2}$$

$$p_I = 500 \text{ kPa}$$

Determine p_{II}.

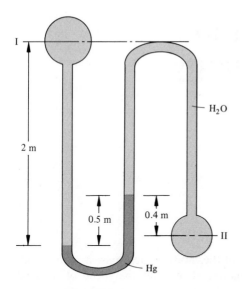

18. The piston shown below is held in equilibrium by the pressure of the gas flowing through the pipe. The piston has a mass of 21 kg; $p_I = 600$ kPa; $p_{II} = 170$ kPa. Determine the pressure of the gas in the pipe, p_{III}.

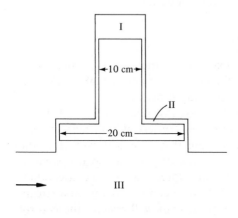

19. Consider the same problem as above, but $p_I = 350$ kPa, $p_{II} = 130$ kPa, $p_{III} = 210$ kPa. Find the mass of the piston.

Chapter 3
Conservation of
Mass and Energy

For most applications in the area of thermodynamics, mass is considered constant. This precludes the variation of mass as the speed of light is approached and the conversion of mass to energy in a nuclear reaction. These topics may be covered by thermodynamic analyses, but they will not be covered in this text. This chapter develops the laws of conservation of mass and conservation of energy for open and closed systems. Various energy forms, such as heat, work, and potential energy, are defined prior to inclusion in conservation equations.

3.1 CONSERVATION OF MASS

The law of the conservation of mass states that the total mass is a constant. As with all laws this is a deduction of experimental evidence; it can only be demonstrated, but not proved, to be true.

In thermodynamics there are two systems: open and closed. For a closed system, a mass of fixed identity, the conservation of mass is true; no equation is necessary to demonstrate this. It is self-evident in the definition of the system. However, in the case of an open system, a system of fixed space, we can develop an expression for the conservation of mass. Consider Figure 3.1, which shows a control volume in a stream of moving fluid. At some time t a given mass will occupy the control volume. This will be a mass of fixed identity. We will follow this mass of fixed identity as it moves across the boundaries of the control volume and from this determine the conservation of mass, or continuity equation, for the control volume.

After a time Δt, some of the mass in the control volume has moved across the control surface. This is represented as the system boundary at time $t + \Delta t$. The control surface boundary and the system boundary are coincident at time t.

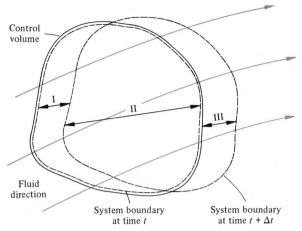

Figure 3.1   *A control volume located in a moving fluid.*

The regions of space that the mass occupies at times t and $t + \Delta t$ may be visualized as three regions, I, II, and III. Since the system mass is constant with time, we may write

$$m_{It} + m_{IIt} = m_{IIt+\Delta t} + m_{IIIt+\Delta t} \tag{3.1}$$

where the subscripts on each mass term represent the spatial region and the time. This equation may be rearranged as follows:

$$m_{IIt+\Delta t} - m_{IIt} - m_{It} + m_{IIIt+\Delta t} = 0 \tag{3.2}$$

Let us add zero to Equation (3.2), but zero expressed in a mathematically convenient form. What is the mass in region III at time t? Zero. What is the mass in region I at time $t + \Delta t$? Zero. Add these quantities to Equation (3.2) and divide by Δt.

$$\frac{m_{IIt+\Delta t} - m_{IIt}}{\Delta t} + \frac{m_{It+\Delta t} - m_{It}}{\Delta t} + \frac{m_{IIIt+\Delta t} - m_{IIIt}}{\Delta t} = 0 \tag{3.3}$$

The first term of Equation (3.3) represents the change in mass of region II. As Δt approaches zero, region II coincides with the control volume, so the change in mass of region II is. identical to the change in mass of the control volume.

$$\lim_{\Delta t \to 0} \frac{m_{IIt+\Delta t} - m_{IIt}}{\Delta t} = \frac{dm_{II}}{dt} = \frac{dm_{cv}}{dt}$$

If the limit is taken for the other two terms in Equation (3.3), the result is

$$\frac{dm_{cv}}{dt} - \frac{dm_{I}}{dt} + \frac{dm_{III}}{dt} = 0 \tag{3.4}$$

where the negative sign is due to the fact that $m_{I_{t+\Delta t}} - m_{I_t}$ is a decreasing, or negative, term physically. The mass in region I is decreasing, not increasing, with time. As Δt approaches zero, the change of mass of region I represents the mass entering the control volume, and the change of mass of region III represents the mass leaving the control volume. Let $dm/dt \equiv \dot{m}$, then $dm_I/dt = \dot{m}_{in}$ and $dm_{III}/dt = \dot{m}_{out}$, and considering $\dot{m}$ as positive Equation (3.4) becomes

$$\frac{dm_{cv}}{dt} = \dot{m}_{in} - \dot{m}_{out} \tag{3.5}$$

Equation (3.5) may be interpreted physically this way: The change of the mass within the control volume is equal to the difference between the mass entering and the mass leaving. We would probably have deduced this, but it is comforting to have a rigorous development as support.

It is possible to express the conservation of mass, Equation (3.5), in terms of area, velocity, and density, all measurable quantities. The volume will be denoted as $\mathcal{V}$ in the next few paragraphs to distinguish it from the mass velocity, v. Let $d\mathcal{V}$ be a differential volume in the total control volume. The fluid in $d\mathcal{V}$ has a density ρ. Thus the total mass in the control volume is the volume integral of $\rho\, d\mathcal{V}$:

$$m_{cv} = \int_{cv} \rho\, d\mathcal{V} \tag{3.6}$$

The time rate of change of m_{cv} is

$$\frac{dm_{cv}}{dt} = \frac{d}{dt}\int_{cv} \rho\, d\mathcal{V} \tag{3.7}$$

The volume flow across a differential surface is the product of the area, dA, times the fluid velocity normal to the area, v_n. The mass flow is found by multiplying this term by the fluid density, ρ, at area dA. The total mass flow across the entire area is found by integrating each differential flow area over the entire area A. Thus

$$\dot{m} = \int_A \rho v_n\, dA \tag{3.8}$$

Equation (3.5) may now be written as

$$\frac{d}{dt}\int_{cv} \rho\, d\mathcal{V} = \int_A \rho v_n\, dA_{in} - \int_A \rho v_n\, dA_{out} \tag{3.9}$$

For steady flow, the change in mass within the control volume is zero; hence the mass rates of flow in and out are identical:

$$\int_A \rho v_n\, dA_{in} = \int_A \rho v_n\, dA_{out} \tag{3.10}$$

For steady, one-dimensional flow, where the fluid enters at state i and leaves at state o, as illustrated by Figure 3.2, the conservation of mass becomes

$$\frac{A_i \mathsf{v}_i}{v_i} = \frac{A_o \mathsf{v}_o}{v_o} \tag{3.11}$$

Thus the expression for steady, one-dimensional flow, the continuity equation, or conservation of mass, is

$$\dot{m} = \frac{A\mathsf{v}}{v} = \rho A \mathsf{v} \frac{\text{mass}}{\text{time}} \tag{3.12}$$

Equation (3.12) expresses the conservation of mass for open systems in terms of readily measured or determined properties.

3.2 ENERGY FORMS

We have mentioned the energy of a system, but we have not defined energy. This was no accident. The total energy, E, cannot be simply defined, although parts of the total energy can be, as we will do later in the chapter. There is a conservation law for energy, the first law of thermodynamics. It states that whenever energy transfer occurs, energy must be conserved. This is stated simply enough, but before we develop the first law for open and closed systems, let us examine some of the various energy forms that we will be using in thermodynamics.

Work and heat are two of the most fundamental energy forms with which we will deal and it is essential that we understand them thoroughly.

Work

Work is usually defined as a force, F, acting through a distance, x, in the direction of that force.

$$W = \int F \, dx \tag{3.13}$$

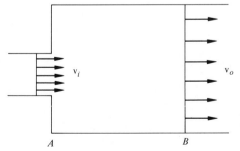

Figure 3.2   One-dimensional slug flow in a pipe.

This force, F, can appear in many forms, as we shall see. It can be the force acting on a mass to raise it, or it can be the force necessary to move a charged particle in a magnetic field. When work is considered in thermodynamics, a system is involved. Either the system will be doing work on the surroundings, everything external to the system, or the surroundings will be doing work on the system. To distinguish these two cases, work done by a system is, conventionally, positive and work done on a system is negative. Notice that the units (N · m) are units of energy, for work is an energy form.

Consider the system in Figure 3.3(a), (b), and (c). What is the direction of the system work in these three cases? Referring to the definition of work, we note that in case (a) the falling mass in the gravitational field provides the energy for work to be done on the system. Note that the work crosses the system boundary, going into the system, and hence the work is negative. In case (b) the mass and pulley are considered the system, and we note that the work is done by the system to the surroundings. In this case the work is positive as it goes from the system to the surroundings, the box with paddle wheels. In case (c) the paddle wheel and the mass and pulley are considered as the system. Is any work done on the system by the surroundings? No. Does the system do any work on the surroundings? No. So the work is zero. What happens within the system does not matter. It becomes important, then, to draw the system boundaries judiciously, or parts of a problem may not be apparent. In the reverse situation unnecessary complications may also

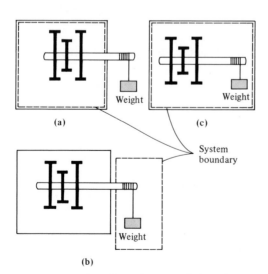

System boundary

Figure 3.3   The effect of a system's boundary on the work.

occur; both types of difficulties are eliminated with practice in problem solving.

Let us find the work done by a closed system, a system in which there is no mass flow. Again the piston-cylinder provides a ready tool for our analysis (Figure 3.4). There is a piston of uniform area, A, acted upon by a pressure, p_1; the pressure pushes the piston to the final position, where the pressure has dropped to p_2. The system mass has remained constant, but the volume has changed from V_1 to V_2, where $V_2 > V_1$. From our previous experience we note that the system has done work on the surroundings and the work is positive.

We would find this qualitative description somewhat lacking if we were asked how much work was done. We know that work was performed by the system; now we must attempt to model, with a mathematical equation, what is actually happening. We must remember that this is a model, not a description.

Looking back to Chapter 2 we find that properties, such as pressure, are defined only for equilibrium states. The pressure must be uniform throughout the system. This is handy, because the pressure acting on the piston face will be uniform, and a pressure multiplied by an area yields a force. Consider what happens when the piston moves a distance dL. There will be a drop in pressure, and, furthermore, by our model the

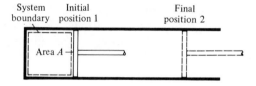

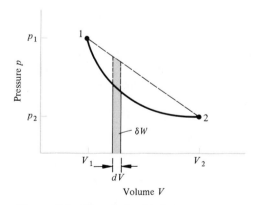

Figure 3.4   *The relationship between pressure and volume on a p–V diagram and in a piston–cylinder.*

pressure must drop uniformly throughout the system. Such a process is called a "quasi-equilibrium process"—we are assuming the system is in equilibrium at each point going from its initial to final states. Our model will be valid or not depending on whether or not this assumption can be made. It turns out that this assumption is reasonable for many processes, because the time the substance requires to establish equilibrium is usually much less than the time of macroscopic event, the work done.

Using the quasi-equilibrium model, the force $F = pA$ and the work done by moving a distance dL are

$$\delta W = pA \, dL \tag{3.14}$$

however,

$$A \, dL = dV$$

so

$$\delta W = p \, dV \tag{3.15}$$

The total work is found by integrating between 1 and 2:

$$W = \int_1^2 \delta W = \int_1^2 p \, dV \tag{3.16}$$

The reason for the inexactness of the work differential is that work is a path function. This may be seen graphically in Figure 3.4, where δW is denoted by the shaded area on the diagram. This, however, is for a given distribution of pressure with volume. If the distribution were changed to that of a straight line between 1 and 2, the δW would be greater, indicating that work is a function of how the pressure varies in going from state 1 to state 2.

The δ operator indicates an inexact differential, which approaches an infinitesimal, rather than the exact differential, which approaches zero.

Example 3.1

Let the pressure in the cylinder in Figure 3.4 be the following function of volume, $p = C/V$. (a) If the initial pressure is 400 kPa, the initial volume is 0.02 m³, and the final volume is 0.08 m³, find the work done. (b) Is the sign correct?

(a) $\quad W = \int_1^2 p \, dV = C\int_1^2 \dfrac{dV}{V} = C \ln\left(\dfrac{V_2}{V_1}\right)$

$\quad C = pV = p_1 V_1$

$\quad W = p_1 V_1 \ln\left(\dfrac{V_2}{V_1}\right)$

$\quad W = (400 \text{ kN/m}^2)\,(0.02 \text{ m}^3)\ln\left(\dfrac{0.08}{0.02}\right) = 11.09 \text{ kJ}$

(b)   The work is positive, indicating that the system did work on the surroundings, which corresponds to the physical situation.

It is very important to use a unit balance on all the equations, so the terms will be expressed in compatible units.

Example 3.2

The pressure in the cylinder in Figure 3.4 varies in the following manner with volume: $p = C/V^2$. If the initial pressure is 500 kPa, the initial volume is 0.05 m³, and the final pressure is 200 kPa, find the work done by the system.

$$W = \int_1^2 p \, dV = C \int_1^2 \frac{dV}{V^2} = C \left[\frac{1}{V_1} - \frac{1}{V_2} \right]$$

$$C = p_1 V_1^2 = p_2 V_2^2 \qquad V_2 = V_1 \left(\frac{p_1}{p_2} \right)^{1/2}$$

$$W = p_1 V_1 - (p_2 p_1)^{1/2} V_1$$

$$W = (500 \text{ kN}/\text{m}^2)(0.05 \text{ m}^3) - [(500)(200)]^{1/2} \text{ kN}/\text{m}^2 (0.05 \text{ m}^3)$$

$$W = 9.19 \text{ kN} \cdot \text{m} = 9.19 \text{ kJ}$$

Let us now consider the quasi-equilibrium process in more detail. Consider the piston-cylinder arrangement in Figure 3.5(a). The piston is held in place by a mass acted upon by the gravitational field, hence a force. If we move one block from the piston as in Figure 3.5(b), the

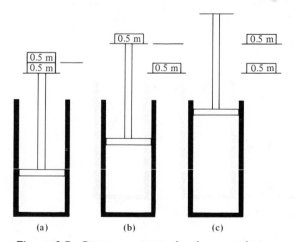

Figure 3.5   *Piston movement when large weight increments are removed.*

piston will shoot upward and oscillate up and down before reaching an equilibrium position. Clearly the pressure is not uniform during such a process, and the process cannot be called a quasi-equilibrium process. The next step toward reaching a quasi-equilibrium process is to divide the mass into smaller and smaller quantities. Figure 3.6 illustrates the mass, being replaced by a stack of cards. As each card is removed, the piston moves upward a slight amount and after a very slight oscillation reaches its equilibrium state. To improve this process even further, we must make the mass of each card infinitesimal, so the change in the piston height would occur in differential steps. This process is imaginary, but it represents the ideal, the maximum work possible; all the energy goes into moving the piston and is not dissipated by the piston's oscillation. Note also that the addition of an infinitesimal card to the piston would start the process in reverse. This process is called an "equilibrium" or "reversible" process. In Figure 3.5 we could not slide the mass onto the piston to start the process over. This process is called an "irreversible" process; it cannot return to its original state along the same path. There are many factors that render a process irreversible. Nonuniform pressure in the system causes the mass to move within the system, using energy that then will not be available for work. Frictional effects, too, are apparent irreversibilities. The energy used to overcome mechanical friction is lost for useful purposes. Also, fluid viscous forces, a fluid friction, dissipates useful energy. There are other causes of irreversibilities, but the effect is always to decrease the useful energy output or increase the energy required.

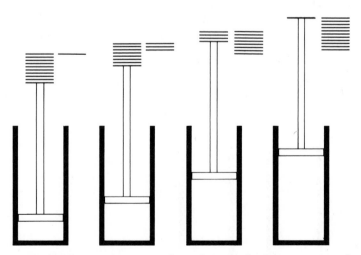

Figure 3.6   *Piston movement when infinitesimal weights are removed modelling a quasi-static process.*

Measuring how quickly work is accomplished gives us another quantity—power, $\dot{W}$.

$$\dot{W} = \frac{\delta W}{dt} \text{ energy/unit time} \tag{3.17}$$

The SI unit of power is the watt. Thus

$$1 \text{ J/s} = 1 \text{ N} \cdot \text{m} \cdot \text{s}^{-1} = 1 \text{ W}$$

Associated with work and power is torque. Torque is the turning moment exerted by a tangential force acting at a distance from the axes of rotation. Torque represents the capacity to do work; whereas, power represents the rate at which work may be done. Section 16.8 explores torque, work, and power in more detail.

There are forms of work other than mechanical work. These forms are seldom dominant, but neglecting them can lead to error. For instance, a film on the surface of a liquid has a surface tension, which is a property of the liquid and the surroundings. The surface tension of a film is a function of the surrounding medium. A simple experiment illustrates the surface tension action. In Figure 3.7(a), the entire loop is covered with a film for instance, a soap film. The film is punctured within the thread loop, and the surface tension of the film acts to make the remaining area a minimum. The surface tension, σ, has units of force per unit length

$$W = -\int_1^2 \sigma \, dA (\text{force/length}) \times \text{area} = \text{force} \times \text{length} \tag{3.18}$$

The minus sign indicates that an increase in area results in work being done on the system. The units of work are correct. Thus we have seen that when pressure with units of force per unit area acts on a volume, work is performed; and now we see that surface tension acting on an area is work. Surface tension is an area phenomenon, whereas pressure is a volume phenomenon.

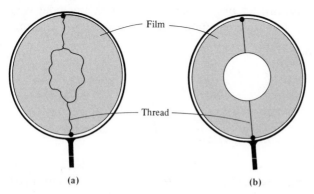

(a)   (b)

Figure 3.7 A hoop with a soap film. This demonstrates the effect of surface tension.

Example 3.3

Let the soap film in Figure 3.7 have a surface tension of 25×10^{-5} $N \cdot m^{-1}$, a hoop diameter of 0.05 m, and an inner-circle diameter in Figure 3.7(b) of 0.02 m. Find the work done by the film.

$$W = -\sigma \int_1^2 dA = -\sigma(A_2 - A_1)$$

$$W = (-25 \times 10^{-5} \, N/m)\left(\frac{\pi}{4}(21 - 25) \times 10^{-4} \, m^2\right)$$

$$= 25\pi \times 10^{-9} \, N \cdot m$$

The system performed $25\pi \times 10^{-9}$ J of work.

To demonstrate another form of work, let us consider a wire as the system. Stretch the wire, within the elastic limits, and the work is done on the wire by the surroundings; F is the tension, then

$$W = -\int_1^2 F \, dL \tag{3.19}$$

where the minus sign indicates that a positive displacement results from work being supplied to the system. If we limit the problem to within the elastic limit, where E is the modulus of elasticity, s is the stress, ε is the strain, and A is the cross-sectional area, then

$$F = sA = E\varepsilon A$$
$$\varepsilon = dL/L_1$$
$$\delta W = -F \, dL = -E\varepsilon AL_1 \, d\varepsilon$$

$$W = -AEL_1 \int_1^2 \varepsilon \, d\varepsilon = -\frac{AEL_1}{2}\left(\varepsilon_2^2 - \varepsilon_1^2\right)$$

There are other, less visible, modes of work—electrical work, magnetic work—but these do not play a significant role in this course in thermodynamics. The following equations summarize some work forms:

compressible fluid $\qquad W = \int_1^2 p \, dV$

surface film $\qquad W = -\int_1^2 \sigma \, dA \tag{3.20}$

stretched wire $\qquad W = -\int_1^2 F \, dL$

What is common to all these work expressions? That work is equal to an intensive property times the change in its extensive property. In advanced thermodynamics the intensive properties are defined in terms of the partial derivatives of the extensive property on which they act. If $\bar{F}$ is an intensive property and $d\overline{X}$ is the change in the related extensive property, then

$$W = \overline{F} \cdot d\overline{X} \tag{3.21}$$

Equation (3.21) is the general work expression for any reversible work. It might be possible to have a system in which many different forms of work are occurring; thus

$$W = \sum_i \overline{F}_i \cdot d\overline{X}_i \tag{3.22}$$

and for Equation (3.20), this becomes

$$\delta W = p\, dV - \sigma\, dA - F\, dL \tag{3.23}$$

Heat

Having looked at work in some of its many forms, let us consider now a transfer of thermal energy between a system and its surroundings. Heat is defined as the energy crossing a system's boundary because of a temperature difference between the system and the surroundings. Heat and work are similar in that they are both energy fluxes and both must cross a system's boundary to have any meaning. This definition differs from the colloquial usage of heat, in that bodies do not have any property called "heat." Heat exists only as energy crossing a boundary; once inside the system it has no meaning.

Just as there are conventions attached to the direction in which work crossed a system's boundary, so there are with heat. Heat flowing into the system is positive; heat flow from the system is negative. Heat is represented by the symbol Q. If there is a process in which $Q = 0$, then the process is called "adiabatic." For instance, if a pipe is well insulated from the surroundings, then any heat flow between the system and the surroundings is very small, negligible in most cases. If it were zero, the pipe would be adiabatically insulated. If we were to model the pipe as the system, then it may be possible to assume the walls were adiabatically insulated. This simplifies the analysis to be made without introducing errors of any great magnitude. Of course, if there actually were substantial heat flow through the walls, then to assume that the walls are adiabatic would be extremely inaccurate.

Since heat occurs as a result of energy flow between the system and its surroundings, it is a function of the method in which this energy is transferred. This means heat is a path function; it depends on how the energy is transferred.

$$Q = \int_1^2 \delta Q \tag{3.24}$$

Q is the heat transferred when a system goes from state 1 to state 2 for any given process. Another convenient notation is heat per unit mass, q.

$$q = \frac{Q}{m} \tag{3.25}$$

where m is the system mass. The integral of δQ may only be evaluated when the process is reversible (continuous), the only time a functional relationship may be determined. Note that the δ operator must be used.

Convertibility of Energy

We have seen that both newton meters (N · m) and joules (J) are energy forms, but how are they related? Starting in 1845, an Englishman, James Prescott Joule, initiated experiments to find the equivalences between heat (J) and mechanical work (N · m). Although he used the English unit system, the principle is the same. Figure 3.8 illustrates a simplified version of his experiment. A closed, insulated container is fitted with a spindle to which paddles are attached. A wire is wound around the external spindle shaft and led over a pulley to a weight (force). The weight was dropped at constant velocity, eliminating all acceleration but local gravity. There was a temperature increase in the water after the weight was dropped. From the definition of thermal energy (heat energy), the ability to increase temperature of a substance, Joule was able to calculate the equivalence between mechanical and thermal energies.

The joule is a basic unit of energy; that is,

$$1 \, N \cdot m = 1 \, J$$

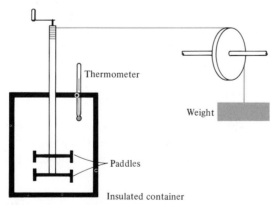

Figure 3.8   Joule's experiment on the equivalence of heat and work.

Potential, Kinetic, and Internal Energy

There are many types of energy, energy forms, but we will discuss three: potential, kinetic, and internal energies—the ones that occur in the typical situations we will be analyzing.

The potential energy of a system mass depends on its position in the gravitational force field. There must be a reference datum and a distance from the datum to the system mass; let this distance be z. Let the system be acted upon by a force F, and assume that this force raises the system. The change in potential energy is equal to the work (force through a distance) necessary to move the system.

$$d(\text{P.E.}) = F\,dz = mg\,dz$$

$$\int_1^2 d(\text{P.E.}) = \int_1^2 mg\,dz = m\int_1^2 g\,dz \tag{3.26}$$

The limits are from the initial to the final position. If the distance from 1 to 2 is not too great, so the gravitational acceleration, g, is essentially constant, Equation (3.26) becomes

$$(\text{P.E.})_2 - (\text{P.E.})_1 = mg(z_2 - z_1)\ \text{N}\cdot\text{m} \tag{3.27}$$

Thus we see that our assumption about the form of this energy equation is realistic; the potential energy of a system increases as the height increases.

The kinetic energy of a system is developed in an analogous manner. Again let us consider a system of mass m. Let a horizontal force act on the system and move it a distance dx. Since it moves horizontally, there is no change in potential energy. The change in kinetic energy is defined as the work in moving the system a distance dx.

$$d(\text{K.E.}) = F\,dx$$

$$F = ma = m\frac{dv}{dt}$$

$$\frac{dv}{dt} = \frac{dv}{dx}\frac{dx}{dt} = v\frac{dv}{dx} \tag{3.28}$$

$$d(\text{K.E.}) = mv\frac{dv}{dx}dx = mv\,dv$$

Integrating,

$$\int_1^2 d(\text{K.E.}) = m\int_1^2 v\,dv$$

$$(\text{K.E.})_2 - (\text{K.E.})_1 = \frac{m}{2}\left(v_2^2 - v_1^2\right)\ \text{N}\cdot\text{m} \tag{3.29}$$

Note that this is the translational velocity of the system, and if the velocity of the system increases, there is a positive change of kinetic energy. If a car is moving at 40 kilometers per hour and is accelerated (force through a distance) to 80 kilometers per hour, the kinetic energy increases. If the car's velocity is zero, its kinetic energy is zero.

One of the less tangible forms of energy of a substance is its internal energy. This is the energy associated with the substance's molecular structure. Although we cannot measure internal energy, we can measure changes of internal energy.

The internal energy of a diatomic molecule can be visualized as a function of four vibratory energies and the energy of position. Energy is stored in the molecular structure by these five modes. The more complex the molecular structure, the more modes there are available for storing energy. For a monatomic gas there are not these five modes of energy storage, but we will take the general case first. In Figure 3.9 the different forms of a diatomic molecule are illustrated. There is an energy associated with the translational motion of the molecule, called the "translational kinetic energy." There is the back-and-forth movement of the atoms in the molecule, toward and away from each other; this is a vibrational energy. The atoms, as a pair, are visualized as rotating around their center of mass; this is called the "first" kind of rotational energy. Also, an atom may rotate around its own center of mass, and this is called the "second" kind of rotational energy. As the number of atoms in

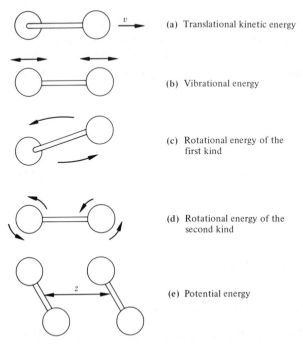

Figure 3.9   Modes of storing internal energy in a di-atomic molecule.

the molecule increases, the complexity of the rotational energy increases. The last term to be included in the internal energy model is that of the potential energy of attraction between adjacent molecules. This force is quite large, as well you realize by the energy it takes to convert water to steam by overcoming the attractive force of the liquid molecules.

The symbol for specific internal energy is u; for total internal energy, U.

$$u = \text{specific internal energy, J/kg} \tag{3.30a}$$

$$U = mu \text{ total internal energy, J} \tag{3.30b}$$

There are many other energies that actually contribute to the internal energy, such as electron spin, but we are more interested in using internal energy than in building models of it.

Several energy forms have now been defined and discussed: kinetic energy, potential energy, internal energy, work, and heat. There is also chemical energy; that will be discussed in detail in the chapter on reactive systems. It may be possible that for a given system all work types, energies, and heat must be included, but that is very rare. Most often, only a few are important for each process.

3.3 FIRST COROLLARY OF THE FIRST LAW

There are two types of systems: fixed mass (closed) and fixed space (open). The first corollary of the first law of thermodynamics, that energy is conserved, is the application of the conservation of energy to closed systems; the second corollary is the application of the conservation of energy to open systems. We will consider the closed system first.

final energy − initial energy = energy added to system

$$E_2 \quad - \quad E_1 \quad = \quad Q - W \tag{3.31}$$

Where heat has been added $(+Q)$ and work has been done on the system $(-W)$.

Let us examine the components of the initial energy of the system, E_1. What energy forms cannot be in E_1? Since work and heat exist only when crossing the boundary, they cannot exist as energy at a system state. Kinetic, potential, and internal energies can all exist. Thus

$$E = U + \frac{mv^2}{2} + mgz \text{ J} \tag{3.32}$$

or dividing by the system mass, m,

$$e = u + \frac{v^2}{2} + gz \text{ J/kg} \tag{3.33}$$

In solving problems it is often easier to find the solution per unit mass and then multiply by the system mass at the conclusion.

Let us get heat and work into the energy equation by having a change occur; assume the system does work W and receives heat Q as in Figure 3.10. Let the piston expand, doing work, and the system receives heat from the surroundings. If the system operates between initial and final states, then the first law of thermodynamics, Equation (3.31), becomes

$$Q = E_2 - E_1 + W \tag{3.34}$$

The signs on Q and W reflect the fact that they were both positive in the thermodynamic sense. A way to avoid this, if it seems confusing, is to perform an energy balance on the system:

energy initial + energy in = energy final + energy out

$$E_1 \quad + \quad Q \quad = \quad E_2 \quad + \quad W \tag{3.35}$$

Solve Equation (3.35) for Q and divide by the system mass, m

$$q = (e_2 - e_1) + w$$

$$q = (u_2 - u_1) + \frac{(v_2^2 - v_1^2)}{2} + g(z_2 - z_1) + w \text{ J/kg} \tag{3.36}$$

Usually the kinetic and potential energies are neglected, especially for closed systems, so

$$q = (u_2 - u_1) + w \tag{3.37}$$

or

$$q = \Delta u + w$$

If the derivative of Equation (3.37) is taken,

$$\delta q = du + \delta w \tag{3.38}$$

the total system change is

$$\delta Q = dU + \delta W \tag{3.39}$$

Boundary

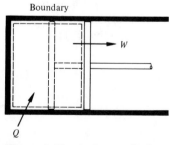

Figure 3.10 *A piston–cylinder system receives heat and does work.*

Equations (3.38) and (3.39) are commonly referred to as the first law equations for a closed system. You have seen they are nothing more than an energy balance.

Example 3.4

In a tank similar to the one that Joule used in determining the mechanical-thermal energy equalities (Figure 3.8) there are 10 kg of water. The total work input is 20 N · m.

(a) Find the change in specific and total internal energy. (Assume the system is adiabatic.)

(b) If a heat leak of 0.1 J/kg is noted, what is the internal energy change?

Since ΔK.E. and ΔP.E. are zero for both cases, the energy balance for the system is

First Law: $Q = \Delta U + W$

(a) Adiabatic: $Q = 0$

$\Delta U = -W = -(-20 \text{ Nm}) = 20 \text{ J}$

$\Delta u = \Delta U / m = 2 \text{ J/kg}$

(b) $Q = (-0.1 \text{ J/kg})(10 \text{ kg}) = -1.0 \text{ J}$

$-1.0 = \Delta U - 20$

$\Delta U = 19.0 \text{ J}$

$\Delta u = 1.9 \text{ J/kg}$

Heat flow out is negative; work in is negative.

3.4 ENERGY AS A PROPERTY

In the preceding analysis we have tacitly assumed that the system energy is a property. The use of an exact differential in describing energy, for instance, implies this. In the following development we will prove that this assumption is valid. First, a cycle analysis of a closed system will be made. Referring to the cycle definition we see that the system state is the same before and after a cyclic process. We assume that both heat and work enter and leave the system as it completes the cyclic process. From Equation (3.35), noting that $E_1 = E_2$, we may write

$$Q_{in} + W_{in} = Q_{out} + W_{out}$$

$$Q_{in} - Q_{out} = W_{out} - W_{in}$$

$$\Sigma Q = \Sigma W \qquad\qquad (3.40)$$

This may be written as a cyclic integral

$$\oint \delta Q = \oint \delta W \tag{3.41}$$

or for any cyclic process

$$\oint \delta Q - \oint \delta W = 0 \tag{3.42}$$

If

$$\delta Q = dE + \delta W \qquad \text{or} \qquad \delta Q - \delta W = dE$$

then

$$\oint \delta Q - \oint \delta W = \oint dE = 0$$

Hence, the system energy depends only on the system state and not on the path followed to arrive at the state; it is a property. Also, the components of the total energy are properties, so

$$\oint dU = \oint d(\text{K.E.}) = \oint d(\text{P.E.}) = 0 \tag{3.43}$$

It is also important to note from Equation (3.40) that $\Sigma Q = \Sigma W$ for any system undergoing a cyclic process. The individual work and heat terms may be positive or negative, depending on direction. The algebraic sum of these terms, then, is the net heat and work, the conclusion being that the net heat is equal to the net work for a closed system operating on a cycle.

3.5 SECOND COROLLARY OF THE FIRST LAW

We will consider now the systems in which there is a mass flow, the open system. Here the volume is fixed, and the mass flows into and out of this fixed space, or control volume. This is a very common situation: In a jet engine, air enters and leaves the engine; in a steam turbine, the steam enters and leaves the turbine. In both cases work is done, heat is exchanged, and energy is transferred.

The second corollary of the first law is the conservation-of-energy principle applied to open systems. Consider the control volume shown in Figure 3.11 to be located in a stream of moving fluid. Heat may be transferred and work may be done, as illustrated.

At some time t there is a control mass, which is the sum of the masses in regions I and II. At some time $t + \Delta t$, some of the mass has moved into region III, and heat and work interactions may have occurred to the control mass. An energy balance on the control mass for the time

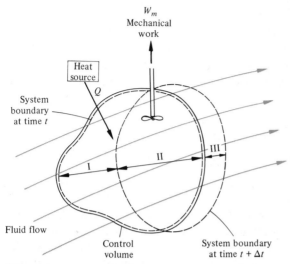

Figure 3.11 *Control volume located in a moving fluid with heat and work being transferred.*

interval Δt is

$$Q - W = \Delta E_{CM} = E_{CMt+\Delta t} - E_{CMt} \tag{3.44}$$

where Q and W are the heat and work done on the control mass.

The energy of the control mass may be expressed in terms of the control volume

$$E_{CMt} = E_{IIt} + E_{It} \tag{3.45}$$

$$E_{CMt+\Delta t} = E_{IIt+\Delta t} + E_{IIIt+\Delta t} \tag{3.46}$$

Subtract Equation (3.45) from Equation (3.46) and add zero to the right-hand side. In this case $E_{IIIt} = 0$ and $E_{It+\Delta t} = 0$.

$$E_{CMt+\Delta t} - E_{CMt} = E_{IIt+\Delta t} - E_{IIt} + E_{It+\Delta t} - E_{It} + E_{IIIt+\Delta t} - E_{IIIt}$$

Divide by Δt and take the limit as $\Delta t \to 0$. Again region II coincides with the control volume as $\Delta t \to 0$, hence $dE_{II}/dt = dE_{cv}/dt$.

$$\frac{dE_{CM}}{dt} = \frac{dE_{cv}}{dt} - \frac{dE_I}{dt} + \frac{dE_{III}}{dt} \tag{3.47}$$

The negative sign on dE_I/dt is due to the difference $(E_{It+\Delta t} - E_{It})$ being negative in the physical sense, the energy flow in region I is decreasing. Furthermore, we let

$$\lim_{\Delta t \to 0} \frac{Q}{\Delta t} = \frac{\delta Q}{dt} = \dot{Q} \tag{3.48}$$

and

$$\lim_{\Delta t \to 0} \frac{W}{\Delta t} = \frac{\delta W}{dt} = \dot{W} \qquad (3.49)$$

The derivative dE_{I}/dt represents the energy entering the control volume, dE_{in}/dt; and dE_{III}/dt represents the energy leaving, dE_{out}/dt. Thus Equation (3.44) becomes

$$\frac{dE_{cv}}{dt} = \dot{Q} - \dot{W} + \frac{dE_{\mathrm{in}}}{dt} - \frac{dE_{\mathrm{out}}}{dt} \qquad (3.50)$$

The total energy may be represented as $E = em$, and $dE/dt = e\dot{m}$, hence

$$\frac{dE_{cv}}{dt} = \dot{Q} - \dot{W} + e_{\mathrm{in}}\dot{m}_{\mathrm{in}} - e_{\mathrm{out}}\dot{m}_{\mathrm{out}} \qquad (3.51)$$

The development is almost complete, but we must account for the different work terms. There is mechanical work done by the fluid on the machinery (turbine, compressor); however, there is also flow work. This is the energy transmitted across the system boundary as a result of a pumping process occurring outside the system, causing the fluid to enter the system. This is the work by the fluid to overcome the normal stress, pressure, at the boundary. Thus there is a flow-work term entering and leaving the system boundary.

Let us consider a differential element on the control surface with area A. The fluid has a pressure p at the area A and the fluid moves a distance dx normal to the area. The flow work is

$$\delta W_{\mathrm{flow}} = pA\, dx$$

$$A\, dx = v\, dm$$

$$\delta W_{\mathrm{flow}} = pv\, dm$$

$$\dot{W}_{\mathrm{flow}} = pv\, \dot{m}$$

The net flow work is the difference between the flow work entering and leaving. Equation (3.51) becomes

$$\frac{dE_{cv}}{dt} = \dot{Q} - \dot{W}_m + (e + pv)_{\mathrm{in}}\dot{m}_{\mathrm{in}} - (e + pv)_{\mathrm{out}}\dot{m}_{\mathrm{out}} \qquad (3.52)$$

For steady state, $dE_{cv}/dt = 0$, and the first law may be stated as an energy balance,

Energy In = Energy Out

$$\dot{Q} + (e + pv)_{\mathrm{in}}\dot{m}_{\mathrm{in}} = (e + pv)_{\mathrm{out}}\dot{m}_{\mathrm{out}} + \dot{W}_m \qquad (3.53)$$

and for steady-flow conditions, $\dot{m}_{\mathrm{in}} = \dot{m}_{\mathrm{out}} = \dot{m}$,

$$\dot{Q} + (e + pv)_{\mathrm{in}}\dot{m} = (e + pv)_{\mathrm{out}}\dot{m} + \dot{W}_m \qquad (3.54)$$

Separating the total energy into three components—internal, kinetic, and potential energies—yields for the energy equation on the unit mass basis,

$$q + u_1 + p_1 v_1 + (\text{K.E.})_1 + (\text{P.E.})_1$$

$$= u_2 + p_2 v_2 + (\text{K.E.})_2 + (\text{P.E.})_2 + w \tag{3.55a}$$

$$q = \Delta u + \Delta(pv) + \Delta(\text{K.E.}) + \Delta(\text{P.E.}) + w \tag{3.55b}$$

$$\delta q = du + d(pv) + d(\text{K.E.}) + d(\text{P.E.}) + \delta w \tag{3.55c}$$

and so

$$\dot{Q} + \dot{m}\left[u_1 + p_1 v_1 + (\text{K.E.})_1 + (\text{P.E.})_1 \right]$$

$$= \dot{m}\left[u_2 + p_2 v_2 + (\text{K.E.})_2 + (\text{P.E.})_2 \right] + \dot{W} \tag{3.56}$$

We can observe that two energy terms appear together on both sides of the equation in the energy term and both are properties. It is easier to write once than twice, so we let

$$h \equiv u + pv$$

$$H \equiv U + pV \tag{3.57}$$

The sum of these two properties, h, is called "enthalpy." Note that u and pv should have compatible units. At various times there may be a temptation to associate enthalpy with heat or work. This is not true; enthalpy is a property having no function physically other than being the sum of $U + pV$.

Equations (3.55b) and (3.56) become, respectively,

$$q = \Delta h + \Delta(\text{K.E.}) + \Delta(\text{P.E.}) + w \tag{3.58a}$$

and

$$\dot{Q} + \dot{m}\left[h_1 + (\text{K.E.})_1 + (\text{P.E.})_1 \right]$$

$$= \dot{m}\left[h_2 + (\text{K.E.})_2 + (\text{P.E.})_2 \right] + \dot{W} \tag{3.58b}$$

The first law may be written on a differential basis as

$$\delta q = dh + d(\text{K.E.}) + d(\text{P.E.}) + \delta w \tag{3.59}$$

Equation (3.59) has been developed from the control volume's point of view. We will consider the unit mass of the system; the first law for the mass is

$$\delta q = du + p\, dv$$

From the definition of enthalpy,

$$dh = du + p\, dv + v\, dp$$

Therefore

$$dh = \delta q + v\, dp \tag{3.60}$$

Substituting Equation (3.60) into Equation (3.59),

$$\delta w = -v\, dp - d(\text{K.E.}) - d(\text{P.E.}) \tag{3.61}$$

If the kinetic and potential energy variations are zero,

$$\delta w = -v\, dp \tag{3.62a}$$

$$\dot{W} = -\dot{m}\int v\, dp \tag{3.62b}$$

The work in Equation (3.61) is valid for reversible processes, as that assumption was used in expressing $\delta w = p\, dv$ for constant mass.

Example 3.5

A steam turbine, illustrated in Figure 3.12, receives steam with a flow rate of 15 kg/s and experiences a heat loss of 14 kW. Using the steam inlet and exit properties listed below, find the power produced.

	Inlet	Exit
Pressure	6205 kPa	9.859 kPa
Temperature	811.1 K	318.8 K
Velocity	$30.48\ \text{m}\cdot\text{s}^{-1}$	$274.3\ \text{m}\cdot\text{s}^{-1}$
Specific internal energy	3150.3 kJ/kg	2211.8 kJ/kg
Specific volume	$0.05789\ \text{m}^3/\text{kg}$	$13.36\ \text{m}^3/\text{kg}$

Open System, Steady-State

First Law: Energy In = Energy Out

$$q + u_1 + p_1 v_1 + (\text{K.E.})_1 + (\text{P.E.})_1$$

$$= u_2 + p_2 v_2 + (\text{K.E.})_2 + (\text{P.E.})_2 + w$$

$$q = \frac{14\ \text{kW}}{15\ \text{kg/s}} = \frac{14\ \text{kJ/s}}{15\ \text{kg/s}} = 0.933\ \text{kJ/kg}$$

$$w = (u_1 - u_2) + (p_1 v_1 - p_2 v_2) + \left(\frac{v_1^2 - v_2^2}{2}\right) + g(z_1 - z_2) + q$$

$$w = (3150.3 - 2211.8) + [(6205)(0.05789) - (9.859)(13.36)]$$

$$+ \left(\frac{30.48^2 - 274.3^2}{2}\right)\left(\frac{1}{1000}\ \frac{\text{kJ}}{\text{J}}\right) + (9.8)(3.0)\left(\frac{1}{1000}\right)$$

$$+ (-0.933)$$

$$w = 1127.93\ \text{kJ/kg} \qquad W = (1127.93)(15) = 16\,919\ \text{kW}$$

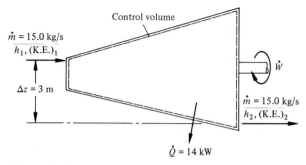

Figure 3.12   *The control volume for turbine (Example 3.5).*

Example 3.6

A nozzle is a device that converts enthalpy into kinetic energy. The kinetic energy is then usually used to drive a mechanical device such as a turbine wheel. The fluid energy is converted into mechanical work. Figure 3.13 shows a typical nozzle. Air enters the nozzle at a pressure of 2700 kPa, at a velocity of 30 m/s, and with an enthalpy of 923.0 kJ/kg, and leaves with a pressure of 700 kPa, and enthalpy of 660.0 kJ/kg. (a) If the heat loss is 0.96 kJ/kg, and if the mass flow rate is 0.2 kg/s, find the exit velocity. (b) Find the exit velocity for adiabatic conditions.

Open System, Steady-State

First Law: Energy In = Energy Out

$$q + h_1 + (K.E.)_1 = h_2 + (K.E.)_2$$

$$(K.E.)_2 = \frac{v_2^2}{2} = h_1 - h_2 + \frac{v_1^2}{2} + q$$

$$\frac{v_2^2}{2} = (923.0 - 660.0) + \frac{30^2}{(2)(1000)} + (-0.96)$$

$$= 262.94 \text{ kJ/kg}$$

(a) $v_2 = [(2)(1000)(262.94)]^{1/2} = 725.1 \text{ m/s}$

(b) $q = 0$

$$v_2 = [(2)(1000)(263.9)]^{1/2} = 726.5 \text{ m/s}$$

The previous examples have all illustrated the use of the first law of thermodynamics applied to open systems. A sketch of the physical system is drawn with the assumed direction for heat and work (dictated by common sense or experience or both). For a steady state, the energy terms entering and leaving are equated and the unknown term solved for

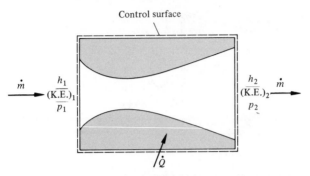

Control surface

Figure 3.13 *The control surface for the nozzle (Example 3.6).*

from this equation. Different situations, such as a nozzle and turbine, are analyzed in a similar manner. The procedure is to (a) draw a sketch; (b) label energy flows; (c) decide on the system (open or closed); (d) write the first law for the system in symbols, then with numbers, making certain that all terms have the same units; and (e) solve the equation for the unknown term.

Example 3.7

Figure 3.14 illustrates a windmill with blade diameter D, air velocity v, and density ρ entering the blades. Consider the windmill to be adiabatic and the exit air velocity from the blades to be negligible. Find the power produced.

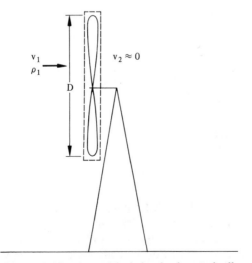

Figure 3.14 *A simplified sketch of a windmill.*

The system in this instance is an open one, the volume enclosing the blades, with the air passing through the volume.

First Law: Energy In = Energy Out

$$\dot{Q} + \dot{m}\left[h_1 + (\text{K.E.})_1 + (.\text{P.E.})_1\right] = \dot{m}\left[h_2 + (\text{K.E.})_1 + (\text{P.E.})_2\right] + \dot{W}$$

Since the elevation did not change, the temperature and pressure did not change and the system was adiabatic

$$\dot{Q} = \Delta h = \Delta(\text{P.E.}) = 0$$

$$\dot{W} = \dot{m}\left[(\text{K.E.})_1 - (\text{K.E.})_2\right] = \dot{m}\frac{v_1^2}{2}$$

From conservation of mass,

$$\dot{m} = \rho A v$$

$$A = \frac{\pi}{4}D^2$$

Therefore,

$$\dot{W} = \frac{\pi}{8}\rho D^2 v_1^3$$

We see that the power produced is a function of the velocity cubed. The above expression represents the ideal or maximum power that can be produced by a windmill.

3.6 SYSTEM BOUNDARIES

We have thus far looked at both open and closed systems and have seen that the system's boundaries are whatever we make them to be, but we should choose them as wisely as possible. Let us consider the case illustrated in Figure 3.15, where a force pushes on block over another block. If block A is considered the system, moving at a uniform velocity, we write the energy equation for it. The opposing friction force, F', is equal and opposite to the applied force, so the net force is zero. The work therefore is zero (net force through a distance); thus

$$Q = \Delta U$$

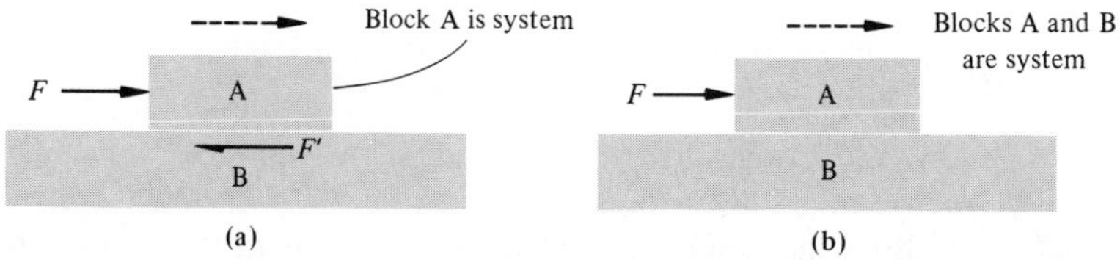

Figure 3.15   *Blocks illustrate the importance of choosing system boundaries wisely.*

In this case ΔU is the internal energy change of the block, and Q is the heat transferred from the surroundings to the block because of a temperature difference. No temperature difference occurs, however, and heat does not cross a boundary. The reason is that friction acts on the system boundary; thus either the equation or the definition of heat is in error. The problem arises because heat, by definition, is supposed to cross a boundary; in this case it does not.

This does not mean that the problem defies solution. We redefine the system to be both blocks A and B. In this case the friction and frictional force are within the system's boundaries. The surroundings do work by exerting a force F and moving block A a distance Δx. The internal energy of both blocks is raised because of frictional energy, and heat flows from the system to the surroundings. An energy balance on the combined block system yields

$$\text{energy in} + \text{energy initial} = \text{energy out} + \text{energy final}$$

$$W + U_1 = Q + U_2$$

$$(U_2 - U_1) = W - Q$$

Thus a realistic energy equation tells us that the change in the internal energy is equal to the energy in minus the energy out. The preceding example demonstrated that system boundaries are arbitrary, but useful if they are chosen wisely. It also illustrated that other physical laws, such as Newton's law, must be used in conjunction with thermodynamic laws to solve real problems.

We have just finished examining the first law for open and closed systems in which certain terms have been included in our system model.

Humans have tried to discover many parts of a substance's energy and catalog them. When we write the conservation of total energy equation in terms of the individual energies, we will neglect some energies on purpose or by chance. Thus what is derived is not the energy equation but a human approximation of what is happening physically. It should be noted that although these are approximations, their accuracy is very high.

Thus far we have been able to express the conservation of energy in terms of specific quantities but not the means to evaluate these quantities. The next two chapters will give us the tools to do so.

PROBLEMS

1. The weight of a bridge crane plus its load must exceed 100 metric tons (1 metric ton = 1000 kg). It is driven by a motor and travels at 1.17 m/s along the crane rails. Determine the energy that must be absorbed by the brakes in stopping the crane.

2. Determine the work required to accelerate a unit mass between the following velocity limits: (a) 10 m/s to 110 m/s; (b) 50 m/s to 150 m/s; (c) 100 m/s to 200 m/s.

3. It is required to lift five people on an elevator a distance of 100 m. The work is found to be 341.2 kJ and the gravitational acceleration is 9.75 m/s². Determine the average mass per person.

4. In the preceding problem the initial potential energy of the elevator was 68.2 kJ vis-à-vis the earth's surface. What is the height above the ground that the people were lifted?

5. A student is watching pilings being driven into the ground. From the size of the pile driver the student calculates the mass to be 500 kg. The distance which the pile driver is raised is measured to be 3 m. Determine the potential energy of the pile driver at its greatest height (the piling is considered the datum). Find the driver velocity just prior to impact with the piling.

6. A child (25-kg mass) is swinging on a play set. The swing's rope is 2.3 m long. Determine: (a) the change in potential energy from zero degrees with the horizontal when the swing goes from 0° to an angle of 45° with the vertical; (b) the velocity when the swing reaches the 0° angle.

7. A piston/cylinder contains air at a pressure of 500 kPa. The piston movement is resisted by a spring and atmospheric pressure of 100 kPa. The air moves the piston and the volume changes from 0.15 m³ to 0.60 m³. Determine the work when: (a) the force of the spring is directly proportional to the displacement; (b) the force of the spring is proportional to the square root of the displacement.

8. An automobile produces 20 kW when moving with a velocity of 50 km/h. Determine (a) the resisting force; (b) the resisting (drag) force if it is proportional to the velocity cubed and the automobile travels at 100 km/h.

9. An elastic sphere of 0.5 m diameter contains a gas at 115 kPa. Heating of the sphere causes it to increase to 0.62 m and during this process the pressure is proportional to the sphere diameter. Determine the work done by the gas.

10. A force, F, is proportional to x^2 and has a value of 133 N when $x = 2$. Determine the work done as it moves an object from $x = 1$ to $x = 4$, where x is in meters.

11. A fluid at 700 kPa, with a specific volume of 0.25 m³/kg and a velocity of 175 m/s enters a device. Heat loss from the device by radiation is 23 kJ/kg. The work done by the fluid is 465 kJ/kg. The fluid exits at 136 kPa, 0.94 m³/kg, and 335 m/s. Determine the change in internal energy.

12. An air compressor handles 8.5 m³/min of air with a density of 1.26 kg/m³ and a pressure of 1 atm and it discharges at 445 kPa (gage)

with a density of 4.86 kg/m³. The change in specific internal energy across the compressor is 82 kJ/kg and the heat loss by cooling is 24 kJ/kg. Neglecting changes in kinetic and potential energies, find the work in kilowatts.

13. A centrifugal pump compresses 3000 liter/min of water from 98 kPa to 300 kPa. The inlet and outlet temperatures are 25°C. The inlet and discharge piping are on the same level, but the diameter of the inlet piping is 15 cm, whereas that of the discharge piping is 10 cm. Determine the pump work in kilowatts.

14. Two gaseous streams containing the same fluid enter a mixing chamber and leave as a single stream. For the first gas the entrance conditions are $A_1 = 500$ cm², $v_1 = 730$ m/s, $\rho_1 = 1.60$ kg/m³. For the second gas the entrance conditions are $A_2 = 400$ cm², $\dot{m}_2 = 8.84$ kg/s, $v_2 = 0.502$ m³/kg. The exit stream condition is $v_3 = 130$ m/s and $v_3 = 0.437$ m³/kg. Determine (a) the total mass flow leaving the chamber; (b) the velocity of gas 2.

15. An insulated 2-kg box falls from a balloon 3.5 km above the earth. What is the change in internal energy of the box after it has hit the earth's surface?

16. A 4-mm steel wire with the Young's modulus of the material (E) equal to 2.067×10^8 kPa, a length of 4 m, is gradually subjected to an axial force of 5000 N. Determine the work done.

17. A soap bubble of a 15 cm radius is formed by blowing through a 2.5-cm diameter wire loop. Assume that all the soap film goes into making the bubble. The surface tension of the film is 0.02 N/m, find the total work required to make the bubble.

18. A closed system containing a gas expands slowly in a piston/cylinder from 600 kPa and 0.10 m³ to a final volume of 0.50 m³. Determine the work done if the pressure distribution is determined to be (a) $p = C$; (b) $pV = C$; (c) $pV^{1.4} = C$; (d) $p = -300V + 630$, where V is in cubic meters and p in kilopascals.

19. An elevator is on the thirtieth floor of an office building when the supporting cable shears. The elevator drops vertically to the ground where several large springs absorb the impact of the elevator. The elevator mass is 2500 kg and it is 100 m above the ground. Determine:
(a) the potential energy of the elevator before its fall.
(b) the velocity and kinetic energy the instant before impact.
(c) the energy change of the springs when they are fully compressed.

20. A closed gaseous system undergoes a reversible process in which 30 kJ of heat are rejected and the volume changes from 0.14 m³ to 0.55 m³. The pressure is constant at 150 kPa. Determine (a) the change in internal energy of the system; (b) the work done.

21. A fluid enters with a steady flow of 3.7 kg/s and an initial pressure of 690 kPa, initial density of 3.2 kg/m³, initial velocity of 60 m/s, and an initial internal energy of 2000 kJ/kg. It leaves at 172 kPa, $\rho = 0.64$ kg/m³, v = 160 m/s, and $u = 1950$ kJ/kg. The heat loss is found to be 18.6 kJ/kg. Find the work in kilowatts.

22. A 32-cm cube of ice at 0°C melts while being used to cool beer and soda at the beach. The specific volume of liquid water at 0°C is 1.002 cm³/gm and of ice at 0°C is 1.094 cm³/gm. Is there any work done by the surroundings, the atmosphere, on the ice?

23. Air and fuel enter a furnace used for home heating. The air has an enthalpy of 302 kJ/kg and the fuel an enthalpy of 43 027 kJ/kg. The gases leaving the furnace have an enthalpy of 616 kJ/kg. There are 17 kg air/kg fuel. Water circulates through the furnace wall receiving heat. The house requires 17.6 kW of heat. What is the fuel consumption per day?

24. An air compressor compresses air with an enthalpy of 96.5 kJ/kg to a pressure and temperature that have an enthalpy of 175 kJ/kg. There are 35 kJ/kg of heat lost from the compressor as the air passes through it. Neglecting kinetic and potential energies, determine the power required for an air mass flow of 0.4 kg/s.

25. A steam condenser receives 9.47 kg/s of steam with an enthalpy of 2570 kJ/kg. The steam condenses to a liquid and leaves with an enthalpy of 160.5 kJ/kg. (a) Find the total heat transferred from the steam. (b) Cooling water passes through the condenser with an unknown flow rate; however, the water temperature increases from 13°C to 24°C. Also, it is known that 1 kg of water will absorb 4.2 kJ of energy per degree temperature rise. Find the cooling water flow rate.

26. Steam enters a turbine with a pressure of 4826 kPa, $u = 2958$ kJ/kg, $h = 3263$ kJ/kg, and with a flow of 6.3 kg/s. Steam leaves with $h = 2232$ kJ/kg, $u = 2102$ kJ/kg, and $p = 20.7$ kPa. There is radiative heat loss equal to 23.3 kJ/kg of steam. Determine (a) the power produced, (b) the adiabatic work, (c) the inlet specific volume, (d) the exit velocity if the exit area is 0.464 m².

27. Steam with a flow rate of 1360 kg/h enters an adiabatic nozzle at 1378 kPa, 3.05 m/s, with a specific volume of 0.147 m³/kg, and with a specific internal energy of 2510 kJ/kg. The exit conditions are $p = 137.8$ kPa, specific volume = 1.099 m³/kg, and internal energy = 2263 kJ/kg. Determine the exit velocity.

28. An injector was used on steam locomotives as a means of providing water to the boiler at a pressure higher than the steam pressure. The injector looks like a T with steam entering horizontally and water entering vertically from the bottom. The fluids mix and discharge

horizontally; 18.2 kg/min of steam enters at 1378 kPa and an internal energy of 2590 kJ/kg, and with a specific volume of 0.143 m³/kg. The water enters 1.52 m below the horizontal with an enthalpy of 42 kJ/kg. The mixture has a pressure of 1550 kPa, an internal energy of 283.8 kJ/kg, and a specific volume of 0.001024 m³/kg. The ratio of water to steam is 10.5 to 1. Neglect changes in kinetic energy. Determine the heat transfer in kilowatts.

Chapter 4
Properties of
Pure Substances

Thus far in the discussion of thermodynamic principles, the substance in the system has not been detailed. We have assumed that it is a gas and that it is homogeneous. Although this is often the case, there are other phases of a substance, and we must be able to analyze these. Homogeneity means that the substance is the same in any part of the system as it is in any other part. This means that if we sample the system here and there and analyze the various samples, they will be the same; we could not distinguish one from the other. In the previous chapter we developed the conservation of energy equations for open and closed systems. There are several properties such as enthalpy, internal energy, and specific volume that must be evaluated; that is a purpose of this chapter.

4.1 PURE SUBSTANCE

Let us pick water as the substance in the system. The water can be a liquid, solid, or vapor. If we sample the system now, the samples will be distinguishable by the various phases of water. Water is an example of a pure substance, which may exist as liquid, solid, or vapor, phases by which all pure substances are characterized. Recall that the definition of a pure substance is that it is homogeneous by nature, does not undergo chemical reactions, and is not a mechanical mixture of different species. The substances in our systems considered heretofore have been phases of a substance. The phases are physically homogeneous. There is usually a sharp distinction between phases, inasmuch as the properties of one phase are decidedly different from those of another phase. There is a difference between ice and liquid water, for example.

4.2 LIQUID–VAPOR EQUILIBRIUM

Where do these phase changes occur and how can these changes be quantified and used? To answer this, let us consider the piston–cylinder arrangement in Figure 4.1(a). This is just a scientific way of considering a pot of water on a kitchen stove with the weightless top sitting on the water instead of the edge of the pot. Let us return to Figure 4.1. Heat is added at constant pressure, $q = \Delta h$, to the liquid water initially at 40°C. As heat is added, the water temperature rises until it reaches 100°C; see Figure 4.1(b). Then what happens? The water boils, as shown in Figure 4.1(c). What is the boiling process? It is a phase transition. The water is going from liquid to vapor. In Figure 4.1(d) all the liquid has just changed to vapor and finally, as more heat is added, the temperature of the vapor increases.

4.3 SATURATED PROPERTIES

Let us plot a $T–v$ diagram showing this process for one atmosphere of pressure (Figure 4.2). When the water is at point b, it is called a "saturated liquid." This means that it is at the highest temperature at which, for this pressure, it can remain liquid. If more heat is added, some of the liquid changes to vapor and a mixture of vapor and liquid occurs, such as at point c. This vapor is called a "saturated vapor." At point d, all the water exists as a saturated vapor. Any addition of heat results in the vapor's being superheated, such as the vapor at point e. This is called a "superheated vapor." The vapor has a temperature greater than the saturation temperature, the temperature of the water when it is saturated liquid and vapor, for a given pressure. Note that the temperature of the water does not change until it has all changed phase. All the heat added

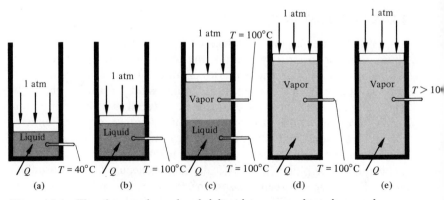

Figure 4.1 The change of a subcooled liquid to a superheated vapor by constant-pressure heat addition.

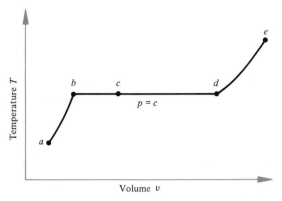

Figure 4.2   Constant-pressure heat addition to water.

during this phase transition goes into forcing the water in the liquid phase to change into the vapor phase.

What about point *a*? The water at point *a* is called a "subcooled liquid" because its temperature is less than the saturated temperature for this pressure.

4.4 CRITICAL PROPERTIES

Now let us run the same test, but at different pressures, and plot them on a *T–v* diagram (Figure 4.3). Furthermore, let us connect the locus of points *b* and the locus of points *c*. These are called the "saturated liquid" line and the "saturated vapor" line, respectively. What about the point of

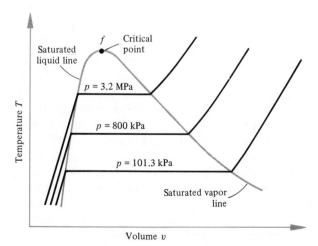

Figure 4.3   The T–v diagram for water at various pressures.

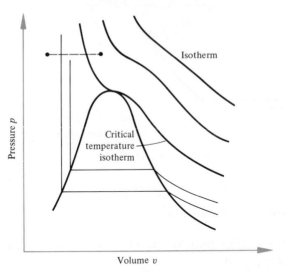

Figure 4.4 *A p–v diagram for water showing lines of constant temperature.*

inflection, point f on Figure 4.3? This is called the "critical point." This point has a unique temperature and pressure known as, not surprisingly, the critical temperature and critical pressure. At pressures higher than the critical pressure, the liquid could be heated from a low temperature to a high temperature without a phase transition occurring. This is illustrated by the dotted line on Figure 4.4, which illustrates the $p–v$ diagram for the liquid–vapor phases of water. At temperatures that are greater than the critical temperature, the pressure may be increased to very high values and no liquefaction will occur.

4.5 SOLID–LIQUID–VAPOR EQUILIBRIUM

Let us consider the solid phase of water, ice. We take a piece of ice at $-17.7°C$ and put it in a vacuum chamber until the pressure is 348 Pa. We now heat the ice. The temperature of the ice will rise to $-6.67°C$ and then further addition of heat will cause the solid water to go directly to water vapor at the same temperature and pressure. This is called "sublimation." The pressure of the system for the next case is raised to 610.8 Pa. Again we start with the ice at $-17.7°C$ and heat it. The temperature will rise to $0.0°C$. Further heating causes some of the solid water to turn to liquid and some to vapor. This point, characterized by this temperature and pressure, is called the "triple point" for water. It is the only point at which all three phases may coexist. If the pressure is further increased from the initial setting to 1.0 kPa and heat is supplied, the ice will rise to $0.0°C$ and change to liquid. This is called "melting." These processes are illustrated in Figure 4.5.

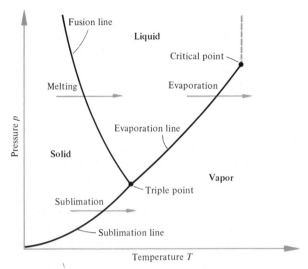

Figure 4.5 *A p–T diagram showing phase equilibrium lines, the triple point, and the critical point.*

The amount of heat added to effect the various phase changes is equal to the change in enthalpy, as noted earlier. These various enthalpy differences across the phase boundaries have certain names. The change of enthalpy between a solid and liquid phase is called the "latent heat of fusion." This is somewhat of a misnomer as heat refers only to thermal energy crossing a boundary, but the term was developed and adopted before this classification. The change of enthalpy between a liquid and vapor phase is called the "latent heat of vaporization." Finally, the change of enthalpy in going from a solid to a vapor phase is called the "latent heat of sublimation."

Referring to Figure 4.2, it is possible to denote the state of water by knowing the temperature and pressure if the water is a subcooled liquid or superheated vapor. This means the lines of constant temperature and constant pressure cross at some unique point. This point is the state of the system at this pressure and temperature. However, what if the state lies somewhere between points *b* and *d*, in the two-phase region? The lines of temperature and pressure are coincident, so they do not uniquely locate the system state. How could the state be located? Knowing the temperature and pressure gives us one line. If the fraction of vapor were known, then we could find out how much water had been evaporated and where along the *a-b* line the system existed. It is important to realize that we need two independent intensive properties to determine the state of a pure substance. In the saturated mixture region, temperature and pressure are not independent and thus do not define a state. This is demonstrated graphically by their being coincident and not intersecting.

4.6 QUALITY

We now define a quantity, x, called the "quality," as the mass of vapor in the system divided by the total system mass (the mass of vapor plus the mass of liquid). Note that the definition of quality presumes a homogeneous mixture of vapor and liquid.

$$x = \frac{\text{mass vapor}}{\text{mass vapor} + \text{mass liquid}} \qquad (4.1)$$

Let us see then whether we can find the value of some extensive property on a unit mass basis. The specific enthalpy, h, is desired at point c on Figure 4.2, and point c is characterized by a quality x. Let the enthalpy at state b be h_b, and at state d let it be h_d. The change of enthalpy in going from b to d is $h_{bd} = h_b - h_d$. To find the enthalpy of state c, we consider first that all the water was initially at state b, with an enthalpy h_b. Then a fraction of the water was evaporated increasing its enthalpy by h_{bd}. The total enthalpy of the mixture at state c is

$$h_c = h_b + x h_{bd} \qquad (4.2a)$$

or

$$h_c = h_d - (1 - x) h_{bd} \qquad (4.2b)$$

This was illustrated for specific enthalpy, but it is valid for all specific extensive properties.

4.7 THREE-DIMENSIONAL SURFACE

Water is not the only pure substance with which thermodynamics concerns itself. We can and do investigate any substance. Another pure substance is carbon dioxide, which is used in refrigeration cycles and has a T–p diagram different from that of water. The point is that one should not expect all phase diagrams to be identical to those of water. The various phases and phase transitions are present, and that is the similarity.

Any two-dimensional diagram is really a projection of that three-dimensional surface on that plane. Figure 4.6(a) illustrates the three-dimensional surface from which the pressure-temperature and pressure–volume diagrams are projected.

The surface and projections in Figure 4.6(a) are for a substance that contracts on freezing. The volume contracts on freezing, which means that the freezing temperature increases as the pressure increases. This is not true with water, as we know; the volume of solid water is greater than that of the same mass of liquid. The result of this is that ice floats and we can enjoy ice skating. Furthermore, all aquatic life would be destroyed if the solid phase were denser than the liquid phase. The three-dimensional

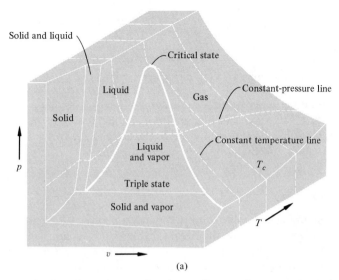

Figure 4.6(a) A PvT surface for a substance that contracts on freezing.

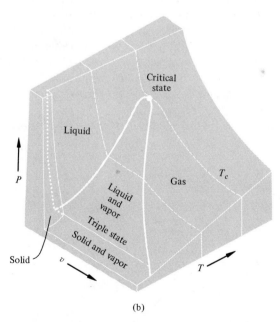

Figure 4.6(b) A PvT surface for a substance that expands on freezing.

diagram for water is illustrated in Figure 4.6(b). Note the difference in the solid–liquid interface.

4.8 TABLES OF THERMODYNAMIC PROPERTIES

As we said before, it certainly would be convenient if tables or charts, or both, existed listing the values of the thermodynamic functions. Fortunately, all this work has been done for us. In the appendixes of the text there are several tables and diagrams for various substances.

Saturated Steam

We shall consider the steam tables first (Appendix Tables A.5 to A.8). These are an abridged edition of the Steam Tables by Keenan, Keyes, Hill, and Moore. How are these tables organized? Tables A.5 and A.6 are concerned only with the properties in the saturated region, region II, on Figure 4.7. The difference in the tables is that in Table A.5 the temperature is the independent variable; it is in even increments, and in Table A.6 the pressure is in even increments. Let us consider an example using Table A.6. The specific volume, specific internal energy, specific enthalpy, and specific entropy are tabulated. Let the symbol r stand for any of these properties. The following symbols are used: r_f is the value of the property as a saturated liquid, denoted by point f on Figure 4.7; r_g is the value of the property as a saturated vapor, denoted by point g on Figure 4.7; r_{fg} is the difference between r_g and r_f, $r_{fg} = r_g - r_f$. This is sometimes called the change in the property due to evaporation.

We already have seen how to use quality, so why not test our knowledge on an example?

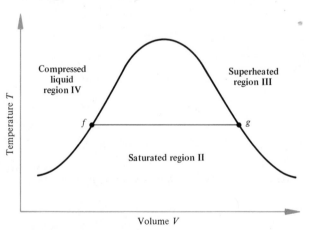

Figure 4.7 A T–v diagram illustrating three regions included in the steam tables.

Example 4.1

Find the enthalpy and specific volume of steam at 250 kPa and 50% quality.

Note the values of the properties as saturated liquid and vapor.

$h_f = 535.37 \text{ kJ/kg} \qquad v_f = 0.001\ 067 \text{ m}^3/\text{kg}$

$h_{fg} = 2181.5 \text{ kJ/kg} \qquad v_g = 0.7187 \text{ m}^3/\text{kg}$

$h_g = 2716.9 \text{ kJ/kg}$

At 50% quality,

$$h = h_f + xh_{fg} \tag{4.3a}$$

$$h = h_g - (1 - x)h_{fg} \tag{4.3b}$$

Using Equation (4.3a),

$$h = 535.37 + (0.50)(2181.5) = 1626.1 \text{ kJ/kg}$$

Thus the value of enthalpy is determined. The other properties are similarly calculated. The specific volume is slightly different in that v_{fg} is not given.

$$v_{fg} = 0.7187 - 0.001\ 067 = 0.7176$$

$$v = v_f + xv_{fg}$$

$$v = 0.001\ 067 + (0.50)(0.7176) = 0.3598 \text{ m}^3/\text{kg}$$

If we were asked to calculate the internal energy, how would we proceed? If we had the steam tables available, all we would have to do is to look up the values of u for the particular temperature or pressure or for both. Using the conditions stated in the previous example,

$$u = u_f + xu_{fg}$$

$$u = 535.1 + (0.50)(2002.1) = 1536.1 \text{ kJ/kg}$$

However, what would happen if we had the results of the previous example and did not have access to the steam tables? How is the enthalpy defined?

$$h = u + pv$$

We know all the quantities except u, so u may be calculated:

$$u = h - pv = 1626.1 - (250)(0.3598)$$

$$u = 1536.1 \text{ kJ/kg}$$

The units of h, pv, and u must be the same, pressure must be in kilopascals for common units of kilojoules. The internal energy is the same, as should be expected.

Superheated Steam

Having a certain amount of confidence in finding the values of properties in the saturated region, region II, let us consider properties in the superheated region, region III. Water vapor existing in this region must be defined by two independent properties before the state can be determined. Usually one of these properties is the pressure, and frequently the other is the temperature.

Example 4.2

Find the enthalpy and specific volume of steam at 200 kPa and 300°C.

Where do we find steam with these properties? In Table A.7. Why? Because the temperature of the steam, 300°C, is greater than the saturated steam temperature, 120.2°C, for the pressure of 200 kPa. If Tables A.5 and A.6 are to be used, the steam must lie in region II and must have saturated temperatures and pressures.

We proceed to the 200-kPa column and go vertically down the column until it intersects with the 300°C line. At this line we read the values of h and v

$$h = 3071.8 \text{ kJ/kg}$$

$$v = 1.3162 \text{ m}^3/\text{kg}$$

We now consider the situation in which we are given steam with a pressure of 200 kPa and internal energy of 2966.7 kJ/kg. We want to find the temperature enthalpy and the specific volume.

Why would Table A.7 be used? We check the maximum value of internal energy that steam could have if it were in region II. That value of u_g at 200 kPa is less than the value of u that is given. Since u is greater than u_g, and steam with this value of internal energy cannot lie in region II. It must be in region III. If steam had a value of u between u_f and u_g for a given pressure, then it would lie in region II. The solution to this problem is simple. We proceed vertically down the 200-kPa column in the superheat tables until the value of u is attained and then read the property values at this level:

$$h = 3276.6 \text{ kJ/kg}$$

$$v = 1.5493 \text{ m}^3/\text{kg}$$

$$T = 400°C$$

Subcooled Liquid Water

This time let us consider water at 10 MPa and 60°C. We want to find the enthalpy and specific volume at this condition. We might suspect that the water is a subcooled liquid in region IV on Figure 4.7 and having property values located in Table A.8. Why? Again we check the saturated water temperature at 10 MPa. It is 311.06°C. If the temperature is less than this saturated value for water with a pressure of 10 MPa, the water is subcooled and its properties may be found in Table A.8. Since the conditions in the problem meet this criteria, Table A.8 will be used for the solution. We go to the column for 10 MPa and proceed vertically down until it intersects with the 60°C line. At this level we read the property values as

$h = 259.49 \text{ kJ/kg}$

$v = 0.001\ 012\ 7 \text{ m}^3/\text{kg}$

Determination of Saturated Vapor State

There are tabulated values for vapor properties, but it is sometimes very difficult to measure the vapor state. Let us consider a situation in which steam is driving a turbine and we want to know the steam state as it enters the turbine.

This means finding the state of a flowing vapor. If the vapor is superheated, then measuring the temperature and pressure would determine the state. What happens if the steam is a saturated mixture? The temperature and pressure are not independent in the saturated region, so more information is needed to determine the state. There is no ready means to measure other properties, such as specific volume, enthalpy, or internal energy. There are several methods to resolve this problem, one of the most common being a throttling calorimeter.

Figure 4.8 schematically illustrates a throttling calorimeter. Notice that the steam supply is sampled from a vertical pipe. If the pipe were horizontal, separation of liquid and vapor would be more apt to occur. A throttling process is one in which the pressure is adiabatically decreased by use of a valve. This is a totally irreversible process. A first-law analysis across the adiabatic valve shows that the initial and final enthalpies have the same value. If there is a sufficient decrease in pressure, the steam will be superheated at a lower pressure and temperature.

The temperature and pressure measurement can be made for the steam leaving the valve (in the cup) and the enthalpy determined. It is important that there be negligible heat loss and negligible velocity at the point of temperature and pressure measurement. By knowing the enthalpy and the initial pressure or temperature, the initial steam state may

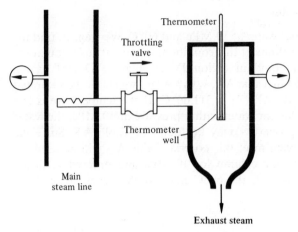

Figure 4.8   Schematic diagram of a throttling steam calorimeter.

be determined. This is the average value of the steam state, as it is based on a representative sample.

There are other types of calorimeters, which operate with varying degrees of success. A problem with the throttling calorimeter is that the initial steam state must not be very wet or throttling will not produce superheated steam. A separating calorimeter mechanically separates the liquid and vapor, volumetrically measuring the liquid and measuring the remaining vapor flow rate. An electric calorimeter, through resistance heating, superheats a sample of steam steadily withdrawn through a sampling tube. By measuring the electrical energy added and the superheated steam state and steam flow rate, the steam quality is determined.

Example 4.3

A throttling calorimeter is connected to the desuperheated steam line supplying steam to the auxiliary feed pump on a ship. The line pressure measures 2.5 MPa. The calorimeter pressure is 110 kPa and the temperature is 150°C. Determine (a) the line steam quality; (b) the line steam enthalpy.

Interpolating to find state 2 in the superheat tables,

State 2

$p_2 = 110 \text{ kPa} \qquad T_2 = 150°C$

$h_2 = 2775.6 \text{ kJ/kg}$

State 1

$$h_1 = h_2 = h_f + xh_{fg} \qquad p_1 = 2.50 \text{ MPa}$$
$$2775.6 = 962.11 + x(1841.0)$$
$$x = 0.985 \quad \text{or} \quad 98.5\%$$

Ammonia and Refrigerant-12 Tables

Two other substances have property tables that are listed in the appendix. These are ammonia and Refrigerant 12 (R 12), dichlorodifluromethane. Both are used as refrigerants, especially R 12. Although it is not necessary to repeat all the problem examples that can be done with these tables, as the techniques developed in using the steam tables are now equally valid, it is worthwhile to do some. The thermodynamic properties of ammonia are listed in Tables A.9 and A.10 and the thermodynamic properties of Refrigerant 12 are listed in Tables A.11 and A.12.

Example 4.4

Ammonia is in a container and found to have a pressure of 250 kPa and a temperature of 30°C. How much ammonia is in the 0.5 m³ container?

First, we note that the ammonia vapor is superheated. Why? Because the temperature of 30°C is greater than the saturated temperature of ammonia at 250 kPa, which is $-13.67°C$, approximately. This is found by looking at the saturated Table A.9. The specific volume at 30°C and 250 kPa is

$$v = 0.5780 \text{ m}^3/\text{kg}$$

The total volume is 0.50 m³.
   The mass is

$$m = \frac{V}{v} = \frac{0.5}{0.5780} = 0.865 \text{ kg}$$

Example 4.5

Refrigerant 12 is used in a refrigeration unit. At one stage in the process, it is a hot gas leaving the compressor at 1.0 MPa and 70°C. Determine the velocity in a 2.5-cm-diameter pipe, if the flow rate is 3 kg/min.

Is the R 12 superheated or saturated? Why? It is superheated because its temperature, 70°C, is greater than the saturation temperature at 1.0 MPa, 41.6°C. We can determine the properties of the R 12 at this state. In particular, we are interested in the specific volume

$$v = 0.020\ 397\ \text{m}^3/\text{kg}$$

The specific volume is known; how can the velocity be determined? Since the mass flow rate is given and the pipe area may be calculated, this should provide the solution. Either remembering the equation for one-dimensional mass flow, or figuring it out by a unit balance as follows, yields

$$\dot{m} = \frac{A\text{v}}{v}$$

$$\text{v} = \frac{\dot{m}v}{A} = \frac{(3.0)(0.020\ 397)(4)}{\pi(0.025)^2}$$

$$\text{v} = 124.6\ \text{m/min} = 2.07\ \text{m/s}$$

Example 4.6

A component in a refrigeration system is the evaporator. It is here that the cold refrigerant picks up heat from the refrigerated space. Figures 4.9(a) and 4.9(b) illustrate the process. The flow rate through the evaporator is 25 kg/h. Determine the heat received from the refrigerated space. The refrigerant is R 12 entering at −30°C as a saturated liquid and leaving at +10°C.

This is an open system (why?). Perform an energy balance on the system, in this case the evaporator.

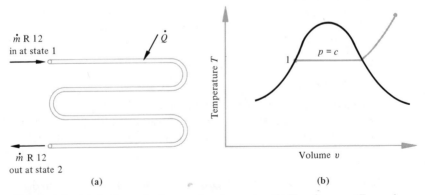

$\dot{m}$ R 12 in at state 1

$\dot{Q}$

$\dot{m}$ R 12 out at state 2

Temperature T

$p = c$

1

Volume v

(a)   (b)

Figure 4.9   (a) A schematic diagram for evaporator. (b) T–v diagram (Example 4.6).

First Law: Energy In = Energy Out

$$\dot{m}h_1 + \dot{Q} = \dot{m}h_2$$

$$\dot{Q} = \dot{m}(h_2 - h_1)$$

Find h_1 from Table A.11. It is the saturated liquid enthalpy at $-30°C$.

$$h_1 = 8.854 \text{ kJ/kg}$$

Find h_2 at $+10°C$. There is not enough information given to locate state 2. What other information can be used? The type of process. Heat exchangers are constant-pressure devices. So the pressure at state 2 is the same as the pressure at state 1. What is the pressure at state 1? It is the saturated pressure for that temperature, $-30°C$, since the problem specified that the R 12 was a saturated liquid when it entered the evaporator.

$$p_1 = 0.1004 \text{ MPa} \qquad p_2 = 0.1004 \text{ MPa}$$

Now there is enough information to find h_2 uniquely. It is a superheated vapor and has a value of

$$h_2 = 197.628 \text{ kJ/kg}$$

$$\dot{Q} = \dot{m}(h_2 - h_1) = \frac{(25.0)(197.628 - 8.854)}{3600}$$

$$\dot{Q} = 1.31 \text{ kW}$$

PROBLEMS

1. A 2-m^3 tank contains a saturated vapor at 40°C. Determine the pressure and mass in the tank if the substance is (a) steam; (b) ammonia; (c) R 12.
2. Determine for R 12 the following:
   (a) h if $T = 85°C$ and $p = 1000$ kPa.
   (b) x if $h = 100$ kJ/kg and $T = 0°C$.
   (c) u if $T = 100°C$ and $p = 800$ kPa.
   (d) p if $T = 20°C$ and $v = 0.001\ 020\ m^3/kg$.
3. Complete the following table for ammonia.

	T, °C	p, kPa	x, %	h, kJ/kg	u, kJ/kg	v, m^3/kg
(a)	10			1225.5		
(b)	50	700				
(c)		752.79	80		1200.0	
(d)	12	1000				
(e)	0		90			
(f)		1554.3	80			

4. Determine the volume occupied by 2 kg of steam at 1000 kPa and 500°C.
5. Complete the following table for water.

	T, °C	p, kPa	x, %	h, kJ/kg	u, kJ/kg	v, m³/kg
(a)	200			852.4		
(b)		150			1000.0	
(c)	300	800				
(d)	200	5000				
(e)		300				0.8500
(f)	300		80			
(g)		1000	90			

6. R 12 is contained in a storage bottle with a diameter of 20 cm and a length of 120 cm. The weight of the R 12 is 370 N ($g = 9.8$ m/s²) and the temperature is 20°C. Determine (a) the ratio of mass vapor to mass liquid in the cylinder; (b) the height of the liquid–vapor line if the bottle is standing upright.
7. Steam has a quality of 90% at 200°C. Determine (a) the enthalpy; (b) the specific volume.
8. A refrigeration system uses R 12 as the refrigerant. The system is evacuated, then charged with refrigerant at a constant 20°C temperature. The system volume is 0.018 m³. Determine (a) the pressure and quality when the system holds 0.8 kg of R 12; (b) the mass of R 12 in the system when the pressure is 200 kPa.
9. A rigid steel tank contains a mixture of vapor and liquid water at a temperature of 65°C. The tank has a volume of 0.5 m³, the liquid phase occupying 30% of the volume. Determine the amount of heat added to the system to raise the pressure to 3.5 MPa.
10. Steam enters an isothermal compressor at 400°C and 100.0 kPa. The exit pressure is 10 MPa, determine the change of enthalpy.
11. Steam enters an adiabatic turbine at 300°C and 400 kPa. It exits as a saturated vapor at 30 kPa. Determine (a) the change of enthalpy; (b) the work; (c) the change of internal energy.
12. A 0.5-m³ tank contains saturated steam at 300 kPa. Heat is transferred until the pressure reaches 100 kPa. Determine (a) the heat transferred; (b) the final temperature; (c) the final steam quality; (d) the process on a T–v diagram.
13. A 500-liter tank contains a saturated mixture of steam and water at 300°C. Determine (a) the mass of each phase if their volumes are equal; (b) the volume occupied by each phase if their masses are equal.

14. A 1-kg steam–water mixture at 1.0 MPa is contained in an inflexible tank. Heat is added until the pressure rises to 3.5 MPa and the temperature to 400°C. Determine the heat added.

15. A rigid vessel contains 5 kg of wet steam at 0.4 MPa. After the addition of 9585 kJ the steam has a pressure of 2.0 MPa and a temperature of 708°C. Determine the initial internal energy and specific volume of the steam.

16. Two kilograms per minute of ammonia at 800 kPa and 70°C are condensed at constant pressure to a saturated liquid. There is no change in kinetic or potential energy across the device. Determine (a) the heat; (b) the work; (c) the change in volume; (d) the change in internal energy.

17. Three kilograms of steam initially at 2.5 MPa and a temperature of 400°C have 2460 kJ of heat removed at constant temperature until the quality is 90%. Determine (a) $T–v$ and $p–v$ diagrams; (b) pressure when dry saturated steam exists; (c) work.

18. You want 400 liters/min of water at 80°C. Cold water is available at 10°C and dry saturated steam at 200 kPa (gage). They are to be mixed directly. Determine (a) steam and water flow rates required; (b) the pipe diameters, if the velocity is not to exceed 2 m/s.

19. Steam condensate at 1.75 MPa leaves a heat exchanger trap and flows at 3.8 kg/s to an adjacent flash tank. Some of the condensate is flashed to steam at 175 kPa and the remaining condensate is pumped back to the boiler. There is no subcooling. Determine (a) the condensate returned at 175 kPa; (b) the steam produced at 175 kPa.

20. A chemical process requires 2000 kg/h of hot water at 85°C and 150 kPa. Steam is available at 600 kPa and 90% quality, and water is available at 600 kPa and 20°C. The steam and water are mixed in an adiabatic chamber, with the hot water exiting. Determine (a) the steam flow rate; (b) the steam-line diameter if the velocity is not to exceed 70 m/s.

21. The main steam turbine of a ship is supplied with steam from two steam generators. One steam generator delivers steam at 6.0 MPa and 500°C, and the other delivers steam at 6.0 MPa and 550°C. Determine the steam entrance enthalpy and temperature to the turbine.

22. An adiabatic feed pump in a steam cycle delivers water to the steam generator at a temperature of 200°C and a pressure of 10 MPa. The water enters the pump as a saturated liquid at 180°C. The power supplied to the pump is 75 kW. Determine (a) the mass flow rate; (b) the volume flow rate leaving the pump; (c) the percentage of error if the exit conditions are assumed to be a saturated liquid at 200°C.

23. An adiabatic rigid tank has two equal sections of 50 liters separated by a partition. The first section contains steam at 2.0 MPa and 95% quality. The second section contains steam at 3.5 MPa and 350°C.

Determine the equilibrium temperature and pressure when the partition is removed.

24. A throttling calorimeter is attached to a steam line where the steam temperature reads 210°C. In the calorimeter the pressure is 100 kPa and the temperature is 125°C. Determine the quality of the steam using the steam tables.

25. A throttling calorimeter is connected to a main steam line where the pressure is 1750 kPa. The calorimeter pressure is 100 mm Hg vacuum and 105°C. Determine the main steam quality.

26. A piston–cylinder containing steam at 700 kPa and 250°C undergoes a constant-pressure process until the quality is 70%. Determine per kilogram (a) the work done; (b) the heat transferred; (c) the change of internal energy; (d) the change of enthalpy.

27. An electric calorimeter samples steam with a line pressure of 0.175 MPa. The calorimeter adds 200 W of electricity to sampled steam having a resultant pressure of 100 kPa, temperature of 140°C, and a flow rate of 11 kg/h. Determine the main steam quality.

28. Three kilograms of ammonia are expanded isothermally in a piston —cylinder from 1400 kPa and 80°C to 100 kPa. The heat loss is 495 kJ/kg. Determine (a) the system work; (b) the change of enthalpy; (c) the change of internal energy.

29. Refrigerant 12 is expanded steadily in an isothermal process. The flow rate is 13.6 kg/min with an inlet state of wet saturated vapor with an 80% quality to a final state of 70°C and 200 kPa. The change of kinetic energy across the device is 3.5 kJ/kg and the heat added is 21.81 kW. Determine the system power.

Chapter 5
Ideal Gas and
Specific Heat

In this chapter the working substance is considered to be a gas. This is only one of three phases—solid, liquid, or vapor (gas)—of a substance, but it is a frequently found fluid state, such as air at atmospheric pressure, for example, and deserves special attention. The vapor or gas phase may be best modeled mathematically, thus permitting more complete analysis.

5.1 IDEAL-GAS EQUATION OF STATE

There are several equations of state for gases, the most common being the ideal-gas equation of state, which relates the dependency of pressure, volume, temperature, and mass at a state. Not all states are allowed, so we consider Figure 5.1, a simplified version of the equilibrium surfaces in Chapter 4.

States A, B, and C are shown. States A and B are on the surface; these represent all possible equilibrium states for a substance. Remember that it is only the equilibrium states for which properties are defined. State C is not an equilibrium state; it is not possible for a substance to remain in that state for any period of time. The system could pass through state C in going from A to B, but it could not exist there. The ideal-gas equation of state represents all the states on the AB surface, relating the various properties to each other. The ideal-gas equation of state, often called the ideal-gas law, is

$$pV = mRT \tag{5.1}$$

where p is the absolute pressure, V is the total volume, m is the mass, T is the absolute temperature, and R is the individual gas constant. This equation comes in many forms; dividing by the mass yields

$$pv = RT \tag{5.2}$$

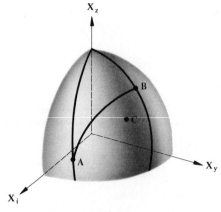

Figure 5.1   *Diagram of an equilibrium surface with points A and B on the surface and point C below the surface.*

Equations (5.1) and (5.2) are the most frequently used forms of the ideal-gas equation of state. The value of R is tabulated in Appendix Table A.1 for several different gases.

The gas constant R is not a magical quantity and may be calculated for any gas if its molecular weight is known. Avogadro's law states that equal volumes of an ideal gas at the same temperature and pressure have equal numbers of molecules. One gram mole of a substance has 6.023×10^{23} molecules (Avogadro's number). If M is the molecular mass, $Mv = \bar{v}$, multiply Equation (5.2) by it:

$$p\bar{v} = MRT \tag{5.3a}$$

$$p\bar{v} = \bar{R}T \tag{5.3b}$$

where $\bar{R} = MR$.

The constant $\bar{R}$ is called the "universal gas constant," having a value of

$$\bar{R} = 8.3143 \text{ kJ/kgmol} \cdot \text{K} \tag{5.4}$$

Note that $n\bar{v} = V$, where n is the number of moles, and Equation (5.3b) becomes

$$pV = n\bar{R}T \tag{5.5}$$

Equations (5.1) and (5.2) are the ones most often used, but Equation (5.5) is used when dealing with chemical reactions. The abbreviation "kgmol" stands for the molecular weight expressed in kilograms.

Lest we forget that the ideal gas equation of state represents not all states, but all possible equilibrium states, we should consider the following example.

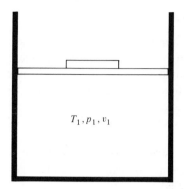

Figure 5.2   *A closed-system piston–cylinder arrangement (Example 5.1).*

Example 5.1

Air is considered to be an ideal gas with a value of $R = 0.287$ kJ/kg · K. If there are 2 kg of air in a piston–cylinder (Figure 5.2) at a temperature of 280 K, a volume of 0.2 m³, find the pressure.

$$p = \frac{mRT}{V} = \frac{(2 \text{ kg})(0.287 \text{ kJ/kg} \cdot \text{K})(280 \text{ K})}{(0.2 \text{ m}^3)}$$

$p = 803.6$ kPa

On the other hand, if the pressure were 1200 kPa and the temperature were unknown, we could find the temperature as follows:

$$T = \frac{pV}{mR} = \frac{(1200 \text{ kN/m}^2)(0.2 \text{ m}^3)}{(2 \text{ kg})(0.287 \text{ kJ/kg} \cdot \text{K})}$$

$T = 418.1$ K

Example 5.2

A student states that he has calculated the pressure, temperature, and specific volume of the gas in Example 5.1 to be 500 kPa, 600 K, and 0.12 m³/kg, respectively. Is this possible?

$pv = RT$ $RT = (0.287)(600)$
$pv = (500)(0.12)$ $RT = 172.2$ kJ/kg
$pv = 60$ kJ/kg $\therefore pv \neq RT$ not possible

There are other gas equations of state, most of which work well when the gas is not near liquefaction. When a phase transition is necessary, the results are not accurate.

5.2 ACTUAL-GAS EQUATIONS OF STATE

All gases that we analyze are actual gases, not ideal gases. In actual gases the molecular collisions are inelastic; at high densities in particular there are intermolecular forces that the simplified equations of state do not account for. There are many gas equations of state that attempt to correct for the nonideal behavior of gases. The disadvantage of all methods is that the equations are more complex and require the use of experimental coefficients.

One of the equations is the Van der Waals equation of state for a gas, which was developed in 1873 as an improvement on the ideal gas law. The Van der Waals equation of state is

$$p = \frac{\bar{R}T}{\bar{v} - b} - \frac{a}{\bar{v}^2} \tag{5.6}$$

The coefficients a and b compensate for the nonideal characteristics of the gas. The constant b accounts for the finite volume occupied by the gas molecules, and the a/v^2 term accounts for intermolecular forces. Constants for selected gases are given in Table 5.1.

A second equation of state is the Beattie-Bridgeman equation of state for a gas:

$$p = \frac{\bar{R}T(1 - \varepsilon)}{\bar{v}^2}(\bar{v} + B) - \frac{A}{\bar{v}^2} \tag{5.7}$$

where

$$A = A_0(1 - a/\bar{v})$$
$$B = B_0(1 - b/\bar{v})$$
$$\varepsilon = \frac{c}{\bar{v}T^3}$$

and the constants A_0, B_0, a, b, c are determined for individual gases. Table 5.2 includes these for certain gases.

Table 5.1   Van der Waals Constants

Substance	a kPa $(m^3/kgmol)^2$	b $(m^3/kgmol)$
Air	135.8	0.0364
Ammonia (NH_3)	423.3	0.0373
Carbon dioxide (CO_2)	364.3	0.0427
Carbon monoxide (CO)	146.3	0.0394
Helium (He)	3.41	0.0234
Hydrogen (H_2)	24.7	0.0265
Methane (CH_4)	228.5	0.0427
Nitrogen (N_2)	136.1	0.0385
Oxygen (O_2)	136.9	0.0315

Table 5.2   Constants for the Beattie-Bridgeman Equation of State

Gas	A_0	a	B_0	b	$10^{-4}c$
Helium	2.1886	0.05984	0.01400	0.0	0.0040
Argon	130.7802	0.02328	0.03931	0.0	5.99
Hydrogen	20.0117	−0.00506	0.02096	−0.04359	0.0504
Nitrogen	136.2315	0.02617	0.05046	−0.00691	4.20
Oxygen	151.0857	0.02562	0.04624	0.004208	4.80
Air	131.8441	0.01931	0.04611	−0.001101	4.34
Carbon dioxide	507.2836	0.07132	0.10476	0.07235	66.00

Pressure in kilopascals; specific volume in m³/kgmol; temperature in kelvin; $\bar{R} = 8.31434$ kJ/kgmol · K

Example 5.3

Calculate the pressure of 2 moles of air at 400 K, with a total volume of 0.5 m³. Use the ideal gas law and the Van der Waals equation of state.

$$\bar{v} = \frac{V}{n} = \frac{0.5}{2} = 0.25 \text{ m}^3/\text{kgmol}$$

$$a = 135.8$$

$$b = 0.0364$$

$$p = \frac{(8.3143)(400)}{(0.25 - 0.0364)} - \frac{135.8}{(0.25)^2} = 13.4 \text{ MPa}$$

$$p = \frac{n\bar{R}T}{V} = \frac{(2.0)(8.3143)(400)}{(0.5)} = 13.3 \text{ MPa}$$

Notice that the difference in the results is only 0.7%, and hence the use of the ideal gas equation of state is a good approximation in many instances.

Compressibility Factor

The ideal gas law works well for gases at low densities. As the pressure of the gas is increased for a given temperature, the molecules are packed closer and closer together. This brings about nonideal behavior due to additional forces acting on the molecule. The other equations of state account for this deviation by introducing empirical constants. The form of Equation (5.2) is very convenient to work with, so there has developed a method to account for nonideal gas behavior using this form. If we divide Equation (5.2) by RT, it yields

$$\frac{pv}{RT} = 1$$

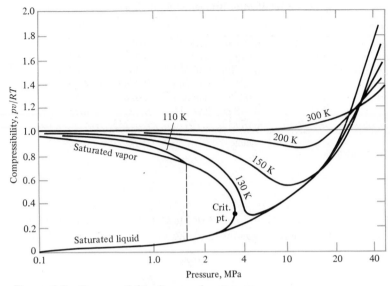

Figure 5.3   Compressibility diagram for nitrogen.

for an ideal gas. However, if the gas is not ideal, this will not be equal to one, but some other number, Z.

$$\frac{pv}{RT} = Z$$

The symbol Z is called the "compressibility factor." It is equal to one for an ideal gas, but will have a value other than one for an actual gas. Figure 5.3 illustrates how the compressibility factor varies with pressure along different isotherms. At room temperature, 300 K, the compressibility factor is unity up to 6890 kPa. Note that as the temperature rises, it requires greater and greater pressures to cause non ideal-gas behavior. In many application areas the temperature and pressure limits allow us to use the ideal-gas equation of state.

Generalized Compressibility Factor

The form of the equation of state using the compressibility factor is simple. At this point the only difficulty lies in acquiring charts for all the gases. Fortunately, this task may be reduced to that of developing only one chart.

This is accomplished by using reduced equations of state. The critical pressure, temperature, and specific volume are unique at the critical point for each gas. Table 5.3 lists the critical properties of some

Table 5.3   Critical Properties

Substance	Formula	Critical Pressure, MPa	Critical Temperature, K	Critical Specific Volume, m^3/kg
Air		3.77	132.7	0.002 859
Alcohol (methyl)	CH_4O	7.97	513.2	0.003 689
Ammonia	NH_3	11.28	405.5	0.004 264
Argon	A	4.87	151.2	0.001 879
Carbon Dioxide	CO_2	7.39	304.2	0.002 135
Carbon Monoxide	CO	3.49	133.0	0.003 321
Helium	He	0.229	5.26	0.0146
Hydrogen	H_2	1.297	33.3	0.0330
Oxygen	O_2	5.04	154.4	0.002 316

substances. The reduced coordinates are

reduced pressure $p_r = p/p_c$, where p_c = critical pressure

reduced temperature $T_r = T/T_c$, where T_c = critical temperature

reduced specific volume $v_r = v/v_c$ where v_c = critical specific volume

A generalized chart of compressibility factors for reduced temperatures and pressures, Figure 5.4, was developed by Nelson and Obert. This

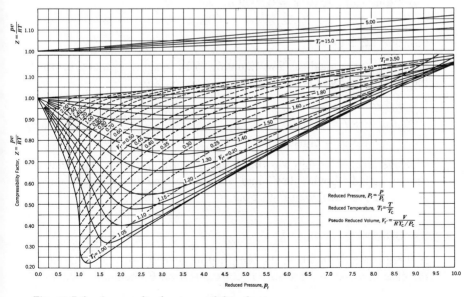

Figure 5.4   A generalized compressibility chart.

may be used to find the properties of a gas, knowing the critical properties. The reduced volume, V_r', is defined as

$$V_r' = \frac{V}{RT_c/p_c} = \frac{ZRT/p}{RT_c/p_c} = Z\frac{T_r}{p_r}$$

by Nelson and Obert.[1]

Example 5.4

Determine the specific volume of ammonia at 2000 kPa and 160°C.

$$p_r = \frac{2000}{11\,280} = 0.177$$

$$T_r = \frac{433}{405.5} = 1.068$$

From Figure 5.4,

$Z = 0.94$

$R = 0.4882 \text{ kJ/kg} \cdot \text{K}$

$$v = Z\frac{RT}{P} = \frac{(0.94)(0.4882)(433)}{2000}$$

$v = 0.0993 \text{ m}^3/\text{kg}$

From Table A.10, $v = 0.0999 \text{ m}^3/\text{kg}$.

5.3 BOYLE'S LAW

Referring to the ideal gas law, Equation (5.2), imagine a process in which the temperature was constant. Then

$$p_1v_1 = RT_1$$
$$p_2v_2 = RT = RT_1 \tag{5.8}$$
$$p_1v_1 = p_2v_2 = \text{constant}$$

This special relationship was first noticed by Robert Boyle. The relationship is called "Boyle's law" and a curve joining the points 1 and 2, $pv = C$, is represented in Figure 5.5.

5.4 CHARLES' LAW

If there is a case in which the temperature is held constant, what about the case in which the pressure is constant? Two Frenchmen, Jacques A.

---

[1]L. C. Nelson and E. F. Obert, General p-V-T Properties of Gases, *Trans. ASME*, Oct. 1954, 1057–1066.

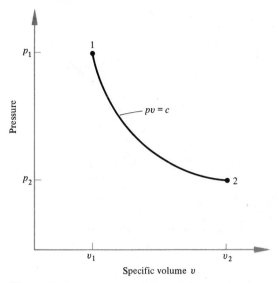

Figure 5.5   A constant-temperature process on a p–v diagram.

Charles and Joseph L. Gay-Lussac, independently discovered the constant-pressure relationship. They did not know about the ideal gas law, as they were investigating air at fairly low temperatures and pressures. Under conditions such as this, air behaves as an ideal gas. For constant pressure,

$$p_1 = \frac{RT_1}{v_1} = p_2 = \frac{RT_2}{v_2}$$

$$\frac{T_2}{v_2} = \frac{T_1}{v_1} = \frac{T}{v} = C \tag{5.9}$$

What if the volume was constant?

$$v_1 = \frac{RT_1}{p_1} = v_2 = \frac{RT_2}{p_2}$$

$$\frac{T_1}{p_1} = \frac{T_2}{p_2} = \frac{T}{p} = C \tag{5.10}$$

Equations (5.9) and (5.10) are called "Charles' law," one for constant pressure, Equation (5.9), and one for constant volume, Equation (5.10). These curves are illustrated in Figures 5.6(a) and 5.6(b).

Let us think about what we found so far. By knowing the equation of state we can predict what one gas property will be if the others are known. Also, we have a function that tells us how the gas varies in going from one state to another state, a process.

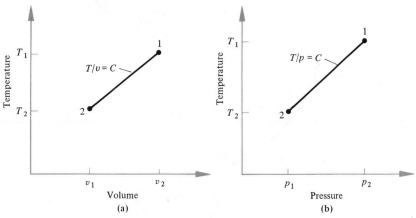

Figure 5.6   *Paths illustrating (a) constant-pressure and (b) constant-volume processes.*

Example 5.5

(a) Calculate the work per kilogram in kJ/kg of air in a piston–cylinder (Figure 5.7) that expands at constant pressure from 278 K and 2100 kPa to 500 K.

$$\frac{W}{m} = w = \int_1^2 p\, dv = p(v_2 - v_1)$$

$$v_1 = \frac{RT_1}{p_1} = \frac{(0.287)(278)}{2100} = 0.038 \text{ m}^3/\text{kg}$$

$$v_2 = \frac{RT_2}{p_2} = 0.0683 \text{ m}^3/\text{kg}$$

$$w = (2100)(0.0683 - 0.038)$$

$$w = 63.63 \text{ kJ/kg}$$

(b) If the heat added is 223.0 kJ/kg, find the change in internal energy.

Closed System
First Law: $q = (u_2 - u_1) + w$
$$223.0 = \Delta u + 63.63$$
$$\Delta u = 159.37 \text{ kJ/kg}$$

By knowing the equation of state for a substance, for example, the ideal gas law, and something about the process path, we are

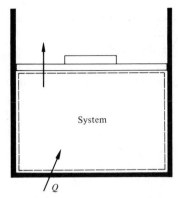

System

Q

Figure 5.7   A closed-system piston–cylinder, where heat is added and the piston moves, doing work (Example 5.5).

able to calculate the work done by the system. This is definitely an asset, but it would also be convenient to calculate the heat transfer and the internal energy change.

5.5 SPECIFIC HEAT

Specific Heat at Constant Volume

Specific heat at constant volume, c_v, is defined as

$$c_v \equiv \left(\frac{\partial u}{\partial T} \right)_v \ \text{kJ/kg} \cdot \text{K} \tag{5.11}$$

An experiment by Joule, Figure 5.8, verified the equation's premise; that thermal energy (heat) transfer at constant volume is only a function of temperature. In Figure 5.8 a vessel with two chambers, I and II,

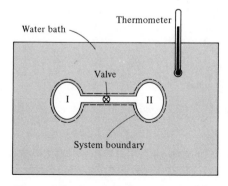

Water bath

Thermometer

Valve

I   II

System boundary

Figure 5.8   Diagram illustrating Joule's experiment regarding internal energy.

connected by a valve, was placed in a constant-temperature bath. When the system achieved thermal equilibrium with the bath, the valve was opened. No temperature change occurred in the bath, the system work was zero, and hence the internal energy change was zero. This leads to the conclusion that specific internal energy is not a function of volume for an ideal gas.

Mathematically, we may say that

$$u = u(v, T)$$

$$du = \left(\frac{\partial u}{\partial v}\right)_T dv + \left(\frac{\partial u}{\partial T}\right)_v dT$$

Since $dv > 0$ from Joule's experiment and the internal energy did not change, we may conclude that $u = u(T)$ only.

From the definition of specific heat at constant volume, Equation (5.11), and from the fact that the specific internal energy of an ideal gas is not a function of specific volume (Joule's experiment), the partial derivative in Equation (5.11) becomes a total derivative; hence

$$c_v = \frac{du}{dT}$$

$$du = c_v \, dT \tag{5.12}$$

$$dU = mc_v \, dT \tag{5.13}$$

If we consider a constant-volume heating process, as in Figure 5.9(a), the first law states

$$\delta Q = dU + \delta W$$

but

$$\delta W = \int p \, dV = 0$$

so

$$\delta Q = dU$$

$$Q = \int_1^2 \delta Q = \int_1^2 dU = U_2 - U_1$$

for an ideal gas,

$$U_2 - U_1 = \int_1^2 mc_v \, dT \tag{5.14}$$

For constant-volume processes, with constant mass and constant specific heat,

$$Q = \Delta U = mc_v(T_2 - T_1)$$

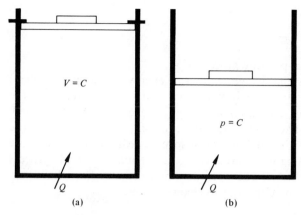

Figure 5.9 *Piston–cylinder arrangements illustrating constant-volume and constant-pressure processes.*

The change of internal energy for an ideal gas with constant specific heat is always

$$U_2 - U_1 = \Delta U = mc_v(T_2 - T_1) \tag{5.15}$$

The change of internal energy for an ideal gas is always denoted by Equation (5.13). The process may be reversible or irreversible, constant pressure, constant volume, or any other process. Internal energy, specific heat at constant volume, temperature, and mass are properties and do not depend on path. The equation of state for internal energy, Equation (5.13), follows from the definition of specific heat at constant volume.

Specific Heat of Constant Pressure

The coefficient c_p is the specific heat at constant pressure, and is rigorously defined as

$$c_p \equiv \left(\frac{\partial h}{\partial T} \right)_p \text{ kJ/kg} \cdot \text{K} \tag{5.16}$$

From the definition of specific heat at constant pressure and from $h = u + pv = u + RT$ for an ideal gas,

$$dh = c_p \, dT \tag{5.17a}$$

$$dH = mc_p \, dT \tag{5.17b}$$

We now consider heat addition at constant pressure, as in Figure 5.9(b). Let us analyze the first-law equation for this system.

$$\delta Q = dU + p \, dV = dU + d(pV)$$

$$\delta Q = dH \qquad \text{for } p = c \tag{5.18}$$

hence

$$\delta Q = dH = mc_p \, dT$$

for ideal-gas constant-pressure processes.

For an ideal gas, the changes in internal energy and enthalpy depend only on the temperature; so if the initial and final temperatures are known, the enthalpy and internal energy changes may be calculated. Thus Equations (5.17a) and (5.13) are true only for ideal gases, the reason following from the definitions of specific heat. We have therefore a method of calculating the internal energy and enthalpy for any process if we know the end states. The work may be calculated if we know the process path, that is, how the system changes state.

Further Considerations

An important ratio that we shall use extensively is the ratio of the specific heats, k.

$$k = \frac{c_p}{c_v} \tag{5.19}$$

Heretofore the specific heat has been written on a mass basis; it can also be written on the mole basis.

We denote the molal constant-volume specific heat as

$$\bar{c}_v \equiv \left(\frac{\partial \bar{u}}{\partial T} \right)_v \text{ kJ/kgmol} \cdot \text{K} \tag{5.20}$$

and the molal constant-pressure specific heat as

$$\bar{c}_p \equiv \left(\frac{\partial \bar{h}}{\partial T} \right)_p \text{ kJ/kgmol} \cdot \text{K} \tag{5.21}$$

(The bar over any property denotes that it is on the mole basis, hence both enthalpy $\bar{h}$ and internal energy $\bar{u}$ have units of kJ/kgmol · K.)

Another relation between the specific heats and the gas constant may be developed by considering the enthalpy of an ideal gas.

$$h = u + pv$$

$$pv = RT$$

$$h = u + RT$$

We now differentiate, and substitute Equations (5.13) and (5.17a) into the resulting equation,

$$dh = du + R \, dT$$

$$c_p \, dT = c_v \, dT + R \, dT$$

$$c_p - c_v = R \tag{5.22}$$

Thus even though the specific heats may be a function of temperature, for an ideal gas their difference is always constant.

The specific heats used in Appendix Table A.1 are measured experimentally. Specific heats may be developed mathematically, but using experimental values adds to the accuracy with which we may calculate, and hence predict, system changes.

It is interesting to note that the idea of heat capacity developed from the days when the caloric theory of heat was in vogue. A body, when heated, was viewed to be gaining "caloric." Heat capacity was the caloric necessary to raise the substance one degree in temperature. To make heat capacity an intensive property, we divide by the system mass, yielding specific heat capacity.

Variation of Specific Heat with Temperature

It may have been inferred that the specific heats were constant for all temperatures for a gas. This is not the case. The specific heat of a gas varies with temperature. The functional relationships denoting this variation are determined from experimental tests. Table 5.4 lists several formulas that predict the specific heat of a gas at a given temperature. Figure 5.10 shows a diagram of the specific heat variation with temperature. If increased accuracy is desired, these equations may be used. In many applications an average value of specific heat is used for the temperature range under consideration. The justification is based on the fact that the relative error involved in doing so is less than the error involved in the initial model assumptions. The reason the specific heats

Table 5.4   Formulas for Specific Heat Variation with Temperature

Substance and Temperature Range	c_p, kJ/kg·K
Air, 280–1500 K	$0.9167 + 2.577 \times 10^{-4}T - 3.974 \times 10^{-8}T^2$
Ammonia, 300–1000 K	$1.5194 + 1.936 \times 10^{-3}T - 1.789 \times 10^{-7}T^2$
Sulfur dioxide, 300–1000 K	$0.7848 + 0.7113 \times 10^{-4}T - 1.73 \times 10^4 T^{-2}$
Hydrogen, 300–2200 K	$11.959 + 2.160 \times 10^{-3}T + 30.95T^{-1/2}$
Oxygen, 300–2750 K	$1.507 - 16.77T^{-1/2} + 111.1T^{-1}$
Nitrogen, 300–5000 K	$1.415 - 287.9T^{-1} + 5.35 \times 10^4 T^{-2}$
Carbon monoxide, 300–5000 K	$1.415 - 273.3T^{-1} + 4.96 \times 10^4 T^{-2}$
H_2O, 300–3000 K	$4.613 - 103.3T^{-1/2} + 967.5T^{-1}$
Carbon dioxide, 300–3500 K	$1.540 - 345.1T^{-1} + 4.13 \times 10^4 T^{-2}$
Methane, 300–1500 K	$0.8832 + 4.71 \times 10^{-3}T - 1.123 \times 10^{-6}T^2$
Ethylene, 300–1500 K	$0.4039 + 4.35 \times 10^{-3}T - 1.35 \times 10^{-6}T^2$
Ethane, 300–1500 K	$0.306 + 5.34 \times 10^{-3}T - 1.53 \times 10^{-6}T^2$
n-butane (C_4H_{10}), 300–1500 K	$0.314 + 5.23 \times 10^{-3}T - 1.60 \times 10^{-6}T^2$
Propane, 300–1500 K	$0.214 + 5.48 \times 10^{-3}T - 1.67 \times 10^{-6}T^2$
Acetylene (C_2H_2) 280–1250 K	$1.921 + 7.06 \times 10^{-4}T - 3.73 \times 10^4 T^{-2}$
Octane, 225–610 K	$0.290 + 3.97 \times 10^{-3}T$

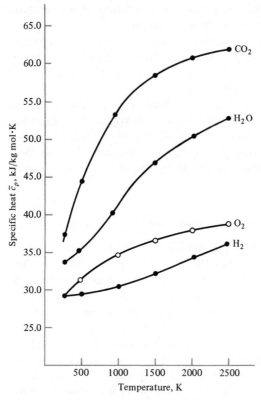

Figure 5.10 *The variation of constant-pressure specific heat with temperature for four gases, plotted from equations in Table 5.4.*

vary with temperature is that the energy associated with each vibrational mode becomes greater, especially at high temperatures. This is illustrated in Figure 5.10 for water and carbon dioxide, where there are more vibrational modes available in the atomic structure than in the simpler diatomic molecules, hydrogen and oxygen. Hence the specific heat variation is more pronounced.

5.6 GAS TABLES

We have examined two methods of analyzing gas behavior; the first is the use of an ideal gas equation of state and the second is the use of the compressibility chart for real gas behavior. There is a third resource available to us, the Gas Tables by Keenan and Kaye. Tables A.2 to A.4 in the Appendix are adapted from these gas tables. These tables account for the variation of specific heat with temperature for an ideal gas. Temperature is the independent variable, and tabulated values for internal energy and enthalpy are given on the same line. Some other symbols are also given values, and we will learn about them when we deal with process paths, how the system changes state. Note that equations using constant specific heats cannot be used, as the substance is now a gas with variable specific heats.

Example 5.6

Air with a temperature of 27°C receives heat at constant volume until the temperature is 927°C. Find the heat added per kilogram and use the ideal gas law and the air tables.

Closed System

First Law: $q = \Delta u + w$

$\quad w = 0 \quad$ for $v = c$

$\quad q = \Delta u$

$\Delta u = c_v(T_2 - T_1) = (0.7176)(927 - 27) = 645.84 \text{ kJ/kg}$

From Table A.2,

$$u_1 = 214.07 \text{ kJ/kg}$$

$$u_2 = 933.33 \text{ kJ/kg}$$

$$\Delta u = 719.26 \text{ kJ/kg}$$

$$\text{percent error} = \frac{73.42}{719.26} \times 100 = 10.2\%$$

The difference between the answers lies in the fact that the specific heat is not constant but varies with temperature. If accurate calculations are to be performed, gas tables are preferable. An alternative is integration over temperature ranges where c_v is essentially constant.

PROBLEMS

1. An unknown gas has a mass of 1.5 kg and occupies 2.5 m³ while at a temperature of 300 K and a pressure of 200 kPa. Determine the ideal gas constant for the gas.
2. A motorist equips his automobile tires with a relief-type valve so that the pressure inside the tire never will exceed 240 kPa (gage). He starts a trip with a pressure of 200 kPa (gage) and a temperature of 23°C in the tires. During the long drive the temperature of the air in the tires

reaches 83°C. Each tire contains 0.11 kg of air. Determine (a) the mass of air escaping each tire; (b) the pressure of the tire when the temperature returns to 23°C.

3. A 6-m³ tank contains helium at 400 K and is evacuated from atmospheric pressure to a pressure of 740 mm Hg vacuum. Determine (a) mass of helium remaining in the tank; (b) mass of helium pumped out; (c) the temperature of the remaining helium falls to 10°C. What is the pressure in kPa?

4. A 1.5-kg quantity of ethane is cooled at constant pressure from 170°C to 65°C. Determine (a) the change of enthalpy; (b) the change of internal energy; (c) the heat transferred; (d) the work.

5. A 5-m³ tank contains chlorine at 300 kPa and 300 K after 3 kg of chlorine has been used. Determine the original mass and pressure if the original temperature was 315 K.

6. Carbon dioxide at 25°C and 101.3 kPa has a density of 1.799 kg/m³. Determine (a) the gas constant; (b) the molecular weight based on the gas constant.

7. Given that a carbon monoxide gas has a temperature of 500 K and a specific volume of 0.4 m³/kg, determine the pressure using the Van der Waals equation of state and the ideal-gas equation of state.

8. Determine the specific volume of helium at 200 kPa and 300 K using the Van der Waals equation of state and the ideal-gas equation of state.

9. Helium is assumed to obey the Beattie-Bridgeman equation of state. Determine the pressure for a temperature of 500°C and a specific volume of 5.2 m³/kg. Compare with the ideal-gas equation of state.

10. Given a pressure of 500 kPa and a temperature of 500 K for carbon dioxide, compute the specific volume using the Beattie-Bridgeman and ideal-gas equations of state.

11. Determine the specific volume of the following gases using the generalized compressibility chart:
    (a) Methyl alcohol at 6000 kPa and 600 K.
    (b) Carbon dioxide at 15 MPa and 300°C.
    (c) Helium at 500 kPa and 60°C.

12. Air exists at 38°C and 4200 kPa. Determine the specific volume using the ideal gas law and the Van der Waals equation of state.

13. For a particular ideal gas, the value of R is 0.280 kJ/kg · K and the value of k is 1.375. Determine the values of c_p and c_v.

14. For a certain ideal gas, $R = 0.270$ kJ/kg · K and $k = 1.25$. Determine (a) c_p; (b) c_v; (c) M.

15. For a certain ideal gas, $c_p = 1.1$ kJ/kg · K and $k = 1.3$. Determine (a) M; (b) R; (c) c_v.

16. A 5-kg quantity of oxygen is heated from 250 K to 400 K at constant pressure. Determine (a) the change of enthalpy; (b) the change of internal energy; (c) the heat added; (d) the work done.

17. The specific heat of carbon dioxide may be found in Table 5.4 as a function of temperature. Find the change of enthalpy of carbon dioxide when its temperature is increased from 325 K to 1100 K. Compare this value with that calculated for constant specific heat found in Table A.1.

18. For a given gas the specific heat at constant-pressure specific heat varies 0.003 kJ/kg $\cdot$ K^2. At $93°C$ the specific heat at constant pressure is equal to 1.214 kJ/kg $\cdot$ K. Then 2 kg of this gas is heated at constant pressure from $93°C$ to $653°C$. Determine (a) the heat added; (b) the change of enthalpy; (c) the change of internal energy for a value of $R = 0.293$ kJ/kg $\cdot$ K.

19. A mountain is to be measured by finding the change in pressure at constant temperature. A barometer at the base of the mountain reads 730 mm Hg, while at the top it reads 470 mm Hg. The local gravitational acceleration is 9.6 m $\cdot$ s^{-2}. Find the mountain's height. Assume $T = 298$ K.

20. An empty, opened can is 30 cm high with a 10-cm diameter. The can, with the open end down, is pushed under water with a density of 1000 kg/m^3. Find the water level in the can when the top of the can is 50 cm below the surface. Thermal equilibrium exists at all times.

21. Compute the average specific heat at constant pressure using the equations found in Table 5.4 between temperature limits of 350 K and 1200 K for (a) nitrogen; (b) methane; (c) propane.

22. Determine the specific heats, c_p and c_v, of steam at 7 MPa and $500°C$. Compare this to the specific heats calculated at 7 MPa and $350°C$.

23. An insulated, constant-volume system containing 1.36 kg of air receives 53 kJ of paddle work. The initial temperature is $27°C$. Determine (a) the final temperature; (b) change of internal energy.

24. A 1-kg gaseous system is in a piston–cylinder and receives heat at a constant pressure of 350 kPa. The internal energy increases 200 kJ and the temperature increases 70 K. If the work done is 100 kJ, determine (a) c_p; (b) the change in volume.

25. An ideal gas occupies a volume of 0.5 m^3 at a temperature of 340 K and a given pressure. The gas undergoes a constant-pressure process until the temperature decreases to 290 K. Determine (a) the final volume; (b) the work if the pressure is 120 kPa.

26. Using Table A.2, calculate the specific heats at constant pressure and constant volume for air at 1000 K. (*Hint:* Use $c_p \approx \Delta h/\Delta T$.)

27. A spherical balloon measures 10 m in diameter and is filled with helium at 101 kPa and 325 K. The balloon is surrounded by air at 101 kPa and 320 K. Determine the lifting force.

28. Repeat Problem 27, except that now the balloon contains hydrogen. Note that the molecular weight of hydrogen is one-half that of helium. Does this halve the lifting force? Why?

29. Determine the size of a spherical balloon required to lift a payload of 1360 kg. The gas to be used is helium at 101.3 kPa and 23°C. The surrounding air is 101.3 kPa and 10°C.

30. A nitrogen cylinder of 0.1 m³ originally has a pressure of 17.25 MPa and a temperature of 20°C. The nitrogen is gradually used until the pressure is 2.75 MPa, the temperature remaining at 20°C. What is the mass of nitrogen used? Use compressibility chart and ideal gas and compare results.

Chapter 6
Processes Using Tables and Gas Equations of State

The purpose of this chapter is to synthesize the concepts of Chapter 3, the first-law analysis, with those of Chapters 4 and 5, property calculations for pure substances and gases. The concept of reversible processes is further illustrated.

6.1 EQUILIBRIUM AND NONEQUILIBRIUM PROCESSES

We know that a process occurs whenever a system changes from one state to another state. It would be convenient if we also knew the path along which this process occurs. However, as soon as the process path is mentioned, an equilibrium or reversible process is assumed to occur. The equilibrium surface illustrated in Figure 6.1 shows the path denoting the process from state 1 to state 2. Imagine that a light is shown on the surface; then a shadow, or projection, of the path will be made on the plane surface. This is denoted by the solid, continuous, line on the $p-V$ plane. Most often the two-dimensional surface is used in drawing a system path. It may be helpful, however, to remember how this projection occurs; a projection of an equilibrium process is implied as soon as a continuous function is drawn.

What if the process were an irreversible one, going again between states 1 and 2? Let us assume that as it crosses the surface in Figure 6.1, a dot is marked on the surface of each crossing. Figure 6.2 might denote such a process, with its projection on the $p-V$ plane.

From mathematics we know that a dotted line is a discontinuous function. To find work, we must be able to integrate $p\,dV$ along a continuous path from state 1 to state 2. Hence when the process is irreversible, the path is discontinuous and we cannot integrate to find the work; the work must be found from the first-law analysis in these cases.

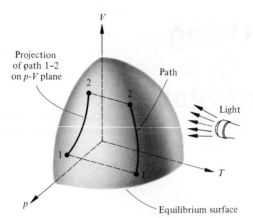

Figure 6.1   *Projection of an equilibrium path on a p–V plane.*

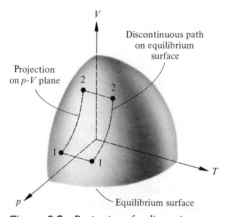

Figure 6.2   *Projection of a discontinuous, or irreversible, path on a p–V plane.*

As soon as we integrate a function, we are defining a continuous function and therefore a reversible, or equilibrium, process.

6.2 CLOSED SYSTEMS

Science evolves by taking individual cases and relating their common aspects to form a general case. The temptation is to present the general case and assume the special cases will be understood. However, past experience indicates that people, present and past, grasp the more complex or general case when the simpler cases are understood.

Before we proceed in solving some problems, we must first decide whether an open- or a closed-system analysis is to be used. Although here the sample problems and discussion will indicate the system, in home-work problems and in life situations you must decide.

Constant-Pressure Process

We consider first a constant-pressure process for a closed system, il-lustrated in Figure 6.3(a). The system receives an amount of heat, Q, performs work, W, and experiences changes in internal energy, $\Delta U = U_2 - U_1$. The first law of thermodynamics for a closed system at rest, valid for all processes, is

$$Q = (U_2 - U_1) + W \tag{6.1}$$

For reversible processes we may write the first law as

$$Q = (U_2 - U_1) + \int_1^2 p\, dV$$

or

$$\delta Q = dU + p\, dV \tag{6.2}$$

Also,

$$H = U + pV$$

$$dH = dU + p\, dV + V\, dp \tag{6.3}$$

$$dp = 0 \qquad \text{for } p = C$$

$$\delta Q = dH \qquad \text{for } p = C \tag{6.4}$$

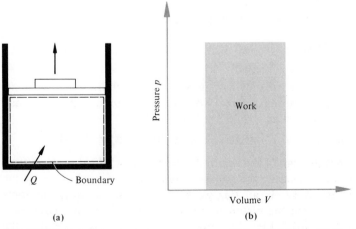

Figure 6.3   (a) A constant-pressure process with heat addition. (b) Graphical interpretation of mechanical work for a constant-pressure process.

For an ideal gas with constant specific heats

$$\Delta H = mc_p(T_2 - T_1) \text{ kJ} \qquad (6.5a)$$

$$\Delta U = mc_v(T_2 - T_1) \text{ kJ} \qquad (6.5b)$$

For pure substances, the values of enthalpy and internal energy may be looked up in tables.

For a reversible constant-pressure process in a closed system, the work is

$$W = \int_1^2 p \, dV = p(V_2 - V_1)$$

$$W = p_2V_2 - p_1V_1 \qquad (6.6a)$$

Using the ideal-gas equation of state, $pV = mRT$, yields

$$W = mR(T_2 - T_1) \qquad (6.6b)$$

Example 6.1

A piston–cylinder containing air expands, at a constant pressure of 150 kPa, from a temperature of 285 K to a temperature of 550 K. The mass of air is 0.05 kg; find the heat, work, change in enthalpy, and the change in internal energy.

Closed System

First Law: $Q = \Delta U + W;$ $Q = \Delta H$ for $p = C$

$$\Delta H = mc_p(T_2 - T_1) = (0.05)(1.0047)(550 - 285)$$

$$\Delta H = 13.31 \text{ kJ}$$

$$\Delta U = mc_v(T_2 - T_1) = (0.05)(0.7176)(550 - 285)$$

$$\Delta U = 9.51 \text{ kJ}$$

$$W = \Delta H - \Delta U = 3.8 \text{ kJ}$$

This is true with the assumption of reversibility, as a constant-pressure process was denoted. For a reversible process the work may also be expressed as

$$W = mR(T_2 - T_1) = (0.05)(0.287)(500 - 285) = 3.8 \text{ kJ}$$

Note that the sign of the heat indicates that it is added to the system and work is done by the system. The amount of work done by the system is 3.8 kJ. Is all this work available to do work on some object? Assume the piston to be in a room where the barometric pressure is 101.3 kPa. Looking at Figure 6.4, we note that the

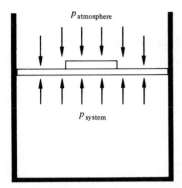

$P_{\text{atmosphere}}$

P_{system}

Figure 6.4   *The pressure in a cylinder resisted by the pressure of the surroundings.*

atmospheric pressure acts against the piston. As the piston expands, it is pushing against the surrounding atmospheric pressure and has to perform work to overcome this resistance.

The work against the atmosphere, W_a, is

$$W_a = p_a \, dV = p_a(V_2 - V_1) \tag{6.7}$$

$$V_2 = \frac{mRT_2}{P_2}$$

$$V_1 = \frac{mRT_1}{P_1}$$

$$P_2 = P_1$$

$$W_a = mRp_a\left(\frac{T_2}{P_2} - \frac{T_1}{P_1}\right) \tag{6.8}$$

Does the work against the atmosphere depend on the system process? No. The atmosphere is always at constant pressure, so the work against the atmosphere is a function of only the initial and final system volume. Thus we may find the system work from the first-law analysis of the system; but to find the net work available to run a piece of equipment, we must subtract the work against the surroundings from the system work.

$$W_{\text{net}} = W_{\text{sys}} - W_a \tag{6.9}$$

The net work does not influence the system's work. It provides us with additional information. In most problems it is only system work that we are interested in.

Example 6.2

For the system noted in Example 6.1, find the net work where the surroundings are at 101.3 kPa.

$$W_a = mRp_a \left(\frac{T_2}{p_2} - \frac{T_1}{p_1} \right)$$

$$p_2 = p_1 = 150 \text{ kPa}$$

$$W_a = \frac{(0.05)(0.287)(101.3)(550 - 285)}{150} = 2.57 \text{ kJ}$$

$$W_{net} = W_{sys} - W_a = 3.8 - 2.57 = 1.23 \text{ kJ}$$

The implied assumption in the previous examples is that heat is added along a reversible, constant-pressure path.

Irreversible Work—Paddle Work

There is a typical case in which energy is added by a paddle wheel and dissipated by fluid friction (Figure 6.5). In this case energy is added to the system irreversibly; that is, there is no direct way in which the system could lift the weight, once the weight dropped, turning the paddles. Consider this irreversible work, called "paddle work," W_p; then the first law for a closed system is

$$Q = \Delta U + W - W_p \qquad (6.10)$$

Usually, the system is considered to be adiabatic, so $Q = 0$,

$$0 = \Delta U + W - W_p \qquad (6.11)$$

Let us consider the following example.

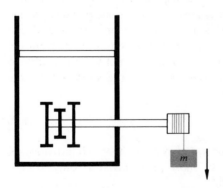

Figure 6.5 A closed system receiving paddle work (Example 6.3).

Example 6.3

A constant-pressure, adiabatic system contains 0.2 kg of steam at 200 kPa and 200°C and is similar to the system depicted in Figure 6.5. The system receives 40.26 kJ of paddle work. The final steam temperature is 300°C. Find the mechanical work, and the changes of internal energy and enthalpy.

$$W = \int_1^2 mp \, dv = mp(v_2 - v_1)$$

From Appendix Table A. 7,

$$v_1 = 1.0803 \text{ m}^3/\text{kg} \qquad v_2 = 1.3162 \text{ m}^3/\text{kg}$$

Hence

$$W = (0.2)(200)(1.3162 - 1.0803) = 9.43 \text{ kJ}$$

$$\Delta U = m(u_2 - u_1) = (0.2)(2808.6 - 2654.4) = 30.84 \text{ kJ}$$

$$\Delta H = m(h_2 - h_1) = (0.2)(3071.8 - 2870.5) = 40.26 \text{ kJ}$$

Constant-Volume Processes

Let us consider next a constant-volume process, a closed container, as is Figure 6.6(a). Let heat be added, causing a change in internal energy. The first law for a closed system shows that

Closed System

First Law: $Q = \Delta U + \int p \, dV; \qquad dV = 0 \qquad$ for $V = C$

$Q = \Delta U$ (6.12)

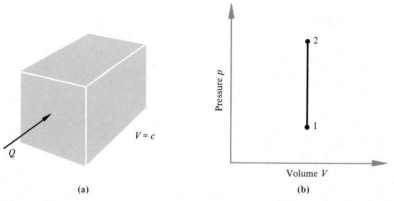

(a)   (b)

Figure 6.6 *Constant-volume heat addition with the p–V diagram for the process.*

For an ideal gas

$$\Delta U = mc_v(T_2 - T_1)$$

hence

$$Q = mc_v(T_2 - T_1) \tag{6.13}$$

Does it matter whether the heat is added reversibly in this process? No. The heat is equal to the change in a property, which depends only on the system state. It does not matter whether there is a path on the equilibrium surface or not, as points 1 and 2 are fixed. As soon as a p–V diagram is drawn illustrating the process, Figure 6.6(b), the implication that the process is reversible is made. The equation for the mechanical work, $\int p\, dV$, is zero whether the process is reversible or irreversible, as dV is zero in both cases, so the continuity of pressure on the equilibrium surface does not matter.

Example 6.4

A closed rigid container has a volume of 1 m³ and holds steam at 300 kPa and with a quality of 90 percent. If 500 kJ of heat is added, find the final temperature.

Closed System

First Law: $Q = \Delta U + W$; $W = 0$ for $V = C$
From Table A.6,

$$u_1 = u_f + xu_{fg} = 561.5 + (0.90)(1982.4) = 2345.31 \text{ kJ/kg}$$

$$v_1 = v_f + xv_{fg} = 0.001\,073 + (0.90)(0.6048) = 0.5454 \text{ m}^3/\text{kg}$$

$$m = \frac{V}{v_1} = 1.833 \text{ kg}$$

$$Q = m(u_2 - u_1)$$

$$500 = 1.833(u_2 - 2345.31)$$

$$u_2 = 2618.1 \text{ kJ/kg}$$

Interpolating from Table A.7,

$$T_2 = 180°C$$

Consider the case in which energy is added in a totally irreversible manner through paddle work (Figure 6.7). The energy balance for a closed system (first law of thermodynamics) yields for a constant volume system

$$Q = \Delta U + W - W_p$$

$$W = 0 \quad \text{for } V = C$$

$$Q = \Delta U - W_p \tag{6.14}$$

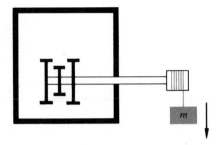

Figure 6.7   Paddle work in a constant-volume system.

If the system is adiabatic, $Q = 0$, and

$$W_p = \Delta U \tag{6.15}$$

Example 6.5

A closed constant-volume system receives 10.5 kJ of paddle work. The system contains oxygen at 344 kPa, 278 K, and occupies 0.06 m³. Find the heat (loss or gain) if the final temperature is 400 K.

Closed System

First Law: $Q = \Delta U + W - W_p;$ $W = 0$ for $V = C$

$$\Delta U = mc_v(T_2 - T_1)$$

$$m = \frac{p_1 V_1}{RT_1} = 0.285 \text{ kg}$$

$$\Delta U = (0.285)(0.6585)(400 - 278) = 22.9 \text{ kJ}$$

$$Q = \Delta U - W_p = 22.9 - 10.5 = 12.4 \text{ kJ}$$

added to the system.

Constant-Temperature Processes

The third type of special process we can envision is a reversible constant-temperature process. Because an ideal-gas equation of state is used in this discussion, the results must be limited to ideal gases. It is, of course, possible to have constant-temperature processes using tables of properties, but the following equations do not apply.

Consider again a piston–cylinder in Figure 6.8(a), which receives heat and expands doing work at constant temperature. Figure 6.8(b)

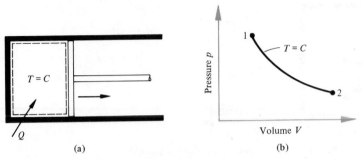

Figure 6.8   *A constant-temperature process in a piston–cylinder and illustrated -on a p–V diagram.*

illustrates the process on a p–V diagram. The first law for a closed system is

$$Q = \Delta U + W = \Delta U + \int p \, dV$$

We need p as a function of V for a constant-temperature process. The ideal gas law shows that

$$p_1 V_1 = mRT_1 = mRT_2 = p_2 V_2 = C$$

or

$$pV = C \qquad \text{for } T = C$$

$$W = \int_1^2 p \, dV = \int_1^2 C \frac{dV}{V} = C \ln \left(\frac{V_2}{V_1} \right)$$

$$W = p_1 V_1 \ln \left(\frac{V_2}{V_1} \right) \tag{6.16}$$

and since $U = U(T)$, then $\Delta U = 0$. This may be readily seen for an ideal gas

$$U = mc_v (T_2 - T_1) = 0 \qquad \text{for } T = C$$

and

$$Q = W \qquad \text{for reversible } T = C \tag{6.17}$$

Example 6.6

A piston–cylinder containing 0.25 kg of helium at 278 K receives heat at constant temperature until the pressure is one half its initial value. Find the heat added, and the work.

Closed System

First Law: $Q = \Delta U + W;$ $\quad \Delta U = 0 \quad$ for $T = C$

$$V_2 = \frac{C}{p_2} \qquad V_1 = \frac{C}{p_1}$$

$$\frac{V_2}{V_1} = \frac{p_1}{p_2} = 2$$

$$Q = W = p_1 V_1 \ln\left(\frac{p_1}{p_2}\right)$$

$$p_1 V_1 = mRT_1$$

$$Q = W = (0.25)(2.077)(278)\ln(2) = 100.0 \text{ kJ}$$

6.3 OPEN SYSTEMS

When dealing with open systems in which there are steady-state, steady-flow conditions, the analysis is much easier. One simply equates all the energy into the system to all the energy leaving the system and solves that equation for the unknown, usually the heat or work of the system. Figure 6.9 illustrates a typical open system.

From Chapter 3 we learned that heat may be expressed as (Equation (3.60))

$$\delta q = dh - v\, dp \tag{6.18}$$

and work may be expressed as (Equation (3.61))

$$\delta w = -v\, dp - d(\text{K.E.}) - d(\text{P.E.}) \tag{6.19}$$

Equation (6.19) reduces to

$$\delta w = -v\, dp \tag{6.20}$$

$$\dot{W} = -\dot{m} \int v\, dp \tag{6.21}$$

for negligible potential and kinetic energy changes.

Figure 6.9 An open system with various energy forms.

At this point let us consider the various processes. Of course, constant volume is meaningless for open systems, but constant specific volume may be considered.

Constant Pressure

Referring to Equation (6.20), we note that in the absence of kinetic and potential energy changes, work is zero. The heat transfer, from Equation (6.18), is equal to the change of enthalpy. The typical situation where open-system, constant-pressure processes occur are in heat exchangers. There is no work, and the analysis is concerned with determining the heat transferred from one fluid to the other.

Example 6.7

A simple tube-in-tube heat exchanger (Figure 6.10) has air flowing in the outside tube, cooling the helium that flows in the inside tube. Assume the change in kinetic and potential energies is negligible. The work is zero. Determine the heat loss from the exchanger given the following information. Air, flowing at 200 kg/h, enters at 280 K and leaves at 340 K; helium, flowing at 44 kg/h, enters at 450 K and leaves at 390 K.

Open System

First Law: Energy In = Energy Out

$$\dot{Q} + \dot{m}_a h_{a\,in} + \dot{m}_{He} h_{He\,in} = \dot{m}_a h_{a\,out} + \dot{m}_{He} h_{He\,out}$$

$$-\dot{Q} = \dot{m}_a(h_{a\,in} - h_{a\,out})$$

$$+ \dot{m}_{He}(h_{He\,in} - h_{He\,out})$$

$$(h_{a\,in} - h_{a\,out}) = c_p(T_{a\,in} - T_{a\,out}) = (1.0047)(280 - 340)$$

$$= -60.28 \text{ kJ/kg}$$

$$(h_{He\,in} - h_{He\,out}) = c_p(T_{He\,in} - T_{He\,out})$$

$$= (5.1954)(450 - 390)$$

$$= 311.72 \text{ kJ/kg}$$

$$-\dot{Q} = (200)(-60.28) + (44)(311.72)$$

$$= 0.461 \text{ kW}$$

In this case the heat exchanger is not adiabatic, so there is a heat flow from the exchanger to the surroundings.

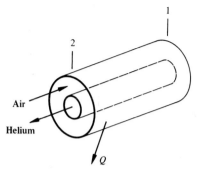

Figure 6.10 *A simple tube-in-tube counterflow heat exchanger.*

Constant-Temperature Processes—Ideal Gas

Let us calculate the power produced by a constant-temperature process. This is found by integrating Equation (6.21). For an ideal gas the expression for the isothermal open-system work (neglecting kinetic and potential energy) may be found by

$$pv = RT = C$$

$$v = \frac{C}{p} \qquad \dot{W} = \dot{m} \int_1^2 v \, dp$$

$$\dot{W} = -\dot{m} \int_1^2 C \frac{dp}{p} = -\dot{m} p_1 v_1 \ln\left(\frac{p_2}{p_1}\right)$$

Note that the assumption of reversibility was incorporated into the development when the work was expressed as $v \, dp$. This is true only for reversible flow process. This is not a terrible circumstance, as a constant-temperature process is not physically realizable, and assuming constant temperature is at least as great a restriction as assuming reversibility. The heat flow is found from Equation (6.18); note that it is equal to the power when $dh = 0$; which for an ideal gas, $dh = c_p \, dT$, occurs in isothermal processes.

Example 6.8

Air flows steadily through an engine at constant temperature, 400 K. Find the work per kilogram if the exit pressure is one-third the inlet pressure and the inlet pressure is 207 kPa. Assume that the kinetic and potential energy variation is negligible.

Open System

$T = C$

$$w = -p_1 v_1 \ln\left(\frac{p_2}{p_1}\right)$$

$$v_1 = \frac{RT_1}{p_1} = \frac{(0.287)(400)}{207} = 0.5546 \text{ m}^3/\text{kg}$$

$$w = -(207)(0.5546)\ln\left(\tfrac{1}{3}\right)$$

$$w = +126.1 \text{ kJ}$$

Constant Specific Volume

This type of process occurs when we introduce the constraint of incompressibility. The specific volume, hence density, remains constant. Very often in the pumping of liquids, the assumption of incompressibility is a realistic one to make. The following example illustrates the considerations involved in this type of problem.

Example 6.9

Water enters a pump at a rate of 5 kg/s, at 200 kPa and 100°C and leaves at 1000 kPa. Determine the power required if the pump is adiabatic. The changes in kinetic and potential energies are negligible.

From equation (6.21)

$$\dot{W} = -\dot{m}\int_1^2 v \, dp$$

$v_1 \approx v_f$ at $100°C = 0.001\ 044 \text{ m}^3/\text{kg}$

$$\dot{W} = -(5.0)(0.001\ 044)(1000 - 200) = -4.176 \text{ kW}$$

This sign is negative since power is put into the system. Should the pump have a heat loss or gain, the heat transferred may be evaluated from either a first-law analysis or from Equation (6.18). In this case the exit enthalpy must also be determined. Additional information, such as the actual exit temperature, would have to be provided to calculate the exit enthalpy. Most often, however, a pump is considered adiabatic.

6.4 POLYTROPIC PROCESS—IDEAL GAS

Closed System

The previous process types for ideal gases in open and closed systems are special cases of a more general type of process, called the "polytropic" process,

$$pV^n = C \tag{6.22a}$$

$$pv^n = C \tag{6.22b}$$

On a $p-V$ diagram this process is illustrated in Figure 6.11, as well as by the special constant-property processes.

The constant-temperature process is a case when $n = 1$; the constant-pressure process $n = 0$; and the constant-volume process $n = \infty$. The polytropic processes are all assumed to be reversible; note that $pV^n = C$ is the equation of the line projected from the equilibrium surface.

Relationships may be developed between the temperature, pressure, and volume for a polytropic process between state 1 and state 2. Using

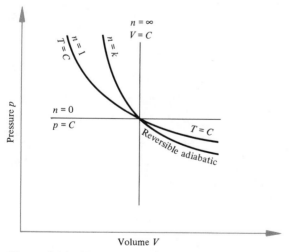

Figure 6.11   *The variation of pressure with volume for different values of n.*

the ideal-gas law,

$$pv = RT$$

$$p_1v_1^n = p_2v_2^n$$

$$p_1 = \frac{RT_1}{v_1} ; \qquad p_2 = \frac{RT_2}{v_2}$$

$$\frac{p_1}{p_2} = \left(\frac{T_1}{T_2}\right)\left(\frac{v_2}{v_1}\right) \tag{6.23}$$

For a polytropic process between states 1 and 2,

$$\frac{p_1}{p_2} = \left(\frac{v_2}{v_1}\right)^n \tag{6.24a}$$

We substitute Equation (6.23) into (6.24a):

$$\left(\frac{T_1}{T_2}\right)\left(\frac{v_2}{v_1}\right) = \left(\frac{v_2}{v_1}\right)^n$$

$$\frac{T_1}{T_2} = \left(\frac{v_2}{v_1}\right)^{n-1} = \left(\frac{V_2}{V_1}\right)^{n-1} \tag{6.24b}$$

$$\frac{V_2}{V_1} = \left(\frac{T_1}{T_2}\right)^{1/(n-1)} \tag{6.24c}$$

On eliminating the volumes the result is

$$\frac{T_1}{T_2} = \left(\frac{p_2}{p_1}\right)^{(1-n)/n} = \left(\frac{p_1}{p_2}\right)^{(n-1)/n} \tag{6.25a}$$

$$\frac{p_1}{p_2} = \left(\frac{T_1}{T_2}\right)^{n/(n-1)} \tag{6.25b}$$

The work may also be calculated from the pressure–volume functional relationship, as follows:

$$W = \int_1^2 p \, dV; \qquad p = CV^{-n}$$

$$W = \int_1^2 CV^{-n} \, dV = \frac{CV_2^{1-n} - CV_1^{1-n}}{1-n}$$

$$C = p_2V_2^n = p_1V_1^n$$

$$W = \frac{p_2V_2 - p_1V_1}{1-n} \tag{6.26a}$$

Using the ideal-gas law,

$$p_2 V_2 = mRT_2$$

$$p_1 V_1 = mRT_1$$

$$W = \frac{mR(T_2 - T_1)}{1 - n} \tag{6.26b}$$

To derive Equation (6.26a) the ideal-gas equation of state is not required; however, to derive Equation (6.26b) the ideal-gas equation of state is necessary. Reversibility is common to both.

One type of process that is frequently used as a standard of comparison for any actual process is the reversible adiabatic process. For a closed system the first law of thermodynamics for a reversible change may be written as

$$\delta q = du + p \, dv \tag{6.27a}$$

$$\delta q = dh - v \, dp \tag{6.27b}$$

Since $\delta q = 0$ for a reversible adiabatic process and defining du and dh in terms of an ideal gas yields

$$c_v \, dT = -p \, dv \tag{6.27c}$$

$$c_p \, dT = v \, dp \tag{6.27d}$$

Dividing Equation (6.27d) by Equation (6.27c),

$$\frac{c_p}{c_v} = k = -\frac{v}{p} \frac{dp}{dv}$$

$$k \frac{dv}{v} = -\frac{dp}{p} \tag{6.27e}$$

Integrating Equation (6.27e),

$$k \ln\left(\frac{v_2}{v_1}\right) = \ln\left(\frac{p_1}{p_2}\right).$$

We then take antilogs

$$\left(\frac{v_2}{v_1}\right)^k = \left(\frac{p_1}{p_2}\right)$$

or

$$p_2 v_2^k = p_1 v_1^k = p v^k = C \tag{6.28a}$$

$$p V^k = C \tag{6.28b}$$

Equation (6.28b) is the same as Equation (6.22a) except that $n = k$; hence the relationships among temperature, pressure, and volume are

$$\left(\frac{V_2}{V_1}\right) = \left(\frac{T_1}{T_2}\right)^{1/(k-1)} \tag{6.29a}$$

$$\left(\frac{p_1}{p_2}\right) = \left(\frac{T_1}{T_2}\right)^{k/(k-1)} \tag{6.29b}$$

Example 6.10

Helium is compressed polytropically from 100 kPa and 0°C to 1000 kPa and 165°C.
  (a) Determine the work and the heat for 0.11 kg of helium.
  (b) If the process is reversible adiabatic, find the work.

From Equation (6.25b) the value of n is determined for this particular process

$$\frac{100}{1000} = \left(\frac{273}{438}\right)^{n/n-1}$$

Take logs of both sides and solve for n

$n = 1.258$

From Equation (6.26b)

$$W = \frac{(0.11)(2.077)(165)}{(-0.258)} = -146.1 \text{ kJ}$$

The heat is calculated by using the equation for the first law of thermodynamics for a closed system.

$Q = \Delta U + W$

$\Delta U = mc_v(T_2 - T_1) = (0.11)(3.1189)(165) = 56.6 \text{ kJ}$

$Q = 56.6 - 146.1 = -89.5 \text{ kJ}$

  For a reversible adiabatic process $n = k$, and

$$W = \frac{mR(T_2 - T_1)}{(1 - k)} = \frac{(0.11)(2.077)(165)}{(-0.666)} = -56.6 \text{ kJ}$$

  Let us start with the problem from the beginning and observe how it fits together. The first calculation shows that the exponent n is positive. The work is calculated and found to be negative. This indicates that the system, the helium gas, is having work done on it by the surroundings. The calculation for the heat shows that it is negative. This indicates that the heat is leaving the system. Note that for the reversible adiabatic process, $n = k$ and $Q = 0$, the work is negative, as we originally thought it should be. In this case the piston does work on the gas, raising the gas temperature and pressure. Incidentally, the work in the adiabatic case is equal to the change of internal energy.

Open System (Steady-State, Steady-Flow)

In this case an expression for the power may be determined because the process is known. Neglecting variations in kinetic and potential energies, from Equation (6.21) and the process $pv^n = C$,

$$\dot{W} = -\dot{m}\int_1^2 v\, dp = -\dot{m}\int_1^2 C^{1/n}p^{-1/n}\, dp$$

$$C^{1/n} = p_1^{1/n}v_1 = p_2^{1/n}v_2$$

$$\dot{W} = \frac{n}{n-1}\dot{m}[-p_1v_1 - p_2v_2]$$

Noting that

$$\frac{p_2v_2}{p_1v_1} = \left(\frac{p_2}{p_1}\right)^{(n-1)/n}$$

$$\dot{W} = \frac{n}{n-1}\dot{m}RT_1\left[1 - \left(\frac{p_2}{p_1}\right)^{(n-1)/n}\right] \tag{6.30}$$

Example 6.11

In a turbine 4500 kg/min of air expands polytropically from 425 kPa and 1360 K to 101 kPa. The exponent n is equal to 1.45 for the process. Find the work and the heat. Figure 6.12 illustrates the problem.

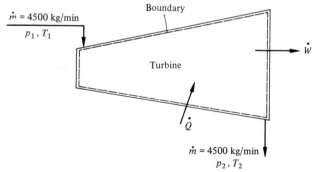

Figure 6.12 *A gas turbine receiving hot compressed air and expanding it polytropically.*

Open System

First Law: Energy In = Energy Out

$$\dot{m}h_1 + \dot{Q} = \dot{m}h_2 + \dot{W}$$

From Equation (6.30),

$$\dot{W} = \frac{1.45}{0.45}\left(\frac{4500}{60}\right)(0.287)(1360)\left[1 - \left(\frac{101}{425}\right)^{0.45/1.45}\right]$$

$$\dot{W} = 33\,939 \text{ kW}$$

Solve for $\dot{Q}$, as follows:

$$\dot{Q} = \dot{m}(h_2 - h_1) + \dot{W}$$

$$T_2 = 1360\left(\frac{101}{425}\right)^{0.45/1.45} = 870.7 \text{ K}$$

$$\dot{Q} = \left(\frac{4500}{60}\right)(1.0047)(870.7 - 1360) + 33\,939$$

$$\dot{Q} = -2927 \text{ kW}$$

The negative sign on the heat indicates that it is out of the system. This is reasonable as the turbine casing must be at a higher temperature than the surroundings.

6.5 THREE-PROCESS CYCLES

Until this point we have analyzed processes, however, by combining two or more processes together it is possible to construct a cycle. In later chapters we will analyze several applications of cycles, but for now let us consider three processes, joined together, forming a cycle. Three is the minimum practical number of processes to create a cycle. In Chapter 3 we noted some interesting behavior for cycles, namely, that $\oint \delta Q = \oint \delta W$ or $\Sigma Q = \Sigma W$. With this in mind, let us consider the following example.

Example 6.12

The following cycle involves 3 kg of air: polytropic compression from 1 to 2, where $p_1 = 150$ kPa, $T_1 = 360$ K, $p_2 = 750$ kPa and $n = 1.2$; constant-pressure cooling from 2 to 3; and constant-temperature heating from 3 to 1. Figure 6.13 illustrates the p–V

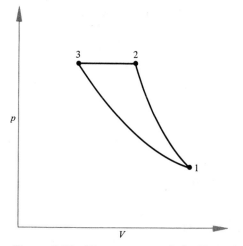

Figure 6.13   Three-process cycle for Example 6.12.

diagram for the cycle. Find the temperatures, pressures, and volumes at each state and determine the net work and heat.

The key to solving three-process-cycle problems is to draw a sketch. Once that is done, follow the processes around the cycle.
   State 1:

$$V_1 = \frac{mRT_1}{p_1} = \frac{(3.0)(0.287)(360)}{150} = 2.066 \text{ m}^3$$

Process 1–2, $pV^n = C$:

$$\frac{T_2}{T_1} = \left(\frac{p_2}{p_1}\right)^{n-1/n} \qquad T_2 = 360\left(\frac{750}{150}\right)^{0.2/1.2} = 470.7 \text{ K}$$

$$V_2 = \frac{MRT_2}{p_2} \qquad V_2 = \frac{(3.0)(0.287)(470.7)}{750} = 0.540 \text{ m}^3$$

Process 2–3, $p = C$ and $T_3 = T_1 = 360$ K, since the process 3–1 is at a constant temperature:

$$p_3 = p_2 = 750 \text{ kPa}$$

$$V_3 = \frac{mRT_3}{p_3} = 0.413 \text{ m}^3$$

$$T_3 = 360 \text{ K}$$

At this point let us calculate Q and W for each process.

$$Q_{1-2} = U_2 - U_1 + W_{1-2}$$

$$W_{1-2} = \frac{mR(T_2 - T_1)}{1 - n} = -476.5 \text{ kJ}$$

$$U_2 - U_1 = mc_v(T_2 - T_1) = 238.3 \text{ kJ}$$

$$Q_{1-2} = -238.2 \text{ kJ}$$

$$Q_{2-3} = H_3 - H_2 = mc_p(T_3 - T_2) = -333.6 \text{ kJ}$$

$$Q_{2-3} = U_3 - U_2 + W_{2-3}$$

$$U_3 - U_2 = mc_v(T_3 - T_2) = -238.3$$

$$W_{2-3} = -95.3 \text{ kJ}$$

The work could also be evaluated by $\int p \, dV$. For constant temperature processes,

$$Q_{3-1} = W_{3-1} = p_3 V_3 \ln\left(\frac{V_1}{V_3}\right) = mRT_3 \ln\left(\frac{V_1}{V_3}\right)$$

$$Q_{3-1} = W_{3-1} = (3.0)(0.287)(360) \ln\left(\frac{2.066}{0.413}\right) = 499 \text{ kJ}$$

$$\sum Q = -72.8 \text{ kJ} \qquad \sum W = -72.8 \text{ kJ}$$

In this case the cycle required a net amount of work to operate, hence the negative sign.

6.6 TRANSIENT FLOW

The typical situations studied in undergraduate thermodynamics are closed systems and steady-state open systems. However, the unsteady open system must be considered if we are to consider transient flow; the two cases which we will consider are charging and discharging a tank.

Consider Equation (3.52), which may be rewritten as

$$dE_{cv} = \delta Q - dW + \left(h + \frac{v^2}{2} + gz\right)_{in} dm_{in} - \left(h + \frac{v^2}{2} + gz\right)_{out} dm_{out}$$

(6.31)

where $e = u + v^2/2 + gz$. Equation (6.31) may be integrated over the control volume for any instant of time, yielding

$$\Delta E_{cv} = Q - W + \int \left(h + \frac{v^2}{2} + gz\right)_{in} dm_{in}$$

$$- \int \left(h + \frac{v^2}{2} + gz\right)_{out} dm_{out}$$

(6.32)

Furthermore, the change of energy within the control volume is equal to the change of internal energy. Thus, Equations (6.31) and (6.32) become

$$dU_{cv} = \delta Q - \delta W + \left(h + \frac{v^2}{2} + gz\right)_{in} dm_{in}$$

$$- \left(h + \frac{v^2}{2} + gz\right)_{out} dm_{out} \quad (6.33)$$

and

$$\Delta U_{cv} = Q - W + \int \left(h + \frac{v^2}{2} + gz\right)_{in} dm_{in}$$

$$- \int \left(h + \frac{v^2}{2} + gz\right)_{out} dm_{out} \quad (6.34)$$

Discharging a Tank

Consider Figure 6.14, which illustrates a tank connected to a supply line. The volume of the connection is considered negligible. Apply the first law to the control volume (Equation (6.33))

$$dV_{cv} = \delta Q + \left(h + \frac{v^2}{2} + gz\right)_{out} dm_{out} \quad (6.35)$$

Note that $\delta W = 0$ and $dm_{in} = 0$. Furthermore, let us restrict the velocity of the exit stream to less than 50 m/s and the change of z to be small, thus kinetic and potential energies may be neglected. This yields

$$dU_{cv} = \delta Q + (h\, dm)_{out} \quad (6.36)$$

The typical assumption at this point is that the tank is adiabatic; if it is not, δQ must be evaluated by other means (heat transfer analysis,

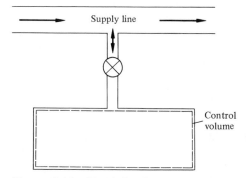

Figure 6.14   Sketch for charging and discharging a tank.

experimental data). Thus, for $\delta Q = 0$, Equation (6.36) becomes

$$d(mu) = h \, dm \tag{6.37}$$

However, $d(mu) = u \, dm + m \, du$, and upon substituting into Equation (6.37),

$$u \, dm + m \, du = h \, dm$$

$$\frac{dm}{m} = \frac{du}{(h-u)}$$

For any substance, $h = u + pv$; hence

$$\frac{dm}{m} = \frac{du}{pv} \tag{6.38}$$

A separate relationship between specific volume and mass may be written by considering the total control volume, $V = vm = $ constant, hence

$$dV = 0 = v \, dm + m \, dv$$

or

$$\frac{dm}{m} = -\frac{dv}{v} \tag{6.39}$$

Substituting Equation (6.39) into Equation (6.38) yields

$$\frac{du}{pv} = -\frac{dv}{v}$$

or

$$du + p \, dv = 0$$

Since $du + p \, dv = \delta q = 0$, the process is reversible adiabatic. In order to evaluate properties at any instant of time, the substance in the tank is assumed to be in equilibrium at that instant.

By using the reversible adiabatic properties for an ideal gas,

$$\frac{T_2}{T_1} = \left(\frac{v_1}{v_2} \right)^{k-1}$$

$$\ln\left(\frac{T_2}{T_1} \right) = (1-k)\ln\left(\frac{v_2}{v_1} \right)$$

$$\frac{dT}{T} = (1-k)\frac{dv}{v}$$

Substituting in Equation (6.39) yields

$$\left(\frac{1}{k-1} \right) \frac{dT}{T} = \frac{dm}{m}$$

and integrating yields

$$\frac{m_2}{m_1} = \left(\frac{T_2}{T_1}\right)^{1/(k-1)} \tag{6.40}$$

In a similar fashion

$$\frac{m_2}{m_1} = \left(\frac{p_2}{p_1}\right)^{1/k} = \frac{v_1}{v_2} \tag{6.41}$$

Charging a Tank

Let us consider the opposite situation, that of charging a tank. Figure 6.14 still applies, but the flow is into the control volume. As before, we will neglect changes in kinetic and potential energies, note that the work is zero, and let the tank be adiabatic, Equation (6.34) becomes

$$\int (h \, dm)_{in} = \Delta U_{cv} = m_2 u_2 - m_1 u_1 \tag{6.42}$$

At this point, let us assume that the properties in the line are constant with time (charging the tank does not affect them). Integrate from 0 to $m_L(m_L = m_{line})$, with $h_L = $ constant. Thus

$$m_L h_L = m_2 u_2 - m_1 u_1 \tag{6.43}$$

Let us apply this to the situation where the tank is initially evacuated ($m_1 = 0$), thus $m_L = m_2$ and

$$m_2 h_L = m_2 u_2$$

or

$$h_L = u_2$$

Consider an ideal gas as the substance entering the tank ($h = c_p T$, $u = c_v T$); thus

$$T_2 = kT_L$$

Hence, when charging a tank, the final equilibrium temperature in the tank is k times the line temperature. Again equilibrium at any instant is assumed to exist in the tank, so properties are defined.

Example 6.13

An adiabatic tank containing air is used to power an air turbine during times of peak power demand. The tank has a volume of 500 m^3 and contains air at 1000 kPa and 500 K. Determine: (a) the mass remaining when the pressure reaches 100 kPa; (b) the temperature at this instant.

(a) For air $k = 1.4$ and, from Equation (6.41),

$$\frac{m_2}{m_1} = \left(\frac{100}{1000}\right)^{1/1.4} = 0.193$$

$$m_2 = 0.193\ m_1$$

$$m_1 = \frac{p_1 V_1}{RT_1} = 3484.3\ \text{kg}$$

$$m_2 = 672.5\ \text{kg}$$

(b) The temperature may be calculated several ways:

$$T_2 = \frac{p_2 V_2}{m_2 R} = \frac{(100)(500)}{(672.5)(0.287)} = 259\ \text{K}$$

$$T_2 = T_1 \left(\frac{p_2}{p_1}\right)^{(k-1)/k} = 259\ \text{K}$$

$$T_2 = T_1 \left(\frac{m_2}{m_1}\right)^{k-1} = 259\ \text{K}$$

PROBLEMS

1. An insulated box containing carbon dioxide gas falls from a balloon 3.5 km above the earth's surface. Determine the temperature rise of the carbon dioxide when the box hits the ground.

2. Air from the discharge of a compressor enters a 1-m³ storage tank. The initial air pressure in the tank is 500 kPa and the temperature is 600 K. The tank cools, and the internal energy decreases by 213 kJ/kg. Determine (a) the work done; (b) the heat loss; (c) the change of enthalpy; (d) the final temperature.

3. A rigid, perfectly insulated system contains 0.53 m³ of helium at 1000 kPa. The system receives 1000 kJ of paddle work. Determine the final pressure.

4. A constant-pressure, adiabatic system contains 0.15 kg of air at 150 kPa and is comparable to the system depicted in Figure 6.5. The system receives 20.79 kJ of paddle work. The temperature of the air is initially 278 K and finally 416 K. Find the mechanical work and the changes of internal energy and enthalpy.

5. A closed rigid container has a volume of 1 m³ and holds air at 344.8 kPa and 273 K. Heat is added until the temperature is 600 K. Determine the heat added and the final pressure.

6. A piston–cylinder containing air receives heat at a constant temperature of 500 K and an initial pressure of 200 kPa. The initial volume is 0.01 m³ and the final volume is 0.07 m³. Determine the heat and work.

7. Steam enters an adiabatic turbine at 1 MPa and 600°C and exits at 50 kPa and 150°C. Determine the work per kilogram of steam.

8. A steam turbine receives 5 kg/s of steam at 2.0 MPa and 600°C and discharges the steam as a saturated vapor at 100 kPa. The heat loss through the turbine is 6 kW. Determine the power.

9. A nozzle receives 5 kg/s of steam at 0.6 MPa and 350°C and discharges it at 100 kPa and 200°C. The inlet velocity is negligible, the heat loss is 250 kJ/kg, determine the exit velocity.

10. Air is compressed polytropically from 101 kPa and 23°C and delivered to a tank at 1500 kPa and 175°C. Determine per kilogram of air (a) the heat removed during compression; (b) the reversible work.

11. A tank with a volume of 50 m³ is being filled with air, an ideal gas. At a particular instant, the air in the tank has a temperature of 400 K and 1380 kPa. At this instant the pressure is increasing at a rate of 138 kPa/s and the temperature is increasing at a rate of 25 K/s. Calculate the air flow into the tank at this instant in kg/s.

12. Air at a pressure of 100 kPa has a volume of 0.32 m³. The air is compressed in a reversible adiabatic manner until the temperature is 190°C. The reversible work is −63 kJ. Determine (a) the initial temperature; (b) the air mass; (c) the change of internal energy.

13. Air in a piston-cylinder occupies 0.12 m³ at 552 kPa. The air expands in a reversible adiabatic process, doing work on the piston until the volume is 0.24 m³. Determine (a) the work of the system; (b) the net work if the atmospheric pressure is 101 kPa.

14. Three moles of oxygen are compressed in a piston-cylinder in a reversible adiabatic process from a temperature of 300 K and a pressure of 102 kPa until the final volume is one tenth the initial volume. Determine (a) the final temperature; (b) the final pressure; (c) the system work.

15. A scuba tank contains 1.5 kg of air. The air in the tank is initially at 15°C. The tank is left near an engine exhaust line, and the tank's pressure doubles. Determine (a) the final temperature; (b) the change in internal energy; (c) the heat added.

16. In a reversible adiabatic manner, 17.6 m³/min of air are compressed from 277 K and 101 kPa to 700 kPa. Determine (a) the final temperature; (b) the change of enthalpy; (c) the mass flow rate; (d) the power required.

17. An adiabatic device looks like an inverted T with 3.03 kg/s of steam at 4 MPa and 600°C entering from the top, and two streams, one at 0.5 kg/s, 0.2 MPa and 6°C exiting horizontally and the other at 0.2

MPa and an unknown temperature also exiting horizontally. Determine the unknown temperature.

18. A control volume receives three streams of steam $(1, 2, 3)$, power is produced or consumed, 800 kW of heat leaves the control volume, and two streams $(4, 5)$ leave. Given that

$$m_1 = 0.5 \text{ kg/s}, v_1 = 400 \text{ m/s}, p_1 = 1.5 \text{ MPa, dry saturated vapor}$$

$$\dot{m}_2 = 5.0 \text{ kg/s}, v_2 \approx 0 \text{ m/s}, p_2 = 1.0 \text{ MPa}, v_2 = 0.100 \text{ m}^3/\text{kg}$$

$$\dot{m}_3 = 3.0 \text{ kg/s}, v_3 \approx 0 \text{ m/s}, p_3 = 600 \text{ kPa}, T_3 = 450°C$$

$$\dot{m}_4 = 4.0 \text{ kg/s}, v_4 = 500 \text{ m/s}, p_4 = 2.0 \text{ MPa}, T_4 = 300°C$$

$$\dot{m}_5 = 4.5 \text{ kg/s}, v_5 \approx 0 \text{ m/s}, p_5 = 800 \text{ kPa}, T_5 = 500°C$$

Determine the power.

19. An adiabatic condenser receives 100 kg/s of steam at 92% quality and 60 kPa. The steam leaves at 60 kPa and 60°C. The cooling water enters at atmospheric pressure and 40°C and discharges at 60°C. Determine (a) the heat transferred; (b) the cooling water flow rate.

20. An adiabatic rigid container, holding water initially at 100 kPa and 20°C, receives 209 kJ/kg of paddle work. Determine the final temperature.

21. A pneumatic lift system is being demonstrated at a sales show. The total load is 70 kg, and the lift piston is 15.2 cm in diameter and has an 20.2-cm stroke. A portable air bottle with an initial pressure of 20 MPa and a temperature of 23°C is to be used as the pneumatic supply. A regulator reduces the pressure from the bottle to the lift system. Neglecting all volume in the lines from the bottle to the piston, determine the number of times the piston can operate per air bottle if the air in the bottle remains at 23°C and the volume of the bottle is 0.05 m³.

22. One kilogram per second of air initially at 101 kPa and 300 K is compressed polytropically according to the process $pV^{1.3} = C$. Calculate the power necessary to compress the air to 1380 kPa. An aftercooler removes 100 kW from the air before it enters a storage tank. Determine the change in internal energy of the air across the aftercooler.

23. Oxygen expands in a reversible adiabatic manner through a nozzle from an initial pressure and initial temperature and with an initial velocity of 50 m/s. There is a decrease of 38 K in temperature across the nozzle. Determine (a) the exit velocity; (b) for inlet conditions of 410 kPa and 320 K, find the exit pressure.

24. An ideal gas having a mass of 2 kg at 465 K and 415 kPa expands in a reversible adiabatic process to 138 kPa. The ideal gas constant is 242 J/kg · K and $k = 1.40$. Determine (a) the final temperature; (b) the change in internal energy; (c) the work; (d) c_p, c_v.

25. An ideal gas with a molecular weight of 6.5 kg/kgmol is compressed in a reversible manner from 690 kPa and 277 K to a final specific volume of 0.47 m³/kg according to $p = 561 + 200\,v + 100v^2$, where p is the pressure in kPa, v is the specific volume in m³/kg. The specific heat at constant volume is 0.837 kJ/kg · K. Determine (a) the work; (b) the heat; (c) the final temperature; (d) the initial specific volume.

26. Three kilograms of neon are in a constant-volume system. The neon is initially at a pressure of 550 kPa and at a temperature of 350 K. Its pressure is increased to 2000 kPa by paddle work plus 210 kJ of heat addition. Determine (a) the final temperature; (b) the change of internal energy; (c) the work input.

27. A 2-kg mass oxygen expands at a constant pressure of 172 kPa in a piston-cylinder system from a temperature of 32°C to a final temperature of 182°C. Determine (a) the heat required; (b) the work done; (c) the change of enthalpy; (d) the change of internal energy.

28. One kilogram of carbon dioxide is contained in a constant-pressure piston-cylinder system. The carbon dioxide receives paddle work and rejects heat while changing from an initial temperature of 417 K to a final temperature of 277 K. The heat rejected is found to be three times the work input. Determine (a) the heat; (b) the paddle work; (c) the change of enthalpy; (d) the net work if the surroundings are at 101 kPa.

29. One kilogram of air expands at a constant temperature from a pressure of 800 kPa and a volume of 2 m³ to a pressure of 200 kPa. Determine (a) the work; (b) the heat; (c) the change of internal energy; (d) the change of enthalpy.

30. An ideal compressor compresses 12 kg/min of air isothermally from 99 kPa and a specific volume of 0.81 m³/kg to a final pressure of 600 kPa. Determine (a) the work in kW; (b) the heat loss in kW.

31. A vertical piston-cylinder system is constructed so the piston may travel between two stops. The system is surrounded by air at 100 kPa. The enclosed volume is 0.05 m³ at the lower stop and 0.12 m³ at the upper stop. Carbon dioxide is contained in the system at a temperature of 300 K and a pressure of 200 kPa. Heat is added and the piston rises, the system changing at constant pressure, until the piston reaches the upper stop. Heat continues to be added until the temperature is 900 K. Determine (a) the heat added; (b) the change of enthalpy; (c) the system work; (d) the net work; (e) the final pressure.

32. Air is contained in a piston-cylinder and is compressed in a reversible adiabatic manner from a temperature of 300 K and a pressure of 120 kPa to a final pressure of 480 kPa. Determine (a) the final temperature; (b) the work per kilogram.

33. Two cubic meters of helium at 277 K and 101 kPa are compressed to 404 kPa in a reversible adiabatic manner. Determine (a) the final temperature; (b) the power required.

34. In a natural gas pipeline compressor 110 m^3/min propane are compressed polytropically. The inlet pressure is 101 kPa and the temperature is 38°C. The process follows $pV^{1.08} = C$. The exit pressure is 510 kPa. Determine (a) the exit temperature; (b) the heat loss; (c) the work required; (d) the mass flow rate.

35. Air is compressed polytropically in a cylinder according to $pV^2 = C$. The work required is 180 kJ. Determine (a) the change of internal energy; (b) the heat transferred.

36. A piston–cylinder contains 2 kg of steam. The piston is frictionless and rests on two stops. The steam inside is initially at 1000 kPa and 75% quality. Heat is added until the temperature is 500°C. However, the piston does not move until the system pressure reaches 2 MPa and then it moves at constant pressure. Determine (a) the system work; (b) the total heat transferred.

37. Helium expands polytropically through a turbine according to the process $pV^{1.5} = C$. The inlet temperature is 1000 K, the inlet pressure is 1000 kPa, and the exit pressure is 150 k Pa. The turbine produces 1×10^5 kW. Determine (a) the exit temperature; (b) the heat transferred (kW); (c) the mass flow rate.

38. Carbon dioxide flows steadily at 1 kg/s through a device where the pressure, specific volume, and velocity of the gas are tripled according to $pV^n = C$. The inlet conditions are $p_1 = 200$ kPa, $v_1 = 0.4$ m^3/kg, $v_1 = 100$ m/s. Determine (a) n; (b) the initial and final temperatures; (c) the change of enthalpy; (d) the work; (e) the heat.

39. Nitrogen expands through a turbine in a reversible adiabatic process. The nitrogen enters at 1100 K and 550 kPa and exits at 100 kPa. The turbine produces 37 MW. Using Table A.1 determine (a) the exit temperature; (b) the exit enthalpy (kJ/kg); (c) the flow rate required.

40. Three kilograms of dry saturated steam expands at a constant pressure of 300 kPa to a final temperature of 400°C. Determine (a) the heat required; (b) the work done; (c) the change of internal energy.

41. In the previous problem the atmospheric pressure is 100 kPa. Determine the net system work.

42. In a piston–cylinder, 0.5 kg of air expands polytropically, $n = 1.8$, from an initial pressure of 5000 kPa and an initial volume of 0.07 m^3 to a final pressure of 500 kPa. Calculate the system work and heat.

43. A cycle, composed of three processes, is: polytropic compression ($n = 1.5$) from 137 kPa and 38°C to state 2; constant pressure from

state 2 to state 3; and constant volume from state 3 to state 1. The heat rejected is 1560 kJ/kg and the substance is air. Determine (a) the pressures, temperatures, and specific volumes around the cycle; (b) the heat added; (c) the heat rejected; (d) the work for each process.

44. A three-process cycle operating with nitrogen as the working substance has: constant temperature compression 1–2 ($T_1 = 40°C$, $p_1 = 110$ kPa); constant volume heating 2–3; and polytropic expansion 3–1 ($n = 1.35$). The isothermal compression requires -67 kJ/kg of work. Determine (a) p, T, and v around the cycle; (b) the heats in and out; (c) the net work.

45. Two kilograms of helium operate on a three-process cycle where the processes are: constant volume (1–2); constant pressure (2–3); and constant temperature (3–1). Given that $p_1 = 100$ kPa, $T_1 = 300$ K, and $v_1/v_3 = 5$, determine (a) the pressure, volume, and temperature around the cycle; (b) the work for each process; (c) the heat added.

46. An insulated tank initially evacuated with a volume of 3 m³ is charged with helium from a constant-temperature supply line at 300 K and 1000 kPa. If the final tank pressure is 900 kPa, determine the final helium temperature.

47. The tank in the preceding problem has an initial charge of helium at 300 K and 100 kPa. For the same final pressure of 900 kPa, determine the final helium temperature.

48. An adiabatic tank contains 2 kg of air at 3000 kPa and 325 K. The tank is discharged until the pressure reaches 500 kPa. Determine (a) the mass remaining; (b) the final tank temperature.

49. The tank in the preceding problem is now heated so that the temperature remains constant at 325 K. Determine the heat added.

Chapter 7
The Second Law of Thermodynamics and the Carnot Cycle

7.1 THE SECOND LAW OF THERMODYNAMICS

The second law of thermodynamics may be written in several ways. Regardless of the terminology used, however, the purpose of the second law is to give a sense of direction to energy-transfer processes. Combining this with the first law gives us the information necessary to analyze energy-transfer processes. The second law of thermodynamics states:

> Whenever energy is transferred, the level of energy cannot be conserved and some energy must be permanently reduced to a lower level.

When this is combined with the first law of thermodynamics, the law of energy conservation, the following results:

> Whenever energy is transferred, energy must be conserved, but the level of energy cannot be conserved and some energy must be permanently reduced to a lower level.

We understand the conservation of energy, but what is the energy level? What follows is a wholly qualitative description of the combined first and second laws of thermodynamics. It is intended to give a feeling, a sense of appreciation, for the laws and their implications.

7.2 ENERGY LEVEL

Let us consider two identical blocks of material with different temperatures. Let the temperature of one block be 1000 K, the temperature of the second one be 500 K. The block at 1000 K has a higher thermal energy level than the one at 500 K. The internal energy of the blocks is

$$U_1 = m_1 c T_1 \text{ kJ}$$
$$U_2 = m_2 c T_2 \text{ kJ}$$

If these blocks are brought together and reach thermal equilibrium at temperature T_3, then the energy of the two blocks is

$U_3 = (m_1 + m_2)cT_3$

However, by the first law,

$$U_3 = U_1 + U_2$$
$$(m_1 + m_2)cT_3 = m_1cT_1 + m_2cT_2$$

let

$m_1 = m_2$

$$T_3 = \frac{T_1 + T_2}{2} = 750 \text{ K}$$

Thus the final temperature is less than one of the initial temperatures. We assume the energy level to be proportional to temperature, so we see that there has been an energy-level decrease. There has also been an energy-level increase, but the second law states only that there will be a decrease. The energy level finally is less than the highest energy level initially. Note that the total energy of either block has nothing to do with the energy level. The mass of block 2 could be 10 times larger than that of block 1, but there will still be an energy-level decrease. Check this and you will find that $T_3 = 545.4$ K.

Actually the energy level is more than the temperature level of a system. A block of material of a small mass at a distance above a zero datum would have a higher energy level than a block of large mass of the same material resting at the zero datum. In this case the energy level is indicated by the potential energy. The potential energy of the small mass is greater than that of the larger mass. The energy level is a function of all the thermodynamic forces in a system that will cause energy transfer from the system. The energy level of the system is lowered when the values of these thermodynamic forces within the system decrease. The direction of energy transport is from a high energy level to a low energy level.

7.3 SECOND LAW FOR A CYCLE

The first law of thermodynamics gives us a technique for energy analysis, but it does not describe how the energy will flow. The second law of thermodynamics gives direction to the energy flow.

As with all laws, there is no way to verify the second law. Sadi Carnot first observed that in steam engines there could be no work produced unless there was heat flow from a high-temperature reservoir to a low-temperature reservoir. He further noted that the work produced was a function of the temperature difference between the reservoirs and was greater for a greater temperature difference.

Let us consider the simple heat engine in Figure 7.1, which could be the basic power plant of a ship. In this figure, fuel is burned in a boiler, and the thermal energy released (heat) by the process converts the liquid water into high-temperature steam. The boiler is the high-temperature reservoir. The steam expands through the turbine, the work reservoir, to a low temperature. The steam is condensed at a low temperature in the condenser, the low-temperature reservoir, and the liquid water is pumped back to the boiler, completing the cycle. The Kelvin–Planck statement of the second law is

> No cyclic process is possible whose sole result is the flow of heat from a single heat reservoir and the performance of an equivalent amount of work.

In relating this statement of the second law to the simple power cycle in Figure 7.1, could we eliminate the condenser from the system and still maintain a cycle? No. There is no way to return the steam to the boiler. If the condenser were at the same temperature as the boiler, could work be accomplished? No. If work were accomplished, then there would be a reduction in temperature, and heat would have to flow from cold to hot to leave the turbine and enter the condenser. This is a violation of the second law. The Kelvin–Planck statement further disallows the conversion of all the heat into work. Only a portion of the heat may be converted into work. This corresponds to the statement of the second law in terms of the energy level.

Consider a constant volume container which receives paddle work, as shown in Figure 7.2(a). The substance will increase in temperature, but cooling will return the system to its initial condition. Let us reverse the situation and add heat to the system. Can the weight be raised? No. This

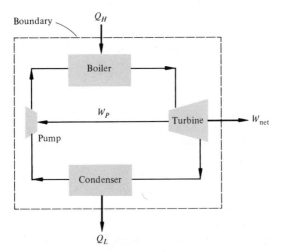

Figure 7.1   A simple heat engine with heat and work denoted.

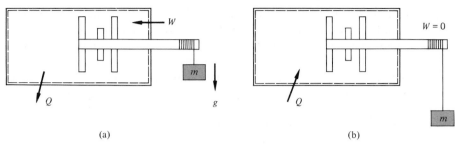

Figure 7.2 *Demonstration of effect of the second law.*

is analogous to trying to produce work with only a single heat reservoir, which violates the second law. Notice that all the work can be converted into heat, but not all the heat can be converted into work.

Thus the second law tells us that not all heat may be converted into work, but it does not tell us the maximum amount of work that can be produced for heat flow between two temperatures. That is the purpose of the next section. The concept of the second law will be more completely understood after the next chapters, which discuss entropy and available energy.

7.4 CARNOT CYCLE

A thermodynamic cycle occurs when a system undergoes two or more processes and returns to its initial state. Many machines are analyzed by use of cyclic investigation.

The first person to recognize the energy-transfer processes in machines was Sadi Carnot. A retired army corporal at the age of 24, he had the time and inclination to think about these transfer processes. In 1824 he published *Reflections on the Motive Power of Fire*, the document that established the basis for the second law of thermodynamics and introduced the concepts of "cycle" and "reversible processes." It remained for Lord Kelvin and Rudolf Clausius to amplify the principles in modern thermodynamic form.

Necessity is the mother, not only of invention, but also of discovery. The basis for many scientific discoveries has been a practical need to understand some physical phenomena. Although the steam engine was becoming the motive force of the industrial revolution, no one had the understanding necessary to advance its design, and its manufacture proceeded on a hit-and-miss basis.

7.5 CARNOT ENGINE

Carnot noted that the higher the temperature of the steam entering the engine and the lower the temperature of the steam leaving the engine, the greater the engine's work output. He devised a theoretical engine that would operate on a closed cycle; it would receive heat at a constant temperature and reject heat at a constant temperature. The boiling and condensing of water takes place at nearly a constant temperature. The engine would be perfectly insulated, and the work would be done reversibly. Hence there would be a reversible adiabatic expansion by the engine to produce work, and to complete the cycle there would be a reversible adiabatic compression. Figure 7.3 illustrates the Carnot cycle on pressure–volume coordinates.

How does the Carnot engine produce work? We can see that there is on the p–V diagram a net area equal to the net work produced, but this does not always give a physical appreciation of the work performed. If the engine could turn a shaft, as an internal-combustion engine turns a crankshaft, then a physical appreciation could be attached to the use of the p–V diagram.

Consider a Carnot heat engine, as depicted in Figure 7.4(a)–(e). This engine can provide work to turn the shaft as shown. Let us follow the engine through a cycle and compare the piston movement to the points on Figure 7.3. The cylinder arrangement starts with the fluid in the cylinder at state 1. A perfect conductor is connected between the engine and the heat source, and heat is transferred at constant temperature until the fluid reaches state 2, Figure 7.4(b). A perfect insulator is now placed between the heat reservoirs, and the engine and the piston expand in a reversible adiabatic process until state 3 is reached, Figure 7.4(c). A perfect conductor is now inserted·between the engine and the heat sink, and heat is rejected from the engine to the heat sink at constant temperature. This brings the engine to state 4, Figure 7.4(d). In order to return the fluid to state 1, some of the energy developed as work in going

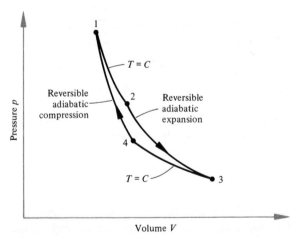

Figure 7.3 *The p–V diagram for the Carnot cycle.*

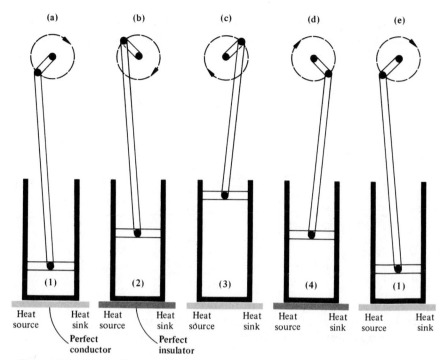

Figure 7.4   A visualization of the Carnot engine.

from state 1 to state 3 must be returned to the engine, compressing the fluid from state 4 to state 1. This compression must be reversible adiabatic, so a perfect insulator is inserted between the engine and the heat reservoirs and the fluid is compressed to state 1, Figure 7.4(e). The cycle is completed as the engine returns to its starting point, and work has been produced.

The Carnot engine is important because it converts thermal energy (heat) into the maximum amount of mechanical energy (work). No other engine or device works more efficiently between two temperature reservoirs, although it may be possible to develop engines with the same efficiency as the Carnot engine. It is desirable to have an expression for the thermal efficiency of the engine and to see what factors influence the efficiency. Later we will be able to compare other engines to the Carnot engine and determine the relative merit of the particular engine.

Carnot Engine Efficiency

The thermal efficiency, η_{th}, of any engine is the net work produced divided by the energy supplied to produce that work; that is,

$$\eta_{th} = \frac{W_{net}}{Q_{in}} \tag{7.1}$$

The net work can be evaluated by using a cyclic integral. Ideal gas relationships will be used for this analysis, but later in the text (Chapter 8) it will be shown that the results are valid for any substance as the working fluid.

$$W_{net} = \oint \delta W = \oint p \, dV = \int_1^2 p \, dV + \int_2^3 p \, dV + \int_3^4 p \, dV + \int_4^1 p \, dV$$

$$\int_1^2 p \, dV = p_1 V_1 \ln\left(\frac{V_2}{V_1}\right) = mRT_H \ln\left(\frac{V_2}{V_1}\right)$$

$$\int_2^3 p \, dV = \frac{p_3 V_3 - p_2 V_2}{1 - k} = \frac{mRT_C - mRT_H}{1 - k} \tag{7.2}$$

$$\int_3^4 p \, dV = p_3 V_3 \ln\left(\frac{V_4}{V_3}\right) = mRT_C \ln\left(\frac{V_4}{V_3}\right)$$

$$\int_4^1 p \, dV = \frac{p_1 V_1 - p_4 V_4}{1 - k} = \frac{mRT_H - mRT_C}{1 - k}$$

where T_H is the temperature of the high-temperature heat source and T_C is the temperature of the low-temperature heat sink.

If these terms are added, the net work reduces to

$$W_{net} = mRT_H \ln\left(\frac{V_2}{V_1}\right) + mRT_C \ln\left(\frac{V_4}{V_3}\right) \tag{7.3}$$

The expression for the heat supplied must be deduced. Heat is supplied at constant temperature from state 1 to state 2. This is a closed system, so the first law of thermodynamics tells us

$$Q_{1-2} = U_2 - U_1 + W_{1-2}$$

$$U_2 - U_1 = 0 \qquad \text{since } T = C \tag{7.4}$$

$$Q_{1-2} = W_{1-2} = mRT_H \ln\left(\frac{V_2}{V_1}\right)$$

The heat rejected may be found in a similar manner to be

$$Q_{3-4} = mRT_C \ln\left(\frac{V_4}{V_3}\right) \tag{7.5}$$

An important conclusion regarding theoretical heat engines has been demonstrated when Equations (7.5), (7.4), and (7.3) are compared. That is,

$$W_{net} = \Sigma Q = Q_{in} + Q_{out} \tag{7.6}$$

Substituting Equations (7.3) and (7.4) into Equation (7.1) yields

$$\eta_{th} = \frac{mRT_H \ln(V_2/V_1) + mRT_C \ln(V_4/V_3)}{mRT_H \ln(V_2/V_1)} \tag{7.7}$$

Using the fact that processes 2 to 3 and 4 to 1 are reversible adiabatic processes,

$$\frac{T_2}{T_3} = \left(\frac{V_3}{V_2}\right)^{k-1} = \frac{T_H}{T_C} \qquad \frac{T_1}{T_4} = \left(\frac{V_4}{V_1}\right)^{k-1} = \frac{T_H}{T_C}$$

$$\frac{V_2}{V_1} = \frac{V_3}{V_4}$$

Equation (7.7) may now be written

$$\eta_{th} = \frac{mRT_H \ln(V_2/V_1) - mRT_C \ln(V_2/V_1)}{mRT_H \ln(V_2/V_1)}$$

$$\eta_{th} = \frac{T_H - T_C}{T_H} \tag{7.8}$$

Thus we discover that the thermal efficiency of a Carnot engine with ideal gas as the working fluid is dependent only on the absolute temperature of the heat supplied and the heat rejected. The greater the difference between the two, the more efficient the engine. In this development the heat was transferred at constant temperature. This is ideal and usually a temperature difference exists between the heat reservoir and an actual heat engine.

Example 7.1

An engine operates on the Carnot cycle. It produces 50 kW while operating between temperatures of 800°C and 100°C. Determine the efficiency and the heat added.

$$\eta_{th} = \frac{T_H - T_C}{T_H} = \frac{1073 - 373}{1073} = 0.652$$

$$\dot{Q}_{in} = \frac{\dot{W}_{net}}{\eta_{th}} = \frac{50}{0.652} = 76.6 \text{ kW}$$

7.6 MEAN EFFECTIVE PRESSURE

The work for the Carnot engine may be evaluated by Equation (7.3). However, it is sometimes convenient to have another means of evaluating an engine. One such indicator is the mean effective pressure, an imaginary pressure developed by equating the cycle work to an equivalent work:

$$W_{net} = \int_1^3 dV = p_m \int_1^3 dV \tag{7.9}$$

$$W_{net} = p_m \times \text{piston displacement}$$

$$p_m = \frac{W_{net}}{\text{piston displacement}} \tag{7.10}$$

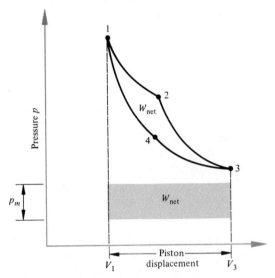

Figure 7.5   A graphical illustration of mean effective pressure, p_m.

The mean effective pressure, p_m, is equal to the net cycle work divided by the piston displacement. Figure 7.5 illustrates the p–V diagram for the Carnot engine and the shaded rectangular area is the same area as that enclosed by the Carnot engine, but it is characterized by a mean effective pressure. The higher the mean effective pressure, the greater will be the force per cycle and the greater the work output per cycle.

Example 7.2

A six-cylinder engine with a 10-×10-cm bore and stroke operates on the Carnot cycle. It receives 54 kJ per cycle of heat at 833 K and rejects heat at 555 K while running at 300 rev/min. Determine the mean effective pressure and the power of the engine.

$$\eta_{th} = \frac{T_H - T_C}{T_H} = \frac{833 - 555}{833} = 0.333$$

$$W_{net} = (0.333)Q_{in} = (0.333)(54) = 18 \text{ kJ/cycle}$$

$$\text{total piston displacement} = \frac{6\pi}{4}(0.1)^2(0.1) = 0.004\,712 \text{ m}^3$$

$$p_m = \frac{W_{net}}{\text{piston displacement}} = \frac{18 \text{ kJ/cycle}}{0.004\,712 \text{ m}^3/\text{cycle}}$$

$$p_m = 3819 \text{ kPa}$$

$$\dot{W} = (18 \text{ kJ/cycle})(5 \text{ cycles/s}) = 90 \text{ kW}$$

Example 7.3

The working substance in a Carnot engine is 0.05 kg of air. The maximum cycle temperature is 940 K, and the maximum pressure is 8.4×10^3 kPa. The heat added per cycle is 4.2 kJ. Determine the maximum cylinder volume if the minimum temperature during the cycle is 300 K.

From Figure 7.5 we note that the maximum temperature and pressure occur at state 1 and the maximum volume occurs at state 3. To find V_3 we must follow the processes 1–2–3.

$$V_1 = \frac{mRT_1}{p_1} = \frac{(0.05)(0.287)(940)}{8.4 \times 10^3} = 1.606 \times 10^{-3} \text{ m}^3$$

$$Q_{in} = p_1 V_1 \ln\left(\frac{V_2}{V_1}\right)$$

$$4.2 = (8.4 \times 10^3)(1.606 \times 10^{-3}) \ln\left(\frac{V_2}{V_1}\right)$$

$$V_2 = 2.193 \times 10^{-3} \text{ m}^3 \qquad T_2 = 940 \text{ K}$$

$$p_2 = \frac{mRT_2}{V_2} = \frac{(0.05)(0.287)(940)}{2.193 \times 10^{-3}}$$

$$p_2 = 6.15 \times 10^3 \text{ kPa}$$

$$\frac{V_3}{V_2} = \left(\frac{T_2}{T_3}\right)^{1/(k-1)} \qquad \text{for reversible adiabatic processes}$$

$$V_3 = 3.811 \times 10^{-2} \text{ m}^3$$

7.7 REVERSED CARNOT ENGINE

The Carnot engine is a power-producing engine that uses heat supplied as an energy source and delivers mechanical work as an energy output. When the Carnot cycle is reversed, it means that mechanical work is supplied as an energy input and heat may be moved from one energy level (temperature) to another energy level. Why do this? For instance, this is the manner in which refrigerators operate. Energy is supplied by an electric motor, driving a compressor, and the refrigerant (the working substance) picks up heat from inside the refrigerator at a low temperature and discharges it at a high temperature to the condensing coils on the outside of the refrigerator. We will discuss actual refrigeration systems later, but for now let us limit the discussion to the reversed Carnot cycle.

The reversed Carnot cycle has all the same processes as the power-producing Carnot cycle, but the cycle operates in the counterclockwise direction. The p–V for the reversed Carnot engine is given in Figure 7.6

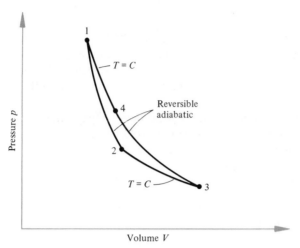

Figure 7.6   A p–V diagram for the reversed Carnot cycle.

The purpose of the reversed Carnot engine is to remove a quantity of heat at a low temperature, Q_{in}, by supplying work. This results in a rejection of heat at a high temperature Q_{out}. The performance of reversed engines is denoted by the coefficient of performance (COP), rather than an efficiency term. This has, however, the same purpose as efficiency—the desired effect divided by the cost of achieving the desired effect. For a reversed Carnot engine using an ideal gas, the working substance, the coefficient of performance is

$$\text{COP} = \frac{Q_{in}}{W_{net}} = \frac{p_1 V_1 \ln(V_2/V_1)}{-p_1 V_1 \ln(V_2/V_1) - p_3 V_3 \ln(V_4/V_3)}$$

$$\text{COP} = \frac{T_C}{T_H - T_C} \tag{7.11}$$

where the net work has been written as a positive quantity. Thus the coefficient of a performance of a reversed Carnot engine depends only on the absolute temperatures of the heat reservoirs. The coefficient of performance of most reversed cycles is greater than unity.

A device operating on this cycle would have the capability of providing either cooling or heating by locating the desired reservoir in the space requiring temperature conditioning. Machines do operate in this manner and provide year-round conditioning of buildings. This will be covered in greater detail in the chapter on refrigeration.

Example 7.4

A reversed Carnot engine removes 40 000 W from a heat sink. The temperature of the heat sink is 260 K and the temperature of the heat reservoir is 320 K. Determine the power required of the engine.

$$COP = \frac{Q_{in}}{W_{net}} = \frac{T_C}{T_H - T_C} = \frac{260}{320 - 260}$$

$$COP = 4.33$$

$$W_{net} = \frac{Q_{in}}{COP} = \frac{40\,000}{4.33} = 9230.7 \text{ W}$$

7.8 FIRST COROLLARY OF THE SECOND LAW

The second law gives a sense of direction to a process whereas the first law does not. It also tells us that it is impossible to have a perpetual machine of the second kind, one that violates the second law. There are two corollaries to this law, the first one:

> It is impossible to construct an engine to operate between two heat reservoirs, each having a fixed and uniform temperature, which will exceed the efficiency of a reversible engine operating between the same reservoirs.

Figure 7.7 illustrates such a situation. Note that the reversible engine is a Carnot engine, whereas the other engine may have irreversibilities associated with it.

Assume that the irreversible engine has an efficiency greater than that of the reversible engine. Let W_I be the work produced by the irreversible engine, while receiving heat Q_1, and let W_R be the work produced by the reversible engine (Carnot) when receiving the same amount of heat, Q_1. The assumption tells us that $W_I > W_R$.

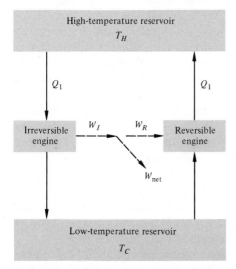

Figure 7.7 *Two heat engines, one reversible and one irreversible, operating between the same temperature limits.*

Now run the reversible engine as a reversed Carnot engine so it will supply the amount of heat Q_1 to the high-temperature reservoir. It would be possible to run the reversible engine by supplying work from the irreversible engine and still have some net work remaining. This is a perpetual-motion machine of the second kind and cannot exist because it violates the second law of thermodynamics. Thus the assumption that $W_I > W_R$ is not valid and the first corollary must be valid.

7.9 SECOND COROLLARY OF THE SECOND LAW

The second corollary follows:

All reversible engines have the same efficiency when working between the same two constant-temperature heat reservoirs.

The proof of the proposition follows that of the previous one. Assume that the proposition is not true in a particular case, and then show that a perpetual-motion machine of the second kind is a consequence. In this case the irreversible engine is replaced by a Carnot engine of higher efficiency than the other reversible engine.

7.10 THERMODYNAMIC TEMPERATURE SCALE

When discussing the zeroth law of thermodynamics, we discovered that temperature could be measured by observing a change in magnitude of a property. Usually this is the volume change of mercury or another liquid in a thermometer. Temperature scales have been developed using various properties of different substances. The scales may be standardized at the boiling and ice points of water, but they will differ at other temperature levels. It is possible to develop a temperature scale that is independent of substance properties, but that relies on a reversible engine.

We are at the point where we can discuss a thermodynamic temperature scale by considering reversible engines operating between constant-temperature heat reservoirs. Figure 7.8 illustrates this. Three reversible engines that produce work, W_{12}, W_{23}, and W_{13}, are placed between constant-temperature reservoirs at temperatures T_1, T_2, and T_3, respectively. Engine 12 receives heat Q_1 at temperature T_1 and rejects heat Q_2 at temperature T_2 while producing work W_{12}. Engine 23 receives heat Q_2 at temperature T_2 and rejects heat Q_3 at temperature T_3 and produces work W_{23}. Engine 13 receives heat Q_1 at temperature T_1 and rejects heat Q_3 at temperature T_3 while producing work W_{13}. Heat Q_3 is the same for engine 23 and engine 13 as demonstrated by the first corollary.

The thermal efficiency of a reversible engine operating between two constant-temperature reservoirs has been shown to be a function of only the temperatures of the reservoirs. The thermal efficiency for engine 12,

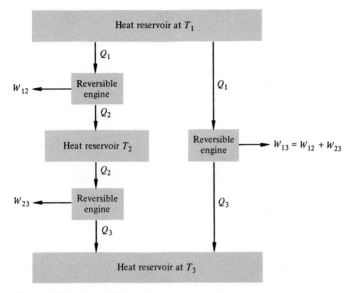

Figure 7.8   Reversible engines operating between two temperature reservoirs.

η_{12}, is then

$$\eta_{12} = f(T_1, T_2) \tag{7.12}$$

and the thermal efficiency is also

$$\eta_{12} = 1 - \frac{Q_2}{Q_1} \tag{7.13}$$

and hence a new functional relationship is shown to exist between the heat ratio and the temperatures of the heat reservoir by comparing Equations (7.12) and (7.13).

$$\frac{Q_1}{Q_2} = \phi(T_1, T_2)$$

In the same manner, relationships for the other heat ratios and temperatures may be developed or deduced as

$$\frac{Q_2}{Q_3} = \phi(T_2, T_3)$$

$$\frac{Q_1}{Q_3} = \phi(T_1, T_3)$$

A relationship between the three heat ratios is

$$\frac{Q_1}{Q_2} = \frac{Q_1/Q_3}{Q_2/Q_3}$$

and substitution of the temperature equivalents yields

$$\phi(T_1, T_2) = \frac{\phi(T_1, T_3)}{\phi(T_2, T_3)}$$

Temperature T_3 must cancel from the right-hand side if the equation is to maintain a functional identity. This yields a new relationship

$$\phi(T_1, T_2) = \frac{\psi(T_1)}{\psi(T_2)} \tag{7.14}$$

where ψ denotes the functional relationship.

There are many functional relationships that will satisfy Equation (7.14). The simplest, and the one used throughout this text, was proposed by Kelvin:

$$\psi(T) = T \tag{7.15}$$

Thus

$$\frac{Q_1}{Q_2} = \frac{T_1}{T_2} \tag{7.16}$$

and Equation (7.12) becomes

$$\eta_{12} = 1 - \frac{Q_2}{Q_1} = 1 - \frac{T_2}{T_1}$$

This is the Kelvin temperature scale with which the reader is familiar. The limits are from 0 to $+\infty$.

Kelvin also proposed another functional relationship, which would permit the lower temperature limit to be $-\infty$, rather than zero. To do this he introduced a logarithmic scale with temperature θ. The relationship between the scales may be shown.

$$\psi(\theta) = e^\theta \tag{7.17}$$

$$\frac{Q_1}{Q_2} = \frac{e^{\theta_1}}{e^{\theta_2}} \tag{7.18}$$

Equating Equations (7.16) and (7.18) yields

$$\frac{e^{\theta_1}}{e^{\theta_2}} = \frac{T_1}{T_2}$$

$$\theta_1 - \theta_2 = \ln(T_1) - \ln(T_2)$$

and therefore

$$\theta = \ln(T) + C$$

where C is a constant that determines the level of temperature corresponding to zero on the logarithmic temperature scale.

The Kelvin temperature scale showed that it was possible to have an absolute-zero temperature that is independent of the properties of any substance. We must resort, in the final analysis, to a "thermometer" type instrument to measure temperature, as it is impossible to design a heat engine running on the Carnot cycle.

PROBLEMS

1. A Carnot engine operates with 0.136 kg of air as the working substance. The pressure and volume at the beginning of isothermal expansion are 2.1 MPa and 9.6 liters, respectively. The air behaves as an ideal gas and the sink temperature is 50°C and the heat added is 32 kJ. Determine (a) the source temperature; (b) cycle efficiency; (c) pressure at the end of isothermal expansion; (d) heat rejected to sink per cycle.

2. A Carnot engine produces 25 kW while operating between temperature limits of 1000 K and 300 K. Determine (a) heat supplied per second; (b) heat rejection per second.

3. A Carnot engine uses nitrogen as the working fluid. The heat supplied is 53 kJ and the adiabatic expansion ratio is 16:1. The receiver temperature is 295 K. Determine (a) the thermal efficiency; (b) the heat rejected; (c) the work.

4. A Carnot engine uses air as the working substance and receives heat at a temperature of 315°C and rejects it at 65°C. The maximum possible cycle pressure is 6.0 MPa and minimum volume is 0.95 liter. When heat is added, the volume increases by 250%. Determine the pressure and volume at each state in the cycle.

5. A Carnot engine operates between temperatures of 1000 K and 300 K. The engine operates at 2000 rev/min and develops 200 kW. The total engine displacement is such that the mean effective pressure is 300 kPa. Determine (a) the cycle efficiency; (b) the heat supplied (kW); (c) the total engine displacement (m^3).

6. A Carnot cycle uses nitrogen as the working substance. The heat supplied is 54 kJ. The temperature of the heat rejected is 21°C, and $V_3/V_2 = 10$. Determine (a) the cycle efficiency; (b) the temperature of heat added; (c) the work.

7. Helium is used in a Carnot engine where the volumes beginning with the constant-temperature heat addition are $V_1 = 0.3565$, $V_2 = 0.5130$, $V_3 = 8.0$, $V_4 = 5.57$ m^3. Determine the thermal efficiency.

8. Air is used in a Carnot engine where 22 kJ of heat is received at 560 K. Heat is rejected at 270 K. The displacement volume is 0.127 m^3. Determine (a) the work; (b) the mean effective pressure.

9. A Carnot engine operates between temperature limits of 1200 K and 400 K, using 0.4 kg of air and running at 500 rev/min. The pressure

at the beginning of heat addition is 1500 kPa and at the end of heat addition is 750 kPa. Determine (a) the heat added per cycle; (b) the heat rejected; (c) the power; (d) the volume at the end of heat addition; (e) the mean effective pressure; (f) the thermal efficiency.

10. A reversed Carnot engine receives 316 kJ of heat. The reversible adiabatic compression process increases by 50% the absolute temperature of heat addition. Determine (a) the COP; (b) the work.

11. Two reversible engines operate in series between a high-temperature (T_h) and a low-temperature (T_c) reservoir. Engine A rejects heat to engine B, which in turn rejects heat to the low-temperature reservoir. The high-temperature reservoir supplies heat to engine A. Let $T_h = 1000$ K, $T_c = 400$ K, and the engine thermal efficiencies are equal. The heat received by engine A is 500 kJ. Determine (a) the temperature of heat rejection by engine A; (b) the works of engine A and engine B; (c) the heat rejected by engine B.

12. Use the engine arrangement described in Problem 11 with engines of equal thermal efficiency; let the work of engine B be equal to 527.5 kJ, the low temperature be equal to 305 K, and the heat received by engine A be equal to 2110 kJ. Determine (a) the thermal efficiency of each engine; (b) the temperature of heat supplied to engine A; (c) the work of engine A; (d) the heat rejected by engine B; (e) the temperature of heat addition to engine B.

13. A nonpolluting power plant can be constructed using the temperature difference in the ocean. At the surface of the ocean in tropical climates, the average water temperature year-round is 30°C. At a depth of 305 m, the temperature is 4.5°C. Determine the maximum thermal efficiency of such a power plant.

14. A heat pump is used to heat a house in the winter months. When the average outside temperature is 0°C and the indoor temperature is 23°C, the heat loss from the house is 20 kW. Determine the minimum power required to operate the heat pump.

15. A heat pump is used for cooling in summer and heating in the winter. The house is maintained at 24°C year-round. The heat loss is 0.44 kW per degree difference between outside and inside temperature. The average outside temperature is 32°C in the summer and −4°C in the winter. Determine (a) the power requirements for both heating and cooling; (b) which condition determines the unit size that must be purchased.

16. The Carnot efficiency of a heat engine is a function of the high-temperature heat source, T_h, and the low-temperature heat sink, T_c. It is possible to change one heat reservoir by an amount ΔT to improve the engine efficiency. Which reservoir should be changed?

17. A Carnot refrigerator rejects 2500 kJ of heat at 80°C while using 1100 kJ of work. Find (a) the cycle low temperature; (b) the COP; (c) the heat absorbed.

18. A Carnot heat engine rejects 230 kJ of heat at 25°C. The net cycle work is 375 kJ. Determine the thermal efficiency and the cycle high temperature.

19. A Carnot refrigerator operates between temperature limits of −5°C and 30°C. The power consumed is 4 kW and the heat absorbed is 30 kJ/kg. Determine (a) the COP; (b) the refrigerant flow rate.

20. The Novel Air Conditioning Company claims to have developed a new air conditioner which will maintain a room at 22°C while rejecting 105 kJ of heat at 42°C. The unit requires 10 kJ of work. Is this device believable?

21. A Carnot heat pump is being considered for home heating in a location where the outside temperature may be as low as −35°C. The expected COP for the heat pump is 1.50. To what temperature could this unit provide heat?

Chapter 8
Entropy

In this chapter we will develop a property called "entropy." Although it is impossible to attach a physical meaning to entropy, as we can with internal energy, it helps us quantitatively understand the second law.

In the late 1800s, Clausius was investigating the equilibrium conditions for an isolated system. He knew that the total energy of the system was a constant, and he wanted to determine whether or not there would be a change of state in the isolated system. Experience had shown that a change of state might occur, but could a method be devised by which the change could be predicted to occur? If the system were in equilibrium, then no change would occur. It would be especially helpful if there were a system property that would denote whether or not the system was in equilibrium.

8.1 CLAUSIUS INEQUALITY

Clausius developed a method to evaluate a closed system operating as a heat engine, as shown in Figure 8.1. The system receives heat Q_{S1} and Q_{S2} at temperatures T_1 and T_2, respectively. The system produces work W_S and rejects heat Q_{S3} at temperature T_3. All the processes are assumed to be reversible. The work from the system drives two Carnot engines, A and B, which act as heat pumps. What Clausius did, and we shall do, is assume that the system develops a slight irreversibility. What happens provides the answer to his original question: Does a property exist to determine system equilibrium?

The energy into the heat reservoirs I and II is equal to the energy out; hence

$$Q_{A1} = Q_{S1} \quad \text{and} \quad Q_{B2} = Q_{S2}$$

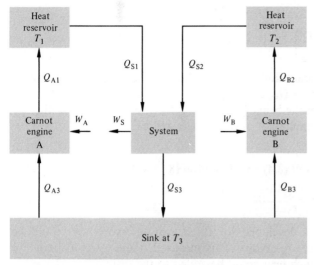

Figure 8.1 A system used to derive two reversed Carnot engines.

Since the Carnot engines are reversible, all the work leaving the system is utilized as work by them:

$$W_S = W_A + W_B$$

The system rejects heat Q_{S3}. Since all processes are reversible,

$$Q_{S3} = Q_{A3} + Q_{B3} \qquad (8.1)$$

We let the system act in an irreversible manner, causing rejected heat to be greater than the reversible heat supplied to the Carnot engines.

$$Q_{S3} > Q_{A3} + Q_{B3} \qquad (8.2)$$

Equations (8.1) and (8.2) may be combined as

$$Q_{S3} \geqslant Q_{A3} + Q_{B3} \qquad (8.3)$$

Analyzing Carnot engine A, we find

$$Q_{A3} + W_A = Q_{A1}$$

From the Carnot engine efficiency,

$$W_A = Q_{A1}\left(1 - \frac{T_3}{T_1}\right)$$

hence,

$$Q_{A3} = Q_{A1}\left(\frac{T_3}{T_1}\right) = Q_{S1}\left(\frac{T_3}{T_1}\right) \qquad (8.4)$$

and similarly, for Carnot engine B;

$$Q_{B3} = Q_{B2}\left(\frac{T_3}{T_2}\right) = Q_{S2}\left(\frac{T_3}{T_2}\right) \tag{8.5}$$

Consider the algebraic summation for the heat flowing into and out of the sink. Let the heat Q_{S3} entering the sink be negative (as it is leaving the system), and let the heats Q_{A3} and Q_{B3} be positive (as they will enter the system). The summation of these heats will be

$$Q_{A3} + Q_{B3} + Q_{S3} \leqslant 0 \tag{8.6}$$

Substituting Equation (8.4) and Equation (8.5) in Equation (8.6) yields

$$\frac{Q_{S1}}{T_1} + \frac{Q_{S2}}{T_2} + \frac{Q_{S3}}{T_3} \leqslant 0 \tag{8.7}$$

Equation (8.7) was developed for three heat reservoirs and could be expanded for n heat reservoirs. This means that there would be $n - 1$ temperatures at which heat could be added to the system:

$$\sum_n \frac{Q}{T} \leqslant 0$$

As the number of heat reservoirs becomes large, the heat transferred at any given reservoir becomes small. Taking the limit as n becomes infinitely large, then Q becomes infinitesimally small, and

$$\lim_{n \to \infty} \sum_n \frac{Q}{T} = \int \frac{dQ}{T} \leqslant 0$$

$$\oint \frac{\delta Q}{T} \leqslant 0 \tag{8.8}$$

Equation (8.8) is called the "Clausius inequality."

8.2 DERIVATION OF ENTROPY

The definition of entropy was developed by applying the Clausius inequality to a cyclic process. Figure 8.2 illustrates a two-process cycle, composed of reversible paths A and B. Since the cycle is reversible, it is possible to proceed in the clockwise or counterclockwise direction.

Point P is located on path A. If the system proceeds an infinitesimal distance from P in the clockwise direction, an amount of heat δQ_{cl} will be added. If the direction is reversed—and the system returns that infinitesimal distance to point P—an infinitesimal amount of heat δQ_{cc} is removed.

Heat added is positive, and heat removed is negative; so

$$\delta Q_{cl} = -\delta Q_{cc}$$

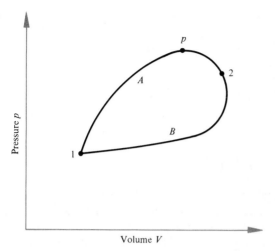

Figure 8.2 *Two paths, A and B, joining states 1 and 2.*

Applying the Clausius inequality to the cycle,

$$\oint \frac{\delta Q_{cl}}{T} \leqslant 0 \qquad (8.9)$$

or

$$\oint \frac{\delta Q_{cc}}{T} \leqslant 0 \qquad (8.10)$$

but

$$-\delta Q_{cl} = +\delta Q_{cc}$$

hence

$$-\oint \frac{\delta Q_{cl}}{T} \leqslant 0 \qquad (8.11)$$

There is only one solution that satisfies both Equations (8.9) and (8.10):

$$\oint \frac{\delta Q_{cl}}{T} = 0$$

Since it would be possible to obtain the same result for the counterclock-wise direction, the expression may be written in general as

$$\oint \frac{\delta Q}{T} = 0 \qquad (8.12)$$

We apply Equation (8.12) to the cycle illustrated in Figure 8.2

$$\oint \frac{\delta Q}{T} = \int_1^2 \left(\frac{\delta Q}{T} \right)_A + \int_2^1 \left(\frac{\delta Q}{T} \right)_B = 0$$

$$\int_1^2 \left(\frac{\delta Q}{T} \right)_A = \int_1^2 \left(\frac{\delta Q}{T} \right)_B$$

Since the paths A and B are arbitrary paths, for any path between the points 1 and 2

$$\int_1^2 \frac{\delta Q}{T} = \int_1^2 \left(\frac{\delta Q}{T} \right)_A = \int_1^2 \left(\frac{\delta Q}{T} \right)_B = C$$

The integral must be equal to some constant value because it does not depend on a particular system path. This indicates that the quantity $(\delta Q/T)_{rev}$ is an exact differential. In the thermodynamic sense the quantity is not a function of the system but depends only the system states 1 and 2. This is the definition of a property. Clausius called this property "entropy," S. Hence

$$S_2 - S_1 = \int_1^2 \left(\frac{\delta Q}{T} \right)_{rev} \tag{8.13}$$

Although the quantity $(\delta Q/T)_{rev}$ may denote a differential change in the property S, the property difference is described by Equation (8.13). This poses no problem in undergraduate thermodynamics, since the analysis of problems utilizes the entropy difference. However, the resolution of this dilemma remains a point of interest.

Let us now apply the Clausius inequality to a system operating on a cycle composed of a reversible path A and an irreversible path B.

$$\oint \frac{\delta Q}{T} = \int_1^2 \left(\frac{\delta Q}{T} \right)_A + \int_2^1 \left(\frac{\delta Q}{T} \right)_B \leq 0$$

$$\int_1^2 \left(\frac{\delta Q}{T} \right)_A = S_2 - S_1$$

However, for an irreversible path $\delta Q/T$ cannot be evaluated. The entropy change for an irreversible process between states 1 and 2 is

$$S_2 - S_1 > \int_1^2 \left(\frac{\delta Q}{T} \right)_B$$

The equality sign may be included if we write the previous equation for both reversible and irreversible situations.

$$S_2 - S_1 \geq \int_1^2 \left(\frac{\delta Q}{T} \right)_{rev} \tag{8.14a}$$

or

$$dS \geq \left(\frac{\delta Q}{T} \right)_{rev} \tag{8.14b}$$

We may reach several conclusions regarding the entropy change of a system. For adiabatic closed systems, the entropy will increase due to internal irreversibilities. Only for reversible adiabatic processes will $dS = 0$.

Let us consider an isolated system, one that is closed, is adiabatic, and has no energy flow across its boundary. We know from Equation (8.14b) that

$$(dS)_{\text{isolated}} \geqslant 0 \tag{8.15a}$$

or

$$(\Delta S)_{\text{isolated}} \geqslant 0 \tag{8.15b}$$

The system may have several parts to it, i subsystems, so if we let Equation (8.15b) apply to the overall system, then

$$\Delta S_{\text{total}} = \sum_{i} (\Delta S)_{i} \geqslant 0 \tag{8.16}$$

8.3 THIRD LAW OF THERMODYNAMICS

The third law of thermodynamics allows the calculation of absolute entropy. The Nernst postulate of the third law is

The absolute entropy of a pure crystalline substance in complete internal equilibrium is zero at zero degrees absolute.

Although a complete discussion of the third law and its implications is beyond the scope of this text, it can be seen that the absolute entropy of a substance at some state 2 may be determined as follows:

$$S_2 - S_1 = \int_{T=T_1}^{T=T_2} \frac{\delta Q}{T} \tag{8.17}$$

however,

$$S_1 - S_0 = \int_{T=0}^{T=T_1} \frac{\delta Q}{T} \tag{8.18}$$

but $S_0 = 0$, so the absolute value of S_1 may be determined by Equation (8.18) and the absolute value of S_2 may be from Equation (8.17). If S_0 is a constant, then the absolute value of S_2 may be found in the same manner.

8.4 EQUILIBRIUM STATE

Clausius started his investigation of systems and developed, or discovered, the property of entropy in order to determine whether or not an isolated system was in equilibrium. Noting that $dS = (\delta Q/T)_{\text{rev}}$, there are two methods by which dS will be nonzero. One method can be an exchange of heat, δQ, between the system and the surroundings. However, the system is isolated, and so there can be no energy transfer between the system and the surroundings. The other method requires a temperature differential within the system. This would cause an irreversible energy flow within the system and result in a change in the system temperature. Thus it is only by an irreversible process that the entropy of the system will increase.

The following conclusions about the change in entropy and system equilibrium may be made.

1. When the entropy of an isolated system is at its maximum value, no change in state can occur.
2. When it is possible for the entropy of an isolated system to increase, the system cannot be in a state of equilibrium and it is possible for a change of system state to occur.

8.5 ENTROPY CHANGE OF A CLOSED SYSTEM

Figure 8.3 illustrates an entropy equilibrium surface. The system is characterized by temperature T_1 and pressure p_1. If the system is in equilibrium, then it is denoted by point A on the equilibrium surface. However, if the system is not in equilibrium, then it has to be below the equilibrium surface, as illustrated by point B. A change may occur that would bring the system to the equilibrium surface; however, we cannot determine whether or not it will occur. Note that the entropy is a maximum for its state; if the temperature or pressure were different, then another maximum value of entropy for that state would be the equilibrium value, such as point C.

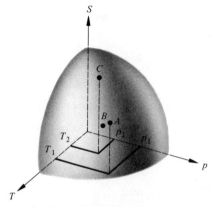

Figure 8.3 An equilibrium surface with states A and C on the surface and state B below the surface.

The following statements summarize entropy changes of a closed system.

1. The entropy will decrease when heat is removed from the system, all processes being reversible ones.
2. The entropy will remain constant when reversible adiabatic processes occur within the system.
3. The entropy will increase when heat is added to the system, reversibly or irreversibly.
4. The entropy of an isolated system will increase when irreversible processes occur within it.

Statement 1 requires the restriction of reversibility because irreversibilities within the system may cause an entropy increase; and although the heat removal results in an entropy decrease, the net entropy change could be positive or negative, depending on the magnitude of each term.

8.6 CALCULATION OF ENTROPY CHANGE FOR IDEAL GASES

Thus far we have not calculated the value of entropy, or the difference in entropy between two states for a closed system. We will first calculate the entropy of gases, using the ideal gas laws and the gas tables, and then calculate the entropy of pure substances.

The first law for a closed system is

$$\delta Q = dU + \delta W$$

For a reversible process, and considering only mechanical work, this becomes

$$T\,dS = dU + p\,dV \tag{8.19}$$

Equation (8.19) is a very important thermodynamic relation. A second important relation is

$$dH = dU + p\,dV + V\,dp$$

and for reversible processes, with only mechanical work, this becomes

$$T\,dS = dH - V\,dp \tag{8.20}$$

If the ideal gas equation of state is used, $pV = mRT$, and the change of enthalpy and internal energy is written for an ideal gas, Equations (8.19) and (8.20) become, respectively,

$$dS = mc_v\frac{dT}{T} + mR\frac{dV}{V} \tag{8.21}$$

and

$$dS = mc_p\frac{dT}{T} - mR\frac{dp}{p} \tag{8.22}$$

Integrating between state 1 and state 2 with constant specific heats and mass yields

$$S_2 - S_1 = mc_v \ln\left(\frac{T_2}{T_1}\right) + mR \ln\left(\frac{V_2}{V_1}\right) \qquad (8.23)$$

$$S_2 - S_1 = mc_p \ln\left(\frac{T_2}{T_1}\right) - mR \ln\left(\frac{P_2}{P_1}\right) \qquad (8.24)$$

Thus there are two equations describing the change of entropy for an ideal gas in a closed system. The particular equation to use is the one that will make the solution simplest.

Example 8.1

Three kilograms of air are heated from 300 K to 800 K while the pressure changes from 100 kPa to 500 kPa. Determine the change in entropy. Show that the two ideal gas equations yield the same result.

$$V_1 = \frac{mRT_1}{P_1} = \frac{(3)(0.287)(300)}{100} = 2.583 \text{ m}^3$$

$$V_2 = \frac{mRT_2}{P_2} = \frac{(3)(0.287)(800)}{500} = 1.378 \text{ m}^3$$

$$S_2 - S_1 = mc_v \ln\left(\frac{T_2}{T_1}\right) + mR \ln\left(\frac{V_2}{V_1}\right)$$

$$S_2 - S_1 = (3)(0.7176)\ln\left(\frac{800}{300}\right) + (3)(0.287)\ln\left(\frac{1.378}{2.583}\right)$$

$$S_2 - S_1 = 1.57 \text{ kJ/K}$$

$$S_2 - S_1 = mc_p \ln\left(\frac{T_2}{T_1}\right) - mR \ln\left(\frac{P_2}{P_1}\right)$$

$$S_2 - S_1 = (3)(1.0047)\ln\left(\frac{800}{300}\right) - (3)(0.287)\ln\left(\frac{500}{100}\right)$$

$$S_2 - S_1 = 1.57 \text{ kJ/K}$$

The gas tables account for the variation of specific heat with temperature. If Equation (8.22) is partially integrated as follows,

$$S_2 - S_1 = m\int_1^2 c_p \frac{dT}{T} - mR \ln\left(\frac{P_2}{P_1}\right)$$

and in the gas tables,

$$\int_1^2 c_p \frac{dT}{T} = \phi_2 - \phi_1$$

then

$$S_2 - S_1 = m(\phi_2 - \phi_1) - mR \ln \left(\frac{p_2}{p_1} \right) \tag{8.25}$$

Example 8.2

Calculate the change in entropy for air at the conditions indicated in Example 8.1 using Table A.2 in the Appendix.

At 300 K, $\phi_1 = 2.5153$;   at 800 K, $\phi_2 = 3.5312$

$$S_2 - S_1 = (3)(3.5312 - 2.5153) - (3)(0.287) \ln(5)$$
$$= 1.66 \ \text{kJ/K}$$

Thus we see that the variation of specific heat with temperature has a significant effect in our calculating the change of entropy. In this case the difference between the results of Examples 8.1 and 8.2 is nearly 6 percent. Use of the gas tables results in a more accurate analysis.

8.7 RELATIVE PRESSURE AND RELATIVE SPECIFIC VOLUME

In the gas tables there is also the means of calculating an isentropic process from state 1 to state 2. For an isentropic process, $\Delta S = 0$; hence

$$\ln \left(\frac{p_2}{p_1} \right) = \frac{\phi_2 - \phi_1}{R}$$

The relative pressure p_r is defined as

$$\ln(p_r) = \ln \left(\frac{p}{p_0} \right) = \frac{\phi}{R}$$

where p_0 is a reference state. Then

$$\ln \left(\frac{p_{r2}}{p_{r1}} \right) = \frac{\phi_2 - \phi_1}{R} \tag{8.26}$$

and hence

$$\frac{p_{r2}}{p_{r1}} = \left(\frac{p_2}{p_1} \right)_S \tag{8.27}$$

There is also a relative specific volume, v_r

$$v_r = \frac{RT}{p_r}$$

Hence

$$\left(\frac{v_2}{v_1} \right)_S = \frac{v_{r2}}{v_{r1}} \tag{8.28}$$

Example 8.3

Air expands isentropically in a piston–cylinder from 1100 K and 420 kPa to 100 kPa. Calculate the work per kilogram of air. Use the gas tables.

First Law, Closed System

$$q = \Delta u + w$$

$$q = 0 \qquad \text{for } s = c$$

$$u_1 = 845.33 \text{ kJ/kg} \qquad p_{r1} = 167.1$$

$$p_{r2} = p_{r1}\left(\frac{p_2}{p_1}\right) = 167.1\left(\frac{100}{420}\right) = 39.78$$

$$u_2 = 561.96 \text{ kJ/kg}$$

$$w = u_1 - u_2 = 283.37 \text{ kJ/kg}$$

8.8 ENTROPY OF A PURE SUBSTANCE

Entropy has been defined as a property of a system, and the specific entropy is tabulated in the tables of thermodynamic properties. Since entropy is a property, it may be calculated for any quality within the saturated region. It is listed in the compressed liquid and superheat vapor tables. Figure 8.4 illustrates a T–s diagram for steam. Another frequently used chart for steam is the h–s diagram, called the "Mollier diagram," illustrated in Figure 8.5. A disadvantage of the Mollier diagram is that it cannot be used for steam with a quality of 50 percent or less. However, at times, as we will see in later chapters, the Mollier diagram is the only

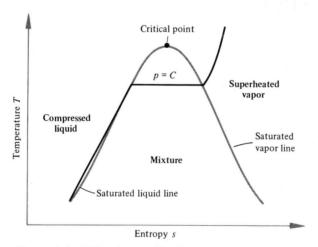

Figure 8.4   A T–s diagram for water.

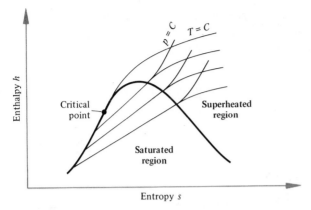

Figure 8.5   An h–s or Mollier diagram for water.

means available for the solution of steam turbine problems. The T–s diagram usually does not list specific volume, so there is a disadvantage in its use. However, both charts are very useful for particular problems and will be used for them. The charts are listed in the Tables B.1 and B.2 of the Appendix.

Example 8.4

A steam engine receives steam at 3.0 MPa and 300°C and expands isentropically to 1.0 MPa. Determine the work done if the cylinder volume at the start of expansion is 3.2 liters. Figure 8.6 illustrates the process on a T–s diagram.

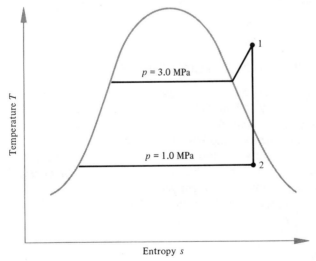

Figure 8.6   A T–s diagram.

Solve the problem in general and then substitute the property values from the steam tables.

This is a closed system; the first law for a closed system is

$$Q = \Delta U + W$$

but $Q = 0$ since the process is isentropic; hence

$$W = -\Delta U = U_1 - U_2$$

From Table A.7,

$$u_1 = 2750.1 \text{ kJ/kg}$$

$$s_1 = 6.5390 \text{ kJ/kg} \cdot \text{K}$$

$$v_1 = 0.08114 \text{ m}^3/\text{kg}$$

We need $U_1 = m_1 u_1$, where

$$m_1 = \frac{V_1}{v_1} = \frac{0.0032}{0.08114} = 0.0394 \text{ kg}$$

The property at point 2 must be determined. What information is available? The pressure is 1.0 MPa, and the entropy is the same as it was at state 1, $s_1 = 6.5390$. At state 2, we find that the steam lies in the saturated region because $s_2 = 6.5390$, which is less than $s_g = 6.5865$ at 1 MPa. We need to find the internal energy, but first we need to calculate the quality.

$$s_2 = s_f + x s_{fg}$$

$$x = 0.988$$

$$u_2 = u_f + x u_{fg} = 2564.2 \text{ kJ/kg}$$

$$W = m(u_1 - u_2) = 7.32 \text{ kJ}$$

8.9 CARNOT CYCLE USING T–S COORDINATES

One of the great advantages of a T–S diagram is that the reversible heat transfer may be represented. Since $Q_{rev} = \int T \, dS$, these are the natural coordinates for reversible heat flow. When talking about the Carnot cycle, we had to invoke the ideal gas law to determine the efficiency. This is no longer necessary. Figure 8.7 illustrates the Carnot cycle on the T–S coordinates. Notice that the reversible adiabatic process, a constant entropy or isentropic process, is a straight vertical line.

From the equation for reversible heat transfer,

$$\delta Q = T \, dS$$

$$Q = T \int_1^2 dS \qquad \text{for } T = C$$

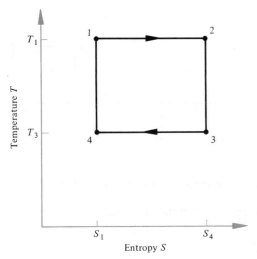

Figure 8.7   *A T–S diagram for the Carnot cycle.*

Hence

$$Q_{1-2} = T_H(S_2 - S_1) \tag{8.29}$$

and

$$Q_{3-4} = T_C(S_4 - S_3) \tag{8.30}$$

but

$$(S_4 - S_3) = -(S_2 - S_1)$$

and recalling that

$$\eta_{th} = \frac{Q_{in} + Q_{out}}{Q_{in}}$$

$$\eta_{th} = \frac{T_H - T_C}{T_H} \tag{8.31}$$

Not surprisingly, the result is obtained more directly when using the proper thermodynamic coordinates. In this case the ideal-gas law did not have to be invoked, thus demonstrating that Equation (8.31) is valid regardless of the working fluid in the Carnot engine.

8.10 HEAT AND WORK AS AREAS

That the work is equal to the algebraic sum of the heat supplied and heat rejected may be demonstrated graphically. The enclosed area on Figure 8.8(a) represents the heat supplied. The enclosed area on Figure 8.8(b) represents the heat rejected. Since heat in is positive and heat out is

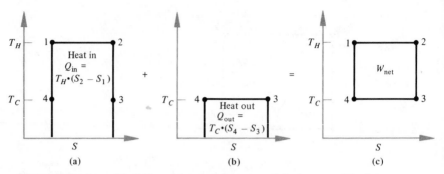

Figure 8.8   A graphical illustration that the net work is the sum of the heats.

negative, the two areas, when added, yield the net work, shown in Figure 8.8(c).

8.11 THE SECOND LAW FOR OPEN SYSTEMS

We have considered the second law of thermodynamics for closed systems; let us extend this to open systems. Figure 8.9 illustrates a general open system, a control volume located in a moving fluid with heat and work being done.

The second law may be stated for constant mass as

$$dS_{CM} \geq \frac{\delta Q}{T} \qquad\qquad (8.32)$$

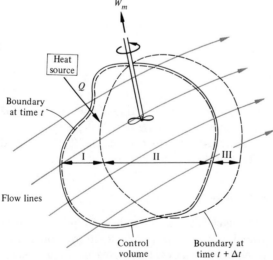

Figure 8.9   A control volume located in a moving fluid.

The mass in the control volume at time t, the control mass, is constant. We will develop another expression for the entropy of the control mass and from this develop the expression of the second law for open systems.

The entropy of the control mass at time t is

$$S_{CM,t} = S_{II,t} + S_{I,t} + S_{III,t} \tag{8.33}$$

The term $S_{III,t}$ is zero, so it may be included. At time $t + \Delta t$,

$$S_{CM,t+\Delta t} = S_{II,t+\Delta t} + S_{I,t+\Delta t} + S_{III,t+\Delta t} \tag{8.34}$$

Again, $S_{I,t+\Delta t}$ is a convenient zero to have in the equation. Take the difference between Equations (8.33) and (8.34) and divide by Δt, taking the limit as Δt approaches zero. This yields

$$\frac{dS_{CM}}{dt} = \frac{dS_{II}}{dt} - \frac{dS_{I}}{dt} + \frac{dS_{III}}{dt} \tag{8.35}$$

The negative sign on dS_I/dt is due to a physical decrease in the quantity $(S_{I,t+\Delta t} - S_{I,t})$. Let us adopt the following convention:

$$S = ms$$

$$\frac{dS}{dt} = \frac{dm}{dt}s = \dot{m}s$$

Thus Equation (8.32) becomes

$$\frac{dS_{CM}}{dt} \geqslant \frac{\delta \dot{Q}}{T} \tag{8.36}$$

when expressed as a rate equation, where $\delta \dot{Q}$ is the infinitesimal heat flux.

As Δt approaches zero, region II coincides with the control volume and regions I and III represent flow in and out of the control volume, respectively.

Equation (8.36) becomes

$$\frac{\delta \dot{Q}}{T} \leqslant \frac{dS_{CM}}{dt} = \frac{dS_{CV}}{dt} - \dot{m}_{in}s_{in} + \dot{m}_{out}s_{out} \tag{8.37}$$

For steady, one-dimensional flow, where $\dot{m} = \dot{m}_{in} = \dot{m}_{out}$, Equation (8.37) reduces to

$$\frac{\delta \dot{Q}}{T} \leqslant \dot{m}(s_{out} - s_{in}) \tag{8.38}$$

For no heat transfer across the control surface, the surface of the control volume, Equation (8.38) becomes

$$s_{out} \geqslant s_{in} \tag{8.39}$$

Thus for steady, one-dimensional, adiabatic flow, the second law tells us that the entropy increases or remains constant; it cannot decrease.

Example 8.5

Steam flows steadily through a turbine with a mass flow rate of 2.52 kg/s. The inlet steam conditions are 7.0 MPa and 500°C. The exit steam pressure is 20 kPa and the expansion is isentropic. Determine the turbine work in kilowatts.

Open System

First Law: Energy In = Energy Out

$$\dot{m}h_1 = \dot{m}h_2 + \dot{W}$$

$$\dot{W} = \dot{m}(h_1 - h_2)_s$$

$$h_1 = 3410.3 \text{ kJ/kg}$$

$$s_1 = 6.7975 \text{ kJ/kg} \cdot \text{K}$$

$$s_2 = s_1 = s_f + xs_{fg} = 0.8320 + x(7.0766)$$

$$x = 0.843$$

$$h_2 = h_f + xh_{fg} = 251.4 + (0.843)(2358.3) = 2239.4 \text{ kJ/kg}$$

$$\dot{W} = (2.52)(3410.3 - 2239.4) = 2950 \text{ kW}$$

8.12 FURTHER CONSIDERATIONS

Heretofore we have considered a system to be an amount of fixed mas or space, most likely an engine of a relatively small size. Consider now that the system is the universe, which, being of fixed mass, satisfies the definition of a system. If all the processes within any system, in general, and the universe were isentropic or reversible cycles, then there would be no increase in entropy. However, actual processes are irreversible, and so there is a continual increase in the entropy of the universe. This means that the entropy must have been originally constrained, or it could not be increasing now. People are still speculating about how the entropy of the universe was minimized originally. Although it is beyond the scope of this text, it remains a point from which many interesting discussions and speculations may result.

PROBLEMS

1. Two kilograms of a gas are cooled from 500°C to 200°C at constant pressure in a heat exchanger. Determine the change of entropy for (a) air; (b) carbon dioxide; (c) helium.
2. Calculate the change in entropy per kilogram between 250 K and 75 kPa and 750 K and 300 kPa using the ideal-gas law and the gas tables for air.

3. Isentropic compression of a gas occurs. Determine the temperature ratio that allows this if the pressure ratio is six and the gas is (a) air; (b) propane; (c) helium.

4. An isothermal expansion of air occurs in a piston–cylinder from 298 K and 800 kPa to 225 kPa. Determine the change of entropy per kilogram.

5. Air is compressed isothermally at 300 K from 150 kPa to 400 kPa. Determine the change of entropy and the heat transfer per kilogram.

6. A piston–cylinder arrangement has been developed to compress air adiabatically from 30°C and 100 kPa to 450 kPa with 120 kJ/kg of work. Is this possible?

7. An isothermal compression process of 50 kg/s of nitrogen from 305 K and 120 kPa to 480 kPa occurs in a large, water-cooled compressor. Determine (a) the change of entropy; (b) the power required.

8. A piston–cylinder containing 0.2 kg of air at 800 K and 2.1 MPa expands isentropically until the pressure is 210 kPa. Determine (a) the work done (use gas tables); (b) the initial and final specific volume.

9. A closed system undergoes a polytropic process according to $pV^{1.3} = C$. The initial temperature and pressure are 600 K and 200 kPa, respectively. Three kilograms of air are the working substance. Determine the change of entropy for the process if the final temperature is 900 K.

10. When air is throttled, there is an entropy increase. For 2 kg/s of air the entropy increases by 0.06 kW/K. Determine the pressure ratio of final to initial for this to occur.

11. In a constant-volume process 1.5 kg of air is cooled from 1200 K. The change of entropy is -1.492 kJ/K. Determine the final temperature.

12. A cooling process at constant pressure involves the removal of 3.5 kW from the system. Compute the mass flow required and the change in entropy of the following substances: (a) air decreasing in temperature from 0°C to -20°C; (b) refrigerant 12 condensing at -20°C; (c) ammonia condensing at -20°C.

13. Determine the heat transferred from state (a) to state (b) if $T = 300 + 1.5s^2$ for the process.

14. A Carnot engine operates between 4°C and 280°C. If the engine produces 310 kJ of work, determine the entropy change during heat addition.

15. Two kilograms of an ideal gas, $R = 317$ J/kg K and $k = 1.26$, are contained in a rigid cylinder; 21.1 kJ of heat are added to the gas, which has an initial temperature of 305 K. Determine (a) the final temperature; (b) the change of entropy; (c) the change of enthalpy; (d) the change of internal energy.

16. A gas turbine expands 50 kg/s of helium polytropically, $pV^{1.8} = C$, from 1100 K and 500 kPa to 350 K. Determine (a) the power produced; (b) the entropy change; (c) the final pressure; (d) the heat loss; (e) the power produced by isentropic expansion to the same temperature.

17. A steam turbine receives steam as a dry saturated vapor at 7.0 MPa and expands adiabatically to atmospheric pressure, 100 kPa. The turbine uses 24.3 kg/h of steam for each kilowatt. What is the entropy of the steam leaving the turbine?

18. Hydrogen is compressed isentropically from $p_1 = 750$ kPa and $T_1 = 305$ K to a pressure of 2.25 MPa. Determine per kilogram (a) the final temperature; (b) the work; (c) the change of enthalpy and internal energy.

19. 2.27 kilograms of steam expand adiabatically from a volume of 0.234 m^3 and a temperature of 300°C to a pressure of 125 kPa. Determine (a) for reversible adiabatic expansion the work, the initial pressure, and the final quality; (b) for irreversible expansion, where the final quality is 100%, find the work, the initial pressure, and the change of entropy.

20. A closed system contains an ideal gas, $R = 269$ J/kg · K, $c_v = 0.502$ kJ/kg · K. The system undergoes the following cycle: At state 1 the temperature is 444 K and the pressure is 448 kPa; heat is transferred at constant pressure until the temperature is 889 K, state 2; the gas is compressed at constant temperatures until the value of the entropy equals that at state 1, state 3; and finally there is an isentropic expansion from state 3 to state 1. Determine (a) the p–V and T–s diagrams; (b) the pressure at state 3; (c) the heat transferred from 1 to 2 and from 2 to 3; (d) the cycle work.

21. Steam, contained in a 22.6-liter piston–cylinder at 1.4 MPa and 250°C, expands isentropically until it becomes a saturated vapor. Determine (a) the final pressure; (b) the work required; (c) the final volume.

22. A refrigeration system uses ammonia as the working substance. The compressor receives ammonia as a dry saturated vapor at −22°C and discharges it at 1.4 MPa. The compression process is isentropic. Determine (a) the discharge temperature; (b) the work of compression per kilogram.

23. A natural gas pipeline distributes gas throughout the country, pumped by large gas-turbine-driven centrifugal compressors. Assume that the natural gas is methane, the pipe diameter is 0.2 m, and the gas enters the compressor at 300 K and 105 KPa. The velocity of the methane entering the compressor is 4 m/s. The compression process is isentropic and the discharge pressure is 700 kPa. Determine (a) the discharge temperature; (b) the mass flow rate; (c) the power required.

24. Air is contained in a 2.8-liter piston–cylinder at 37°C and 100 kPa. Isentropic compression occurs until the final volume is one-third the initial volume. Using the gas tables, determine (a) the final temperature; (b) the final pressure; (c) the work required.

25. Two insulated tanks, one containing 1.7 m³ of argon at 1.0 MPa and 38°C, the other completely evacuated and having a volume of 3.4 m³, are connected by a short pipe and valve. The valve is opened and the argon pressure is equalized between the two tanks. Determine (a) the final pressure; (b) the final temperature; (c) the change of entropy; (d) the change of enthalpy; (e) the work done.

26. Develop the expression for the Clausius inequality starting with the premise that the thermal efficiency of a Carnot engine is greater than or equal to the thermal efficiency of a heat engine.

27. A capillary tube is used in lieu of a throttling valve in home refrigerators. Liquid R 12 enters saturated at 40°C and leaves at a pressure of 362 kPa. Determine the entropy change per kilogram across the capillary tube.

28. Two kg/s of saturated steam at 200°C are condensed to a saturated liquid. The coolant is R 12, which is vaporized at 30°C. Determine (a) the mass flow rate of R 12 vaporized; (b) the change of entropy of the steam and of R 12.

29. A thermodynamic cycle is composed of the following reversible processes: isothermal expansion, 1–2; isentropic compression, 2–3; and constant pressure cooling, 3–1. The cycle operates on 1.5 kg of steam. State 1 is dry saturated steam at 200°C; state 2 has a pressure of 100 kPa. Determine (a) the heat addition; (b) the net work; (c) the entropy change from 1 to 2.

Chapter 9
Available Energy
and Availability

The second law has been stated in terms of entropy and of energy level. There must be some manner in which these two methods combine, or in which the equivalence between the two is apparent. The concept of available energy demonstrates the equivalence of the entropic and energy formulations of the second law. Available energy will be defined as that portion of the thermal energy that can be used for work. There will be two formulations: one for systems with heat transfer, the other for adiabatic open systems. Following the discussion of available energy, we will examine the more general concept of availability of energy where the restriction that only thermal energy be considered is not invoked. The concept of availability of energy will reduce to available energy under special conditions. The major portion of this chapter deals with available energy, which is very useful in thermal-energy applications and is easier to understand than availability. Using the precepts of available energy, the concept of second-law efficiency is introduced.

9.1 AVAILABLE ENERGY FOR SYSTEMS WITH HEAT TRANSFER

To determine the available energy of any heat source, it is possible to imagine the heat source's supplying energy to a reversible heat engine. For any engine, even the Carnot engine, only a portion of the heat received is able to be used for work. This is the available portion of the heat supplied. The rest of the heat is unavailable energy, unavailable to do work. Thus

$$\text{energy added} \quad = \quad \text{available energy} \quad + \quad \text{unavailable energy}$$

$$Q \quad = \quad \text{A.E.} \quad + \quad \text{U.E.} \quad (9.1)$$

The Carnot engine produces the maximum possible work for a given heat input, so this will be equal to the available portion of the heat supplied. Let us consider that heat is added to a system over a temperature range from T_1 to T_2. This may be approximated by a sum of infinitesimal amounts of heat, δQ, added at a temperature T. This infinitesimal amount of heat will be added to an imaginary differential Carnot engine. Figure 9.1 illustrates this. The total shaded area is the amount of heat, δQ. The Carnot engine must operate between two temperature limits. The lowest temperature at which it can reject heat is the temperature of the surroundings, T_0. If the temperature of heat rejection were below that of the surroundings, then heat would have to be transferred from a cooler body to a warmer body, an impossibility.

The amount of work possible for the Carnot engine, δW_{max}, may be related to the heat added by the Carnot efficiency; hence using Equations (7.1) and (7.8) yields

$$\delta W_{max} = \delta Q \left(1 - \frac{T_0}{T} \right) \tag{9.2}$$

however,

$$\delta(\text{A.E.}) = \delta W_{max}$$

$$\delta(\text{A.E.}) = \delta Q \left(1 - \frac{T_0}{T} \right)$$

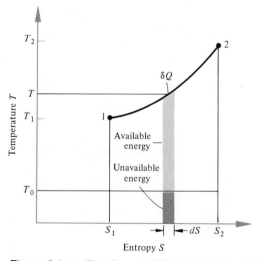

Figure 9.1   *A T–S diagram illustrating the available and unavailable portions of a differential amount of heat.*

We now integrate from state 1 to state 2,

$$\int_1^2 \delta(A.E.) = \int_1^2 \delta Q - T_0 \int_1^2 \frac{\delta Q}{T}$$

$$(A.E.)_{1-2} = Q_{1-2} - T_0(S_2 - S_1) \tag{9.3}$$

The unavailable energy from Equation (9.1) is

$$(U.E.)_{1-2} = +T_0(S_2 - S_1) \tag{9.4}$$

Equations (9.3) and (9.4) apply for a reversible or irreversible process between states 1 and 2.

Example 9.1

If 300 kJ of heat are reversibly added to 0.2 kg of air at constant pressure and the initial temperature is 300°C, find the available and unavailable energies of the heat added. The temperature of the surroundings is 30°C.

First we determine T_2.

$$Q = \Delta H = mc_p(T_2 - T_1)$$

$$300 = (0.2)(1.0047)(T_2 - 300)$$

$$T_2 = 1793°C = 2066 \text{ K}$$

from which

$$(A.E.)_{1-2} = Q_{1-2} - T_0 mc_p \ln\left(\frac{T_2}{T_1}\right)$$

$$(A.E.)_{1-2} = 300 - (303)(0.2)(1.0047) \ln\left(\frac{2066}{573}\right)$$

$$(A.E.)_{1-2} = 221.9 \text{ kJ}$$

$$(U.E.)_{1-2} = Q_{1-2} - (A.E.)_{1-2} = 78.1 \text{ kJ}$$

Example 9.2

We will remove 500 kJ of heat from a constant-temperature heat reservoir at 835 K. The heat is received by a system at a constant temperature of 720 K. The temperature of the surroundings, the lowest available temperature, is 280 K. Find the net loss of available energy as a result of this irreversible heat transfer. Figure 9.2 illustrates this problem graphically.

The total heat is the same in both the heat reservoir and the system, but calculating the available energy will lead to an interesting

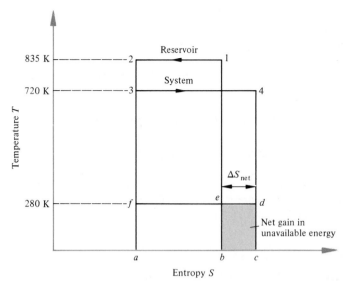

Figure 9.2 *A T–S diagram illustrating the processes in Example 9.2.*

conclusion:

$$(\text{A.E.})_{1-2} = Q_{1-2} - T_0(S_2 - S_1)$$

$$= -500 - 280\left(\frac{-500}{835}\right) = -332.34 \text{ kJ}$$

$$(\text{A.E.})_{3-4} = Q_{3-4} - T_0(S_4 - S_3)$$

$$= +500 - 280\left(\frac{+500}{720}\right) = +305.55 \text{ kJ}$$

$$(\text{A.E.})_{net} = \sum_i (\text{A.E.})_i = -26.78 \text{ kJ}$$

There is a net loss of available energy in the transfer process, and this is equal to the net gain in unavailable energy.

$$(\text{U.E.})_{1-2} = T_0(S_2 - S_1) = +280\left(\frac{-500}{835}\right) = -167.66 \text{ kJ}$$

$$(\text{U.E.})_{3-4} = T_0(S_4 - S_3) = +280\left(\frac{+500}{720}\right) = +194.44 \text{ kJ}$$

$$(\text{U.E.})_{net} = \sum_i (\text{U.E.})_i = +26.78 \text{ kJ}$$

The use of a T–S diagram is very useful in visualizing heat, available energy, and unavailable energy. The first law states that energy is conserved, so the areas $12ab$ and $34ca$ must be equal. We can determine graphically that $(S_4 - S_3)$ must be greater than $|S_2 - S_1|$

for the first law to be valid. Thus there is a net increase in entropy due to the irreversible heat-transfer process. This is denoted as ΔS_{net}. If the heat transfer between the reservoir and the system were done reversibly, the system and reservoir temperatures were equal, there would be no net entropy increase. The available portion of the heat when it leaves the reservoir is denoted by area 12 *fe*, and the available portion of the heat the system receives is denoted by area 34 *df*. The unavailable energy of the heat leaving the reservoir is denoted by area *efab* and the unavailable portion of the heat entering the system is area *fdca*. The net gain in unavailable energy is denoted by area *edcb*; notice that this is equal in magnitude to the net loss of available energy. It is sometimes easier to find the net gain of unavailable energy ($UE_{net} = T_0 \Delta S_{net}$) than the net loss of available energy.

Example 9.3

The exhaust gases from a gas turbine are used to heat water in an adiabatic counterflow heat exchanger. The exhaust temperature of the gases leaving the turbine is 260°C. The exit temperature of the gases from the heat exchanger is 120°C. The water enters at 65°C. The flow rates of the gas and water are 0.38 kg/s and 0.50 kg/s, respectively. The constant-pressure specific heat of the gases is 1.09 kJ/kg · K. The lowest available temperature is 35°C. Find the loss of available energy due to heat transfer. Figures 9.3(a) and 9.3(b) illustrate the physical system and the T–s diagram. Calculate the water exit temperature and heat transfer by a first-law analysis.

Open System, First Law

$$\dot{H}_{w\,in} + \dot{H}_{g\,in} = \dot{H}_{w\,out} + \dot{H}_{g\,out}$$

On rearranging this yields

$$\dot{m}_g c_{pg}(T_{g\,in} - T_{g\,out}) = (h_{w\,out} - h_{w\,in})\dot{m}_w$$

$$(0.38)(1.09)(260 - 120) = (0.50)(h_{w\,out} - 272.06)$$

$$h_{w\,out} = 388.0 \qquad T_{w\,out} = 93C$$

$$Q_{1-2} = \Delta \dot{H}_{gas} = -57.99 \text{ kW}$$

There is a loss of available energy from the gas. The heat leaves the gas; hence it is negative.

$$(\dot{A}.E.)_{1-2} = \dot{Q}_{1-2} - \dot{m}c_p T_0 \ln\left(\frac{T_2}{T_1}\right)$$

$$(\dot{A}.E.)_{1-2} = -57.99 - (0.38)(1.09)(308) \ln\left(\frac{393}{533}\right)$$

$$(\dot{A}.E.)_{1-2} = -19.11 \text{ kW} \qquad \text{(loss from gas)}$$

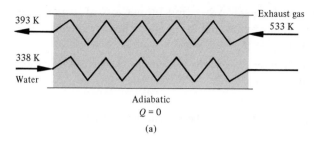

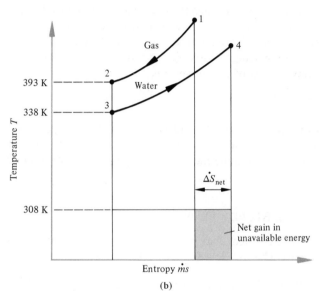

Figure 9.3 *(a) An adiabatic counterflow heat exchanger. (b) A T–s diagram for counterflow heat exchanger (Example 9.3).*

There is a gain in available energy of the water. The heat enters the water; hence it is positive.

$$(\text{Å.E.})_{3-4} = \dot{Q}_{3-4} - \dot{m}T_0(s_4 - s_3)$$

$$(\text{Å.E.})_{3-4} = +57.99 - (0.50)(308)(1.2271 - 0.8935)$$

$$(\text{Å.E.})_{3-4} = +6.61 \text{ kW} \qquad (\text{gain by water})$$

The net loss of available energy is

$$(\text{Å.E.})_{\text{net}} = \sum_i (\text{A.E.})_i = -19.11 + 6.61 = -12.5 \text{ kW}$$

and this is equal to the net gain in unavailable energy, as illustrated in Figure 9.3(b).

Considering Figure 9.2, the equality of areas $12ab$ and $34ca$ demonstrates the first law of thermodynamics. The second law is demonstrated by the net entropy increase, ΔS_{net}, and by net loss of available energy (equal to the net gain of unavailable energy). Thus the second law could be stated as either

> In any energy-transfer process, the net entropy change will be greater than or equal to zero.

or

> In any energy-transfer process, the net change of available energy will be less than or equal to zero.

The entropy increase of the universe may also be visualized. Consider now that the system in Example 9.2 is used to heat air in a room, causing a further entropy increase, the air in the room eventually reaching the surrounding temperature, T_0. The amount of heat is constant, so the entropy change of the universe is

$$(\Delta S)_{universe} = \frac{Q}{T_0} \tag{9.5}$$

The changes in available energy and entropy analyzed so far have been based on heat addition to the Carnot cycle. Carnot-cycle analysis does not explain how to determine the available energy of steam entering a turbine.

9.2 OPEN SYSTEMS, STEADY FLOW[1]

Let us consider an open system that performs work and exchanges heat with the surroundings. The first law for such a system would be

$$\dot{W} = \dot{m}(h_1 - h_2) + \dot{Q} \tag{9.6}$$

where changes in kinetic and potential energies are not included for clarity, but may be included in the final equations. The process for such a device is shown by the expansion 1 to 2 in Figure 9.4.

In addition in Figure 9.4, a reversible path connecting states 1 and 2 is shown, $1-a-b-2$. The process $1-a$ is an isentropic expansion, $a-b$ is an isothermal heat addition, and $b-2$ is an isentropic compression. The reversible work for process $1-2$ may be found by summing the works for the reversible processes connecting states 1 and 2

$$\dot{W}_{1a} = \dot{m}(h_1 - h_a)$$
$$\dot{Q}_{1a} = 0$$
$$\dot{W}_{ab} = \dot{m}(h_a - h_b) + \dot{Q}_{ab}$$
$$\dot{Q}_{ab} = \dot{m}T_0(s_2 - s_1)$$
$$\dot{W}_{b2} = \dot{m}(h_b - h_2)$$
$$\dot{Q}_{b2} = 0$$

[1]M. A. Paolino and J. P. Kuspo, *Mechanical Engineering News*, vol. 17, no. 3, 1980.

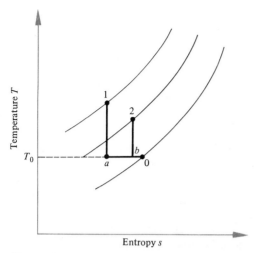

Figure 9.4   *A T–s diagram illustrating reversible and irreversible processes between states 1 and 2.*

The reversible power is

$$\dot{W}_{rev} = \dot{W}_{1a} + \dot{W}_{ab} + \dot{W}_{b2}$$

$$\dot{W}_{rev} = \dot{m}(h_1 - h_2) - \dot{m}T_0(s_1 - s_2) \qquad (9.7)$$

Equation (9.7) represents the maximum power, the available energy, the device can produce when operating between fixed states 1 and 2.

The difference between Equations (9.6) and (9.7) represents the loss of available energy due to irreversibilities in the system. This is called the irreversibility, I, and in this case

$$\dot{I} = \dot{W}_{rev} - \dot{W}$$

or

$$\dot{I} = \dot{m}T_0(s_2 - s_1) - \dot{Q} \qquad (9.8)$$

Comparing Equation (9.8) to Equation (8.38), we see that this is a direct result of the second law.

In determining the reversible work for the process 1–2, the reversible process never reached the surrounding pressure p_0. A question arises as to what the maximum work, reversible work, could be obtained from a substance at state 1 if the final state were in equilibrium with the surroundings at p_0 and T_0, state 0 in Figure 9.4. In this case the reversible work path is 1–a, a–o. The reversible work, referenced to T_0 and p_0 is called availability, ψ. Further discussion follows in the next section for closed systems.

$$\dot{W}_{1-0} = \dot{W}_{1-a} + \dot{W}_{a-0}$$

$$\dot{W}_{a-0} = \dot{m}(h_a - h_0) + \dot{Q}_{a-0}$$

$$\dot{Q}_{a-0} = \dot{m}T_0(s_0 - s_1)$$

$$\dot{W}_{1-0} = \dot{m}(h_1 - h_0) + \dot{m}T_0(s_0 - s_1)$$

Including the kinetic and potential energies and writing the maximum work on a unit mass basis yields

$$\psi_1 = \left(h_1 + \frac{v_1^2}{2} + gz_1\right) - (h_0 + gz_0) - T_0(s_1 - s_0) \tag{9.9}$$

Notice that the final velocity is zero because the system must be in equilibrium with the surroundings. Equation (9.7) may be written as

$$\dot{W}_{rev} = \dot{m}(\psi_1 - \psi_2) \tag{9.10}$$

A Special Case—Turbines

Practice in a field sometimes differs with classical presentations. Such is the case in turbine analysis. In practice a turbine is considered to be adiabatic, with an expansion as illustrated in Figure 9.5. In the ideal turbine the fluid expands isentropically from inlet to exit, 1 to 2. The maximum work is the isentropic enthalpy drop from inlet to exit. This

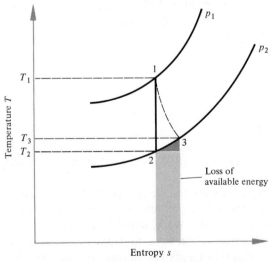

Figure 9.5 A T–s diagram for flow through a turbine.

applies to a turbine stage, as well as to the overall turbine. Thus,

$$(\text{a.e.})_{1-2} = w_{max} = (h_1 - h_2)_{s=c} \text{ kJ/kg} \tag{9.11}$$

However, in actual turbines irreversibilities do not permit isentropic expansion of the fluid. The ideal exit pressure is p_2, and the actual fluid exit is at this pressure but at some temperature greater than T_2. Since the fluid did not expand to the temperature T_2, it did not utilize all its available energy; there is a loss of available energy utilization by the turbine. If we look at the difference between the availabilities at states 2 and 3 we find

$$\psi_3 - \psi_2 = (h_3 - h_2) - T_0(s_3 - s_2)$$

which is the energy that could be used for work. Comparing this to Equation (9.11) we find that the loss of available energy due to irreversible expansion is

$$(\text{a.e.})_{loss} = -T_0(s_3 - s_2) \tag{9.12}$$

The loss of available energy is represented by the rectangular lightly shaded area on Figure 9.5. The darkened triangular area is the available energy of the exit fluid that could drive a Carnot engine.

At this point let us reflect on the adiabatic turbine. The isentropic enthalpy drop across the turbine is the available energy of the entering fluid. In an actual turbine, irreversibilities decrease this amount by an amount $T_0(s_3 - s_2)$. The turbine can only use an amount $(h_1 - h_3)$; the rest is lost from the turbine's point of view.

Example 9.4

Air flows through an adiabatic compressor at 2 kg/s. The inlet conditions are 100 kPa and 310 K, and the exit conditions are 700 kPa and 560 K. Consider T_0 to be 298 K. Determine the change of availability and the irreversibility.

From Equation (9.9),

$$\dot{m}(\psi_2 - \psi_1) = \dot{m}(h_2 - h_1) - \dot{m}T_0(s_2 - s_1)$$

$$= (2.0)(1.0047)(560 - 310) - (2.0)(298)$$

$$\times \left[(1.0047) \ln \left(\frac{560}{310} \right) - (0.287) \ln \left(\frac{7}{1} \right) \right]$$

$$= 481.1 \text{ kW}$$

From Equation (9.8),

$$\dot{I} = \dot{m}T_0(s_2 - s_1) = 21.2 \text{ kW}$$

9.3 FURTHER CONSIDERATIONS OF AVAILABLE ENERGY—AVAILABILITY

The presentation at the beginning of this chapter gives one enough tools to understand, and calculate, available energy. There are instances in which a more general approach to a problem is needed, and this section develops the concept of maximum work for all cases. It illustrates the depth of thermodynamics and the cohesiveness of the theoretical foundations. This section has been adapted, in part, from Keenan's *Thermodynamics*.

Any system operates within a medium, which we have called the surroundings. Since the surroundings are assumed to be in a stable, equilibrium state, they cannot serve as a source of work; neither can they react chemically with the system or diffuse into the system. They must remain distinct. If the system within the surroundings is in equilibrium with the surroundings and is itself in an equilibrium state, then there can be no interaction between the two. This is called the "dead state."

If the system is in some state other than the dead state, it will change spontaneously toward the dead state. In Figure 8.3 this dead state is indicated by the apex of the spheroid. The change by the system from any state other than the dead state to the dead state will occur without work being supplied from an outside source. Hence the work done by the system in changing to the dead state will never be less than zero. It will be zero when the system is initially at the dead state:

$$W_{max} \geq 0 \tag{9.13}$$

where W_{max} is the maximum useful work that can be produced by the system as it interacts with the surroundings in achieving the dead state. Availability of a system state is defined as the "maximum work that can result from the interaction of the system and surroundings when only cyclic changes occur in external things, except the lifting of a weight."

Now let us find an expression for the availability. The symbols p_0, T_0, E_0, V_0, and S_0 will represent the pressure, temperature, energy, volume, and entropy of the surroundings, which are in the dead state. The system will also exist with properties p_0 and T_0, but will have values of E, V, and S for the energy, volume, and entropy.

When the system changes state, an infinitesimal amount of heat, δQ, may be added; if δQ is negative, heat flows from the system. A certain amount of work will be delivered, and the maximum amount of work, δW_{max}, that can be delivered is that work done when the heat flow is reversible. For the heat transfer to be reversible, it is necessary to place a reversible engine between the system and the surroundings (Figure 9.6). Let the temperature of the system rejecting the heat be T and that of the

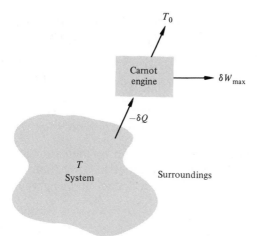

Figure 9.6   Reversible heat transfer from a system.

surroundings be T_0. The work, δW_q,

$$\delta W_q = \delta Q \frac{(T_0 - T)}{T} \tag{9.14}$$

found from the Carnot efficiency, is always positive. The sign of the heat flow must be considered.

There may be other forms of work occurring between the system and the surroundings. There may be a change in surface tension, electrical work, and so on. All these works will be lumped in one term, δW. Consider, however, a change in the system volume, dV. This change of system volume is resisted by the pressure of the surroundings, p_0. The lumped work is resisted by the pressure-volume work, $p_0 \, dV$; the net work that can be delivered by the system is

$$\delta W - p_0 \, dV$$

The net work is the sum of all the individual works

$$\delta W_{net} = \delta W - p_0 \, dV + \frac{(T_0 - T)}{T} \delta Q$$

$$\delta W_{net} = \delta W - \delta Q - p_0 \, dV + T_0 \frac{\delta Q}{T}$$

The first law of thermodynamics for a closed system, $\delta Q = dE + \delta W$, yields the following, where the heat transfer is a reversible process:

$$\delta W_{net} = -dE - p_0 \, dV + T_0 \, dS \tag{9.15}$$

The term E is considered in the most general sense, including kinetic, potential, capillary, and so on, energies. Notice that if the function,

$dE + p_0 \, dV - T_0 \, dS$, decreases, the work increases. The maximum work that can be accomplished by any system at state 1 to state 0 can be represented by the decrease in this function:

$$W_{max} = - \int_1^0 (dE + p_0 \, dV - T_0 \, dS) \qquad (9.16)$$

The function, $E + p_0 V - T_0 \, dS$, is called the "availability," $\mathcal{Q}$, thus

$$W_{max} = \mathcal{Q}_1 - \mathcal{Q}_0 \qquad (9.17)$$

The availability, $\mathcal{Q}$, is a property, but it is bound by the particular values of T_0 and p_0. However, the availability varies as a function of system state; and the change in availability of a system in going from state 1 to state 2 is

$$\mathcal{Q}_2 - \mathcal{Q}_1 = \int_1^2 d\mathcal{Q} = \int_1^0 d\mathcal{Q} + \int_0^2 d\mathcal{Q} = \int_1^0 d\mathcal{Q} - \int_2^0 d\mathcal{Q}$$

$$\mathcal{Q}_2 - \mathcal{Q}_1 = \int_1^0 (dE + p_0 \, dV - T_0 \, dS) - \int_2^0 (dE + p_0 \, dV - T_0 \, dS)$$

$$\mathcal{Q}_2 - \mathcal{Q}_1 = (E_2 + p_0 V_2 - T_0 S_2) - (E_1 + p_0 V_1 - T_0 S_1) \qquad (9.18)$$

The concept of availability may be expressed in terms of another thermodynamic function, and since this function is tabulated, the availability may be calculated. The Gibbs function, or free energy, G, is defined as

$$G = H - TS \qquad (9.19)$$
$$dG = dH - T \, dS - S \, dT$$
$$H = E + pV$$
$$dH = dE + p \, dV + V \, dp$$
$$dG = dE + p \, dV + V \, dp - T \, dS - S \, dT \qquad (9.20)$$

Conditions of constant temperature and pressure, where the temperature is T_0 and the pressure is p_0, yield for the Gibbs function

$$dG = dE + p_0 \, dV - T_0 \, dS \qquad (9.21)$$

Notice that Equation (9.21) is the same as Equation (9.18) on the differential basis; the change in availability of a system therefore is equal to the change in the Gibbs function of the system at constant temperature and pressure:

$$d\mathcal{Q} = (dG)_{T, p} \qquad (9.22)$$

This is particularly useful when evaluating the availability of systems in which chemical reactions occur. These reactions may often be considered at constant temperature and pressure.

9.4 SECOND-LAW EFFICIENCY[2]

It is important to conserve energy and energy quality. Conserving energy is made possible by using less for a given task or improving the first-law efficiency for a process—for example, turning off extra lights or driving cars with higher gas mileage. Additionally, it is possible to conserve energy quality by matching source and system energy quality needs, improving the second-law efficiency of the process.

Processes

The second-law efficiency for a process may be defined as

$$\eta_2 = \frac{\text{the change of available energy of the system}}{\text{the change of available energy of the source}}$$

$$\eta_2 = \frac{\Delta(\text{a.e.})_{system}}{\Delta(\text{a.e.})_{source}} \tag{9.23}$$

We will consider three types of processes: power consuming, power producing, and heat transfer. Figure 9.7 illustrates a T–s diagram for a fluid passing through a pump or compressor from state 1 to state 2. To accomplish this change of state, energy in the form of work had to be supplied to the compressor, w_{source}. This is 100 percent available energy by definition. The change in available energy of the system represents the work required to change from state 1 to state 2 reversibly. From Equation (9.7) this is

$$\Delta(\text{a.e.})_{system} = (h_2 - h_1) - T_0(s_2 - s_1)$$

$$\Delta(\text{a.e.})_{system} = (w_{1-2})_{rev} \tag{9.24}$$

The second law efficiency is then

$$\eta_2 = \frac{(w_{1-2})_{rev}}{w_{source}} \tag{9.25}$$

In a turbine the fluid expands irreversibly from state 1 to state 2 as illustrated in Figure 9.8. In this instance the change of available energy of the system is the actual work output of the turbine, w_{system}. The actual work produced is less than the change of available energy between states 1 and 2 because irreversibilities have diminished the available energy. The change of available energy between states 1 and 2 is equal to the

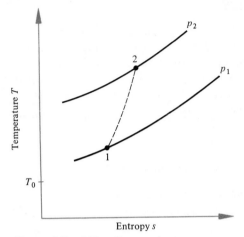

Figure 9.7 *A T–s diagram for a pump or compressor.*

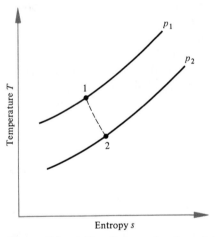

Figure 9.8 *A T–s diagram showing adiabatic expansion in a turbine.*

maximum work found from Equation (9.7).

$$\Delta(\text{a.e.})_{\text{source}} = (h_1 - h_2) - T_0(s_1 - s_2)$$

$$\Delta(\text{a.e.})_{\text{source}} = (w_{1-2})_{\text{rev}}$$

The second-law efficiency is

$$\eta_2 = \frac{w_{\text{system}}}{(w_{1-2})_{\text{rev}}} \tag{9.26}$$

Example 9.5

An adiabatic turbine receives gas ($c_p = 1.09$ kJ/kg·K, $k = 1.3$) at 700 kPa and 1000°C and discharges it at 150 kPa and 665°C. The surrounding temperature is 25°C. Determine the second law and isentropic efficiencies of the turbine.

From Equation (9.7) the reversible work may be determined.

$$w_{rev} = (h_1 - h_2) - T_0(s_1 - s_2)$$

$$w_{rev} = (1.09)(1273 - 938) - 298\left[(1.09)\ln\left(\frac{1273}{938}\right)\right.$$

$$\left. - (0.252)\ln\left(\frac{700}{150}\right)\right]$$

$$w_{rev} = 381.6 \text{ kJ/kg}$$

$$w_{sys} = (h_1 - h_2) = 365.1 \text{ kJ/kg}$$

$$\eta_2 = \frac{w_{sys}}{w_{rev}} = 0.956$$

The isentropic efficiency exit temperature is

$$T_{2s} = 1273\left(\frac{150}{700}\right)^{0.3/1.3} = 892 \text{ K}$$

$$\eta_s = \frac{(h_1 - h_2)_{actual}}{(h_1 - h_2)_s}$$

$$\eta_s = \frac{1273 - 938}{1273 - 892} = 0.879$$

While the isentropic efficiency is the standard of comparison, it does not reflect the energy available from the source because the isentropic exit state is not the actual exit state of the turbine.

The last process we will consider is heating. Let us consider a solar collector, used for heat air or water, which in turn is used for home heating. The collector receives a total amount of energy, Q_A, from the sun at temperature T_A. Of this total energy some is lost to the surroundings, Q_C, and the remainder, Q_B, is transferred to the water at temperature T_B. Figure 9.9 illustrates this process. Note that $Q_A \neq Q_B$. The

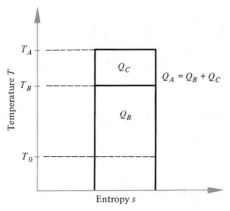

Figure 9.9 *A T–S diagram for a nonadiabatic solar collector.*

change of the system and source available energies are

$$\Delta(\text{A.E.})_{\text{system}} = Q_B \left(1 - \frac{T_0}{T_B} \right)$$

$$\Delta(\text{A.E.})_{\text{source}} = Q_A \left(1 - \frac{T_0}{T_A} \right)$$

and

$$\eta_2 = \frac{1 - (T_0/T_B)}{1 - (T_0/T_A)} \eta_1 \tag{9.27}$$

where

$$\eta_1 = \frac{Q_B}{Q_A}$$

The first-law efficiency of the collector, η_1, reflects the adiabatic effectiveness of the collector.

Cycles

The second-law efficiency for a cycle is slightly different than for a process in that a cycle's purpose is to perform a given task, be it cooling or power production. The second-law efficiency for a cycle is defined as

$$\eta_2 = \frac{\text{minimum available energy necessary to perform a given task}}{\text{actual available energy required to perform a given task}}$$

$$\tag{9.28}$$

We will analyze two cycles, a general heat engine cycle and a refrigeration cycle. Figure 9.10 illustrates a *T–s* diagram for a general

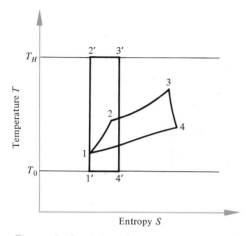

Figure 9.10   A T–S diagram for a general heat engine cycle.

heat engine cycle, 1234. In addition, a Carnot heat engine is illustrated by the cycle $1'2'3'4'$. The task of the heat engine is to perform work, w_{net}. In performing the work the cycle operates between temperature limits T_1 and T_3. The high- and low-temperature heat reservoirs are at T_H and T_0. The energy supplied by the high-temperature reservoir is q_H, the available portion of which is

$$(\text{a.e.})_{actual} = \left(1 - \frac{T_0}{T_H}\right) q_H$$

What is the minimum available energy required? This is the net work of a Carnot engine operating between T_H and T_0, as illustrated in Figure 9.10. Thus

$$(\text{a.e.})_{minimum} = w_{net}$$

The second-law efficiency is

$$\eta_2 = \frac{w_{net}}{\left(1 - \dfrac{T_0}{T_H}\right) q_H} = \frac{w_{net}/q_H}{\left(1 - \dfrac{T_0}{T_H}\right)}$$

$$\eta_2 = \eta_1 / \eta_{Carnot} \tag{9.29}$$

If the actual cycle has many reversibilities, it will cause the first law efficiency to be low and consequently the second law efficiency will likewise be low. The minimum available energy required for the task is the net work of a Carnot engine operating between the reservoir temperature limits.

For a refrigeration system, illustrated in Figure 9.11, the given task is to provide cooling, q_{4-1}. The actual refrigeration cycle is 1234. To

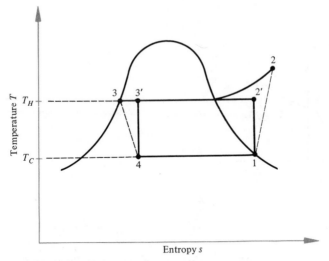

Figure 9.11   A T–s diagram for a refrigeration system.

produce this amount of cooling, work must be supplied, hence the available energy actually used is w_{net}. We must determine the minimum available energy required to provide cooling, q_{4-1}. This is the cooling supplied by a reversed Carnot cycle, 12'3'4. The minimum available energy is equal to the net work of the reversed Carnot cycle.

$$\left(a.e.\right)_{minimum} = \left(\frac{T_H}{T_C} - 1\right) q_{4-1}$$

The second-law efficiency is

$$\eta_2 = \frac{\left(\dfrac{T_H}{T_C} - 1\right) q_{4-1}}{w_{net}}$$

Let

$$\beta = COP \text{ (coefficient of performance)}$$

$$\eta_2 = \frac{\beta}{\beta_{Carnot}} \tag{9.30}$$

Second-law efficiencies may be used in energy conservation programs, evaluating the available energy usage. The first-law efficiency evaluates the energy quantity utilization, whereas the second-law efficiency evaluates the energy quality utilization.

PROBLEMS

1. Determine the available energy of furnace gas, $c_p = 1.046 \text{ kJ/kg} \cdot \text{K}$, when it is cooled from 1260 K to 480 K at constant pressure. The surroundings are at 295 K.

2. A Carnot engine receives heat from a constant-temperature reservoir. The engine has a maximum temperature of 820 K. For each 1100 kJ of heat transferred, compute the net change of entropy when the high-temperature reservoir is at the following temperatures: (a) 1700 K; (b) 1425 K; (c) 1150 K; (d) 875 K.

3. At constant pressure, 138 kPa, 5 kg of air are cooled from 500 K to 300 K. The temperature of the surroundings is 277 K. Determine the available portion of the heat removed and the entropy increase of the universe.

4. Steam is contained in a constant-pressure closed system at 200 kPa and 200°C and is allowed to reach thermal equilibrium with the surroundings, which are at 26°C. Find the loss of available energy per kilogram of steam.

5. Steam exits a power turbine with a quality of 85% and a pressure of 10 kPa and is condensed to a saturated liquid. The lowest available temperature is 21°C. What fraction of the heat rejected to the condenser is available energy?

6. A tank contains dry, saturated steam at 80°C. Heat is added until the pressure reaches 200 kPa. The lowest available temperature is 20°C. Determine the available portion of the heat added per kilogram.

7. If 2 kg of air at 21°C are heated at constant pressure until the absolute temperature doubles, determine for the process (a) the heat required; (b) the entropy change; (c) the unavailable energy; (d) the available energy if $T_0 = 10°C$.

8. A constant-volume container holds air at 102 kPa and 300 K. A paddle wheel does work on the air until the temperature is 422 K. The air is now cooled by the surroundings (at 289 K) to its original state. Determine (a) the paddle work required (adiabatic) per kilogram; (b) the available portion of the heat removed per kilogram.

9. Air is contained within a piston–cylinder initially at 140 kPa and 16°C. An irreversible process occurs during which 2.11 kJ of heat is removed and 2.53 kJ of work is done on the air. The system contains 0.068 kg. (a). If the final pressure is 275 kPa, determine the entropy change. (b) If the sink temperature is 15°C, find the available portion of the heat transferred.

10. Air is contained in a rigid tank at 138 kPa and 335 K. Heat is transferred to the air from a constant temperature reservoir at 555 K until the pressure is 207 kPa. The temperature of the surroundings is 289 K. Determine per kilogram of air (a) the heat transferred; (b) the

available portion of the heat transferred; (c) the loss of available energy due to irreversible heat transfer.

11. Air having a mass of 2.5 kg is cooled from 210 kPa and 205°C to 5°C at constant volume. All the heat is rejected to the surroundings at −4°C. Determine the available portion of the heat removed. Draw a T–s diagram and label the available and unavailable portions of the energy rejected.

12. A tank contains water vapor at a pressure of 75 kPa and a temperature of 100°C. Heat is added to the water vapor until the pressure is tripled. The lowest available temperature is 16°C. Find the available portion of the heat added per kilogram.

13. A boiler produces dry saturated steam at 3.5 MPa. The furnace gas enters the tube bank at a temperature of 1100°C, leaves at a temperature of 430°C, and has an average specific heat, $c_p = 1.046$ kJ/kg · K, over this temperature range. Neglecting heat losses from the boiler and for the water entering the boiler as a saturated liquid with a flow rate of 12.6 kg/s, determine for $T_0 = 21$°C (a) the heat transfer; (b) the loss of available energy of the gas; (c) the gain of available energy of the steam; (d) the entropy production.

14. A steam turbine uses 12.6 kg/s of steam and exhausts to a condenser at 5 kPa and 90% quality. Cooling water enters the condenser at 21°C and leaves at the steam temperature. Determine (a) the heat rejected; (b) the net loss of available energy; (c) the entropy production; (d) the entropy increase of the universe.

15. Hot, vaporized oil is used in a heat exchanger to generate steam on the other side of the tubes. The oil vapor condenses at constant temperature equal to 585 K, and saturated steam is generated at 548 K. The steam is passed through a turbine, doing work, and is rejected to a condenser. The lowest available temperature is 15°C. Find the net loss of available energy in the heat exchanger.

16. An air heater is located in a closed gas-turbine cycle. The air enters the heater at 512 K and 690 kPa and is heated to 1067 K. A pressure drop of 138 kPa occurs across the heater. Determine the percentage loss of available energy due to the pressure drop. $T_0 = 311$ K.

17. Steam enters a turbine at 6.0 MPa and 500°C and exits at 10 kPa with a quality of 89%. The flow rate is 9.07 kg/s. Determine (a) the loss of available energy; (b) the available energy entering the turbine; (c) the available portion of the steam exiting the turbine.

18. An air turbine receives air at 825 kPa and 815°C. The air leaves the turbine at 100 kPa and 390°C. Determine (a) the available energy entering the turbine; (b) the entropy increase across the turbine. The lowest available temperature is 300 K.

19. Nitrogen, initially at 1380 kPa and 700 K, undergoes a constant-pressure process in which 346.7 kJ/kg of heat is removed. The lowest

available temperature is 289 K, and the lowest available pressure is 101 kPa. Determine (a) the change of availability; (b) the maximum work.

20. Air expands from 825 kPa and 500 K to 140 kPa and 500 K. Determine (a) the Gibbs function at the initial conditions; (b) the maximum work; (c) the entropy change.

21. An economizer, a gas–water finned tube heat exchanger, receives 67.5 kg/s of gas, $c_p = 1.0046$ kJ/kg · K, and 51.1 kg/s of water, $c_p = 4.186$ kJ/kg · K. The water rises in temperature from 402 K to 469 K, whereas the gas lowers in temperature from 682 K to 470 K. There are no changes of kinetic energy, and $p_0 = 103$ kPa and $T_0 = 289$ K. Determine (a) the change of the availability of the water; (b) the change of availability of the gas; (c) the entropy production.

22. A Carnot engine receives 2000 kJ of heat at 1100°C from a heat reservoir at 1400°C. Heat is rejected at 100°C to a reservoir at 50°C. The expansion and compression processes are isentropic. Determine (a) the net change of entropy of the universe for one cycle; (b) the second-law efficiencies of the heat addition process and the cycle.

23. In a steam generator the steam generating tubes receive heat from hot gases passing over the outside surface, evaporating water inside the tubes. The flue gas flow rate is 20 kg/s with an average specific heat of 1.04 kJ/kg · K. The gas temperature decreases from 650°C to 400°C while generating steam at 300°C. The water enters the tubes as a saturated liquid and leaves with a quality of 90%. $T_0 = 27°C$. Determine (a) the water flow rate; (b) the net change of available energy; (c) the second-law efficiency.

24. A tank contains 3 kg of air at 300 kPa and 400 K and is cooled to 300 K. The constant-temperature heat reservoir is at 290 K. Determine (a) the net change of available energy; (b) the second-law efficiency of the process.

25. A Carnot cycle engine receives and rejects heat with a 20°C temperature differential between it and the heat reservoirs. The expansion and compression processes have pressure ratios of 50. For 1 kg of air as the substance and cycle temperature limits of 1000 K and 300 K and $T_0 = 280$ K, determine the second-law efficiency.

Chapter 10
Thermodynamic Relationships

The problems in previous chapters were based on the built-in assumption that the property values for the various substances were available. The specific heats, the tables of properties, and so on, all have to be determined from experimental data. The question arises as to how, from an experiment, the necessary properties can be determined. Experimenters cannot blindly set up an experiment and hope to achieve meaningful results. They must know what to vary in the experiment and what to hold constant. They are further limited in that only four properties are measurable: temperature, pressure, volume, and mass. All remaining properties must be calculated from these four or from changes these properties undergo under certain conditions.

In order to achieve the desirable result of useful tests, experimenters must be theoreticians as well. They must develop a mathematical model for thermodynamic properties and use this in determining the functional relationship between the four properties. They can measure and control, on the one hand, and the remaining properties that they want to determine, on the other. This is the purpose of the chapter: to show you the method used and to illustrate how properties may be determined. The model is a good model of a substance, but it is a mathematical model, not the substance itself. This is true of all sciences that wish to predict system behaviors.

10.1 INTERPRETING DIFFERENTIALS AND PARTIAL DERIVATIVES

To develop a model we must first understand differential equations and partial derivatives. Consider $Z = f(X, Y)$; then

$$dZ = \left(\frac{\partial Z}{\partial X} \right)_Y dX + \left(\frac{\partial Z}{\partial Y} \right)_X dY \tag{10.1}$$

where dZ is the total derivative $(\partial Z/\partial X)_y$ tells us how Z varies with respect to X at constant Y through a differential distance dX. The second term, $(\partial Z/\partial Y)_x$, shows how Z varies with respect to Y at constant X through a differential distance dY. The sum of both terms tells how dZ does vary. If Z is assumed to be a continuous function with continuous derivatives, then

$$\frac{\partial^2 Z}{\partial X \partial Y} = \frac{\partial^2 Z}{\partial Y \partial X} \tag{10.2}$$

Equation (10.2) is an important relationship and is valid for exact differentials, differentials that are point, not path, functions, This is true of properties, which depend only on the system state, but it is not true of the inexact differentials, heat and work. To illustrate the inexact differential, consider the differential of mechanical work, $\delta W = p\,dV$. Let us assume momentarily that work is an exact differential. Then

$$W = f(p, V)$$

$$dW = \left(\frac{\partial W}{\partial p}\right)_V dp + \left(\frac{\partial W}{\partial V}\right)_p dV$$

The first term on the right-hand side is zero because no work is accomplished at constant volume: this tells us, then, that the only work is work at constant pressure. This, obviously, is not true, and the hypothesis is invalid.

Physical Interpretation of Partial Derivatives

Before we continue with the mathematical development of our thermodynamic model, let us first consider the physical significance of the partial derivative and how these derivatives may be used in developing a thermodynamic surface. The importance of understanding the difference between independent and dependent variables is a key to the understanding of the mathematical model. Any two variables are independent if one may be held constant while the other is varied. This illustrates that the changing of one variable will not change the other variable, thus demonstrating their independence from one another. Figure 10.1 shows a set of three independent variables, each of which may be varied while the other two remain constant, demonstrating that each is independent of the others.

When X, Y, and Z are used to describe some surface, a functional relationship is developed between the variables, and the concept of dependence and independence appears. Let these variables describe a sphere of radius R; then

$$X^2 + Y^2 + Z^2 = R^2 \tag{10.3}$$

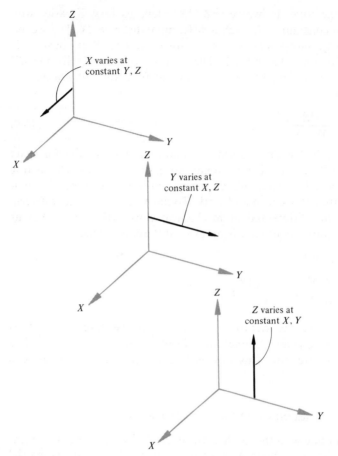

Figure 10.1 *A set of three independent variables with one varying while two remain constant.*

An octant of the sphere is shown in Figure 10.2, and the lines shown on the surface demonstrate how two variables change while the third is held constant.

On the surface only two variables may be independent; the third is dependent. For a point to be on the surface in Figure 10.2, any two values of X and Y may be chosen independently, and that locates the point in the XY plane at $Z = 0$; Z must be a certain value, dependent, to go from that point to the surface. If it is any other value, the point, or state, will not be on the surface. This is shown mathematically by solving Equation (10.3) for any variable, in this case Z:

$$Z = (R^2 - X^2 - Y^2)^{1/2} \tag{10.4}$$

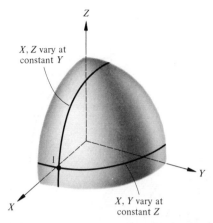

X, Z vary at constant Y

X, Y vary at constant Z

Figure 10.2 *The variation of two variables with the third dependent, keeping the line on the surface.*

Since R is a constant, this may be written as

$$Z = f(X, Y)$$

(10.5)

Hence, from Equation (10.1),

$$dZ = \left(\frac{\partial Z}{\partial X} \right)_Y dX + \left(\frac{\partial Z}{\partial Y} \right)_X Y$$

The partial derivatives are graphically shown in Figure 10.3.

The equilibrium surfaces used in the previous chapters are thermodynamic surfaces. We will be able to develop the mathematical tools used in making these surfaces. The steam tables and other tables of properties rely on these techniques in their development.

10.2 AN IMPORTANT RELATIONSHIP

Another mathematical relationship between derivatives may be developed as follows. Let us consider three variables, X, Y, and Z, and suppose that

$$Z = f(X, Y)$$

then

$$dZ = \left(\frac{\partial Z}{\partial X} \right)_Y dX + \left(\frac{\partial Z}{\partial Y} \right)_X dY$$

(10.6a)

but the relationship may also be written as

$$Y = f(X, Z)$$

$$dY = \left(\frac{\partial Y}{\partial X} \right)_Z dX + \left(\frac{\partial Y}{\partial Z} \right)_X dZ$$

(10.6b)

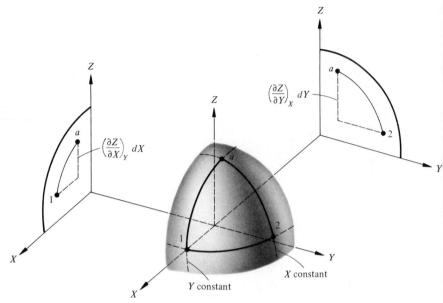

Figure 10.3   *Illustration of true projection of curves a–1 and a–2 in the XZ and ZY planes, respectively.*

Solving for dZ,

$$dZ = \frac{dY - (\partial Y/\partial X)_Z \, dX}{(\partial Y/\partial Z)_X}$$

and substituting in Equation (10.6a) yields

$$dY - \left(\frac{\partial Y}{\partial X}\right)_Z dX = \left(\frac{\partial Y}{\partial Z}\right)_X \left(\frac{\partial Z}{\partial X}\right)_Y dX + \left(\frac{\partial Y}{\partial Z}\right)_X \left(\frac{\partial Z}{\partial Y}\right)_X dY$$

and, rearranging,

$$\left[1 - \left(\frac{\partial Y}{\partial Z}\right)_X \left(\frac{\partial Z}{\partial Y}\right)_X\right] dY = \left[\left(\frac{\partial Y}{\partial X}\right)_Z + \left(\frac{\partial Y}{\partial Z}\right)_X \left(\frac{\partial Z}{\partial X}\right)_Y\right] dX$$

$$(10.7)$$

Since X and Y are independent variables, any values may be assigned to them and Equation (10.7) will be valid. Let $X = $ constant, then $dX = 0$,

$$\left(\frac{\partial Y}{\partial Z}\right)_X \left(\frac{\partial Z}{\partial Y}\right)_X = 1 \qquad (10.8)$$

or

$$\left(\frac{\partial Z}{\partial Y}\right)_X = \frac{1}{(\partial Y/\partial Z)_X} = \left(\frac{\partial Z}{\partial Y}\right)_X \qquad (10.9)$$

Equation (10.9) proves that partial derivatives may be inverted. If Y is held constant in Equation (10.7), then

$$\left(\frac{\partial Y}{\partial X}\right)_Z + \left(\frac{\partial Y}{\partial Z}\right)_X \left(\frac{\partial Z}{\partial X}\right)_Y = 0 \qquad (10.10)$$

If we multiply Equation (10.10) by $(\partial X/\partial Y)_Z$,

$$\left(\frac{\partial X}{\partial Y}\right)_Z \left(\frac{\partial Y}{\partial X}\right)_Z + \left(\frac{\partial X}{\partial Y}\right)_Z \left(\frac{\partial Y}{\partial Z}\right)_X \left(\frac{\partial Z}{\partial X}\right)_Y = 0$$

but the first term is equal to one by Equation (10.8), yielding

$$\left(\frac{\partial Y}{\partial Z}\right)_X \left(\frac{\partial Z}{\partial X}\right)_Y \left(\frac{\partial X}{\partial Y}\right)_Z = -1 \qquad (10.11)$$

Equation (10.11) is very important and will be used later.

10.3 APPLICATION OF MATHEMATICAL METHODS TO THERMODYNAMIC RELATIONS

The first law for a closed system on a unit mass basis is

$$\delta q = du + p\, dv \qquad (10.12)$$

If the process is reversible,

$$\delta q = T\, ds$$

hence

$$du = T\, ds - p\, dv \qquad (10.13)$$

Also,

$$dh = du + p\, dv + v\, dp$$
$$dh = T\, ds + v\, dp \qquad (10.14)$$

Two other thermodynamic relationships that are frequently used are the Helmholtz function, a, and the Gibbs function, g. Both are written on a unit mass basis

$$a = u - Ts \qquad (10.15)$$
$$g = h - Ts \qquad (10.16)$$

Taking derivatives of each yields

$$da = du - T\, ds - s\, dT$$

but, from Equation (10.13),

$$du = T\, ds - p\, dv$$

hence

$$da = p\, dv - s\, dT$$
$$dg = dh - T\, ds - s\, dT \qquad (10.17)$$

but

$$dh - T\,ds = v\,dp$$

hence

$$dg = v\,dp - s\,dT \tag{10.18}$$

Equations (10.13), (10.14), (10.17), and (10.18) are of the form $x = f(x, y)$. Using Equation (10.13) as an example, we let

$$u = f(s, v)$$

then

$$du = \left(\frac{\partial u}{\partial s}\right)_v ds + \left(\frac{\partial u}{\partial v}\right)_s dv \tag{10.19}$$

Comparing Equation (10.19) and (10.13) yields

$$T = \left(\frac{\partial u}{\partial s}\right)_v \qquad -p = \left(\frac{\partial u}{\partial v}\right)_s \tag{10.20}$$

Relationships between the other differentials may be developed from Equations (10.14), (10.17), and (10.18) in a similar manner. They are summarized below:

$$\left(\frac{\partial u}{\partial s}\right)_v = T \qquad \left(\frac{\partial h}{\partial s}\right) = T$$

$$\left(\frac{\partial a}{\partial v}\right)_T = -p \qquad \left(\frac{\partial u}{\partial v}\right)_s = -p$$

$$\left(\frac{\partial g}{\partial p}\right)_T = v \qquad \left(\frac{\partial h}{\partial p}\right)_s = v$$

$$\left(\frac{\partial a}{\partial T}\right)_v = -s \qquad \left(\frac{\partial g}{\partial T}\right)_p = -s \tag{10.21}$$

10.4 MAXWELL'S RELATIONS

Another group of relations, called Maxwell's relations, may be developed in the following manner. Differentiating Equation (10.20),

$$\left(\frac{\partial T}{\partial v}\right)_s = \frac{\partial^2 u}{\partial v\,\partial s} \qquad -\left(\frac{\partial p}{\partial s}\right) = \frac{\partial^2 u}{\partial s\,\partial v}$$

but, from Equation (10.2),

$$\frac{\partial^2 u}{\partial v\,\partial s} = \frac{\partial^2 u}{\partial s\,\partial v}$$

hence

$$\left(\frac{\partial T}{\partial v}\right)_s = -\left(\frac{\partial p}{\partial s}\right)_v \tag{10.22}$$

Equation (10.22) is one Maxwell relation; the others may be derived in a similar manner, and the following will result:

$$\left(\frac{\partial T}{\partial v}\right)_s = -\left(\frac{\partial p}{\partial s}\right)_v$$

$$\left(\frac{\partial T}{\partial p}\right)_s = \left(\frac{\partial v}{\partial s}\right)_p$$

$$\left(\frac{\partial p}{\partial T}\right)_v = \left(\frac{\partial s}{\partial v}\right)_T$$

$$\left(\frac{\partial v}{\partial T}\right)_p = -\left(\frac{\partial s}{\partial p}\right)_T \tag{10.23}$$

It should be recognized that this is not a complete listing; this listing is a function of Equations (10.13), (10.14), (10.17), and (10.18). If more terms, such as surface work, were added to the first law equation, then there would be more Maxwell relations. However, the ones listed in Equation (10.23) will be sufficient for the applications in the text. Should others be needed, they may be readily developed.

10.5 SPECIFIC HEATS, ENTHALPY, AND INTERNAL ENERGY

The specific heat at constant volume was developed by considering a constant-volume process; hence the first law is written as $\delta q = du = c_v dT$, and c_v is defined as

$$c_v \equiv \left(\frac{\partial u}{\partial T}\right)_v = \left(\frac{\partial u}{\partial s}\right)_v \left(\frac{\partial s}{\partial T}\right)_v = T\left(\frac{\partial s}{\partial T}\right)_v \tag{10.24}$$

In a similar manner for a constant-pressure process, $\delta q = dh = c_p dT$, and c_p is defined as

$$c_p \equiv \left(\frac{\partial h}{\partial T}\right)_p = \left(\frac{\partial h}{\partial s}\right)_p \left(\frac{\partial s}{\partial T}\right)_p = T\left(\frac{\partial s}{\partial T}\right)_p \tag{10.25}$$

It is possible to develop an expression for enthalpy and internal energy in terms of the specific heats and temperature and pressure. This is the intent of the chapter, to use the measurable properties to obtain nonmeasurable properties.

Let $u = f(T, v)$; then

$$du = \left(\frac{\partial u}{\partial T}\right)_v dT + \left(\frac{\partial u}{\partial v}\right)_T dv$$

$$T ds = du + p\, dv \tag{10.26}$$

For $T = C$, it follows that

$$T\left(\frac{\partial s}{\partial v}\right)_T = \left(\frac{\partial u}{\partial v}\right)_T + p$$

$$\left(\frac{\partial u}{\partial v}\right)_T = T\left(\frac{\partial s}{\partial v}\right)_T - p \tag{10.27}$$

By use of a Maxwell relation, Equation (10.27) becomes

$$\left(\frac{\partial u}{\partial v}\right)_T = T\left(\frac{\partial p}{\partial T}\right)_v - p \tag{10.28}$$

and Equation (10.26) becomes

$$du = c_v dT + \left[T\left(\frac{\partial p}{\partial T}\right) - p\right]dv \tag{10.29}$$

Notice that the change of internal energy may be calculated in terms of measurable properties.

Let $h = f(T, p)$; then

$$dh = \left(\frac{\partial h}{\partial T}\right)_p dT + \left(\frac{\partial h}{\partial p}\right)_T dp \tag{10.30}$$

and

$$T\,ds = dh - v\,dp \tag{10.31}$$

Hence

$$\left(\frac{\partial h}{\partial p}\right)_T = T\left(\frac{\partial s}{\partial p}\right)_T + v$$

If we use a Maxwell relation,

$$\left(\frac{\partial h}{\partial p}\right)_T = -T\left(\frac{\partial v}{\partial T}\right)_p + v \tag{10.32}$$

Substituting Equation (10.32) into Equation (10.30) yields

$$dh = c_p dT + \left[v - T\left(\frac{\partial v}{\partial T}\right)_p\right]dp \tag{10.33}$$

The functional variables were chosen in each case such that the first term would be the specific heat. All that is needed to integrate Equation (10.33) is a correlation between T, p, and v.

Example 10.1

Assume air has an equation of state $pv = RT$. The specific heat is known. Develop an expression for the change in internal energy.

$$du = c_v dT + \left[T \left(\frac{\partial p}{\partial T} \right) - p \right] dv$$

$$\frac{T}{p} = \frac{v}{R} = \text{constant} \qquad \text{for } v = c$$

Take derivative with respect to T

$$T \left(\frac{\partial p}{\partial T} \right)_v = p$$

$$du = c_v dT + [p - p] dv = c_v dT$$

This is what we would have expected. For other equations of state, this simplification will not result.

Entropy and Specific Heat Relationship

The equation for the change of entropy may be found in the following manner. The results lead to a very important thermodynamic relationship between the specific heat at constant volume and constant pressure.
We let

$$s = f(T, p)$$

$$ds = \left(\frac{\partial s}{\partial T} \right)_p dT + \left(\frac{\partial s}{\partial p} \right)_T dp \tag{10.34}$$

Using Equation (10.25) this becomes

$$ds = c_p \frac{dT}{T} + \left(\frac{\partial s}{\partial p} \right)_T dp$$

Using the Maxwell relation from Equation (10.23),

$$ds = c_p \frac{dT}{T} - \left(\frac{\partial v}{\partial T} \right)_p dp \tag{10.35}$$

We now let

$$s = f(T, v)$$

$$ds = \left(\frac{\partial s}{\partial T} \right)_v dT + \left(\frac{\partial s}{\partial v} \right)_T dv$$

$$ds = c_v \frac{dT}{T} + \left(\frac{\partial s}{\partial v} \right)_T dv$$

$$ds = c_v \frac{dT}{T} + \left(\frac{\partial p}{\partial T} \right)_v dv \tag{10.36}$$

Equations (10.36) and (10.35) both express the change of entropy, but they use different specific heats. Again, an equation of state is needed to determine the partial derivatives.

Specific Heat Difference

Another relationship denoting the difference in specific heats may be found in the following manner. Let us equate Equations (10.35) and (10.36):

$$c_p \frac{dT}{T} - \left(\frac{\partial v}{\partial T} \right)_p dp = c_v \frac{dT}{T} + \left(\frac{\partial p}{\partial T} \right)_v dv$$

$$dT = \frac{T(\partial v / \partial T)_p dp + T(\partial p / \partial T)_v dv}{c_p - c_v} \tag{10.37}$$

Since $T = T(p, v)$,

$$dT = \left(\frac{\partial T}{\partial p} \right)_v dp + \left(\frac{\partial T}{\partial v} \right)_p dv \tag{10.38}$$

We equate the coefficients of dp or dv in Equations (10.38) and (10.37). The following is found by equating the coefficients of dp:

$$(c_p - c_v)\left(\frac{\partial T}{\partial p} \right)_v = T \left(\frac{\partial v}{\partial T} \right)_p$$

$$c_p - c_v = T \left(\frac{\partial v}{\partial T} \right)_p \left(\frac{\partial p}{\partial T} \right)_v \tag{10.39}$$

From Equation (10.11), we may write for p, v, and T

$$\left(\frac{\partial v}{\partial T} \right)_p \left(\frac{\partial T}{\partial p} \right)_v \left(\frac{\partial p}{\partial v} \right)_T = -1$$

from which

$$\left(\frac{\partial p}{\partial T} \right)_v = - \left(\frac{\partial v}{\partial T} \right)_p \left(\frac{\partial p}{\partial v} \right)_T \tag{10.40}$$

Substituting Equation (10.40) into Equation (10.39),

$$c_p - c_v = -T \left(\frac{\partial v}{\partial T} \right)_p \left(\frac{\partial v}{\partial T} \right)_p \left(\frac{\partial p}{\partial v} \right)_T$$

$$c_p - c_v = -T \left(\frac{\partial v}{\partial T} \right)_p^2 \left(\frac{\partial p}{\partial v} \right)_T \tag{10.41}$$

Three conclusions may be made by applying experimental observations to Equation (10.41).

1. Since $(\partial v / \partial T)_p$ is very small for liquids and solids, the difference between c_p and c_v is essentially zero; only one specific heat for a

liquid or solid is usually tabulated with designation of constant pressure or constant volume.

2. As T approaches zero, c_p approaches c_v.
3. For all known substances $(\partial p/\partial v)_T < 0$, and $(\partial v/\partial T)_p^2 > 0$ all the time; hence $c_p \geq c_v$.

10.6 CLAPEYRON EQUATION

It is possible to predict values of enthalpy for changes in phase of a substance in thermodynamic equilibrium. By equilibrium we mean that the system is in mechanical equilibrium (both phases are at the same pressure) and in thermal equilibrium (both phases are at the same temperature). There is a direct correlation between the enthalpy change in vaporization and volume change in vaporization.

The relationship that we want is $(\partial h/\partial v)_p$, but enthalpy cannot be calculated, so

$$\left(\frac{\partial h}{\partial v}\right)_p = \left(\frac{\partial h}{\partial s}\right)_p \left(\frac{\partial s}{\partial v}\right)_p$$

and

$$\left(\frac{\partial h}{\partial s}\right)_p = T$$

Hence

$$\left(\frac{\partial h}{\partial v}\right)_p = T\left(\frac{\partial s}{\partial v}\right)_p \tag{10.42}$$

Fortunately, in the saturated region the temperature is also a constant, so

$$\left(\frac{\partial s}{\partial v}\right)_p = \left(\frac{\partial s}{\partial v}\right)_T = \left(\frac{\partial p}{\partial T}\right)_v \tag{10.43}$$

from Equation (10.23).

Since the pressure is only a function of temperature when two phases are present, the partial derivative may be written as a total derivative in this saturated region. Equation (10.43) becomes

$$\left(\frac{\partial h}{\partial v}\right)_p = T\left(\frac{dp}{dT}\right) \tag{10.44}$$

Denoting the change of a property from a saturated liquid to a saturated vapor phase as fg, Equation (10.44) becomes

$$\frac{h_{fg}}{v_{fg}} = T\frac{dp}{dT}$$

or

$$h_{fg} = v_{fg}T\left(\frac{dp}{dT}\right) \tag{10.45}$$

Equation (10.45), which is the Clapeyron equation, enables us to determine the change in enthalpy during a phase change by measuring the volume change, the temperature, and the slope of the vapor pressure curve. If the assumption is made that the specific volume of the vapor is much larger than that of the liquid and if we employ the ideal gas law to determine the specific volume of the vapor, the enthalpy of vaporization may be readily calculated. Let us assume that

$$v_g \gg v_f$$

and

$$v_g = \frac{RT}{p}$$

then

$$h_{fg} = \frac{RT^2}{p} \frac{dp}{dT} \tag{10.46}$$

or

$$\frac{dp}{p} = \frac{h_{fg}}{R} \frac{dT}{T^2} \tag{10.47}$$

Example 10.2

Determine the pressure of saturated steam at 40°C if at 35°C the pressure is 5.628 kPa, the enthalpy of vaporization is 2418.6 kJ/kg, and the specific volume is 25.22 m³/kg. The enthalpy of vaporization is essentially constant over this temperature range.

Integrating Equation (10.47) between 1 and 2, we obtain

$$\ln\left(\frac{p_2}{p_1}\right) = \frac{h_{fg}}{R}\left(\frac{1}{T_1} - \frac{1}{T_2}\right)$$

$$R = \frac{v_1 p_1}{T_1} = \frac{(25.22)(5.628)}{308} = 0.46 \text{ kJ/kg} \cdot \text{K}$$

$$\ln\left(\frac{p_2}{p_1}\right) = \frac{2418.6}{0.46}\left(\frac{1}{308} - \frac{1}{313}\right)$$

$$\ln\left(\frac{p_2}{p_1}\right) = 0.2727$$

$$p_2 = 7.392 \text{ kPa}$$

From steam tables $p = 7.384$ kPa.

10.7 IMPORTANT PHYSICAL COEFFICIENTS

There are several other important coefficients that are used to determine actual properties of a substance. The coefficients are determined and, from knowledge of the coefficients, property changes may be determined. Let us consider that $V = f(T, p)$; then

$$dV = \left(\frac{\partial V}{\partial T}\right)_p dT + \left(\frac{\partial V}{\partial p}\right)_T dp$$

We divide each partial derivative by V, which results in the following coefficients:

$$\alpha \equiv \frac{1}{V}\left(\frac{\partial V}{\partial T}\right)_p \tag{10.48}$$

and

$$\beta_T \equiv -\frac{1}{V}\left(\frac{\partial V}{\partial p}\right)_T \tag{10.49}$$

where α is called the coefficient of "thermal expansion" and β_T is called the coefficient of "isothermal compressibility." The negative sign is included in Equation (10.49) because the volume decreases with increasing pressure, and vice versa, so the partial derivative is always negative and the minus sign allows β_T always to be positive. The reciprocal of the isothermal compressibility is called the "isothermal bulk modulus," B_T;

$$B_T \equiv -V\left(\frac{\partial p}{\partial V}\right)_T \tag{10.50}$$

The difference between the specific heats may be expressed in terms of these coefficients. From Equation (10.41),

$$c_p - c_v = -T\left(\frac{\partial v}{\partial T}\right)_p^2 \left(\frac{\partial p}{\partial v}\right)_T \tag{10.51}$$

but

$$\frac{B_T}{v} = -\left(\frac{\partial p}{\partial v}\right)_T$$

and

$$\alpha^2 v^2 = \left(\frac{\partial v}{\partial T}\right)_p^2$$

Hence

$$c_p - c_v = \frac{TB_T \alpha^2 v^2}{v}$$

$$c_p - c_v = \frac{T\alpha^2 v}{\beta_T} \tag{10.52}$$

The change in entropy at constant temperature may also be found when entropy is considered as a function of temperature and pressure. Then

$$ds = \left(\frac{\partial s}{\partial p} \right)_T dp \quad \text{for } T = c$$

but from the Maxwell relations,

$$\left(\frac{\partial s}{\partial p} \right)_T = - \left(\frac{\partial v}{\partial T} \right)_P = -v\alpha \tag{10.53}$$

and

$$(ds)_T = -v\alpha(dp)_T$$

$$(s_2 - s_1)_T = -v\alpha(p_2 - p_1)_T \tag{10.54}$$

Thus by knowing tabulated coefficients the property change of certain systems may be calculated. For instance, the *Handbook of Chemistry and Physics* lists these and other coefficients for a great variety of substances.

The homework problems at the end of the chapter will illustrate some typical cases where the coefficients may be calculated if an equation of state is known, as in the case of gases. The ideal gas law will yield much simplified results but these results can be helpful in confirming your understanding of the meaning of the various coefficients.

Joule-Thomson Coefficient

A throttling process produces no change in enthalpy; hence, for an ideal gas, the temperature remains constant. For a real gas, however, the throttling process will cause the temperature to increase or decrease. The Joule-Thomson coefficient, μ, relates this change and is defined as

$$\mu \equiv \left(\frac{\partial T}{\partial p} \right)_h \tag{10.55}$$

A positive value of μ indicates that the temperature decreased as the pressure decreased; a cooling effect is thus observed. This is true for almost all gases at ordinary pressures and temperatures. The exceptions are hydrogen (H_2), neon, and helium, which have a temperature increase with a pressure decrease, hence $\mu < 0$. Even for these gases there is a temperature above which the Joule-Thomson coefficient changes from negative to positive. This is called the "inversion temperature," and at this temperature $\mu = 0$.

Let us analyze the Joule-Thomson coefficient to gain insight into why the coefficient changes sign:

$$dh = T \, ds + v \, dp$$

$$s = f(T, p)$$

$$ds = \left(\frac{\partial s}{\partial T}\right)_p dT + \left(\frac{\partial s}{\partial p}\right)_T dp$$

$$dh = T\left(\frac{\partial s}{\partial T}\right)_p dT + T\left(\frac{\partial s}{\partial p}\right)_T dp + v \, dp$$

For constant T, $dT = 0$

$$dh = \left[T\left(\frac{\partial s}{\partial p}\right)_T + v\right] dp$$

or

$$\left(\frac{\partial h}{\partial p}\right)_T = T\left(\frac{\partial s}{\partial p}\right)_T + v \tag{10.56}$$

From Equation (10.11) for $h = f(T, p)$

$$\left(\frac{\partial T}{\partial p}\right)_h \left(\frac{\partial p}{\partial h}\right)_T \left(\frac{\partial h}{\partial T}\right)_p = -1$$

$$\left(\frac{\partial T}{\partial p}\right)_h = -\frac{1}{(\partial h/\partial T)_p}\left(\frac{\partial h}{\partial p}\right)_T \tag{10.57}$$

and

$$\left(\frac{\partial h}{\partial T}\right)_p \equiv c_p \tag{10.58}$$

and from Equation (10.32)

$$\left(\frac{\partial h}{\partial p}\right)_T = -T\left(\frac{\partial v}{\partial T}\right)_p + v \tag{10.59}$$

Therefore we substitute Equations (10.58) and (10.59) into (10.57)

$$\mu = -\frac{1}{c_p}\left(\frac{\partial h}{\partial p}\right)_T \tag{10.60}$$

$$\mu = -\frac{1}{c_p}\left[-T\left(\frac{\partial v}{\partial T}\right)_p + v\right] \tag{10.61}$$

For an ideal gas, $pv = RT$

$$\left(\frac{\partial v}{\partial T}\right)_p = \frac{R}{P} = \frac{v}{T}$$

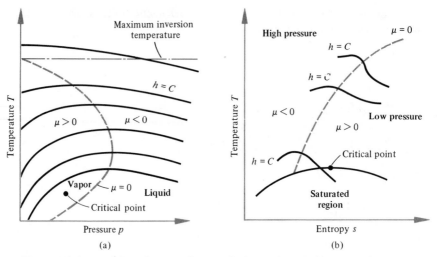

Figure 10.4   (a) A T–p diagram showing the locus of points for $\mu = 0$. (b) A T–s diagram showing the locus of points for $\mu = 0$.

and substituting into Equation (10.61) yields

$$\mu = 0 \tag{10.62}$$

For an ideal gas, $\mu = 0$, but that is not a necessary and sufficient condition that the gas is ideal. Remember, at the inversion temperature, $\mu = 0$ also.

To illustrate why the Joule-Thomson coefficient changes sign, we substitute $h = u + pv$ into Equation (10.60):

$$\mu = +\frac{1}{c_p}\left[-\left(\frac{\partial u}{\partial p}\right)_T - \left(\frac{\partial(pv)}{\partial p}\right)_T\right] \tag{10.63}$$

The first term in the brackets denotes the deviation from Joule's law, which states that the internal energy is a function only of temperature. On expansion, there is an increase in the molecular potential energy, and hence $(\partial u/\partial p)_T$ is negative. This results in a positive μ and a temperature decrease. The second term in the brackets indicates the deviation from Boyle's law (that v varies inversely with p) for a real gas. For most gases at low temperatures and pressures, $(\partial(pv)/\partial p)_T$ is negative; however, it changes sign at higher temperatures and pressures.

Figures 10.4(a) and 10.4(b) illustrate the plot of μ on T–p and T–s diagrams. Lines of constant enthalpy are also noted.

10.8 REAL-GAS BEHAVIOR

Thus far we have had tables for gas properties, or we have used the ideal-gas equations of state to determine the properties of gases, or both.

The ideal-gas law does not accurately portray the behavior of the gas, and the other equations of state of a gas, such as the Van der Waals or Beattie-Bridgeman equations of state offer a more accurate portrayal. All of these are approximations, to be used when tabulated property values are not available. There are additional concepts that are helpful in approximating gas behavior when tabulated data do not exist.

One of these concepts is the compressibility factor, Z, which denotes the deviation of an actual gas from the ideal gas behavior. The compressibility factor is defined as

$$Z = \frac{pv}{RT} \qquad (10.64)$$

where Z is unity for an ideal gas. Use of the generalized compressibility chart enabled us to use one chart for all gases in Chapter 5.

Why is this so? The use of the Van der Waals equation of state will demonstrate that the use of reduced coordinates leads to an equation of state in terms of these coordinates. The constants for the various gases are eliminated.

The Van der Waals equation of state is

$$p = \frac{RT}{v-b} - \frac{a}{v^2} \qquad (10.65)$$

Figure 10.5 shows a p–v diagram with a plot of isotherms for a typical pure substance. The critical isotherm is of importance. Examination of the experimental data has shown that at least the first two derivatives are

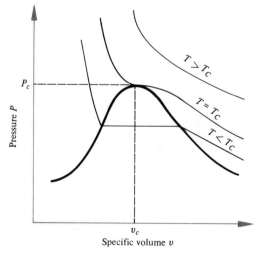

Figure 10.5 A p–v diagram for a typical pure substance.

zero at the critical point:

$$\left(\frac{\partial p}{\partial v}\right)_T = 0 \tag{10.66}$$

and

$$\left(\frac{\partial^2 p}{\partial v^2}\right)_T = 0 \tag{10.67}$$

If Equations (10.66) and (10.67) are applied to Equation (10.65), we may determine the critical properties from the Van der Waals equation:

$$\left(\frac{\partial p}{\partial v}\right)_T = -\frac{RT_c}{(v_c - b)^2} + \frac{2a}{v_c^3} = 0$$

$$RT_c = \frac{2a}{v_c^3}(v_c - b)^2 \tag{10.68}$$

$$\left(\frac{\partial^2 p}{\partial v^2}\right)_T = +\frac{2RT_c}{(v_c - b)^3} - \frac{6a}{v_c^4} = 0$$

$$RT_c = \frac{3a}{v_c^4}(v_c - b)^3 \tag{10.69}$$

Solving Equations (10.68) and (10.69) for v_c yields

$$v_c = 3b \tag{10.70}$$

and substituting this is Equation (10.68) results in

$$T_c = \frac{8a}{27Rb} \tag{10.71}$$

and substitution into Equation (10.65) yields for the critical pressure

$$p_c = \frac{a}{27b^2} \tag{10.72}$$

Using the critical properties, we rewrite the Van der Waals equation of state with reduced coordinates:

$$p = \frac{p_r a}{27b^2} \qquad v = 3bv_r \qquad T = \frac{8aT_r}{27Rb}$$

Substituting these into Equation (10.65) yields

$$p_r = \frac{8T_r}{(3v_r - 1)} - \frac{3}{v_r^2} \tag{10.73}$$

Equation (10.73) illustrates that an equation of state may be expressed in terms of the reduced coordinates and be applicable for any gas. However, no substance obeys Van der Waals equation of state, which is an approximation of actual gas behavior.

PROBLEMS

1. Derive an expression for the change of internal energy of a gas using the Van der Waals equation of state.
2. Derive an expression for the change of enthalpy of a gas using (a) the ideal-gas equation of state; (b) the Van der Waals equation of state.
3. Show that Equation (10.11) is valid when the variables are p, v, and T and are related by the ideal-gas equation of state.
4. Prove that heat is an inexact differential $(Q(T, s))$.
5. Using the Maxwell relation in Equations (10.22) and (10.11), develop the three remaining relations given in Equation (10.23).
6. The following information is obtained from the steam tables of the vaporization process:

1378 kPa	467.7 K
1413 kPa	468.8 K
1448 kPa	469.9 K

   At 1413 kPa, $v_{fg} = 0.1383$ m^3/kg. Determine the latent heat of vaporization at 1413 kPa.
7. Use the Clapeyron relation to determine the change of enthalpy of (a) steam at 2000 kPa; (b) ammonia at 40°C; (c) R 12 at 15°C.
8. Determine the difference between c_p and c_v for a gas that obeys (a) the ideal-gas equation of state; (b) Van der Waals equation of state.
9. Derive an expression for the change of entropy of a gas that obeys the Van der Waals equation of state.
10. The Berthelot equation of state is $(p + a/Tv^2)(v - b) = RT$, where a and b are constants. Show that, at the critical isotherm, (a) $a = (9/8)Rv_cT_c^2$; (b) $b = v_c/3$.
11. The Dieterici equation of state is $p(v - b)e^{a/RTv} = RT$, where a and b are constants. Show that, at the critical isotherm, (a) $a = 2p_cv_c^2$; (b) $b = v_c/2$.
12. Compute a and b for carbon dioxide in Problem 10.
13. Compute a and b for ammonia in Problem 11.
14. Calculate the coefficient of thermal expansion and the coefficient of isothermal compressibility for a gas that obeys (a) the ideal-gas equation of state; (b) the Van der Waals equation of state.
15. Compute the coefficient of thermal expansion for methane at 32°C and 1400 kPa using (a) the ideal-gas equation of state; (b) the Van der Waals equation of state.
16. Derive an expression for the change of enthalpy and entropy at constant temperature using the Van der Waals equation of state.
17. Two kilograms of oxygen are expanded isothermally from 380 K and 700 kPa while obeying the Van der Waals equation of state. Determine the change in internal energy, enthalpy, and entropy for the process.

18. Determine the volume occupied by 3 kg of carbon dioxide at 7.0 MPa and 95°C (use compressibility factor).

29. Oxygen is contained in a 70-liter compressed gas tank at 20.1 MPa and 37°C. Find the mass of oxygen contained in the tank.

20. Helium is compressed in a piston-cylinder until the final temperature is 425°C and the final pressure is 20.1 MPa. Determine the size of the cylinder to contain 0.22 kg of helium at the final state.

21. Determine the Joule-Thomson coefficient for a gas with the following equation of state: (a) $p(v - b) = RT$; (b) $(p + a/v^2)(v - b) = RT$.

22. Determine the average Joule-Thomson coefficient for steam that is throttled from 1.1 MPa and 280°C to 140 kPa.

23. Determine the average Joule-Thomson coefficient for ammonia that is throttled from 2.0 MPa and 140°C to 500 kPa.

24. Show that the Joule-Thomson coefficient equals $\mu = (v/c_p)(T\alpha - 1)$.

Chapter 11
Vapor Power Cycles

Equipment using a vapor as the working fluid operates on either a power-producing or a power-consuming cycle. In either case we must put together two or more processes and develop a cycle that will produce work. In this chapter we will not consider power-consuming cycles, such as the refrigeration cycle. Furthermore, we will limit ourselves, with one or two exceptions, to using water as the working substance, because steam is widely used for running engines, and in the historical as well as the thermodynamic sense the steam engine was the motive force of the industrial revolution.

11.1 CARNOT VAPOR CYCLE

As a first guess we might decide to use a Carnot cycle to produce power. It is the most efficient heat cycle operating between two temperature limits. A T–s diagram (Figure 11.1) illustrates a steam Carnot cycle. Saturated water at state 1 is evaporated, at constant temperature and pressure, to state 2, where it is a saturated vapor. Further heating at constant temperature would necessitate a pressure drop. The steam enters the engine at state 2 and expands isentropically, doing work, until state 3 is reached. The steam-water mixture would then partially condense at constant temperature and pressure until state 4 is reached. At state 4 a compressor would isentropically compress this vapor-liquid mixture to state 1. Some of the work developed in going from state 2 to state 3 would be returned in the compression process.

Difficulties arise in this situation. One is that the turbine will have to handle steam with a low quality. Steam with qualities less than 85 to 90 percent has too much moisture, and the liquid impingement and erosion of the blading is detrimental. For another, there would be difficulties in

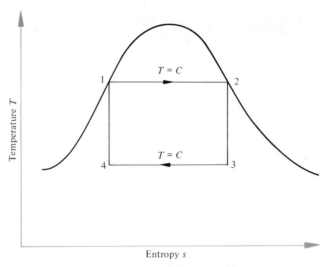

Figure 11.1 *A T–s diagram for a Carnot cycle using steam as the working substance.*

designing a device to compress the liquid-vapor mixture, not to mention the difficulty of controlling a partial condensation process.

11.2 THE RANKINE CYCLE

The Rankine cycle overcomes many of the operational difficulties encountered with the Carnot cycle when the working fluid is a vapor. In this cycle the heating and cooling processes occur at constant pressure. Figure 11.2(a) illustrates the Rankine cycle on a $T–s$ diagram, and Figure 11.2(b) shows the equipment used in the cycle.

Following the cycle from state 1 in Figures 11.2(a) and 11.2(b), we see that the water enters the steam generator as a subcooled liquid at pressure p_2. The energy supplied in the steam generator raises the state of the water from that of a subcooled liquid to that of a saturated liquid and, further, to that of a saturated vapor at state 2. The vapor leaves the steam generator at state 2 and enters a steam turbine, where it expands isentropically to state 3. It enters the condenser at this point and is condensed at constant pressure from state 3 to state 4. At state 4 the water is a saturated liquid at the pressure in the condenser. The liquid cannot enter the steam generator, which is at a higher pressure, until its pressure is raised to that of the steam generator. A pump performs this very easily, in contrast to the compressor in the Carnot vapor cycle, and raises the pressure of the liquid to p_2, the steam-generator pressure. The liquid is now a subcooled liquid at state 1, and the cycle is complete. Problems would still exist if the steam entered the turbine as a saturated

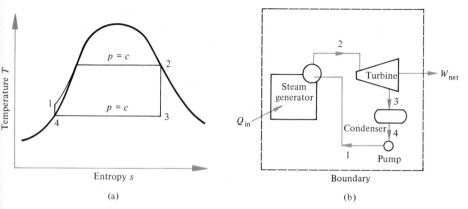

Figure 11.2   (a) A T–s diagram for the Rankine cycle. (b) A schematic diagram for the Rankine cycle.

vapor, in that the moisture content of the steam would be too high as it passed through the turbine, resulting in liquid impingement and erosion of the blading.

Since the Rankine cycle is characterized by constant-pressure heating, there is no reason to stop heating the steam when it reaches the saturated vapor state. The customary practice is to superheat the steam to a much higher temperature. Figure 11.3 illustrates how the superheating shifts the isentropic expansion process to the right, thus preventing a high moisture content of the steam when it exits the turbine. Typical values

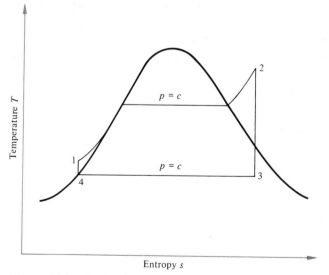

Figure 11.3   The Rankine cycle with superheated steam at state 2.

for the temperature of the steam at state 2 is 485 to 540°C. Metallurgical limitations prevent higher values. The pressure is not limited, and a wide range of pressure will be found.

11.3 RANKINE CYCLE COMPONENTS

Before we can find the overall Rankine cycle efficiency, it is necessary to calculate the values of turbine work, pump work, heat rejected, and heat added. The Rankine cycle is a closed cycle; no mass leaves the system, as indicated by the system boundary in Figure 11.2(b).

Turbine

Let us now narrow the system to its components, one at a time, beginning with the turbine, as illustrated in Figure 11.4. There is, indeed, mass flow across the system boundary, which is fixed, so the turbine constitutes an open system.

There are many forms of energy that must be accounted for. The enthalpy and the kinetic and potential energies must be included, both entering and leaving as well as the turbine work and any heat losses. We write the energy balance across the turbine

$$\dot{H}_2 + (\text{K.\,E.})_2 + (\text{P.\,E.})_2 = \dot{H}_3 + (\text{K.\,E.})_3 + (\text{P.\,E.})_3 + \dot{Q} + \dot{W}_t$$

$$(11.1)$$

where $\dot{W}_t$ is the turbine work and $\dot{Q}$ is the heat loss from the turbine. The heat loss (radiative and convective losses to the surroundings) is usually quite small. If the turbine is considered adiabatic, no significant error results. The potential energy terms also have a negligible effect. The vertical distance across the turbine is short, on the order of 3 m. If the

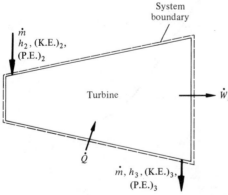

Figure 11.4 *The steam turbine with energy terms noted.*

turbine is adiabatic and the potential energy is neglected, Equation (11.1) becomes

$$\dot{H}_2 + (\text{K.E.})_2 = \dot{H}_3 + (\text{K.E.})_3 + \dot{W}_t \qquad (11.2)$$

Example 11.1

An adiabatic turbine in a steam generating plant receives steam at a pressure of 7.0 MPa and 550°C and exhausts at 20 kPa. The turbine inlet is 3 m higher than the turbine exit, the inlet steam velocity is 15 m/s and the exit velocity is 300 m/s. Calculate the turbine work per unit mass of steam, and find, in terms of percentage, the effect that each term has on the turbine work.

Solving Equation (11.1) for the turbine work under reversible adiabatic conditions,

$$w_1 = (h_2 - h_3)_s + (\text{K.E.})_2 - (\text{K.E.})_3 + (\text{P.E.})_2 - (\text{P.E.})_3$$

$$h_2 = 3530.9 \text{ kJ/kg} \qquad s_2 = 6.9486$$

$$s_3 = s_2 = s_f + x s_{fg}$$

$$6.9486 = 0.8320 + x(7.0766)$$

$$x = 0.864$$

$$h_3 = h_f + x h_{fg} = 251.4 + (0.864)(2358.3) = 2288.9 \text{ kJ/kg}$$

$$w_t = (3530.9 - 2288.9) + \frac{15^2 - 300^2}{2(1000)} + \frac{(3)(9.8)}{1000}$$

$$w_t = 1242 - 44.9 + 0.029 = 1197.1 \text{ kJ/kg}$$

total energy available for work = 1286.9

$$\Delta h = \left(\frac{1242}{1286.9}\right)(100) = 96.5 \text{ percent}$$

$$\Delta \text{K.E.} = \left(\frac{44.9}{1286.9}\right)(100) = 3.49 \text{ percent}$$

$$\Delta \text{P.E.} = \left(\frac{0.029}{1286.9}\right)(100) = 0.01 \text{ percent}$$

Note that the potential energy has a very small influence and is usually not included in the calculation of turbine work. The kinetic energy has a small effect also and need be included only if high accuracy is desired. If the kinetic and potential energies are not included, Equation (11.2) becomes

$$\dot{W}_t = \dot{m}(h_2 - h_3)_s \qquad (11.3)$$

where the s denotes the difference in enthalpies is taken at constant entropy. This is an assumption on which the Rankine cycle was predicated, but denoting the variables that are held constant is a good habit to get into.

Condenser

The next form of energy transfer occurs in the condenser. This is where the heat is rejected by the steam when it condenses from state 3 to state 4, as indicated on Figure 11.2(a). Physically, the steam passes over the outside of tubes in a shell and tube heat exchanger while cooling water passes through the tubes. This accounts for a substantive decrease in the cycle efficiency because the latent heat of the steam is transferred to the cooling water. A substantial portion of the energy that was supplied in the steam generator to convert the water to steam is now lost from the system. An energy balance on the condenser will yield an expression for this loss. Figure 11.5 illustrates the condenser, considered as the system, and the various energy forms that should be considered. The potential energy has been neglected in this case, as the effect of elevation is small. The condenser is adiabatic, so there is no heat flow to the surroundings, and the difference in velocities of the cooling water in and out is very small, hence the change in kinetic energy of the cooling water is essentially zero. Making an energy balance across the condenser yields,

$$\dot{m}_{\text{sys}}(h_3 - h_4) + \dot{m}_{\text{sys}}\left[(\text{K.E.})_3 - (\text{K.E.})_4\right] = \dot{m}_{\text{H}_2\text{O}}(h_o - h_i) \quad (11.4)$$

The exit steam velocity from the turbine to the condenser may be high, and so the entering kinetic energy should be included in the energy balance. The conservation of mass for the system will give us an estimate of the difference between the inlet and exit velocities of the steam and

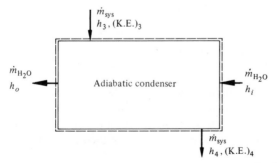

Figure 11.5   *The adiabatic condenser used in the Rankine cycle with mass and energy flows noted.*

saturated liquid:

$$v_4 = v_3 \left(\frac{A_3}{A_4} \right) \left(\frac{v_4}{v_3} \right) \tag{11.5}$$

Picking typical values for the specific volumes at 100 F shows that $v_4 = 0.001\ 006$ and $v_3 = 25.22$. Allowing an area ratio of $A_3/A_4 = 100$, Equation (11.5) becomes

$$v_4 = 0.004 v_3$$

We may conclude that the exit velocity of the saturated liquid from the condenser is very small compared with the inlet velocity of the steam, and thus the exit kinetic energy leaving is also very small compared to the inlet kinetic energy. With these considerations in mind, Equation (11.4) becomes

$$\dot{m}_{\text{sys}} \left[(h_3 - h_4) + (\text{K.E.})_3 \right] = \dot{m}_{\text{H}_2\text{O}} (h_o - h_i) \tag{11.6}$$

where the term on the right-hand side of Equation (11.6) denotes the energy rejected from the system to the surroundings, the cooling water. This energy, about 2100 to 2300 kJ/kg (sys), may cause thermal pollution if the power plant is not correctly situated to minimize thermal-energy impact.

Pump

The feed pump elevates the pressure of the water at state 4 to the steam generator pressure. Figure 11.6(a) is a schematic diagram of the pump, showing the energy terms. Again, potential energy is not considered and the pump is considered to be adiabatic. An energy balance across the pump yields

$$\dot{W}_p = \dot{m}(h_1 - h_4) + \dot{m} \left[(\text{K.E.})_1 - (\text{K.E.})_4 \right] \tag{11.7}$$

but the kinetic energies entering and leaving the pump are essentially equal to each other. We verify this by conservation of mass for equal areas. Hence the work becomes

$$\dot{W}_p = \dot{m}(h_1 - h_4) \tag{11.8}$$

and if we assume the pump is isentropic, the work becomes

$$\dot{W}_p = \dot{m}(h_1 - h_4)_s \tag{11.9}$$

The compressed liquid table, Table A.8, lists temperature and pressure as the ordinates, so the assumption of isentropic compression is not necessary. However, the exit temperature often is not known, so we have to develop another method of determining the pump work.

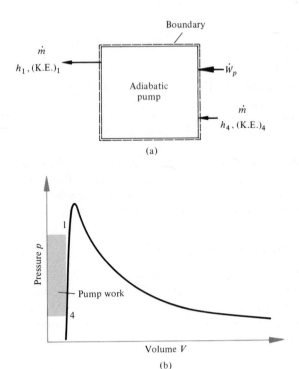

Figure 11.6 (a) The schematic of an adiabatic pump with energy flows noted.
(b) Pump work illustrated on a p–v diagram.

From Equation (11.8), we may deduce that

$$\dot{W}_p = \dot{m} \int_4^1 dh$$

$$dh = v \, dp \qquad \text{for adiabatic case}$$

$$\dot{W}_p = \dot{m} \int_4^1 v \, dp \qquad (11.10a)$$

At low pressures and temperatures, water is essentially incompressible and the specific volume is constant. As the pressure difference between state 4 and state 1 increases, the compressibility of water should be taken into account. For the incompressible case, the work becomes

$$\dot{W}_p = \dot{m} v_4 (p_1 - p_4) \qquad (11.10b)$$

and this is illustrated on Figure 11.6(b) as the shaded area. The next assumption would be a linear variation of specific volume with pressure, such as $v = v_0 + C_1 p$; we would then evaluate v_0, C_1 for the particular case under consideration and use this in the integral of Equation (11.10). It should be remembered that although it is desirable to be as accurate as possible, the effect of the pump work may not demand this accuracy if the exactness of other larger energy terms, such as turbine work, is not so precise. The enthalpy at state 1 is readily determined as follows:

$$h_1 = h_4 + \int_4^1 v \, dp \tag{11.11}$$

Steam Generators

Steam generators, sometimes referred to as boilers, create steam from water. There are many design considerations for the steam generator, but basically the water enters a steam drum at a temperature less than saturated. The water passes from this large drum into small tubes where it is heated to its saturation point with some vaporization occurring. Hot gases, used for heating the water, may be formed by burning fuel or may be the high-temperature exhaust from a diesel or gas-turbine engine. The mixture of steam and water reenters the steam drum where the vapor rises to the top and the water is recirculated. The steam, now a saturated vapor, passes through the superheater tubes where its temperature is increased by further heat transfer from the hot gases. The superheated steam leaves the steam generator and enters the turbine. Figure 11.7 illustrates this schematically.

A first-law analysis, considering only the water, yields

$$\dot{Q} + \dot{m}_s h_1 = \dot{m}_s h_2$$

$$q = h_2 - h_1 \text{ kJ/kg} \tag{11.12}$$

The analysis of boiler is much more complex in practice and entails considering different modes of heat transfer in the various boiler sections. However, from the Rankine cycle viewpoint (that of the water), Equation (11.12) is sufficient at this time.

11.4 EFFICIENCIES

What is the thermal efficiency of the Rankine cycle? the efficiency, η_{th}, is the net work produced, divided by what we had to use to produce this

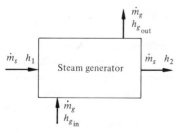

Figure 11.7 Schematic diagram of a steam generator.

work, the heat supplied,

$$\eta_{th} = \frac{\dot{W}_{net}}{\dot{Q}_{in}} = \frac{\dot{W}_t - \dot{W}_p}{\dot{m}(h_2 - h_1)} \tag{11.13}$$

Example 11.2

A steam power plant operates on the Rankine cycle. The steam enters the turbine at 7.0 MPa and 550°C with a velocity of 30 m/s. It discharges to the condenser at 20 kPa with a velocity of 90 m/s. Calculate the thermal efficiency and the net power produced, for a flow rate of 37.8 kg/s. Figure 11.2(a) illustrates the Rankine cycle diagram.

Calculate the thermal efficiency by finding w_p, w_t, q_{in}.

$h_2 = 3530.9$ kJ/kg $s_2 = 6.9486$ kJ/kg · K

Process 2–3, isentropic; $s_3 = s_2$

$\quad x = 0.864 \quad\quad h_3 = 2288.9$ kJ/kg

$h_4 = h_f = 251.4$

$h_1 = 251.4 + (0.001\ 017)(7000 - 20) = 258.5$ kJ/kg

$w_p = (h_1 - h_4) = 7.1$ kJ/kg

$w_t = (h_2 - h_3) + (\text{K.E.})_2 - (\text{K.E.})_3 \quad\quad$ from Equation (11.2)

$w_t = (3530.9 - 2288.9) + \dfrac{30^2 - 90^2}{2(1000)} = 1238.4$ kJ/kg

$w_{net} = \Sigma w = w_t + w_p = 1231.3$ kJ/kg

$q_{in} = (h_2 - h_1) = (3530.9 - 258.5) = 3272.4$ kJ/kg

$\eta_{th} = \dfrac{w_{net}}{q_{in}} = \dfrac{1231.3}{3272.4} = 0.376$ or 37.6 percent

The net power produced, $\dot{W}$,

$\dot{W} = \dot{m}w_{net} = (37.8)(1231.3) = 46\ 543$ kW

The turbine and pump have been considered to be ideal in the case above in which none of the irreversibilities that occur in an actual unit have been considered. To account for these nonideal effects, we introduce the internal turbine and pump efficiencies. The internal efficiency is determined experimentally by the manufacturer; once it is known, the value may be used in computing the actual work produced or consumed. What is the efficiency? All efficiencies are a ratio of what is wanted to what it costs to achieve that end. The turbine produces work, so the internal turbine efficiency, η_t, would be a ratio of the work actually done by the steam to that which could be ideally done. The difference between the two is the loss due to irreversibilities in the turbine:

$$\eta_t = \frac{(h_2 - h_3)\text{actual}}{(h_2 - h_3)_s} \qquad (11.14)$$

The turbine internal efficiency is a determination of how well the available energy is utilized. The denominator, $(h_2 - h_3)_s$, is the available energy of the steam; the numerator, $(h_2 - h_3)_{\text{actual}}$, is the amount of available energy used as work. Once the internal efficiency of the turbine is known, the actual enthalpy of the turbine exit may be calculated.

Example 11.3

A steam turbine with an internal efficiency of 90 percent receives steam at 7.0 MPa and 550°C and exhausts as 20 kPa. Determine the turbine work, exhaust enthalpy, and exit quality of the steam.

$h_2 = 3530.9 \text{ kJ/kg} \qquad s_3 = s_2$

$s_2 = 6.9486 \text{ kJ/kg} \cdot \text{K} \qquad h_3 = 2288.9 \text{ kJ/kg}$

$\eta_t = 0.90 = \dfrac{h_2 - h_{3'}}{(h_2 - h_3)_s}$

$h_{3'} = 2413.1 \text{ kJ/kg}$

$h_{3'} = h_f + x h_{fg} \qquad 2413.1 = 251.4 + x(2358.3)$

$x = 0.916$ or 91.6 percent

$w_t = h_2 - h_{3'} = 1117.8 \text{ kJ/kg}$

The internal efficiency for energy-consuming devices always seems difficult for students to understand, but considering the internal efficiency from the standpoint of available energy should help. The internal efficiency of a pump, η_p, is the ratio of the increase in the available energy of the fluid in passing through the pump to the available energy used to achieve the effect.

$$\eta_p = \frac{\dot{m}(h_1 - h_4)_s}{\dot{W}_p} \qquad (11.15)$$

The internal efficiency for a pump may also be viewed as the cost, the actual pump work, in moving the water to a higher energy state as indicated by $(h_1 - h_4)_s$.

11.5 REGENERATIVE CYCLES

Thus far we have learned how power is produced in a multiphase system and have seen some of the performance factors that must be considered, superheating and internal efficiencies. Let us consider ways in which the thermal efficiency of the cycle may be improved.

As we think about improving the efficiency, we might say that if there were a way to preheat the water before it entered the steam generator, then not as much energy would be added in the boiler. However, this would have to be done so that energy was not added from the surroundings, either. Another problem is the loss of energy in the condenser. If some of the steam were taken from the turbine before it reached the condenser and used to heat the feedwater, then two purposes would be accomplished: one, the preheating would occur with no extra energy input, and two, the latent heat of vaporization would not be lost

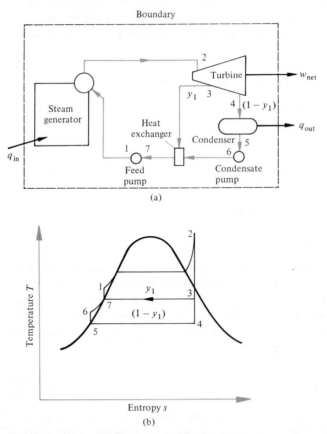

Figure 11.8 *(a) The schematic diagram for a regenerative cycle with one stage of regenerative heating. (b) The T–s diagram for the regenerative cycle with one stage of feed heating.*

from the system in the condenser. A steam cycle using this type of feedwater heating is called a "regenerative cycle." Figure 11.8(a) illustrates such a cycle, and Figure 11.8(b) shows the T–s diagram for the cycle. Note that the only thermal-energy flow across the system boundary occurs in the steam generator and in the condenser. The feedwater heater is within the system boundary, so its effect is only on the heat added and the turbine work.

There are some assumptions implicit in the previous development for regenerative heaters. The first is that all the water leaving the feed heater leaves at the saturated-liquid enthalpy for the pressure of the entering steam and the heater is a direct-contact type. One method that is used in the case of shell and tube heat exchangers is to send the drains back to the previous heat exchanger, in this case, the condenser. Also, a separate drain pump may be employed. However, such detail clouds the issue at this stage in thermodynamic analysis.

When calculating power plant problems, it is easiest to perform the operation per unit mass and, at the conclusion of the problem, incorporate the actual mass flow rates. In Figures 11.8(a) and 11.8(b) the symbol y_1 is the fraction of the total mass in the system, which leaves the turbine to go to the feedwater heater.

$$y_1 = \frac{\text{mass of steam to feed heater}}{\text{total system mass}} \qquad (11.16)$$

How may this value be calculated? We perform an energy balance on the feedwater heater:

First Law, Open System

Energy In = Energy Out

$$y_1 h_3 + (1 - y_1) h_6 = h_7$$

$$y_1 = \frac{h_7 - h_6}{h_3 - h_6} \qquad (11.17)$$

When the turbine work and pump work are calculated, attention must be paid to the fraction of fluid passing through each device. The turbine work per kilogram in the system is

$$w_t = (h_2 - h_3) + (1 - y_1)(h_3 - h_4) \text{ kJ/kg} \qquad (11.18)$$

The pump work per kilogram in the system will have to be modified by the fraction of the total mass that passed through it. There are two pumps, a condensate pump and a feed pump. Sometimes a booster pump is located before the feed pump, but typically two pumps characterized the basic cycle:

$$w_{p1} = (h_6 - h_5)(1 - y_1) \qquad w_{p2} = h_1 - h_7$$

$$w_p = w_{p1} + w_{p2} \qquad (11.19)$$

This provides us with the tools necessary to calculate the cycle efficiency, but where is the heater situated?

Optimum Heater Location

Let us consider two cases, one with a heater using steam before it reaches the turbine, the other using steam after it left the turbine. What is true about both cases? Neither improves the cycle efficiency. In the latter case the temperature of the exhaust steam is that of the liquid leaving the condenser, and no heat transfer can occur. In the former case the heat exchanger acts only as a medium between the steam generator and the feedwater and no gain in efficiency results.

Somewhere between the turbine inlet and exit lies the optimum point for a feed heater. The determination of the location is a trial-and-error procedure. It is sufficient at present to realize that the optimum point occurs after the steam has been used for some work and before it exhausts from the turbine. For one feedwater heater, the optimum point is where the feedwater exit temperature from the heater is halfway between the saturated-steam temperature in the boiler and the condenser temperature.

Example 11.4

Calculate the power plant efficiency and the net work for a steam power plant that has turbine inlet steam conditions of 6.0 MPa and 500°C, bleed steam to the feedwater heater occurring at 800 kPa, and exhaust to the condenser at 15 kPa. The turbine and pump have

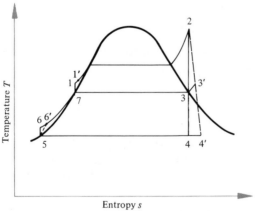

Figure 11.9 *The T–s diagram for one state of regenerative heating where the turbine expansion is adiabatic. This demonstrates the T–s diagram for Example 11.4.*

internal efficiencies of 90 percent. The flow rate is 63.0 kg/s. Compare this to a power plant that does not have regenerative feedwater heating. Figure 11.9 shows the T–s diagram for this situation. Figure 11.8(a) illustrates the schematic diagram.

Determine the enthalpies at states 1 through 7.

$$h_2 = 3422.2 \text{ kJ/kg} \qquad h_{3s} = 2870.2 \text{ kJ/kg}$$
$$h_{4s} = 2229.9 \text{ kJ/kg} \qquad h_5 = 225.94 \text{ kJ/kg}$$
$$h_7 = 721.11 \text{ kJ/kg}$$

$$0.90 = \frac{h_2 - h_{3'}}{h_2 - h_3} \qquad 0.90 = \frac{h_2 - h_{4'}}{h_2 - h_4}$$

$$h_{3'} = 2925.4 \text{ kJ/kg} \qquad h_4 = 2349.1 \text{ kJ/kg}$$

$$h_6 = h_5 + \int_5^6 v\, dp = 225.94 + (0.001\,014)(800 - 15)$$

$$h_6 = 226.7 \text{ kJ/kg}$$

$$h_1 = h_7 + \int_7^1 v\, dp = 721.11 + (0.001\,115)(6000 - 800)$$

$$h_1 = 726.9 \text{ kJ/kg}$$

Apply first-law analysis to the heater, finding y_1.

$$y_1 = \frac{h_7 - h_{6'}}{h_{3'} - h_{6'}} = 0.183 \qquad (1 - y_1) = 0.817$$

$$w_t = (h_2 - h_{3'}) + (1 - y_1)(h_{3'} - h_{4'}) = 967.6 \text{ kJ/kg}$$

$$0.90 = \frac{h_6 - h_5}{h_{6'} - h_5} \qquad h_{6'} = 226.78 \text{ kJ/kg}$$

$$0.90 = \frac{h_1 - h_7}{h_{1'} - h_7} \qquad h_{1'} = 727.5 \text{ kJ/kg}$$

Calculate the total pump work.

$$w_{p1} = (h_{6'} - h_5)(1 - y_1) = 0.68 \text{ kJ/kg}$$
$$w_{p2} = h_{1'} - h_7 = 6.39 \text{ kJ/kg}$$
$$w_p = w_{p1} + w_{p2} = 7.07 \text{ kJ/kg}$$
$$w_{net} = \Sigma w = w_t - w_p = 960.5 \text{ kJ/kg}$$
$$q_{in} = h_2 - h_{1'} = 2694.7 \text{ kJ/kg}$$
$$\eta_{th} = \frac{w_{net}}{q_{in}} = 0.356 \text{ or } 35.6 \text{ percent}$$
$$\dot{W} = \dot{m}w_{net} = (63.0)(960.5) = 60\,511 \text{ kW}$$

For the nonregenerative cycle with 90 percent turbine and pump efficiencies:

$$w_t = h_2 - h_{4'} = 1073.1 \text{ kJ/kg}$$

$$w_p = \int_{p_5}^{p_1} \frac{v_5 \, dp}{\eta_p} = 6.7 \text{ kJ/kg}$$

$$w_{net} = \Sigma w = 1066.4 \text{ kJ/kg}$$

$$q_{in} = h_2 - (h_5 + w_p) = 3189.5 \text{ kJ/kg}$$

$$\eta_{th} = \frac{w_{net}}{q_{in}} = 0.334 \text{ or } 33.4 \text{ percent}$$

The use of the heater improved the basic cycle efficiency by 6.6 percent.

As we have seen, the overall efficiency of the power plant is improved when a feedwater heater is used. If one feedwater heater is good, maybe two are better, and so on. This is true, but only to a point. The efficiency does improve as more feedwater heaters are used, but the gain in efficiency is offset by the increase in capital cost and maintenance. Usually small plants, such as those on ships, have two regenerative heaters, and large stationary plants have six. The location of multiple feed heaters is beyond the scope of the chapter, but, again, it involves trial and error. Ideally the feedwater temperature increases in equal increments across each feed heater located between the condenser and the steam generator. The saturated steam temperature in the boiler and the condenser temperature are the two temperature limits used.

Let us consider a power plant with three feedwater heaters. We will develop expressions to calculate the thermal efficiency for this power plant. Figure 11.10(a) is a schematic drawing for the plant, and Figure 11.10(b) is a T–s diagram of the steam paths. Again the symbols y_1, y_2, and y_3 are the fractions of steam withdrawn at each feed heating state. The drain pumps are included in the schematic, but the assumption that all liquid leaving the feed heater leaves as a saturated liquid at the bleed steam pressure is invoked. The total pumping work is considered to occur with the feed pump and the condensate pump. Ideally the work for each drain pump should be calculated, but the error caused by this assumption is very small. Even if the pumping work was in error by 25 percent, the magnitude of this term, approximately 2.5 kJ/kg, is very small compared with the turbine work of approximately 1000 kJ/kg.

The turbine work per kilogram in the system is

$$w_t = (h_2 - h_3) + (1 - y_1)(h_3 - h_4) + (1 - y_1 - y_2)(h_4 - h_5)$$
$$+ (1 - y_1 - y_2 - y_3)(h_5 - h_6) \tag{11.20}$$

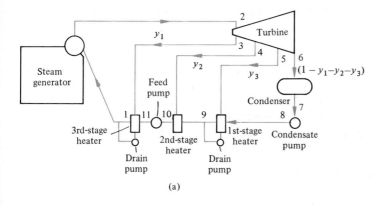

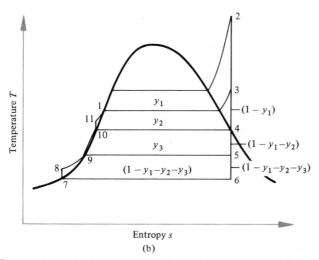

Figure 11.10   (a) The schematic diagram for three stages of regenerative heating.
(b) The T–s diagram for a three-stage regenerative cycle.

The system pump work is found by adding the condensate and feed pump works. To obtain a better approximation for the actual pump work, but not including the drain pumps, let us include the mass fraction of fluid passing through the drain pump in the condensate and feed pumps, respectively. Thus

$$w_{p_1} = v_7(p_{10} - p_7)(1 - y_1 - y_2)$$

$$w_{p_2} = v_{10}(p_1 - p_{10})(1)$$

$$w_p = w_{p_1} + w_{p_2} \tag{11.21}$$

The heat supplied per kilogram is

$$q_{in} = h_2 - h_1 \tag{11.22}$$

and the overall thermal efficiency, from Equation (11.13), is

$$\eta_{th} = \frac{w_{net}}{q_{in}}$$

To solve for the thermal efficiency, we must find the mass flow to each heater. This is found by performing an energy balance on each heater. The easiest method of solution for hand calculations is to take each heater in order and solve for the mass flow, knowing the mass flow from the previous heater. We make an energy balance on the third-state heater:

Open System, First Law

Energy In = Energy Out

$$y_1 h_3 + (1 - y_1)h_{11} = h_1$$

$$y_1 = \frac{h_1 - h_{11}}{h_3 - h_{11}} \tag{11.23}$$

An energy balance on the second-stage heater yields

Open System, First Law

Energy In = Energy Out

$$y_2 h_4 + (1 - y_1 - y_2)h_9 = (1 - y_1)h_{10}$$

$$y_2 = \frac{(1 - y_1)(h_{10} - h_9)}{h_4 - h_9} \tag{11.24}$$

Since y_1 is known from Equation (11.23), y_2 may be calculated. We perform an energy balance on the first-stage heater as follows:

Open System, First Law

Energy In = Energy Out

$$y_3 h_5 + (1 - y_1 - y_2 - y_3)h_8 = (1 - y_1 - y_2)h_9$$

$$m_3 = \frac{(1 - y_1 - y_2)(h_9 - h_8)}{h_5 - h_8} \tag{11.25}$$

Thus all the fractional flow rates may be calculated and the turbine work calculated from Equation (11.20). The reader may ask why the equations for the mass fraction balance are valid physically? What is happening in the heater to allow us to use enthalpies of the fluids entering and leaving? The entering liquid is heated by the condensing steam until it reaches the same temperature as the steam. There is a steam trap, similar to those on steam radiators, which prevents the steam from leaving the heat exchanger until it is a liquid. Thus all the liquid leaving the heat exchanger is at the same temperature. This heat exchanger is an ideal one; actually the water leaves somewhat cooler than the condensed steam temperature.

Example 11.5

A steam power plant runs on a three-state regenerative cycle. The steam enters the turbine at 7.0 MPa and 550°C and fractions are withdrawn at 2.0 MPa, 600 kPa, and 100 kPa for feedwater heating. The remaining steam exhausts at 7.5 kPa to the condenser. Compute the overall thermal efficiency of the power plant.

Calculate the enthalpies for the states noted in Figure 11.10(a).

$h_2 = 3530.9$ $\quad$ $h_3 = 3132.4$ $\quad$ $h_4 = 2841.7$

$h_5 = 2522.3$ $\quad$ $h_6 = 2166.4$ $\quad$ $h_7 = 168.79$

$h_9 = 417.46$ $\quad$ $h_{10} = 670.56$

$h_1 = 908.79$

$h_8 = h_7 + v_7 \Delta p = 168.79 + (0.001\,008)(600 - 7.5) = 169.39$

$h_{11} = h_{10} + v_{10} \Delta p = 670.56 + (0.001\,100)(7000 - 600) = 677.6$

From first-law analysis of the high-pressure heater,

$$y_1 = \frac{h_1 - h_{11}}{h_3 - h_{11}} = 0.0941 \qquad 1 - y_1 = 0.9059$$

Similarly,

$$y_2 = \frac{(1 - y_1)(h_{10} - h_9)}{h_4 - h_9} = 0.0946$$

and

$$y_3 = \frac{(1 - y_1 - y_2)(h_9 - h_8)}{h_5 - h_8} = 0.0855$$

The turbine work is

$$w_t = (h_2 - h_3) + (1 - y_1)(h_3 - h_4) + (1 - y_1 - y_2)(h_4 - h_5)$$
$$+ (1 - y_1 - y_2 - y_3)(h_5 - h_6)$$

$w_t = 1179.3 \text{ kJ/kg}$

The pump works are

$$w_{p_1} = (h_8 - h_7)(1 - y_1 - y_2) = 0.48 \text{ kJ/kg}$$

where the drain pump work is included in w_{p_1} by including the mass fraction y_3, and

$$w_{p_2} = (h_{11} - h_{10})(1) = 7.04 \text{ kJ/kg}$$

where the drain pump work is included in w_{p_2} by including the mass fraction y_1.

$$w_p = w_{p_1} + w_{p_2} = 7.52 \text{ kJ/kg}$$

$$w_{net} = \Sigma w = 1171.78 \text{ kJ/kg}$$

$$q_{in} = h_2 - h_1 = 2622.1 \text{ kJ/kg}$$

$$\eta_{th} = \frac{w_{net}}{q_{in}} = 0.446 \text{ or } 44.6 \text{ percent}$$

11.6 REHEAT CYCLES

The power plant thermal efficiency has been improved by using feedwater heating, but is that all that can be done? No. The steam may be reheated to a high temperature after it has partially expanded through the turbine. A significant portion of the work by the steam is accomplished when the pressure is such that the steam is saturated or nearly saturated. This is the correct place for the vapor to be resuperheated. The steam reenters the turbine and expands to condenser pressure. Figures 11.11(a) and 11.11(b) illustrate the schematic and T–s diagrams for the reheat cycle. The steam expands through the turbine until state 3 is reached, then is removed and reheated at constant pressure to state 4. The temperature of the steam at state 4 is usually within 25°C of the temperature at state 2. The steam reenters the turbine at state 4 and expands to the condenser pressure at state 5. The steam is condensed and pumped back to the steam generator, completing the cycle.

To calculate the thermal efficiency for the reheat cycle, the work and heat added terms must be found. An energy balance on the turbine yields

$$\dot{W}_t = \dot{m}[(h_2 - h_3) + (h_4 - h_5)] \tag{11.26}$$

Equation (11.26) does not need an isentropic expansion but is valid for any turbine efficiency. The pump work is as previously calculated in Equation (11.10a); here,

$$\dot{W}_p = \dot{m} \int_6^1 v \, dp$$

The energy added in the cycle occurs in the steam generator and may be found by an energy balance on the generator:

$$\dot{Q}_{in} = \dot{m}[(h_2 - h_1) + (h_4 - h_3)] \tag{11.27}$$

The thermal efficiency is then

$$\eta_{th} = \frac{[(h_2 - h_3) + (h_4 - h_5)] - \int_6^1 v \, dp}{(h_2 - h_1) + (h_4 - h_3)} \tag{11.28}$$

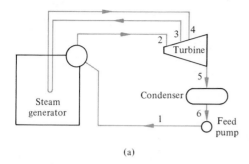

(a)

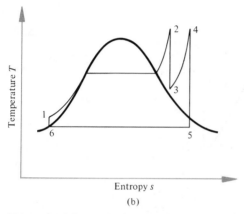

Entropy s

(b)

Figure 11.11   (a) A schematic diagram illustrating one stage of reheating. (b) The
T–s diagram for one stage of reheating.

11.7 REHEAT–REGENERATIVE CYCLE

The reader might deduce that if the reheat and the regenerative cycles
were combined, the overall efficiency would be further improved. The
regenerative–reheat cycle usually has one reheat cycle and two or more
stages of regenerative heating. The power plant has to be sufficiently
large so the increased cost for the reheat piping and maintenance will be
paid for by the improved thermal efficiency. Figures 11.12(a) and 11.12(b)
illustrate a reheat–regenerative cycle with one stage of reheat and two
stages of regenerative heating. The first reheat stage will occur at the
same pressure as the first regenerative feed heater station. The extraction
pressure is determined by optimization of the plant overall efficiency as a
function of the extraction pressure. Results indicate this pressure is 16 to
22 percent of the turbine inlet pressure for a one-stage reheat–regenerative
cycle. The turbine work is

$$w_t = (h_2 - h_3) + (1 - y_1)(h_4 - h_5) + (1 - y_1 - y_2)(h_5 - h_5)$$

$$(11.29)$$

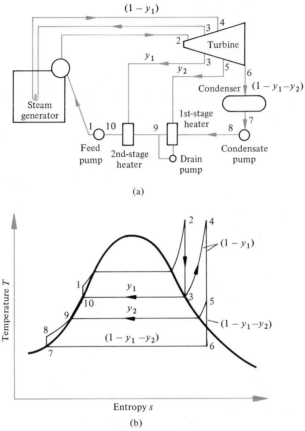

Figure 11.12 (a) The schematic diagram for a regenerative–reheat cycle with two regenerative stages and one reheat stage. (b) The T–s diagram for the two-stage regenerative, one-stage reheat cycle.

The heat added to the steam is

$$q_{in} = (h_2 - h_1) + (1 - y_1)(h_4 - h_3) \qquad (11.30)$$

Before calculating the thermal efficiency, we perform an energy balance on the feed heater to obtain the fraction of steam withdrawn at each state:

$$y_1 = \frac{h_{10} - h_9}{h_3 - h_9} \qquad (11.31)$$

$$y_2 = \frac{(1 - y_1)(h_9 - h_8)}{h_5 - h_8} \qquad (11.32)$$

The thermal efficiency for the cycle may now be calculated as the heat added, and the work produced can be calculated. The following example illustrates the regenerative–reheat cycle analysis.

Example 11.6

A steam power plant operates on a regenerative–reheat cycle with two stages of steam extraction for regenerative heating and one extraction for reheating. The turbine receives steam at 7.0 MPa and 550°C; the steam expands to 2.0 MPa isentropically, where a fraction is extracted for feedwater heating, and the remainder is reheated at constant pressure until the temperature is 540°C. The steam expands isentropically to 400 kPa, where a fraction is withdrawn for low-pressure feedwater heating; the remainder expands isentropically in the turbine to 7.5 kPa, at which time it enters the condenser. Calculate the overall thermal efficiency. The feed pump is located after the second-stage heater.

Calculate the enthalpies for the state points shown in Figure 11.12(a).

$h_2 = 3530.9$ $h_3 = 3132.4$ $h_4 = 3556.1$

$h_5 = 3053.8$ $h_6 = 2352.8$ $h_7 = 168.79$

$h_9 = 604.74$ $h_{10} = 908.79$

$h_8 = h_7 + v_7 \Delta p = 170.79 \text{ kJ/kg}$

$h_1 = h_{10} + v_{10} \Delta p = 914.64 \text{ kJ/kg}$

$$y_1 = \frac{h_{10} - h_9}{h_3 - h_9} = 0.12 \qquad (1 - y_1) = 0.88$$

$$y_2 = \frac{(1 - y_1)(h_9 - h_8)}{h_5 - h_8} = 0.132$$

$w_t = (h_2 - h_3) + (1 - y_1)(h_4 - h_5) + (1 - y_1 - y_2)(h_5 - h_6)$

$w_t = 1364.8 \text{ kJ/kg}$

$w_{p_1} = (h_8 - h_7)(1 - y_1) = 1.76 \text{ kJ/kg}$

$w_{p_2} = (h_1 - h_{10})(1) = 5.85 \text{ kJ/kg}$

$w_p = w_{p_1} + w_{p_2} = 7.61 \text{ kJ/kg}$

$w_{net} = \Sigma w = 1357.2 \text{ kJ/kg}$

$q_{in} = (h_2 - h_1) + (1 - y_1)(h_4 - h_3) = 2989.1 \text{ kJ/kg}$

$$\eta_{th} = \frac{w_{net}}{q_{in}} = 0.454 \text{ or } 45.4 \text{ percent}$$

As you can see from the value of the overall thermal efficiency in the preceding example, the inclusion of reheat and regeneration in the cycle design has significant beneficial results. Less fuel is needed to produce the same power.

11.8 SUPERCRITICAL AND BINARY VAPOR CYCLES

There have been other types of steam power plant cycles developed by industry; supercritical cycles and binary systems are two such examples. We will discuss, but not analyze at this time, these cycles and discover why they are, or were, used.

We have seen that the more reversibly processes occur within a cycle, the greater will be the cycle efficiency. To transfer heat reversibly, the heat should be transferred between two systems at the same temperature, or as near to the same temperature as possible. The temperature inside a furnace is around 1370°C. If heat at this temperature is used to evaporate water at 7.0 MPa, which has a saturated temperature of 286°C, considerable irreversibilities occur in the heat transfer process. One way to overcome this handicap is to have the steam generator operate at a pressure greater than the critical pressure of water. The metallurgical limit of the steam generating tubes is around 600°C. In the supercritical steam generator the heat transfer occurs at constant pressure, but the inefficiency due to exchanging a substantial portion of the heat at a constant low temperature, as occurs in evaporation, is eliminated. The *T–s* diagrams for the supercritical plant and for 7.0 MPa, heated to the same temperature, are shown in Figure 11.13.

Another method that was explored was the use of a fluid other than water, one that evaporated at a temperature near the metallurgical limit. If this fluid could expand to the normal condenser temperature, around 32°C, then the efficiency of the cycle would be improved. One fluid that meets the upper temperature limit is mercury. However, at 32°C the

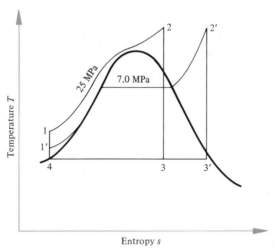

Figure 11.13   A T–s diagram showing a subcritical and supercritical Rankine cycle.

vapor pressure of mercury is on the order of 0.07 Pa. It is not possible to operate at this vacuum in power plants because of air-leakage complications. At a pressure of 7.0 kPa the temperature of mercury is 237°C. If steam is generated by the latent heat released by the condensing mercury, it could expand to the normal condenser pressures. Such a power plant is shown schematically on Figure 11.14(a), and Figure 11.14(b) shows the *T–s* diagram. The latent heat of mercury at 7.0 kPa is about 295 kJ/kg, and for water at 230°C it is 1813.8 kJ/kg. Thus there must be 6.1 kg of mercury per kilogram of water.

There are disadvantages to both the aforementioned cycles. The supercritical cycle demands piping and related equipment able to with-

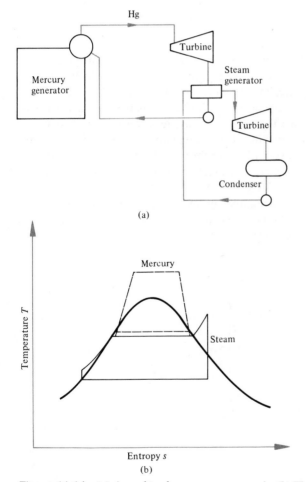

(a)

(b)

Figure 11.14   *(a) A combined mercury–steam cycle. (b) The T–s diagram for the binary mercury–steam cycle.*

stand high pressures, causing structural problems. Mercury's problem is that it is toxic; there are also heat-transfer problems and related equipment complications. The use of mercury in power plants operating on a binary system has not gone beyond a very limited effort made in the mid-1950s. If a fluid that does not pose the difficulties of mercury should be developed, then the binary vapor power cycle would deserve attention. Very high efficiencies are possible, and the capital cost of such plants may be offset by high fuel costs.

At this point you should be able to synthesize Chapters 9 and 11 and perform an energy analysis of a basic steam power plant. The following example illustrates this for isentropic expansions and compressions. The problems will include the use of internal efficiencies.

Example 11.7

A steam power plant operates on a regenerative–reheat cycle where there is one regenerative stage and one reheat stage, which occur at the same pressure. The steam enters the turbine at 7.0 MPa and 550°C, expands isentropically to 2.0 MPa, where steam is extracted for regenerative heating and the remainder is reheated to 540°C. Steam reenters the turbine at this state and expands to 15 kPa. The temperature of the cooling water through the condenser rises from 24°C to 35°C in passing through the condenser. The turbine produces 30 000 kW. Determine the steam flow rate, cooling-water flow rate, change in available energy in the condenser, and the entropy production in the condenser. Figure 11.15 illustrates the T–s diagram for the example.

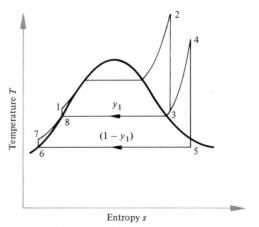

Figure 11.15 *The T–s diagram for Example 11.7.*

Calculate the enthalpies for the cycle state points,

$h_2 = 3530.9 \qquad h_3 = 3132.4 \qquad h_4 = 3556.1$

$h_5 = 2446.9 \qquad h_6 = 225.94 \qquad h_8 = 908.79$

$h_7 = h_6 + v_6 \, \Delta p = 227.95 \text{ kJ/kg}$

$h_1 = h_8 + v_8 \, \Delta p = 914.64 \text{ kJ/kg}$

From a first-law analysis of the heater,

$$y_1 = \frac{h_8 - h_7}{h_3 - h_7} = 0.236 \quad 1 - y_1 = 0.764$$

$w_t = (h_2 - h_3) + (1 - y_1)(h_4 - h_5) = 1245.9 \text{ kJ/kg}$

$w_{p_1} = (1 - y_1)(h_7 - h_6) = 1.53 \text{ kJ/kg}$

$w_{p_2} = h_1 - h_8 = 5.85 \text{ kJ/kg}$

$w_p = w_{p_1} + w_{p_2} = 7.38 \text{ kJ/kg}$

$w_{net} = \Sigma w = 1238.5 \text{ kJ/kg}$

$\dot{W} = \dot{m} w_{net}$

$$\dot{m} = \frac{30\,000}{1238.5} = 24.22 \text{ kg/s}$$

First-law analysis of the condenser yields the following:

$$\dot{m}_{CW}(146.68 - 100.7) = \dot{m}_s(h_5 - h_6)(1 - y_1)$$

$$\dot{m}_{CW} = 893.8 \text{ kg/s}$$

where $\dot{m}_{CW}$ is the cooling water flow rate.

Available energy change of the steam in the condenser is as follows:

$(\text{A.E.})_{5-6} = Q_{5-6} - T_0 \, \Delta S_{5-6}$

$(\text{A.E.})_{5-6} = [\dot{m}_s(h_6 - h_5) - (297)\dot{m}_s(s_6 - s_5)](1 - y_1)$

$(\text{A.E.})_{5-6} = -3789.2 \text{ kW}$

Available energy change of the cooling water in the condenser is as follows:

$(\text{A.E.})_{i-0} = Q_{i-0} - T_0 \Delta S_{i-0}$

$(\text{A.E.})_{i-0} = +41\,096.8 - (297)(893.8)(0.5053 - 0.3534)$

$(\text{A.E.})_{i-0} = +773.6 \text{ kW}$

$(\text{A.E.})_{net} = \Sigma \text{A.E.} = -3015.6 \text{ kW}$

$$\Delta S_{net} = \frac{(\text{U.E.})_{net}}{T_0} = -\frac{(\text{A.E.})_{net}}{T_0} = 10.15 \text{ kW/K}$$

11.9 STEAM-TURBINE REHEAT FACTOR
AND CONDITION CURVE

In the preceding sections we analyzed various vapor power cycles. These cycles utilize a turbine to produce power, and this section analyzes, in more detail, the energy flow through the turbine. Although this analysis uses a steam turbine as the model, the analysis is valid for other turbines, such as a gas turbine. In the case of the gas turbine, the fluid is a mixture of air and oxidized hydrocarbons.

Let us consider an impulse turbine operating in the Rankine cycle. Of the total energy reaching the turbine, only ΔH_s is available for work. The term ΔH_s is the isentropic enthalpy drop between the inlet and exit pressures. Not all the available energy can be used; the following losses can and do occur:

1. Leakage of steam at shaft packings
2. Radiation losses to surroundings
3. Kinetic energy loss to condenser
4. Reheat caused by irreversibilities in fluid flow

Let us consider the turbine as the sum of individual stages. The steam enters the first stage and expands to pressure p_1; the available energy at this pressure is $(\Delta h_s)_1$. Of this available energy, a portion e_1, is turned into mechanical work. The remainder is reheat, Rh_1, rejected to the next stage. This reheat may be broken into two parts: q_{r1}, due to irreversibilities in the fluid flow, friction in the nozzles and blading, and leakage; and $(K.E.)_1$, the exit kinetic energy from the first stage. Thus

$$(\Delta h_s)_1 - e_1 = Rh_1 = q_{r1} + (K.E.)_1 \tag{11.33}$$

The stage efficiency, η_{st}, indicates how well the available energy is converted into mechanical work.

$$(\eta_{st})_1 = \frac{e_1}{(\Delta h_s)_1} \tag{11.34}$$

The reheat from the first stage is passed on to the second stage, where $(\Delta h_s)_2$ is available energy, and e_2 is used as mechanical work. Eventually the reheat from the last stage is passed on to the condenser. The sum of the individual drops in enthalpy will be greater than ΔH_s because one stage's reheat is added to energy entering the successive stage. Thus

$$\sum_i (\Delta h_s)_i = R_f \Delta H_s \tag{11.35}$$

where R_f is the reheat factor, a constant greater than 1 and usually $1.0 \leqslant R_f \leqslant 1.065$. If the stage efficiencies are the same for each stage, which, in general, they are not, then the total mechanical work, E, may be expressed as

$$E = \sum_i e_i = \sum_i \eta_{st}(\Delta h_s)_i = \eta_{st} R_f \Delta H_s \qquad (11.36)$$

The stage efficiency will decrease in stages where moisture is present in the steam. The following example will illustrate several of these turbine concepts.

Example 11.8

Steam enters a six-stage impulse turbine at 3.5 MPa and 450°C. The stage efficiency is 80 percent for all stages, and the isentropic enthalpy drop per stage is 180 kJ/kg. Calculate the reheat factor, the end point. Denote the pressure in each stage. Figure 11.16 illustrates the $h-s$ diagram for the example.

The entering enthalpy, h_0, is 3337.2 kJ/kg and the pressure is 3.5 MPa. In the first stage, 180 kJ/kg of energy is available, so proceed vertically (isentropically) down the $h-s$ diagram until an enthalpy value of 3157 (3337.2 − 180) is reached. The pressure is 2.0 MPa. This would be the enthalpy of the steam entering the next stage if all available energy were used. It is not, however; only 80 percent is used, the remainder, 36 kJ/kg, is reheat passed on to the next stage. The reheat is added at constant pressure ($p = 2.0$ MPa) until a value of 3193 (3157 + 36) is reached. This is the condition of the steam entering the second stage.

The process repeats itself, with the following values denoting the pressure and enthalpy entering such stage:

Stage	Enthalpy	Pressure
1	3337.2	3.5 MPa
2	3193	2.0 MPa
3	3049	1.0 MPa
4	2905	460 kPa
5	2761	180 kPa
6	2617	62 kPa
Condenser	2473	18 kPa

The reheat factor, R_f, is defined as

$$R_f = \sum_i \frac{(\Delta h_s)_i}{\Delta H_s}$$

$$\Delta H_s = 3337.2 - 2294.6 = 1042.6 \text{ kJ/kg}$$

$$R_f = \frac{(6)(180)}{1042.6} = 1.036$$

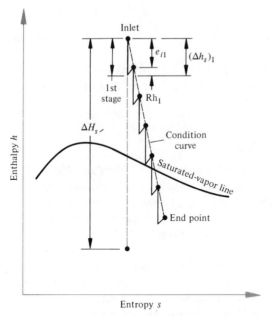

Figure 11.16 The h–s diagram for Example 11.8.

The end point denotes the condition of the steam as it enters the condenser. From the *h–s* diagram, we find that the steam has a quality of 94.2 percent, as well as a pressure of 18 kPa. The condition curve is a line joining the steam states entering the various stages.

11.10 GEOTHERMAL ENERGY

Thermal energy stored below the earth's surface is called geothermal energy. Volcanic activity within the last 3 million years has brought molten rock, magma, to within 8 to 16 km of the earth's surface. Nearby surface rocks contain water in fractures, which when heated by the cooling magma, increases in pressure. At times this heated water erupts as geysers or it may appear as hot springs. In other cases drilling may be necessary to bring it to the surface. The temperature of the water deposits range from 15°C to 300°C, with dissolved solids ranging from 0.1 to 25 percent.

Figure 11.17 illustrates a model of a geothermal system. Surface runoff at state 1 percolates through the pervious rock. The water is heated by the convecting magma and tries to move upward through the rock of low permeability. Typically, the water cannot reach the surface and a well is drilled, which allows the heated water at state 2 to flow to the surface. As the water rises, the pressure drops and flashing occurs (state 3), and a steam–water mixture exits the well at state 4.

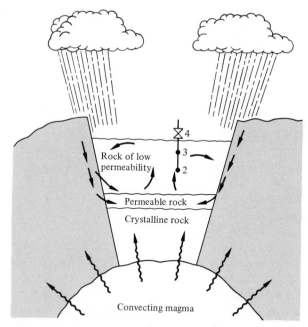

Figure 11.17   Hot-water geothermal system.

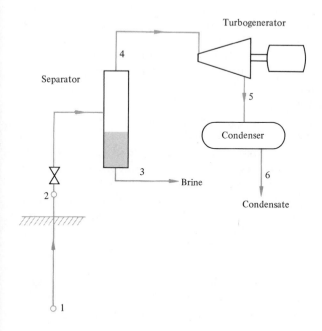

Figure 11.18   Flashed steam system.

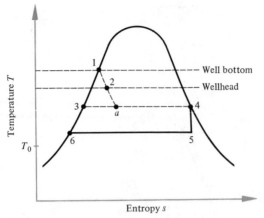

Figure 11.19   A T–s diagram for the flashed steam system.

The flashed steam system, illustrated schematically in Figure 11.18, is used in several countries. The steam–water mixture enters a separator, where the steam flashes and enters a turbine. The brine is discarded. This may not be an easy task as the brine contains minerals and salts harmful to the surface ecology. The steam is condensed, states 5–6. The condensed steam may be recombined with the hot brine and injected into the ground, or it may be stored in a pond and used as cooling water. Figure 11.19 is the *T–s* diagram that accompanies Figure 11.18. The process 1–2-*a* is a throttling process.

There are plans to develop binary cycles, where the hot water is pumped under pressure from the bottom of the well through a heat exchanger. The cooling fluid in the heat exchanger will be a refrigerant, such as R 12, which will operate on a Rankine cycle.

The major areas of difficulty in using geothermal energy are: one, the relatively low pressures and temperatures of the wellhead water (about 700 kPa); two, the extremely corrosive nature of the water; three, the geologically unstable area where plants must be located. However, no new technology is needed to build the plants and the projected cost per kilowatt is significantly less than coal and nuclear plants. As sites for geothermal plants are discovered, more will certainly be built.

PROBLEMS

1. A Carnot cycle uses steam as the working substance and operates between pressures of 7.0 MPa and 7 kPa. Determine (a) the thermal efficiency; (b) the turbine work per kilogram; (c) the compressor work per kilogram.

2. A condenser receives steam from the turbine at 15 kPa, 90 percent quality, and with a velocity of 240 m/s. Determine the increase in circulating water required because of the high exit velocity if the steam flow rate is 38 kg/s and the cooling-water temperature rise is 12°C with an inlet condition of 30°C.

3. With a maximum steam temperature of 560°C, compare the thermal efficiencies of Rankine and Carnot cycles operating with a maximum pressure of (a) 3.5 MPa; (b) 7.0 MPa. The lowest cycle temperature is 30°C.

4. A steam power plant operates on the Rankine cycle, with steam leaving the boiler at 6.0 MPa and 500°C. The plant supervisor wants to know the effect of back pressure (condensing pressure) on the cycle efficiency. (a) Determine and graph the cycle efficiency for back pressures of 7, 15, 30, 45, and 70 kPa. (b) Discuss the results from the available-energy viewpoint.

5. A feed pump in a supercritical power plant raises the pressure of water from 1400 kPa and 175°C to 24.5 MPa isentropically. Determine the pump work using the tables and Equation (11.10). Assume a linear variation of specific volume and recalculate the pump work.

6. A Rankine regenerative–reheat cycle utilizes steam at 8.4 MPa and 560°C entering the high-pressure turbine. The cycle includes one steam-extraction stage for regenerative feedwater heating, the remainder at this point being reheated to 540°C. The condenser temperature is 35°C. Determine (a) the *T–s* diagram for the cycle; (b) optimum extraction pressure; (c) fraction of steam extracted; (d) turbine work, kJ/kg; (e) pump work, kJ/kg; (f) overall thermal efficiency.

7. A two-stage regenerative cycle steam power plant has been selected to power a ship. The requirements are that 37 000 kW be provided. The gears between the turbine (high speed) and shaft (low speed) are 98% efficient. The maximum possible steam pressure and temperature leaving the boiler are 6.0 MPa and 500°C. The condenser has a design vacuum of 722 mm Hg (vacuum). The extraction points are at 150 kPa and 1.4 MPa. Determine (a) the *T–s* diagram; (b) the cycle efficiency; (c) the minimum steam flow rate; (d) the fractions of steam removed for feed heating; (e) the quality of steam entering the condenser.

8. The turbine and pump in problem 7 have internal efficiencies of 90%. Solve Problem 7 with these conditions.

9. A steam power plant produces 1000 MW of electricity while operating on a three-stage regenerative cycle. The steam enters the turbine at 14 MPa and 580°C. Extractions for heating occur at 2.5 MPa, 700

kPa and 150 kPa. The turbine exhausts at 15 kPa and has an internal efficiency of 92 percent. Determine (a) the T–s diagram; (b) the mass flow rate; (c) the heat supplied; (d) the fuel flow rate, if the energy release is 35 000 kJ/kg fuel; (e) the mass fractions y_1, y_2, y_3; (f) the cycle efficiency.

10. A small chemical plant uses steam in its production area. A decision was made to use a reheat cycle with steam entering the turbine at 8.5 MPa and 480°C, being reheated to 440°C at 1.2 MPa and condensing at 7 kPa. Determine (a) the net work per kilogram; (b) what percentage of the total heat supplied is constituted by the reheat; (c) the cycle efficiency; (d) the T–s diagram.

11. Repeat Problem 10 for turbine and pump efficiencies of 92% and 80%, respectively.

12. In the reheat cycle, steam enters the turbine at 12.5 MPa and 560°C. The condenser operates at 7 kPa. Determine the cycle efficiency for the following reheat pressures, if the reheat temperature is always 520°C: (a) 700 kPa; (b) 1.4 MPa (c) 2.1 MPa; (d) 2.8 MPa.

13. A supercritical power plant generates steam at 25 MPa and 580°C. The condenser pressure is 7.0 kPa. Determine the exit quality of the steam if it expands through a turbine in this power plant. Calculate the cycle efficiency.

14. The supercritical power plant in Problem 13 has a reheat stage added at 3.5 MPa with a reheat temperature of 540°C. All other conditions are the same. Determine (a) the quality of steam entering the condenser; (b) the cycle efficiency.

15. A 60 000-kW turbogenerator receives steam at 7.0 MPa and 500°C. There are two steam-extraction stages at 2.0 MPa and 200 kPa. The remainder of the steam at 2.0 MPa is reheated to 480°C. The turbine exhausts at 36 mm Hg (absolute). Determine (a) the T–s diagram; (b) the mass fractions extracted; (c) the cycle efficiency; (d) the mass flow rate of steam entering the turbine.

16. The turbine in Example 11.6 has an internal efficiency of 90% and a pump efficiency of 80%. Recalculate the thermal efficiency. Determine the increased cooling load per kilogram steam in the condenser due to the nonisentropic expansion of steam in the turbine. Calculate the second law efficiency of the turbine.

17. A supercritical power plant produces 800 MW when operating on a regenerative–reheat cycle with three stages of regenerative heating and one state of reheating. The extraction pressures for feed heating are 5.5 MPa, 600 kPa, and 100 kPa. Reheating occurs at 5.5 MPa. The inlet steam condition to the turbine is 30 MPa and 540°C, and the reheat temperature is 520°C. The turbine exhaust pressure is 7 kPa. Determine (a) the T–s diagram; (b) the cycle efficiency; (c) the steam flow rate; (d) the circulating water flow rate if there is a 10°C

temperature rise in the condenser from 20°C; (e) the total power required for pumping.

18. A project engineer wants to determine the turbine internal efficiency of a two-turbine unit operating in a two-stage regenerative steam power plant. He measures the pressure and temperature at the high-pressure turbine inlet and the two regenerative bleed steam stations. The pressures and temperatures are 14 MPa and 520°C, 4.0 MPa and 360°C, and 200 kPa saturated, respectively. The low-pressure turbine receives the nonextracted steam at 4.0 MPa and 360°C. The condenser is at 36°C. Determine (a) the low-pressure and high-pressure turbine internal efficiencies; (b) the quality of steam entering the condenser; (c) the total turbine work per kilogram; (d) the T–s diagram; (e) the cycle second law efficiency.

19. A manufacturing facility uses a steam power plant to generate electricity and provide heating steam to the work spaces. The steam plant operates on the regenerative cycle, with one stage of feed heating. Steam enters the turbine at 5.5 MPa and 500°C, expands to 660 kPa, and extraction for feed heating occurs. The remaining steam expands through the turbine to 150 kPa, where extraction for space heating occurs ($y_2 = y_1$). The remaining steam expands to 15 kPa, where it is condensed. The returns from the space heating enter the condenser as a saturated liquid at the condenser temperature. Determine (a) the cycle efficiency; (b) the T–s diagram; (c) the turbine work per kilogram; (d) for a 50 000-kW load, the mass flow of steam entering the turbine; (e) the heat supplied to the working areas for the conditions in (d).

20. A steam turbine in a Rankine cycle receives steam at 6.0 MPa and 480°C and exhausts at 15 kPa. At reduced loads, the governor valve throttles the steam entering the turbine to 2.2 MPa. Calculate the loss of available energy per kilogram entering the steam turbine.

21. Find the decrease in cycle efficiency for Problem 20.

22. A steam power plant operates on a regenerative–reheat cycle where there is one regenerative stage and one reheat stage that occur at the same pressure. The steam enters the turbine at 8.5 MPa and 540°C and expands isentropically to 2.2 MPa, where steam is extracted for feed heating, the remainder reheated to 500°C. The steam exiting the turbine is at 7 kPa and 240 m/s. There is a 10°C temperature rise in circulating water temperature through the condenser. The water enters the condenser at 20°C. The power plant produces 60 000 kW. Determine (a) the steam flow rate; (b) the cooling-water flow rate; (c) the change of available energy in the condenser; (d) the entropy change of the universe; (e) the T–s diagram.

23. In a binary vapor power cycle, using steam and mercury, saturated mercury vapor enters the mercury turbine at 1250 kPa ($h_g = 361.2$

kJ/kg, $s_g = 0.5024$ kJ/kg · K) and exhausts at 14 kPa ($h_f = 36.0$, $h_{fg} = 294.2$, $s_f = 0.0923$, $s_{fg} = 0.5491$). The condenser–steam generator produces a saturated steam at 3.8 MPa. The steam turbine exhausts at 7 kPa. Determine (a) the mercury turbine work per kilogram mercury; (b) the steam turbine work per kilogram steam; (c) if the steam flow rate is 25.0 kg/s, find the mercury flow rate; (d) the thermal efficiency for the total cycle.

24. An eight-stage impulse turbine receives steam at 7.0 MPa and 550°C. The stage efficiency is 80 percent and the isentropic enthalpy drop per stage is 160 kJ/kg. Determine (a) the exit pressure; (b) the exit quality; (c) the reheat factor; (d) the condition curve; (e) the second law efficiency.

25. The turbine in Problem 24 operates under different steam conditions. The steam enters at 5.5 MPa and 480°C and exhausts at 14 kPa. Determine the turbine work per kilogram.

26. A new power plant is to operate on the Rankine cycle. The following conditions have been established: maximum steam pressure of 9.1 MPa, maximum steam temperature of 460°C, maximum moisture in the exhaust steam of 15%, maximum condenser vacuum of 722 mm Hg (vacuum), and a turbine efficiency of 88%. Determine the optimum inlet temperature and pressure.

27. Steam enters a turbine at 1.4 MPa and 320°C. The turbine internal efficiency is 70%, and the load requirement is 800 kW. The exhaust is to the back pressure system, maintained at 175 kPa. Find the steam flow rate.

28. A turboelectric plant has steam enter the high-pressure turbine at 16.5 MPa and 570°C. The steam expands to 3.5 MPa, where the extraction occurs, and the remainder is reheated to 540°C. The steam enters the low-pressure turbine and exhausts at 14 kPa. Additional regenerative feed heating occurs at 1.2 MPa. The feedwater and condensate leave each heater at the saturated steam temperature. The condensate from the condenser is subcooled 5°C. The turbine internal efficiency is 80% and the generator efficiency is 96%. Determine (a) steam flow rate entering the turbine if the generator load is 25 MW; (b) overall thermal efficiency; (c) heat rejected; (d) heat added; (e) the cycle second law efficiency.

29. An existing steam power plant has the following turbine test data:

    Inlet pressure, 2.1 MPa
    Inlet temperature, 260°C
    Exhaust pressure, 140 kPa
    mechanical efficiency, 95%
    Steam flow, 226.0 kg/s
    Condensate temperature, 70°C
    Turbine power, 1×10^5 kW

The old boiler failed and is to be replaced by a new boiler having exit steam conditions of 4.9 MPa and 320°C. This unit will drive an additional turbine with an internal efficiency of 90%, which exhausts at 2.1 MPa. Part of the exhaust is reheated to the test conditions above, including flow rate, and the remainder heats feedwater to 205°C. Determine (a) total plant power; (b) new total steam flow rate in kilograms per second; (c) old turbine internal efficiency; (d) overall thermal efficiency.

30. A steam turbine carrying a full load of 50.0 MW uses 71.7 kg/s of steam. The turbine efficiency is 75% and it exhausts steam at 25 mm Hg (absolute) and with an enthalpy of 2210 kJ/kg. What is the temperature and pressure of the steam entering the turbine?

31. Steam is admitted to the cylinder of an engine in such a manner that the average pressure is 840 kPa. The diameter of the piston is 25.4 cm and the length of the stroke is 30.5 cm. Determine the work that can be done during one revolution, assuming that the steam is admitted successively to each side (top and bottom) of the piston. What is the power produced when the engine is running at 300 rev/min?

32. A turbine operates in a reheat–regenerative cycle with one reheat stage and two stages of regenerative heating. The heater nearest the condenser is an open type, while the second heater is closed (shell and tube) with its steam drains flowing back to the previous heater. Steam is generated at 4.9 MPa and reaches the turbine at 4.6 MPa and 390°C. Reheating occurs at 760 kPa to a temperature of 370°C. Extractions for feed heating occur at 760 kPa and 100 kPa. The condenser pressure is 7.0 kPa. In the shell and tube type heater the temperature difference between the condensed steam and the feed-water leaving is 5°C. Determine, per kilogram of steam; (a) the T–s diagram; (b) the initial temperature of steam leaving the boiler (the process is adiabatic); (c) the temperature of the feedwater leaving the last heater; (d) the mass fractions of steam extracted for feed heating; (e) the quality of the steam entering the condenser for a turbine efficiency of 80%.

33. A flashed-steam geothermal power plant is located where underground hot water is available as a saturated liquid at 700 kPa. The wellhead pressure is 600 kPa. The flashed steam enters a turbine at 500 kPa and expands to 13 kPa, when it is condensed. The flow rate from the well is 29.6 kg/s. Determine (a) the mass flow rate of steam; (b) the power produced; (c) the cooling water flow if water is available at 30°C and a 10°C rise is allowed through the condenser; (d) the efficiency, where energy input is the available energy of the geothermal water. $T_0 = 27$°C.

34. The same as the previous problem except the turbine has an internal isentropic efficiency of 80%. Additionally, determine the cycle second law efficiency.

35. The earth's temperature has been found to increase 4°C for every 30 m when located over a geothermal reservoir. A well is drilled to a depth of 1000 m. There is a 75-kPa pressure drop from the well to the turbine inlet, including the pressure loss in the separator. The condenser pressure is 14 kPa. Determine (a) the inlet pressure; (b) the work per kilogram of wellhead steam–water mixture; (c) the unit's efficiency (defined above) if $T_0 = 27$°C.

Chapter 12
Refrigeration Systems

Refrigeration in the engineering sense means maintaining a system at a temperature less than the temperature of the surroundings. This will not occur naturally, so a device must be developed that will maintain this condition.

As we saw in dealing with Carnot cycles in Chapter 7, a reversed Carnot engine will remove heat from a low-temperature reservoir and deliver this energy, plus the work necessary to transfer the heat, to a high-temperature reservoir. The refrigerated system in this case is the low-temperature reservoir.

There are several actual refrigeration systems used to perform this function; we shall investigate the mechanical-compression and liquid-absorption systems and look briefly at the low-temperature Linde-Hampson liquefaction system. The first system may be compared to a reversed Carnot cycle operating between the same temperature limits. Therefore the reversed Carnot cycle is worth reviewing briefly.

12.1 REVERSED CARNOT CYCLE

Figures 12.1(a) and 12.1(b) illustrate the physical representation of the reversed Carnot cycle and the T–S diagram representing the cycle, respectively. This is a mechanical-compression type system. The performance ratio of a refrigeration system is not an efficiency, but rather the coefficient of performance, COP, and is defined as the heat supplied (the desired effect) divided by the net work (what it costs):

$$\text{COP} = \frac{Q_{\text{in}}}{W_{\text{net}}} \tag{12.1}$$

For the reversed Carnot cycle, this is

$$\text{COP} = \frac{T_{\text{L}}\,\Delta S}{(T_{\text{H}} - T_{\text{L}})\,\Delta S} = \frac{T_{\text{L}}}{T_{\text{H}} - T_{\text{L}}} \tag{12.2}$$

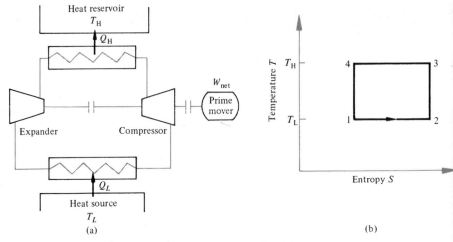

Figure 12.1 *(a) The physical layout of a refrigeration system operating on a reversed Carnot cycle; (b) The T–S diagram for a reversed Carnot cycle.*

Since the reversed Carnot cycle is the best that can be achieved, the COP for the reversed Carnot cycle is often used as a basis of comparison with actual COP values.

12.2 REFRIGERANT CONSIDERATIONS

The next question is what type of working substance, or refrigerant, can be used in the refrigeration system. There are many choices available, as Table 12.1 illustrates, but it is desirable to have the compressor intake pressure equal to or greater than atmospheric pressure, so that air does not leak into the refrigeration system. It is necessary, then, to have a substance that has a boiling temperature less than the surrounding temperature at atmospheric pressure. If the boiling points at atmospheric pressure of various substances are investigated, many will have temperatures low enough to serve as refrigerants. Ammonia, for instance, boils at $-33.3°C$ and Refrigerant 12 at $-29.7°C$, both at atmospheric pressure. If a two-phase substance were used, the T–s diagram for the reversed Carnot vapor-compression system would look similar to Figure 12.2. In this case the heat supplied from the refrigerated region would evaporate the fluid until state 2 was reached. The compressor would compress the wet mixture isentropically to a high temperature until it was a saturated vapor. Heat rejection would occur at constant temperature as the fluid was condensed to a saturated liquid, and then the liquid would enter the constant-entropy expander, producing work, until it reached the pressure and temperature at state 1.

Table 12.1 Physical Properties of Refrigerants

Refrigerant Number	Chemical Formula	Molecular Weight	Boiling Point (1 atm), K	Critical Temperature, K	Critical Pressure, MPa	Latent Heat at Boiling Point, kJ/kg mol	Current Use	Safety Group
729	Air	28.97	78.8	132.6	3.77	—	Gas cycle	1
13	$CClF_3$	104.47	191.7	302.0	3.87	15 503	Yes	1
744	CO_2	44.01	194.6	304.1	7.38	25 306	No	1
13B1	$CBrF_3$	148.9	215.4	340.1	3.96	17 679	Yes	1
22	$CHClF_2$	86.48	232.4	369.1	4.98	20 425	Yes	1
717	NH_3	17.03	239.8	406.1	11.42	23 328	Yes	2
12	CCl_2F_2	120.93	243.4	385.1	4.11	19 969	Yes	1
114	$CClF_2CClF_2$	170.93	276.7	418.9	3.27	23 442	Yes	1
21	$CHCl_2F$	102.93	282.1	451.6	5.17	24 918	No	1
11	CCl_3F	137.38	296.7	471.1	4.38	25 022	Yes	1
113	CCl_2FCClF_2	187.39	320.7	487.3	3.41	27 493	Yes	1

SOURCE: Data extracted from ASHRAE *Guide and Data Book*, 1965–66, by permission.
Safety Group 1: negligible toxicity, nonflammable.
Safety Group 2: toxic, flammable, or both.

249

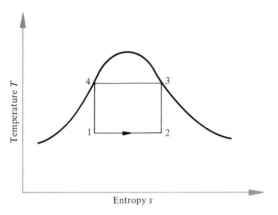

Figure 12.2 *The T–s diagram for a two-phase substance operating on the reversed Carnot cycle.*

12.3 VAPOR-COMPRESSION CYCLE

There are several disadvantages to the aforementioned system. First, reciprocating compressors should not operate in the saturated-mixture region, as lubricating oil may be washed away in the compression process. Second, the work performed by the expander is very small in comparison to the compressor work, and the cost for such an expander would be unnecessarily expensive. The actual or standard vapor-compression system, as illustrated in Figure 12.3(a) and 12.3(b), over-comes these problems in two ways. First, the refrigerant receives heat until it is a saturated vapor at state 2, and, second, the expansion process is an irreversible throttling process for which only an expansion valve is

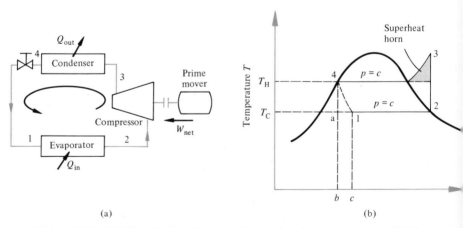

(a) (b)

Figure 12.3 *(a) The schematic diagram of a simple refrigeration system. (b) The T–s diagram for the system in Figure 12.3(a).*

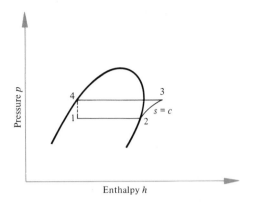

Figure 12.4   The p–h diagram for the system in Figure 12.3(a).

needed. As in the case of power-producing cycles using two-phase substances, the heat addition, states 1 to 2, and heat rejection, states 3 to 4, are constant-pressure processes. The superheat horn, shown in Figure 12.3(b), illustrates the additional work required of dry compression as compared to wet compression. The area *abel* represents the loss of refrigerating effect due to the irreversible throttling process compared with the constant-entropy expansion process. In refrigeration practice, a *p–h* diagram is often used rather than the *T–s* diagram shown in Figure 12.3(b). The same process is shown on a *p–h* diagram (Figure 12.4).

The first means of refrigeration used by humans was the melting of ice blocks. When they developed more sophisticated means of producing a refrigeration effect, it was natural to express this in terms of familiar units, that of melting ice. The latent heat of ice is 334.9 kJ/kg; this is the standard value used, and so the refrigeration produced by 907.18 kg of ice (2000 pounds or 1 ton) is 3.0384×10^5 kJ. A refrigeration system produces the refrigerating effect in a certain time period, and 1 ton of refrigeration is specified as the removal of 3.0384×10^5 kJ in a 24-h period. Hence

$$1 \text{ ton} = 3.516 \text{ kW} \tag{12.3}$$

In the Appendixes, abridged charts and tables are given for several refrigerants. Before proceeding to modifications of the standard vapor-compression cycle, let us put our knowledge to work in analyzing the system in the following example.

Example 12.1

A standard vapor-compression system produces 20 tons of refrigeration using R12 as a refrigerant while operating between a condenser temperature of 41.6°C and an evaporator temperature of $-25°C$. Determine (a) the refrigerating effect in kJ/kg; (b) the circulating rate in kg/s; (c) power supplied; (d) COP; and (e) heat rejected in kW.

The problem is illustrated by Figures 12.3(a), 12.3(b), and (12.4). The refrigerating effect occurs in the evaporator, and performing an energy balance on the evaporator yields

$$q = h_2 - h_1 \text{ kJ/kg}$$

where

$$h_2 = h_g = 176.35 \text{ kJ/kg} \qquad \text{at } -25°C$$
$$h_1 = h_f = 76.17 \text{ kJ/kg} \qquad \text{at } 41.6°C$$

(a) $q = 100.18$ kJ/kg. The refrigeration produced is 20 tons.

$$20 \text{ tons}\left(3.516\frac{\text{kW}}{\text{ton}}\right) = \dot{m}(h_2 - h_1)$$

(b) $\dot{m} = 0.7019$ kg/s. The work of the compressor is found by an energy balance across the compressor

$$w = h_3 - h_2 \text{ kJ/kg}$$

Since h_2 is known, and $s_3 = s_2 = 0.7121$ kJ/kg · K, then in the superheat tables at $p = 1.0$ MPa yields for $h_3 \approx 213.46$ kJ/kg

$$w = 213.46 - 176.35 = 37.11 \text{ kJ/kg}$$

The power $\dot{W}$ is

(c) $\dot{W} = \dot{m}w = (0.7019)(37.11) = 26.05$ kW.

(d) COP $= q_{\text{in}}/w = 100.18/37.11 = 2.70$.

The heat rejected in the condenser is found by an energy balance across the condenser:

$$q_{\text{out}} = h_3 - h_4 = 213.46 - 76.17 = 137.29 \text{ kJ/kg}$$

(e) $\dot{Q}_{\text{out}} = \dot{m}q_{\text{out}} = 96.36$ kW. This size of the compressor depends on the volume flow rate that the compressor must handle at the inlet conditions. For this example, the volume flow rate $\dot{V}$ is

$$\dot{V} = \dot{m}v_2 = (0.7019)(0.131\ 166) = 0.092\frac{\text{m}^3}{\text{s}} = 92\frac{\text{liters}}{\text{s}}$$

Improving the Standard Cycle

It is possible to increase the efficiency of the vapor-compression refrigeration system by subcooling the refrigerant at state 4. Subcooling acts to increase the refrigerating effect. This is accomplished by having greater

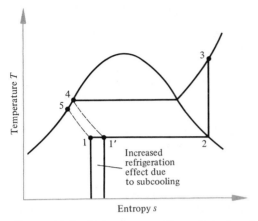

Figure 12.5   A T–s diagram illustrating subcooling.

cooling than is required in the condenser. Figure 12.5 illustrates this and the resulting increase in refrigerating effect.

To locate state 6, we will assume that the saturated liquid properties at the subcooled temperature are the properties at state 6. This assumption is accurate, because the line of constant pressure and the saturated liquid line are close together and the property variation between the two is very small. The oversizing increases the cycle efficiency by increasing the refrigerating effect and by preventing the flashing (vaporizing) of the refrigerant before it enters the throttling valve.

Example 12.2

The refrigerant in Example 12.1 is subcooled by 6.6°C leaving the condenser. Determine the new refrigerating effect and COP.

$$h_5 = h_f @ 35°C = 69.49 \text{ kJ/kg}$$
$$h_1 = h_5$$
$$q = (h_2 - h_1) = 106.86 \text{ kJ/kg}$$
$$w = 37.11 \text{ kJ/kg}$$

COP = 2.88

Thus, the subcooled cycle COP is 6.6 percent higher than the ideal vapor compression cycle.

In actual units the refrigerant leaving the evaporator is also superheated, which increases the refrigerating effect, but increases the work by a greater amount. The superheating is required for the expansion valve control and to prevent liquid refrigerant entering the compressor.

12.4 MULTISTAGE VAPOR-COMPRESSION SYSTEMS

When gas compressors are analyzed in Chapter 15, we will find that the work of compression is reduced by multiple stages of compression. The same is true for refrigeration systems. In this case, rather than just multiple-compressor staging, the entire system is staged, or cascaded. This permits the attaining of lower temperatures for refrigeration with a fixed amount of compressor work.

The advantages will be noted as the system is developed. The optimum interstage pressure, p_i, is found to be

$$p_i = (p_h p_l)^{1/2} \qquad (12.4)$$

where p_h is the maximum system pressure and p_l is the minimum system pressure. Figures 12.6(a), 12.6(b), and 12.6(c) illustrate the schematic diagram and the T–s and p–h diagrams for the cycle.

The evaporator of the high-pressure system serves as the condenser for the low-pressure system. The T–s diagram shows the increased refrigerating effect, area 1abc, due to the cascading. The dotted line 5 to a is an extension of the throttling process from the high pressure to the pressure of the evaporator. Also, the decrease in work compared to that of a single stage is shown by area 3d76.

In this drawing the fluids are assumed to be the same, so the same T–s diagram may be used. As long as a closed cascade condenser is used, the fluids in the high-pressure and low-pressure systems may be different, in which case a correct T–s diagram should be used for each fluid. When the same fluid is used throughout the system, a direct-contact heat exchanger, instead of the closed cascade condenser, is more common. The T–s and p–h diagrams remain the same, but the physical diagram is illustrated by Figure 12.7. The flash gas from the heat exchanger goes to the high-pressure compressor intake and the saturated liquid goes to the low-pressure throttling valve. This type of heat exchanger is more efficient, as all tube resistance to heat transfer is eliminated. The open system also provides some control due to transient changes in the system by allowing the proportion of fluid in each loop to alter slightly. The intermediate pressure, p_i, will vary slightly, depending on conditions.

While the subject of mass flow rates is under discussion, how can the ratio of mass in one loop to that in the other be determined? By performing an energy balance on the cascade condenser, whether it be closed or direct contact, whether using the same fluid or different fluids.

Energy In = Energy Out

$$\dot{m}_2 h_3 + \dot{m}_1 h_5 = \dot{m}_2 h_4 + \dot{m}_1 h_6$$

$$\frac{\dot{m}_1}{\dot{m}_2} = \frac{h_3 - h_4}{h_6 - h_5} \qquad (12.5)$$

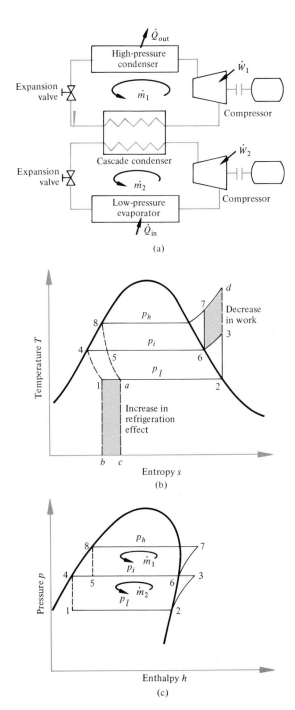

Figure 12.6 (a) The schematic diagram of a two-stage cascade refrigeration system. (b) The corresponding T–s diagram. (c) The corresponding p–h diagram.

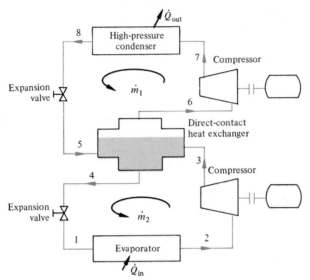

Figure 12.7   An open or direct cascade heat exchanger in a two-stage cascade system.

The desired refrigerating effect determines the flow rate in the low-pressure loop. For instance, X tons of refrigeration is desired:

$$\dot{m}_2(h_2 - h_1) = 3.516X \text{ kW}$$

$$\dot{m}_2 = \frac{3.516X}{h_2 - h_1} \tag{12.6}$$

and $\dot{m}_1$ may be calculated from Equation (12.5).

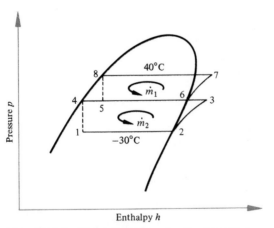

Figure 12.8   The p–h diagram for Example 12.3.

Example 12.3

Calculate the power required by two compressors in an ammonia
system that serves a 30-ton evaporator at $-30°C$. The system uses
a direct-contact cascade condenser, and the condenser temperature
is $40°C$. Figure 12.8 illustrates the p–h diagram for the example and
Figure 12.7 the schematic diagram.

It is necessary to calculate the flow rates in the high- and
low-pressure loops. To do this we must find the intermediate
pressure. Use Tables A.9 and A.10 for ammonia properties.

$$p_i = \sqrt{p_h p_l} \qquad p_h = 1554.3 \text{ kPa (saturated pressure at } 40°C)$$

$$p_l = 119.5 \text{ kPa (saturated pressure at } -30°C)$$

$p_i = 431 \text{ kPa}$

$h_2 = 1404.6$	$s_3 = s_2$	$h_3 = 1574.3$
$h_6 = 1443.5$	$s_7 = s_6$	$h_7 = 1628.1$
$h_8 = 371.7$		$h_5 = 371.7$
$h_4 = 181.5$		$h_1 = 181.5$

Determine the mass flow rate in the low-pressure loop by Equation
(12.6):

$$\dot{m}_2 = \frac{(30)(3.516)}{(1404.6 - 181.5)} = 0.0862 \text{ kg/s}$$

From Equation (12.5),

$$\dot{m}_1 = \dot{m}_2 \left(\frac{h_3 - h_4}{h_6 - h_5} \right) = 0.112 \text{ kg/s}$$

The power of the low-pressure unit is found by an energy balance
on the isentropic compressor:

$$\dot{W}_{LP} = \dot{m}_2(h_3 - h_2) = 14.6 \text{ kW}$$

and similarly for the high-pressure compressor,

$$\dot{W}_{HP} = \dot{m}_1(h_7 - h_6) = 20.6 \text{ kW}$$

The total power is

$$\dot{W}_{total} = 35.2 \text{ kW}$$

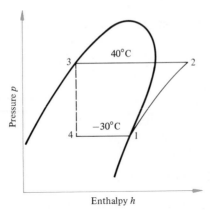

Figure 12.9   *The p–h diagram for one-stage compression refrigeration system in Example 12.3.*

If one stage of compression is used, as in Figure 12.9, then the mass flow determined by Equation (12.6) is

$$h_1 = 1404.6 \qquad h_2 = 1805.1 \qquad h_3 = h_4 = 371.7$$

$$\dot{m} = \frac{(30)(3.516)}{(1404.6 - 371.7)} = 0.1021 \text{ kg/s}$$

$$\dot{W} = \dot{m}(h_2 - h_1) = 40.9 \text{ kW}$$

The power required for one-stage compression is 16.2 percent greater than that required for two-stage compression.

Compare the COP for one and two stages of compression.

One Stage

$$q_{in} = h_1 - h_4 = 1404.6 - 371.7 = 1032.9 \text{ kJ/kg}$$

$$w = h_2 - h_1 = 1805.1 - 1404.6 = 400.5 \text{ kJ/kg}$$

$$\text{COP}_1 = \frac{q_{in}}{w} = 2.58$$

Two Stages

$$q_{in} = h_2 - h_1 = 1404.6 - 181.5 = 1223.1 \text{ kJ/kg}$$

$$w = (h_3 - h_2) + (h_7 - h_6) = 354.3 \text{ kJ/kg}$$

$$\text{COP}_2 = \frac{q_{in}}{w} = 3.45$$

The COP has a significant 33 percent increase when cascading is used.

12.5 ABSORPTION REFRIGERATION SYSTEMS

The largest operating expense in a vapor-compression refrigeration system is due to work, 100 percent available energy, is used to transfer heat from a low temperature to a high temperature. However, the work is transformed into heat and rejected from the system in the condenser. To overcome this use of available energy, the property of absorption of gases by certain fluids may be utilized to transfer heat from a low to high temperature. There is a chemical reaction in the absorption process, heat is released, the heat of reaction. Since the vapor is condensed in another fluid, the latent heat of the vapor, as well as the heat of reaction, must be removed. There are several types of absorption refrigeration systems, among them ammonia–water, water–lithium bromide and water–lithium chloride. We shall look at the ammonia–water system in detail. The other two use water as a refrigerant, which is practical for air conditioning where a temperature below 0°C is not needed. The ammonia–water system, however, is capable of temperatures below 0°C and can achieve temperatures as low as those of the ammonia vapor-compression system.

In this case the ammonia is the refrigerant (R) and the water is the carrier (C). The ammonia vapor is absorbed by liquid water. Figure 12.10 illustrates a schematic diagram for a simple ammonia–water refrigeration system. Some of the complexities encountered in an actual system, such as a rectifier and heat-exchanger modifications, are not included.

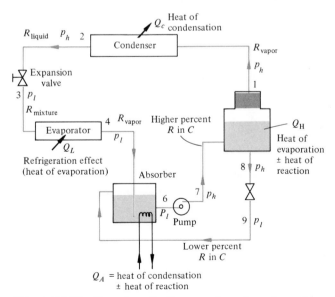

Figure 12.10 *The schematic diagram of an absorption refrigeration system.*

Referring to Figure 12.10, a pump replaces the compressor for changing the pressure level of the system. The work of the pump is very small, so the use of 100 percent available energy is also very small. We start with the vapor leaving the generator at state 1. It is dry saturated ammonia vapor at pressure p_h. The ammonia is condensed in the condenser, leaving as a saturated liquid, state 2; it then passes through a throttling valve, where the pressure is reduced to p_l, leaving at state 3. The ammonia passes through the evaporator, picking up heat from the surroundings, and leaves as a saturated vapor at state 4. This cold vapor then enters the absorber, where it mixes with a hot aqueous solution and is condensed and absorbed. For ammonia the heat of reaction is positive, so a heat exchanger must be located in the absorber to cool the hot aqueous solution, improving its absorbing capability and removing the latent heat of condensation and reaction. This aqueous solution, with a high percentage of ammonia (R) in the water (C), leaves the absorber at state 6 and enters the pump, which it leaves at a high pressure, p_h, at state 7. The high-pressure cool mixture enters the generator, where heat is added to drive the ammonia from solution. The ammonia leaves as a saturated vapor at state 1. Some of the hot liquid, which now has a low percentage of ammonia, leaves the generator at state 8 and is reduced in pressure as it passes through a valve, leaving at state 9. The hot aqueous solution now enters the absorber.

It is possible to calculate the maximum possible and the actual COP for an absorption refrigeration system. The reversed Carnot cycle cannot be used as a basis of comparison in this instance, because the unit does not use a compressor. The refrigeration is accomplished, ideally, by an exchange of heat isothermally. Although this does not actually happen, we can note that condensation and evaporation are isothermal processes.

Figure 12.11 represents the system schematically as various heat sources and sinks. It is possible to represent isothermal heat flow as

$$Q = T\Delta S \tag{12.7}$$

and Figure 12.12 represents these on a $T\text{–}S$ diagram.

COP for Absorption Refrigeration

We know that the COP is the refrigerating effect divided by the cost of producing that effect. This remains unchanged.

$$\text{COP} = \frac{Q_L}{Q_H} \tag{12.8}$$

However, it is possible for the reversible case to reduce this to an equation involving temperature only. An energy balance on the system

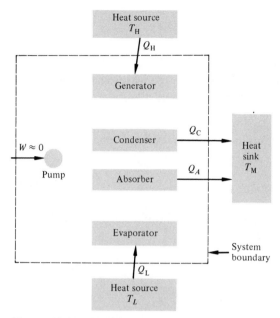

Figure 12.11 *The representation of an absorption refrigeration system as heat sources and sinks.*

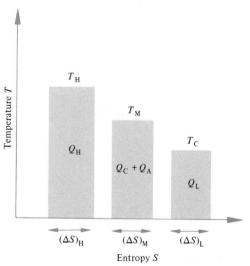

Figure 12.12 *The T–S diagram for the system in Figure 12.11.*

yields the following, with work neglected:

Energy In = Energy Out

$$Q_L + Q_H = Q_C + Q_A = Q_M \tag{12.9}$$

The second law of thermodynamics, when applied to a reversible cycle, tells us that the entropy production is zero, or that

$$(\Delta S)_H + (\Delta S)_L + (\Delta S)_M = 0 \tag{12.10}$$

Using the reversible equation for heat transfer, Equation (12.7), Equation (12.10) becomes

$$\frac{Q_H}{T_H} + \frac{Q_L}{T_L} = \frac{Q_M}{T_M} \tag{12.11}$$

where the direction of the heat flows has been accounted for. We then divide Equation (12.9) by T_M

$$\frac{Q_H}{T_M} + \frac{Q_L}{T_M} = \frac{Q_M}{T_M} \tag{12.12}$$

and equate Equation (12.12) and (12.11)

$$\frac{Q_H}{T_H} + \frac{Q_L}{T_L} = \frac{Q_H}{T_M} + \frac{Q_L}{T_M} \tag{12.13}$$

Simplifying,

$$Q_H\left(\frac{1}{T_H} - \frac{1}{T_M}\right) = Q_L\left(\frac{1}{T_M} - \frac{1}{T_L}\right)$$

and solving for Q_L/Q_H,

$$\text{COP}_{ideal} = \frac{Q_L}{Q_H} = \left(\frac{T_L}{T_M - T_L}\right)\left(\frac{T_H - T_M}{T_H}\right) \tag{12.14}$$

Thus the ideal COP may be determined by knowing the temperature of the various heat-transfer terms. The actual COP is determined by Equation (12.8) after analyzing the actual heat transferred in the absorption system.

Gas–Liquid Equilibrium

Before analyzing the ammonia absorption system, we must first discover what tools are available for the analyses. The system is a mixture of ammonia and water vapor. The system states where the ammonia and water are in equilibrium may be analyzed by using Table B.3 in the Appendix. This chart was developed for analyzing ammonia liquid equilibrium properties. The conditions exist wherever ammonia is in contact with water: in the absorber and generator, condenser, throttling valve, and evaporator. Figure 12.13 illustrates the various properties and specifically denotes them for a temperature of 54.4°C and a pressure of 138 kPa. The concentration fraction of the ammonia in liquid form, x', is

$$x' = \frac{\text{kilograms liquid NH}_3}{\text{kilograms mixture}} \qquad (12.15)$$

and the concentration fraction of ammonia vapor is

$$x'' = \frac{\text{kilograms vapor NH}_3}{\text{kilograms mixture}} \qquad (12.16)$$

The enthalpies, h_L and h_V, are the enthalpies of the liquid mixture and the enthalpy of the vapor, respectively. Thus once the temperature and pressure are known, the other properties, h_V, h_L, x', x'', may be found directly from the chart.

We must be able to determine the enthalpy when two streams of different concentrations are mixed, adiabatically and with heat transfer. These processes occur in the heat exchangers. Figure 12.14(a) denotes a schematic of a mixing chamber, and Figures 12.14(b) and 12.14(c) show the h–x diagram for adiabatic and nonadiabatic mixing, respectively.

The concentration fraction, x, is the percentage of ammonia present in the entering stream. A conservation of mass could be made for the

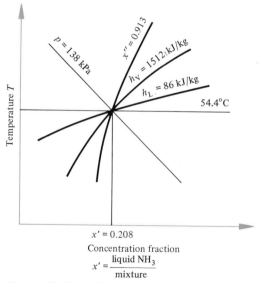

Figure 12.13   A T–x' diagram for an ammonia–water mixture at 54.4°C and 138 kPa.

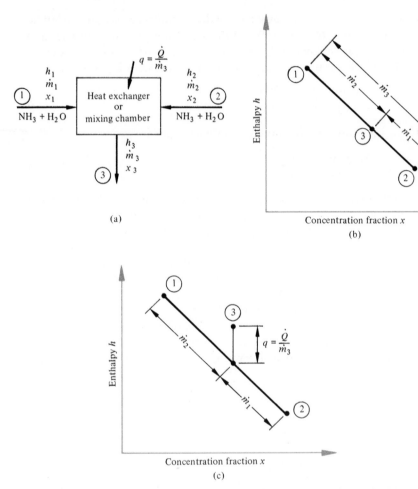

Figure 12.14 (a) The various energy and mass flow into and out of a heat exchanger. (b) The h–x diagram for an adiabatic mixing chamber. (c) The h–x diagram for a mixing chamber with heat transfer.

ammonia, water, or mixture. Let us perform it for the ammonia:

$$(NH_3)_1 + (NH_3)_2 = (NH_3)_3$$
$$\dot{m}_1 x_1 + \dot{m}_2 x_2 = \dot{m}_3 x_3$$
$$\dot{m}_1 + \dot{m}_2 = \dot{m}_3 \qquad (12.17)$$
$$x_3 = \frac{\dot{m}_1 x_1 + \dot{m}_2 x_2}{(\dot{m}_1 + \dot{m}_2)}$$

The reader will notice that the concentration of ammonia at state 3 is independent of heat transfer considerations; it is determined by a mass balance on the heat exchanger. The enthalpy does depend on the heat-transfer process. An energy balance on the exchanger yields

$$\dot{m}_1 h_1 + \dot{m}_2 h_2 + \dot{Q} = \dot{m}_3 h_3$$
$$\dot{m}_1 + \dot{m}_2 = \dot{m}_3$$
$$h_3 = \frac{\dot{m}_1 h_1 + \dot{m}_2 h_2}{\dot{m}_1 + \dot{m}_2} + \frac{\dot{Q}}{\dot{m}_3} \qquad (12.18)$$

and if the exchanger is adiabatic, $Q = 0$,

$$h_3 = \frac{\dot{m}_1 h_1 + \dot{m}_2 h_2}{\dot{m}_1 + \dot{m}_2} \qquad (12.19)$$

The enthalpy at state 3 with heat transfer is equal to the adiabatic enthalpy, Equation (12.19), plus an increase by $\dot{Q}/\dot{m}_3$. Since the mass fraction is the same, this is represented vertically. Figures 12.14(b) and 12.14(c) demonstrate the equations.

Example 12.4

Calculate the quantity of heat transferred in the generator, condenser, absorber per ton of refrigeration for an ammonia–water absorption system when operating at the following temperatures. Generator temperature 104.4°C, condenser pressure 1103 kPa, evaporator pressure 207 kPa, evaporator temperature -1.1°C, absorber temperature 26.6°C. Figure 12.15 shows the schematic with the various temperatures and pressures noted.

Table B.3 will be used to locate the enthalpies at each point.

1. $T = 26.6$°C, $p = 207$ kPa, $x' = 0410$, $h_L = -140$ kJ/kg
2. $T = 104.4$°C, $p = 1103$ kPa, $x'' = 0.920$, $h_V = 1580$ kJ/kg
3. $T = 104.4$°C, $p = 1103$ kPa, $x' = 0.310$, $h_L = 280$ kJ/kg
4. $p = 1103$ kPa, $x_{4'} = x_{2''} = 0.920$

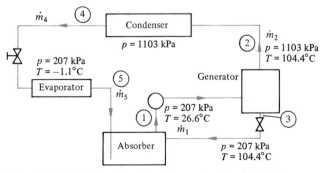

Figure 12.15 The schematic diagram for Example 12.4.

Since p and x' are known, find on the chart $T = 30.5°C$ and $h_L = 84$ kJ/kg.

5. $p = 207$ kPa, $T = -1.1°C$, $x = x_{4'} = x_{2''} = 0.920$

At point 5, the mixture contains 92.0 percent NH_3 and 8.0 percent H_2O. The water cannot be vaporized and must be drained off. This is called "purge liquid." In an actual system a line exists to drain this liquid. To find the enthalpy at 5, it is necessary to find the mass of ammonia and of water, and then to compute the enthalpy of the mixture. For the purge liquid, at $p = 207$ kPa,

$$T = -1.1°C \qquad x_{5'} = 0.620 \qquad h_L = -232 \text{ kJ/kg}$$

For the purge vapor at $p = 207$ kPa

$$T = -1.1°C \qquad x_{5''} = 0.999^+ \qquad h_V = 1284 \text{ kJ/kg}$$

The ammonia mass leaving the evaporator is

$$1(x_5) = m_{p_1} x_{5'} + \left(1 - m_{p_1}\right) x_{5''}$$

where m_{p_1} is the mass of the purge liquid in 1 kg of mixture,

$$1(0.920) = m_{p_1}(0.620) + \left(1 - m_{p_1}\right)(0.999)$$

$$m_{p_1} = 0.208$$

The enthalpy may be found by using Equation (12.19)

$$h_5 = (0.208)(-232) + (1 - 0.208)(1284) = 967 \text{ kJ/kg}$$

All the necessary enthalpies are known, so the refrigerant flow rate per ton of refrigeration may be found.

$$\dot{m}_2 = \dot{m}_4 = \dot{m}_5 = \frac{3.516 \text{ kW/ton}}{(h_5 - h_4) \text{ kJ/kg}} = 0.003\,982 \text{ kg/s/ton}$$

To determine $\dot{m}_1$ and $\dot{m}_3$, make a total mass balance and an ammonia balance around the generator,

(a) Total mass balance: $\dot{m}_1 = \dot{m}_2 + \dot{m}_3$
(b) Ammonia balance: $\dot{m}_1 x_1 = \dot{m}_2 x_2 + \dot{m}_3 x_3$

Substitute (a) into (b) and solve for $\dot{m}_2$ and $\dot{m}_3$:

$$\dot{m}_2 = 0.003\,982 \text{ kg/s}$$

$$\dot{m}_3 = 0.020\,309 \text{ kg/s}$$

and from (a)

$$\dot{m}_1 = 0.024\,291 \text{ kg/s}$$

An energy balance on the generator will yield the heat added, $\dot{Q}_H$

$$\dot{Q}_H + \dot{m}_1 h_1 = \dot{m}_2 h_2 + \dot{m}_3 h_3$$

$$\dot{Q}_H = 15.37 \text{ kW}$$

The heat rejected in the absorber, $\dot{Q}_A$, is found by an energy balance on the absorber.

$$\dot{Q}_A + \dot{m}_1 h_1 = \dot{m}_3 h_3 + \dot{m}_5 h_5$$
$$\dot{Q}_A = 12.94 \text{ kW}$$

The heat rejected in the condenser, $\dot{Q}_C$, is found by an energy balance on the condenser.

$$\dot{Q}_C + \dot{m}_4 h_4 = \dot{m}_2 h_2$$
$$\dot{Q}_C = 5.95 \text{ kW}$$

A check on the accuracy for the heat flows may be made by equating energy supplied and rejected for the cycle.

$$\dot{Q}_H + \dot{Q}_L = \dot{Q}_A + \dot{Q}_C$$
$$15.37 + 3.516 = 12.94 + 5.95$$
$$18.89 = 18.89$$

The COP for the system is

$$\text{COP} = \frac{\dot{Q}_L}{\dot{Q}_H} = \frac{3.516}{15.37} = 0.228$$

Three-Fluid Absorption System

The other absorption systems, such as lithium bromide–water, are analyzed in a similar manner. One interesting variation on the absorption system is the three-fluid system developed by Von-Platen and Munters. In this system no pump is needed, as the entire system is at uniform total pressure. On what is conventionally the high-pressure side, ammonia supports a total pressure of 1240 kPa. On the conventional low-pressure side, the evaporator, ammonia, supports only 210 kPa and hydrogen supports 1030 kPa. A liquid seal separates the two sides, and the pressure drop of the ammonia across the liquid seal is equivalent to throttling. By the use of Dalton's law of partial pressure, refrigeration may occur even through the total system pressure is constant. The heat supply is a gas flame to the generator. This system is used commercially for small gas refrigerators and is currently marketed by Dometic under license of Electrolux of Sweden.

12.6 HEAT PUMP

The heat pump is a reversible refrigerator. It can be used for home heating in cool weather and home cooling in warm weather. Basically it is a vapor-compression refrigeration unit with an external heat sink or source, depending on the climate. The compressor is designed to meet the heating or cooling requirements of the house, whichever is greater.

Let us consider Figure 12.3(a). This is the basic vapor-compression refrigeration system. In the winter it is desirable to use the heat rejected in the condenser for heating. Typically a hot-air heating system is used in homes with heat pumps. The air-cooled condenser is the heat source for the home. The evaporator is the heat source for the refrigeration system. The evaporator is located outdoors. The evaporator coil may be in a well or pond, if available; if not, then it is heated by the outside air. The coefficient of performance for this mode of operation is

$$\text{COP}_h = \frac{\text{desired effect}}{\text{cost of effect}} = \frac{q_{\text{out}}}{w_{\text{net}}} = \frac{h_3 - h_4}{h_3 - h_2} \qquad (12.20)$$

In the summer, a time for home cooling, the evaporator is located in the house and the condenser outdoors. The heat from the living quarters enters the refrigeration system in the evaporator, the refrigerant is compressed, and then cooled. The condenser may be in a pond or well or, if neither is available, the condenser will be air cooled. It should seem apparent with some extra piping and a couple of valves, that the condenser and evaporator could reverse roles depending on the desired effect. The coefficient of performance for the cooling mode of operation is

$$\text{COP}_c = \frac{\text{desired effect}}{\text{cost of effect}} = \frac{q_{\text{in}}}{w_{\text{net}}} = \frac{h_2 - h_1}{h_3 - h_2} \qquad (12.21)$$

The heat pump capacity, motor size, and refrigerant supply must be evaluated by considering both modes of operation. The one with the greater demand determines the system size. Because of the larger size demanded by heating, the utilization of the heat pump has been primarily in those areas where the winter temperature is not extremely low. The popularity of year-round air conditioning has prompted an increase in heat pump application, even in areas of low winter temperatures, such as New England. Another reason is that the cost of fuel oil has risen to a point where the heat pump becomes economically competitive with oil burners for home heating. Problem 16 yields some interesting answers when evaluating the economics of heat pump operations.

A tremendous benefit may be found by using solar heating in conjunction with the heat pump for home heating. Consider the COP_h for the reversed Carnot cycle,

$$\text{COP}_h = \frac{T_{\text{H}}}{T_{\text{H}} - T_{\text{L}}}$$

The solar collector is able to raise T_{L} and by doing so the COP_h increases significantly. Instead of operating between 1.6°C to 48.9°C, the heat pump may only have to operate between 32.2°C and 48.9°C. The increase in the COP_h is from 6.8 to 19.3. Obviously, much less work is required for the same heat output.

12.7 LOW TEMPERATURE AND LIQUEFACTION

The refrigeration processes dealing with the production of very low temperatures is called "cryogenics." Exactly where the cryogenic temperature limit occurs is not rigorously defined; however, several authorities suggest that temperatures from 173 K to 123 K be called cryogenic. To achieve these temperatures, the cascade refrigeration system discussed earlier must be used. The details of such a system will not be analyzed but will be discussed in a general sense. In the cascade system different refrigerants are used in each loop so the evaporator temperature in the lowest-temperature loop will be sufficiently low for whatever purpose is desired, for example, gas liquefaction. Different refrigerants are used so the compressor intake pressure is above or near atmospheric pressure; this prevents air infiltration and also means that the specific volume at the intake will not be great enough to cause a significant increase in the work required for compression.

The gases in the cryogenic region behave similarly to vapors we analyzed in Chapter 4. At room temperature and pressure, gases are very greatly superheated, which is why the ideal gas equations of state may be used. An important thermodynamic phenomenon, the Joule-Thomson effect, is utilized frequency in cryogenic or liquefaction operations.

A throttling process produces no change in enthalpy, and hence for an ideal gas the temperature remains constant:

$$h = u + pv = c_v T + RT$$

In a real gas, however, the throttling process does produce a temperature change, either up or down. The Joule-Thomson coefficient, μ, is defined as

$$\mu \equiv \left(\frac{\partial T}{\partial p} \right)_h \tag{12.22}$$

A positive value of μ indicates that the temperature decreases as the pressure decreases, and a cooling effect is thus observed. This is true for almost all gases at ordinary pressures and temperatures. The exceptions are hydrogen gas, helium, and neon, which have a temperature increase with a pressure decrease, hence $\mu < 0$. Even for these gases there is a temperature above which the Joule-Thomson coefficient changes from negative to positive. This is called the inversion temperature, and at this temperature $\mu = 0$.

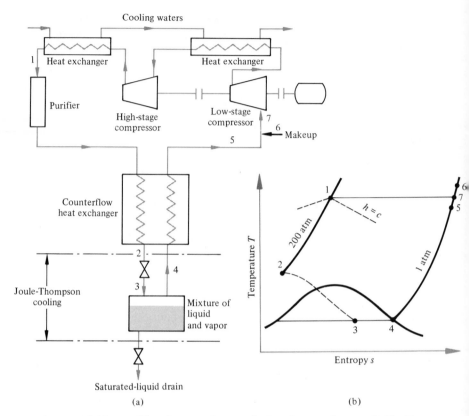

Figure 12.16 *(a) The schematic diagram for liquefaction of gases. (b) The T–s diagram for the liquefaction process.*

The utilization of the Joule-Thomson coefficient in the liquefaction of gases was developed by Linde and Hampson independently. By looking at Figure 10.4(b), we can see that if the gas were at a sufficiently low temperature and high pressure, the throttling process would bring the gas into the saturated-mixture region. Here the vapor and liquid could be separated. Such a system was devised by Linde and Hampson, and it is illustrated schematically in Figure 12.16(a) and the accompanying *T–s* diagram in Figure 12.16(b).

The gas is compressed isothermally in two stages (7 to 1); it is then purified and cooled at constant pressure in a very efficient counterflow heat exchanger (1 to 2); the gas is throttled (2 to 3); and some is liquefied because of the additional Joule-Thomson cooling. The remainder at state 4, a saturated vapor, passes through the counterflow heat exchanger to state 5; makeup gas is added at state 6, and the mixture enters the compressor at state 7.

There are other variations of the liquefaction cycles and refrigeration cycles in general. However, the method of analysis, primarily a first-law analysis, is the same in all cases. Actual refrigeration systems, with their associated control systems, are beyond the scope of this text and must remain an area for further study.

PROBLEMS

1. A reversed Carnot cycle is used for heating and cooling. The work supplied is 10 kW. If the COP = 3.5 for cooling, determine (a) T_2/T_1; (b) the refrigerating effect (tons); (c) the COP for heating.

2. A reversed Carnot cycle is used for refrigeration and rejects 1000 kW of heat at 340 K while receiving heat at 250 K. Determine (a) the T–s diagram; (b) the COP; (c) the power required; (d) the refrigerating effect.

3. Where is the greatest gain in COP derived in a Carnot refrigeration cycle—changing condenser or evaporator temperature?

4. An ideal vapor compression refrigeration cycle uses R12 and has a capacity of 25 tons. The evaporator pressure is 0.22 MPa and the condenser pressure is 1.0 MPa. Find the compressor power.

5. An ideal vapor-compression refrigeration system using R12 operates with an evaporator temperature of $-30°C$ and a condenser exit temperature of 50°C. The compressor requires 75 kW; determine the capacity in tons.

6. An actual vapor-compression refrigerator has a compressor with inlet conditions of $-20°C$ and 130 kPa and exit conditions of 60°C and 900 kPa. The refrigerant is R12. The liquid refrigerant enters the expansion valve at 850 kPa and exits at 160 kPa. Determine the COP.

7. R12 enters an adiabatic compressor as a saturated vapor at $-15°C$ and discharges at 1.0 MPa. If the compressor efficiency is 70%, determine the actual work.

8. An actual vapor-compression refrigerator has a 1.5-kW motor driving the compressor. The compressor efficiency is 70% and has inlet pressure and temperature of 50 kPa and $-20°C$ and a discharge pressure of 900 kPa. The condenser exit pressure is 800 kPa and the expansion valve exit pressure is 60 kPa. Determine (a) the flow rate; (b) the COP; (c) the tons of refrigeration for R 12.

9. A Carnot engine operates between temperature reservoirs of 817°C and 25°C and rejects 25 kW to the low-temperature reservoir. The Carnot engine drives the compressor of an ideal vapor-compression refrigerator, which operates within pressure limits of 190 kPa and 1.2

MPa. The refrigerant is ammonia, determine the COP and the refrigerant flow rate.

10. A 1-ton vapor-compression refrigeration system using R12 as the refrigerant has the following data for the cycle.

> Evaporating pressure 123 kPa
> Condensing pressure 600 kPa
> Temperature of R12 entering the throttling valve 20°C
> Temperature of refrigerant leaving the evaporator −7°C

Assuming isentropic compression, determine (a) the refrigerant flow rate; (b) the power required; (c) the heat rejected; (d) the second law efficiency.

11. A R12 vapor-compression refrigeration system has a condensing temperature of 50°C and an evaporating temperature of 0°C. The refrigeration capacity is 7 tons. The liquid leaving the condenser is a saturated liquid and the compression is isentropic. Determine (a) the refrigerant flow rate; (b) the compressor power requirement; (c) the heat rejected; (d) the coefficient of performance.

12. A vapor-compression refrigeration system uses R12 as the refrigerant. The gaseous refrigerant leaves the compressor at 1200 kPa and 80°C. The heat loss during compression is 14 kJ/kg. The refrigerant enters the expansion valve at 32°C. The liquid leaves the evaporator and enters the compressor as a saturated vapor at −15°C.
The unit must produce 50 tons of refrigeration. Determine (a) the R12 flow rate; (b) the compressor power; (c) the COP.

13. A 15-ton refrigeration system is used to make ice. The water is available at 10°C. Refrigerant 12 is used with saturated temperature limits of −25°C and 75°C. Determine (a) the COP; (b) the refrigerant flow rate; (c) the power required; (d) the maximum kgs of ice manufactured per day.

14. An ideal vapor-compression refrigerating system uses a subcooling–superheating heat exchanger with 10 degrees of superheat added. The evaporator operates at −30°C, the condenser at 1.4 MPa. There is a cooling requirement of 50 tons. If the fluid is R12, determine (a) the mass flow rate; (b) the COP; (c) the refrigerating effect; (d) the degrees of subcooling; (e) the power required; (f) the second law efficiency.

15. The same as Problem 14 except that the fluid is ammonia.

16. A vapor-compression system, using ammonia, is used for home heating. Air is to be heated from 7°C to 30°C at atmospheric pressure. The living requirements demand an air circulation rate of 0.45 m³/s. The evaporator temperature is 0°C, and the condenser pressure is 1.6 MPa. Determine (a) the refrigerant flow rate; (b) the

COP; (c) the power required; (d) the cost, if electricity costs $0.05/kW · h; (e) the cost of fuel, at $0.27/liter, if the heating value of the fuel is 34 900 kJ/liter.

17. A vapor-compression refrigeration system using ammonia as the working substance is required to provide 30 tons of refrigeration at $-16°C$. To accomplish this, the ammonia operates between pressure limits of 144 kPa and 1400 kPa. Determine (a) the quality of ammonia entering the evaporator; (b) the mass flow rate; (c) the change of available energy across the condenser if $T_0 = 26°C$, (d) the volume flow rate at the compressor intake; (e) the bore (D) and stroke (L) ($L = D$) of a double-acting, reciprocating compressor at 600 rev/min. Assume that the fluid is completely discharged from the cylinder each revolution.

18. The same as Problem 17 except that the fluid is R12 and the pressure limits are 123.7 kPa and 1400 kPa.

19. A two-stage cascade refrigeration system uses ammonia as the working substance. The evaporator is at $-30°C$ and the high-pressure condenser is at 1400 kPa. The cascade condenser is a direct-contact type. The cooling load is 100 tons. Determine (a) the optimum cascade condenser pressure; (b) the mass flow rate in the high-pressure loop; (c) the mass flow rate in the low-pressure loop; (d) the COP; (e) the power required; (f) the quality of the fluid entering the evaporator.

20. The same as Problem 19, except that the fluid is R12.

21. For the operating limits described in Problem 19, determine the following for a one-stage vapor-compression system: (a) the COP; (b) the quality of fluid entering the evaporator; (c) the power required.

22. For the operating limits described in Problem 20, determine the following for a one-stage vapor-compression system: (a) the COP; (b) the quality of fluid entering the evaporator; (c) the power required.

23. A two-stage cascade refrigeration system uses ammonia as the working substance. The mass flow rate in the high-pressure loop is 0.10 kg/s. The condenser saturated temperature is 36°C and the evaporator temperature is $-40°C$. The cascade condenser is a direct-contact type. Determine (a) the cascade condenser pressure for minimum work; (b) the refrigerating effect in tons; (c) the COP; (d) the power required.

24. A vapor-compression system, using R12, is to be used for home heating. The maximum heating demand is increasing the temperature of 0.50 m³/s of air at 5°C and atmospheric pressure to 30°C. The minimum condenser temperature must be 10°C greater than the maximum air temperature. The evaporator operates at $-5°C$.

Determine (a) the refrigerant flow rate; (b) the power required; (c) the COP; (d) the cycle second law efficiency.

25. A vapor-compression system, used for home heating, operates between 38°C, saturated condenser temperature, and −5°C, evaporator temperature. The fluid is R12. The home owner decides to increase the evaporator temperature to improve the efficiency of operation. For a heating load of 10 kW, investigate the variation of COP with evaporator temperatures of −5°C, 0°C, +5°C. Do you agree with the home owner?

26. A vapor-compression system is to be used for heating and cooling. The refrigerant is R12 and the system operates between −5°C and 1.6 MPa. The indoor temperature is maintained at 21°C, while the average high outdoor temperature is 32°C in the summer and 5°C in the winter. The building heat loss or gain is 0.33 kW (per degree difference between inside and outside temperature). Determine (a) the mass flow rate required for heating; (b) the COP for heating; (c) the mass flow rate required for cooling; (d) the COP for cooling; (e) which season determines the system size.

27. A vapor-compression system using ammonia is used for heating and cooling. The compressor can handle 6 liters/s of ammonia at −4°C discharging it at 1.8 MPa. Determine (a) the COP for heating and cooling; (b) the maximum cooling load possible; (c) the maximum heating load possible.

28. A refrigeration unit uses the ammonia-absorption system for cooling. The unit is characterized by the following conditions:

Generator temperature 98.8°C
Condenser pressure 7034 kPa
Evaporator pressure 207 kPa
Evaporator temperature 1.6°C
Absorber temperature 23.8°C

Determine (a) the COP; (b) the heat required per ton of refrigeration; (c) the heat rejected in the condenser.

29. An ammonia-absorption refrigeration unit is used for air conditioning a large office building. The cooling requirements are 200 tons at −1.1°C. The vapor leaving the generator is at 115.5°C and 1241 kPa. The evaporator pressure is 138 kPa and the absorber temperature is 23.8°C. Determine (a) the mass flow of ammonia required; (b) the COP; (c) the heat required; (d) the natural gas required if the heat release is 33 500 kJ/m^3.

30. For the liquefaction of gases, the following properties are given for

nitrogen. Refer to Figure 12.16(a).

$$T_5 = T_6 = T_7 = 270 \text{ K}$$
$$h_5 = h_6 = h_7 \text{ kJ/kg}$$
$$h_1 = 405 \text{ kJ/kg at 100 atmospheres}$$
$$T_2 = 150 \text{ K}$$
$$h_2 = 200 \text{ kJ/kg}$$
$$T_3 = 78 \text{ K}$$
$$h_4 = 230 \text{ kJ/kg}$$

Determine (a) the quality of the mixture at state 3 if $h_f = 35$ kJ/kg; (b) the kilograms per minute required at state 7 for 1 kg of N_2 liquefied.

31. A heat pump uses ammonia as the refrigerant and must deliver 30 kW to the building. The discharge pressure is 1.8 MPa and the suction pressure is 300 kPa at 15°C. The compression is isentropic. Determine (a) the ammonia flow rate in liters per second at inlet conditions; (b) the power required; (c) COP; (d) the discharge temperature.

32. An air turbine air at 1400 kPa and 310 K and expands the air to 140 kPa. The turbine internal efficiency is 70% and the power produced is 65 kW. The air exhausting from the turbine will be used for refrigeration. The refrigerated space is to be maintained at 260 K. What is the maximum possible refrigeration in tons?

33. A 25-ton vapor-compression refrigeration system uses ammonia as the working substance. The ammonia enters the compressor as a saturated vapor at −34°C leaves the compressor at 1400 kPa and 180°C, and enters the throttling valve at a temperature of 30°C. Determine (a) the mass flow rate; (b) the COP; (c) the power required.

Chapter 13
Mixtures:
Gas–Gas and
Gas–Vapor

In the thermodynamic analyses in the previous chapters we have assumed the substance to be a pure substance. This is not usually the case, however. Air is a mixture of several pure substances: oxygen, nitrogen, argon, water vapor, and other gases. In chemical reactions, such as combustion, mixtures of gases and vapors are involved.

We start this chapter by analyzing mixtures of ideal gases to establish fundamental laws for gas mixtures and to gain experience in using these concepts. Then we will develop a model for mixtures of gases and vapors and then analyze these mixtures.

13.1 IDEAL-GAS MIXTURES

The gases in a gas mixture are called "components of the mixture." A given component i, will have a mass m_i, and the total mixture will have a mass m, where

$$m = \sum_i m_i \tag{13.1}$$

The mass fraction x_i is defined as

$$x_i = \frac{m_i}{m} \tag{13.2}$$

The component has n_i moles in a total mixture of n moles, where $n = \sum n_i$; we can define the mole fraction y_i as

$$y_i = \frac{n_i}{n} \tag{13.3}$$

Let us consider that we have the components of a gas mixture separated and existing at the same temperature, T, and pressure, p. The equation of state may be written for each of these gases as

$$pV_i = n_i \bar{R} T \tag{13.4}$$

The properties for the mixture of the gases would have a volume V; the properties of the mixture are denoted by the absence of a subscript. For the mixture, the ideal gas equation of state is

$$pV = n \bar{R} T \tag{13.5}$$

and we find the volume fraction of the mixture by dividing Equation (13.4) by Equation (13.5):

$$\frac{V_i}{V} = \frac{n_i}{n} = y_i \tag{13.6}$$

Thus the volume fraction of a gas mixture is equal to the mole fraction of the mixture.

Amagat's Law

Amagat's law of additive volumes is as follows:

> The total volume of a mixture of gases is equal to the sum of the volumes that would be occupied by each component at the mixture temperature, T, and pressure, p.

This applies rigorously to ideal gases:

$$V = \sum_i V_i \tag{13.7}$$

This is illustrated in Figure 13.1 for three components, 1, 2, 3, of an ideal gas mixture.

Another approach in analyzing gas mixtures applies when the components occupy the total mixture volume, V, at the same temperature, T. Figure 13.2 illustrates this case. The ideal-gas equation of state may be

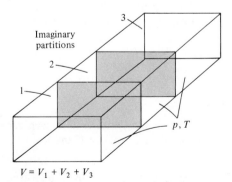

Imaginary partitions

$$V = V_1 + V_2 + V_3$$

Figure 13.1   Visualization of Amagat's law of additive volumes.

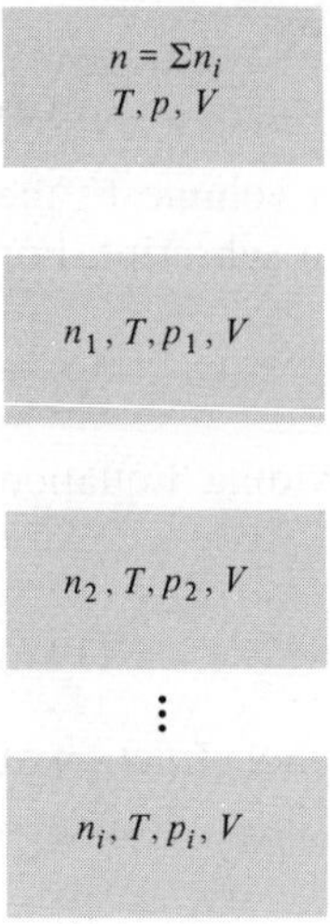

$$n = \Sigma n_i$$
$$T, p, V$$

$$n_1, T, p_1, V$$

$$n_2, T, p_2, V$$

$$\vdots$$

$$n_i, T, p_i, V$$

Figure 13.2   *Visualization of Dalton's law of partial pressures.*

written for each component and for the total mixture:

$$p_i V = n_i \overline{R} T \tag{13.8a}$$

and

$$p V = n \overline{R} T \tag{13.8b}$$

where Equation (13.8a) is the ideal-gas equation of state for the ith component and Equation (13.8b) is the ideal-gas equation of state for the total mixture. Let us divide Equation (13.8a) by Equation (13.8b):

$$\frac{p_i}{p} = \frac{n_i}{n} = y_i \tag{13.9}$$

Thus the ratio of the partial pressure, the pressure of the ith component occupying the mixture volume at the same temperature and volume, is equal to the mole fraction.

Dalton's Law

Dalton's law of partial pressure states:

> The total mixture pressure, p, is the sum of the pressures that each gas would exert if it were to occupy the vessel alone at volume V and temperature T.

This is rigorously true only for ideal gases. Equation (13.9) leads us to the same conclusion:

$$p_i = p \frac{n_i}{n}$$

$$\sum_i p_i = p \sum_i \frac{n_i}{n} = p$$

$$p = \sum_i p_i \tag{13.10}$$

Equation (13.10) is the algebraic statement of Dalton's law of partial pressures.

Mixture Properties

The total mixture properties, such as internal energy, enthalpy, and entropy, may be determined by adding the properties of the components at the mixture conditions. The internal energy and enthalpy are functions of temperature only; hence

$$U = n\bar{u} = \sum_i n_i \bar{u}_i$$

$$H = n\bar{h} = \sum_i n_i \bar{h}_i \tag{13.11}$$

where $\bar{u}_i$ and $\bar{h}_i$ are the internal energy and enthalpy of the component i on the mole basis, energy per unit mole, at the mixture temperature.

The entropy of a component is a function of the temperature and pressure. The entropy of a mixture is the sum of the component entropies,

$$S = n\bar{s} = \sum_i n_i \bar{s}_i \tag{13.12}$$

where $\bar{s}_i$ is the entropy per mole of the ith component at the mixture temperature T and its partial pressure p_i.

Entropy of Mixing

When two gases are mixed together, therefore, there will be an increase in entropy. This is called the "entropy of mixing." The following example will illustrate this phenomenon.

Example 13.1

Figure 13.3 illustrates this example. There are n_a moles of the first component of a gas mixture and n_b moles of the second component;

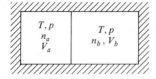

Figure 13.3   Diagram for Example 13.1.

they exist at the same temperature and pressure. The partition separating the components is removed and the gases mix adiabatically. Determine the entropy change of the mixture.

For constant temperature, the change of entropy is

$$S_2 - S_1 = -n\bar{R}\ln\left(\frac{p_2}{p_1}\right)$$

and for the *a* component, the final pressure is p_a, its partial pressure and the initial pressure is p. Therefore

$$(S_2 - S_1)_a = -n_a\bar{R}\ln\left(\frac{p_a}{p}\right) = -n_a\bar{R}\ln(y_a)$$

$$(S_2 - S_1)_b = -n_b\bar{R}\ln\left(\frac{p_b}{p}\right) = -n_b\bar{R}\ln(y_b)$$

and the entropy change of the mixture is the sum of the component entropy changes:

$$(S_2 - S_1) = -n_a\bar{R}\ln(y_a) - n_b\bar{R}\ln(y_b) \tag{13.13}$$

If the components *a* and *b* are the same gas, then the entropy change in mixing is zero. This is because we cannot distinguish between the gases to determine the partial pressures, and the final pressure therefore is also the partial pressure of the gas. In the case discussed in the example for distinguishable components, we may generalize Equation (13.13) for any number of components at the same temperature and pressure:

$$S_2 - S_1 = -\bar{R}\sum_i n_i\ln(y_i) \tag{13.14}$$

We see that there will be an entropy increase by the fact that the sign of the natural logarithm of a fraction is negative.

The specific heats of a mixture may be found by expanding the equations for enthalpy and internal energy of a mixture:

$$c_p = \sum_i x_i c_{pi} \tag{13.15}$$

$$c_v = \sum_i x_i c_{vi} \tag{13.16}$$

The key to analyzing gas mixtures and changing from a mass to a molal analysis, and vice versa, is a unit balance. The following example will illustrate several ways of finding mixture properties.

Example 13.2

Determine the individual gas constant and the molecular weight of a mixture containing 20 percent oxygen, 20 percent nitrogen, and 60 percent carbon dioxide on a volume basis.

	R	M
Oxygen	0.2598	32.0
Nitrogen	0.2968	28.016
Carbon dioxide	0.1889	44.01

$$M = \sum y_i M_i$$

$$\frac{(kg)_{mix}}{(kgmol)_{mix}} = \sum \frac{(kgmol)_i}{(kgmol)_{mix}} \cdot \frac{(kg)_i}{(kgmol)_i}$$

and

$$\sum (kg)_i = (kg)_{mix}$$

$$M = (0.20)(32.0) + (0.20)(28.016) + (0.60)(44.01)$$

$$= 38.409 \ (kg)_{mix}/(kgmol)_{mix}$$

$$R = \sum x_i R_i$$

$$\frac{kJ}{(kg)_{mix} \cdot K} = \sum \frac{(kg)_i}{(kg)_{mix}} \cdot \frac{kJ}{(kg)_i \cdot K}$$

$$x_i = (y_i)\left(\frac{1}{M}\right)(M_i)$$

$$\frac{(kg)_i}{(kg)_{mix}} = \frac{(kgmol)_i}{(kgmol)_{mix}} \cdot \frac{(kgmol)_{mix}}{(kg)_{mix}} \cdot \frac{(kg)_i}{(kgmol)_i}$$

$$x_{O_2} = \frac{(0.2)(32.0)}{38.409} = 0.166$$

$$x_{N_2} = \frac{(0.2)(28.016)}{38.409} = 0.146$$

$$x_{CO_2} = \frac{(0.6)(44.01)}{38.409} = 0.688$$

$$R = (0.166)(0.2598) + (0.146)(0.2968) + (0.688)(0.1889)$$

$$R = 0.2164 \text{ kJ}/ (\text{kg})_{mix} \cdot \text{K}$$

Also, the following may be used.

$$R = \frac{\bar{R}}{M} = \frac{8.3143}{38.409} = 0.2164 \text{ kJ}/ (\text{kg})_{mix} \cdot \text{K}$$

13.2 GAS–VAPOR MIXTURES

The practicing engineer frequently encounters mixtures of vapors and gases. The products of combustion contain water vapor and gas oxides; the carburetor of an automobile has a mixture of gasoline vapor and air. The most common mixture is that of water vapor in air. This is important in heating and cooling problems. The analysis of gas–vapor mixtures may be performed quite easily and accurately if the following assumptions are made: One, the solid or liquid phase contains no dissolved gases; two, the gaseous phase can be treated as an ideal-gas mixture; three, the equilibrium between the condensed phase and vapor phase is independent of the gaseous mixture.

Relative Humidity, Specific Humidity, Saturated Air

There are several frequently used terms that are often misunderstood. Let us define them before proceeding with the analysis.

Consider the cylinder in Figure 13.4(a) and let the substance inside the container be a gas–vapor mixture, where the vapor is superheated at its partial pressure. Water vapor in atmospheric air exists in such a condition. Heat is transferred to the surroundings at constant pressure, as indicated by the line 1–2 on Figure 13.4(b). At point 2, the vapor is a saturated vapor for that pressure; this is called the "dew point." If further heat is transferred, some liquid condensation will occur and the partial pressure will be reduced to state 3, with the saturated liquid at state 4.

When the vapor in the mixture is at state 2, a saturated vapor for its partial pressure, the mixture is called a "saturated mixture"; if the

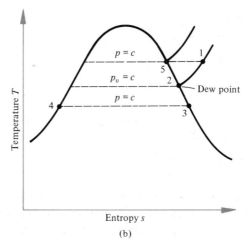

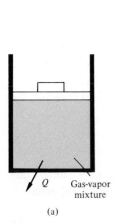

(a)   (b)

Figure 13.4   (a) A constant pressure process with heat transfer. (b) A T–s diagram illustrating dew point.

mixture is air, it is called "saturated air." This is a misnomer, as it is the water vapor in the air, not the air, that is saturated.

The relative humidity, ϕ, of a mixture is the ratio of the mass of vapor in a unit volume to the mass of vapor that the volume could hold if the vapor were saturated at the mixture temperature. The vapor can be considered an ideal gas, and the properties with the subscript g are the saturated vapor properties, then

$$\phi = \frac{m_v}{m_g} = \frac{p_v V/RT}{p_g V/RT} = \frac{p_v}{p_g} \tag{13.17}$$

and from Figure 13.4(b),

$$\phi = \frac{p_1}{p_5}$$

The humidity ratio of an air–water mixture, or specific humidity, ω, is defined as the ratio of the mass of water vapor in a given volume of mixture to the mass of air in the same volume. Let m_v be the mass of water vapor and m_a be the mass of air (without vapor) present in the mixture. Thus

$$\omega = \frac{m_v}{m_a} \tag{13.18}$$

If the ideal gas law is used, then

$$\omega = \frac{R_a p_v}{R_v p_a}$$

where

$$R_a = 0.287 \text{ kJ/kg} \cdot \text{K}$$
$$R_v = 0.4615 \text{ kJ/kg} \cdot \text{K}$$

$$\omega = 0.622 \frac{p_v}{p_a} \tag{13.19}$$

and using the definition of relative humidity, Equation (13.17) relates the two humidity ratios:

$$\phi = \frac{\omega p_a}{0.622 p_g} \tag{13.20}$$

We can use the ideal gas law to describe behavior up to 65°C, above this temperature the saturation of water in air is high and nonideal behavior of the vapor–gas mixture creates too great a discrepancy.

Example 13.3

A room, 6 × 4 × 4 m, is occupied by an air–water vapor mixture at 38°C. The atmospheric pressure is 101 kPa, and the relative humidity is 70 percent. Determine the humidity ratio, dew point, mass of air, and the mass of water vapor.

The total volume is 96 m³. From the steam tables at 38°C,

$$p_g = 6.632 \text{ kPa}$$
$$p_v = (0.70)(6.632) = 4.642 \text{ kPa}$$

The dew point is T_{sat} at $p = 4.642$ kPa

$$T_{\text{dew point}} = 31.5°C$$
$$p_a = p - p_v = 101 - 4.6 = 96.4 \text{ kPa}$$

Solving Equation (13.20) for ω,

$$\omega = 0.622 \frac{p_g \phi}{p_a} = 0.0299$$

The mass of air,

$$m_a = \frac{p_a V}{R_a T} = \frac{(96.4)(96)}{(0.287)(311)} = 103.7 \text{ kg}$$

Since $\omega = m_v / m_a$,

$$m_v = \omega m_a = 3.1 \text{ kg}$$

Example 13.4

The air–water vapor mixture in the previous example is further cooled at constant pressure until the temperature is 10°C. Find the amount of water, in liquid form, that condensed.

To solve this type of problem we must calculate the initial mass of water vapor, as in the previous example, and the final mass of water vapor, and the difference is the water condensed from the mixture.

At 10°C the mixture is saturated because 10°C is less than the dew point, hence

$$p_{v_2} = p_{g_2} = 1.2276 \text{ kPa}$$

$$p_{a_2} = 101 - 1.2 = 99.8 \text{ kPa}$$

$$\omega_2 = (0.622)\frac{p_{v_2}}{p_{a_2}} = 0.00765 \text{ kg water/kg air}$$

mass of water condensed $= m_a(\omega_1 - \omega_2) = 2.31 \text{ kg}$

The previous examples involved closed systems and were aimed at establishing familiarity with the terms used in air–water vapor mixtures. We did not calculate the heat transferred. Let us examine the changes in energy of a mixture when it is cooled. Enthalpy values for steam and air may be found in their respective tables. Air, at times, may be considered an ideal gas if the tables are not available.

Example 13.5

Figure 13.5 illustrates schematically an air-conditioning unit found in many buildings. A coolant passes through the cooling coils while an air-water vapor mixture is blown over the coils. Some water is condensed from the mixture. Calculate the heat transfer per kilogram of air.

Open System, Steady-State

First Law: Energy In = Energy Out

$$\dot{Q} + \sum \dot{m}_i h_i = \sum \dot{m}_j h_j$$

Conservation of mass:

$$\dot{m}_{a_1} = \dot{m}_{a_2} = \dot{m}_a \qquad \dot{m}_{v_1} = \dot{m}_{v_2} + \dot{m}_{l_2}$$

Substitute the mass terms in the first law:

$$-\dot{Q} + \dot{m}_a h_{a_1} + \dot{m}_{v_1} h_{v_1} = \dot{m}_a h_{a_2} + \dot{m}_{v_2} h_{v_2} + \dot{m}_{l_2} h_{l_2}$$

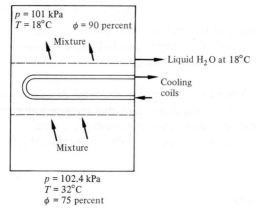

Figure 13.5 *A schematic diagram of an air-conditioning unit as in Example 13.5.*

Divide by $\dot{m}_a$, noting $\omega = \dot{m}_v/\dot{m}_a$,

$$h_{a_1} + \omega_1 h_{v_1} = h_{a_2} + \omega_2 h_{v_2} + (\omega_1 - \omega_2)h_{l_2} + q$$

Use the ideal gas equation of state of enthalpy for air and the steam tables for water vapor enthalpy. Water vapor is considered an ideal gas in this section; so the enthalpy is a function of temperature only and $h = h_g$ at T.

$$p_{v_1} = \phi_1 p_{g_1} = (0.75)(4.759) = 3.57 \text{ kPa}$$

$$\omega_1 = \frac{(0.622)(3.57)}{(102.4 - 3.57)} = 0.0225$$

$$p_{v_2} = \phi_2 p_{g_2} = (0.90)(2.064) = 1.86 \text{ kPa}$$

$$\omega_2 = \frac{(0.622)(1.86)}{(101 - 1.86)} = 0.0117$$

$$q = h_{a_1} - h_{a_2} + \omega_1 h_{v_1} - \omega_2 h_{v_2} - (\omega_1 - \omega_2)h_{l_2}$$

$$q = (1.0047)(305 - 291) + (0.0225)(2559.9) - (0.0117)(2534.4)$$

$$\quad - (0.0108)(75.58)$$

$$q = 41.19 \text{ kJ/kg air}$$

Example 13.6

Consider a tank 1 m in diameter and 2 m high filled with an air–water vapor mixture at condition 1 of the previous example, and cooled until the final condition, condition 2, is reached. Find the mass of liquid condensed and the heat removed from the system.

This is a closed system; hence the first law is

$$\delta Q = dU + \delta W$$

$$\delta W = 0 \qquad \text{for } V = C$$

$$\delta Q = dU \quad \text{or} \quad Q = U_2 - U_1$$

$$Q = \sum m_j u_j - \sum m_i u_i$$

$$m_{a_1} = m_{a_2} = m_a \qquad m_{v_1} = m_{v_2} + m_{l_2}$$

$$Q = m_a u_{a_2} + m_{v_2} u_{v_2} + m_{l_2} u_{l_2} - m_a u_{a_1} - m_v u_{v_1}$$

$$V = \frac{\pi}{4} d^2 L = 1.57 \text{ m}^3$$

$$m_a = \frac{p_a V}{R_a T} = \frac{(102.4 - 3.57)(1.57)}{(0.287)(305)} = 1.772 \text{ kg}$$

$$m_{v_1} = \frac{p_{v_1} V}{R_v T} = \frac{(3.57)(1.57)}{(0.4615)(305)} = 0.0398 \text{ kg}$$

$$m_{v_2} = \frac{p_{v_2} V}{R_v T} = \frac{(1.86)(1.57)}{(0.4615)(291)} = 0.0217 \text{ kg}$$

$$m_{l_2} = m_{v_1} - m_{v_2} = 0.0181 \text{ kg}$$

Use the ideal gas equation of state of internal energy for air and the steam tables for water vapor. Again, $u \approx u_g$ at T.

$$Q = (1.772)(0.7176)(291 - 305) + (0.0217)(2400.2)$$

$$+ (0.0181)(75.58) - (0.0398)(2419.3)$$

$$Q = -60.6 \text{ kJ}$$

Adiabatic Saturation

The adiabatic saturation process is an important process in the study of air–water vapor mixtures. In this process, as the name implies, the mixture is saturated with water vapor adiabatically. To visualize this, consider the schematic diagram in Figure 13.6(a). An unsaturated ($\phi <$ 100 percent) air–water vapor mixture enters an insulated duct. In the bottom of the duct lies water, which evaporates and becomes water vapor in the mixture. The heat from the vaporization must come from the enthalpy of the incoming mixture. Since enthalpy is a function of temperature, the temperature of the mixture decreases. If the mixture and liquid water are in contact for a long enough period of time, the mixture will leave as a saturated mixture at the adiabatic saturated temperature. For this to be a steady-state problem, the changes in kinetic and potential

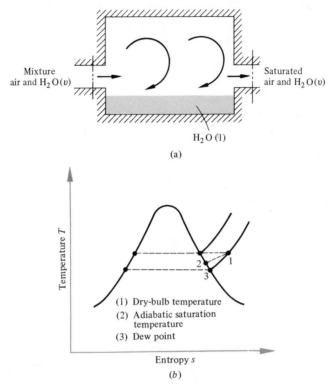

Figure 13.6 (a) An adiabatic saturation unit for steady flow. (b) A T–s diagram illustrating adiabatic saturation temperature.

energies must be zero. The water must be supplied at the adiabatic saturation temperature, or a correction must be made for heating the liquid to the temperature. Since this is an open system, the first law is

Energy In = Energy Out

$$m_a h_{a_1} + m_{v_1} h_{v_1} + m_{l_1} h_{l_1} = m_a h_{a_2} + m_{v_2} h_{v_2}$$

$$h_{a_1} + \omega_1 h_{v_1} + (\omega_2 - \omega_1) h_{l_1} = h_{a_2} + \omega_2 h_{v_2}$$

$$\omega_1 (h_{v_1} - h_{l_1}) = h_{a_2} - h_{a_1} + \omega_2 h_{f_{g_2}}$$

$$\omega_1 (h_{v_1} - h_{l_1}) = c_{pa}(T_2 - T_1) + \omega_2 (h_{f_{g_2}}) \qquad (13.21)$$

The adiabatic saturation temperature, T_2, is a function, then, of the inlet temperature, pressure, relative humidity, and exit pressure. Figure 13.6(b) illustrates the adiabatic saturation temperature on a T–s diagram.

13.3 PSYCHROMETER

To determine the relative humidity of an air–water vapor mixture, a psychrometer uses two thermometers. A cotton wick saturated with water covers the bulb of one. The dry-bulb thermometer measures the temperature of the air–water vapor flow. If the air–water vapor flow is not saturated, then water will evaporate from the wick on the wet bulb. The energy for this evaporation comes, in part, from the internal energy of the mercury in the thermometer, which then decreases, causing a drop in the temperature of the wet-bulb thermometer. Eventually a steady state is reached in which the change of temperature of the air, water vapor, and thermometer is zero with respect to time. This requires that the relative velocity between the air mixture and the wet-bulb thermometer be greater than 3.5 m/s, minimizing the effect of radiative heat transfer and making convective heat transfer a predominant mode.

For air at atmospheric pressure, there is very little actual difference between the wet-bulb temperature and the adiabatic saturation temperature. This is not necessarily true at other pressures or for mixtures other than the air–water vapor mixture.

There are several items that affect the wet-bulb temperature reading: conduction along the thermometer steam; radiant heat transfer from the surroundings to the wet bulb; a boundary layer between the wet bulb and the air. The adiabatic saturation temperature is an equivalent temperature and not affected by these items. The reason the wet- and dry-bulb temperatures determine the state of the mixture is not readily apparent. Two independent properties are necessary for the determination of the state. The dry-bulb temperature is one such property. By knowing the wet-bulb temperature we may determine the vapor pressure for the mixture, which combined with the static pressure, typically atmospheric, defines the second property. Figure 13.7 illustrates a psychrometer; in a sling psychrometer, the two bulbs are in a casing and are whirled around to achieve the necessary relative velocity.

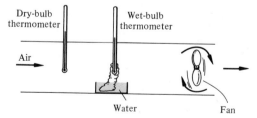

Figure 13.7   Illustration of wet- and dry-bulb thermometers.

13.4 PSYCHROMETRIC CHART

The wet- and dry-bulb temperatures are typically used to determine the relative humidity in conjunction with a pyschrometric chart. It is illustrated in Figure 13.8 and included in Appendix B.4 and may be used in place of the equations developed previously. Either method is correct, but the chart tends to be easier. For pressures other than atmospheric, a correction factor must be used. Many charts indicate what these corrections are, but it is sufficient, at this stage, to know that corrections must be applied.

The pyschrometric charts are widely used in air-conditioning applications, as well as in problems of drying substances. The chart represents the dry-bulb temperature as the abscissa and the wet-bulb temperature, specific volume, relative and specific humidities, vapor pressures, and enthalpy as the other variables. The enthalpy indicated on the chart is per kilogram of dry air. The enthalpy term is the sum of the dry air and water vapor enthalpies in the mixture. Thus $h = h_a + \omega h_f$, where h is the enthalpy given by the chart. The enthalpy of air and water have different reference temperatures: $-17.7°C$ for air, $0°C$ for water. However, the change of enthalpy is most often used and the discrepancy, slight in the first place, is very much minimized.

The enthalpy deviation lines plotted on the chart are the difference between the enthalpies of unsaturated moist air and the saturated air at the same wet-bulb temperature. Thus the enthalpy at the wet-bulb temperature is found and the deviation subtracted from it. This will not typically be used in the text, as the enthalpy may be read directly.

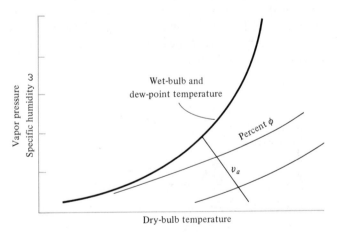

Figure 13.8 The psychrometric chart.

Example 13.7

Find the properties of air at a dry-bulb temperature of 35°C and a wet-bulb temperature of 21°C. Use Table B.4.

$\phi = 29$ percent

$h = 61$ kJ/kg air

$v = 0.885$ m³/kg air

$\omega = 0.010$ kg water/kg air

dew point $= 14$°C

Proceed vertically upward from 35°C until it intersects with the wet-bulb temperature of 21°C. Read the properties; the dew point is found by moving horizontally to the saturated-temperature line. This is the dew point.

Example 13.8

Air at 14°C and 90 percent relative humidity is to be heated until the relative humidity is 50 percent. Determine the heat added per kilogram of dry air. Figures 13.9(a) and 13.9(b) illustrate this problem.

In this process, the humidity ratio is constant, so the process is indicated by a horizontal line between the initial state and the final state. This is also a constant-pressure process, again a horizontal line on the pyschrometric chart. Note that the specific humidity is

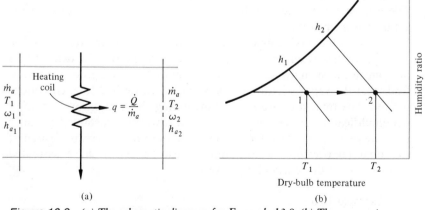

(a)   (b)

Figure 13.9   (a) *The schematic diagram for Example 13.8.* (b) *The process in Example 13.8 on a psychrometric chart.*

defined in terms of the vapor pressure. The heat added per kilogram of air is found from a first-law analysis to be

$$\frac{Q}{m_a} = q = h_2 - h_1$$

$h_1 = h$ at $14°C$ and $\phi = 90$ percent

$h_1 = 36.8 \text{ kJ/kg air}$

$h_2 = 47.0 \text{ kJ/kg air}$

$q = 10.2 \text{ kJ/kg air}$

13.5 AIR-CONDITIONING PROCESSES

The term "air conditioning" implies that the air must be conditioned: heated, cooled, humidified, or dehumidified. Since the heating or cooling process affects the humidity, the air is conditioned in two ways at once. The controls, the actual setup, and all the related equipment used in an actual air conditioner will not be covered here. We will be concerned with the thermodynamic aspects of adjusting the air humidity.

Let us consider dehumidifying the air, removing moisture until the desired humidity is attained. Chemical means are available; substances like silica gel are hygroscopic and absorb water. In most instances, however, the air is cooled below its dew point until the desired amount of liquid is removed; then the remaining air, now cold and humid, is reheated. There are two methods used to cool air. In one, chilled water is sprayed into the air; a fine spray gives a large heat-transfer surface. The air leaves as saturated air at the low-chill water temperature. The second is the direct method employed in room air conditioners, in which the air is passed over the cool or cold evaporator coils of a refrigeration unit. There must be sufficient coils to allow all the air to be cooled to the saturation temperature.

From a thermodynamic viewpoint we are concerned only with the state properties entering and leaving the dehumidifier. In determining how much the air must be cooled in a dehumidification process we must find what the vapor pressure of the water at the desired final conditions will be. The total pressure is constant during the heating process, after dehumidification, so the vapor pressure also will be constant. The vapor pressure leaving the dehumidifier must be the vapor pressure desired at the final temperature, so the vapor must be cooled to a saturated temperature at this particular pressure.

Example 13.9

Moist air at $32°C$ and 60 percent relative humidity enters the refrigeration coils of a dehumidifier with a flow rate of 1.5 kg air/s.

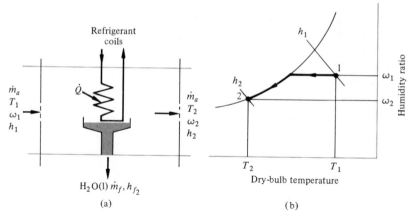

Figure 13.10 *(a) The schematic diagram for Example 13.9. (b) The process in Example 13.9 on a psychrometric chart.*

The air leaves saturated at 15°C. Calculate the condensate removed and the tons of refrigeration required. Figures 13.10(a) and 13.10(b) illustrate the process.

An energy balance across the dehumidifier,

Energy In = Energy Out

$$-\dot{Q} + \dot{m}_a h_1 = \dot{m}_f h_{f_2} + \dot{m}_a h_2$$

$$h_{f_2} = 62.99 \text{ kJ/kg} \qquad \text{(steam tables at 15°C)}$$

By conservation of mass,

$$\dot{m}_f = \dot{m}_a(\omega_1 - \omega_2) = 1.5(0.018 - 0.0106)$$

$$= 0.0111 \text{ kg water/s}$$

$$\dot{Q} = \dot{m}_a(h_1 - h_2) - \dot{m}_f h_{f_2}$$

$$= 1.5(78.2 - 42.0) - 0.0111(62.99)$$

$$\dot{Q} = 53.6 \text{ kW}$$

$$\text{refrigeration} = \frac{53.6}{3.516} = 15.2 \text{ tons}$$

Note that the effect of including the liquid enthalpy is 1.3 percent. Since this magnitude is typical of this term, we can safely neglect the water's enthalpy in our calculations, as the temperature measurements seldom are more than 99 percent accurate.

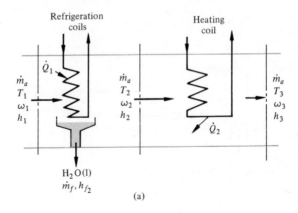

(a)

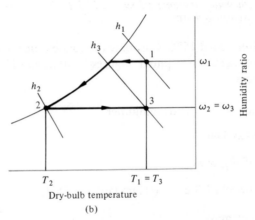

(b)

Figure 13.11   *(a) The schematic diagram for Example 13.10. (b) The processes in Example 13.10 on a psychrometric chart.*

Example 13.10

Moist air enters a humidifier–heater unit at 26°C and 80 percent relative humidity. It is to leave at 26°C and 50 percent relative humidity. For an air flow rate of 0.47 m³/s, find the refrigeration in tons and the heating required in kW. Figures 13.11(a) and 13.11(b) illustrate the process.

In this example, the final conditions are known; from this we can determine the properties at point 2. The energy balance on the dehumidifier yields

$$-\dot{Q} + \dot{m}_a h_1 = \dot{m}_f h_{2f} + \dot{m}_a h_2$$

The flow rate must be determined by finding the specific volume from the psychrometric chart.

$$\dot{m}_a v_{a_1} = \dot{V}_1 \qquad \dot{m}_a = \frac{\dot{V}_1}{v_{a_1}} = \frac{0.47}{0.87} = 0.54 \text{ kg air/s}$$

$$\dot{Q}_1 = \dot{m}_a(h_1 - h_2) - \dot{m}_f h_{2f}$$

$$\dot{m}_f = \dot{m}_a(\omega_1 - \omega_2) \qquad h_{2f} = h_f \qquad \text{at } T_2$$

$$\dot{Q}_1 = (0.54)(69.0 - 49.5) - 0.54(0.0168 - 0.0126)(73.4)$$

$$\dot{Q}_1 = 10.36 \text{ kW}$$

$$\text{refrigeration} = \frac{10.36}{3.516} = 2.95 \text{ tons}$$

Perform an energy balance on the heater to find $\dot{Q}_2$.

$$\dot{m}_a h_2 + \dot{Q}_2 = \dot{m}_a h_3$$

$$\dot{Q}_2 = \dot{m}_a(h_3 - h_2) = 0.54(52.6 - 49.5) = 1.67 \text{ kW}$$

The preceding examples have been devoted to conditioning warm, moist air, the situation normally encountered during the summer when home air conditioners are operating. Air is conditioned in the winter as well, however, by adding moisture and heat to the air. This is important in home heating; very dry air causes wooden furniture to shrink and is responsible for such adverse physiological effects as drying the mucous membranes; it also means that more fuel must be consumed because of the poor heat-conduction properties of dry air. The specific heat and thermal conductivity are affected directly by the moisture in the air.

An example will illustrate the method of solution. Again, the energy balance and mass balances are the keys to problem solving in this area.

Example 13.11

Air enters a humidifier–heater at 10°C and 10 percent relative humidity. The air mixture leaves the heater at a temperature of 24°C and a relative humidity of 50 percent. Determine the heat and water added if the air flow rate is 0.75 kg air/s. The water temperature entering is 15°C. Figures 13.12(a) and 13.12(b) illustrate the process.

Energy In = Energy Out

$$\dot{m}_a h_1 + \dot{Q} + \dot{m}_f h_f = \dot{m}_a h_2$$

where

$$\dot{m}_f = \dot{m}_a(\omega_2 - \omega_1)$$

$$h_f = 62.99 \text{ kJ/kg water}$$

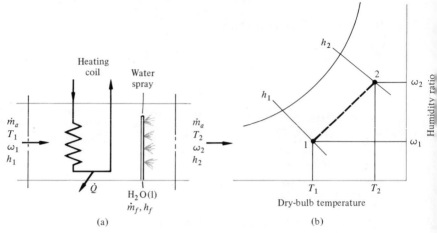

Figure 13.12 (a) *The schematic diagram for Example 13.11.* (b) *The process of Example 13.11 on a psychrometric chart.*

$$\dot{Q} = \dot{m}_a(h_2 - h_1) - \dot{m}_f h_f$$

$$\dot{Q} = 0.75(47.9 - 12) - 0.75(0.0094 - 0.0007)(62.99)$$

$$\dot{Q} = 26.51 \text{ kW}$$

$$\dot{m}_f = \dot{m}_a(\omega_2 - \omega_1) = 0.00652 \text{ kg water/s}$$

13.6 COOLING TOWERS

As we learned in Chapter 9, all energy eventually reaches the dead state, the state of the surroundings. The atmosphere is the ultimate surrounding for all processes on earth. In a heat-transfer process, the final energy will exist at the temperature of the surroundings, the air. We learned that in power plant operations a condenser is a necessary part of the cycle. Usually water is circulated through the condenser, picks up the rejected heat from the system mass, and is discharged. This arrangement works very well when cooling water is in sufficient supply. In many areas, however, cooling water is not available or the environmental effects, thermal pollution, are too damaging to permit such circulation of water.

There are several methods to overcome this problem: dry cooling towers, where air is passed over tubes containing the system mass; cooling ponds; and evaporative cooling towers. In the evaporative cooling tower, circulating water is used in the condenser, but air is used to cool this circulating water. The water is cooled by partially evaporating it, humidifying the air, and thus cooling the remaining circulating water. Figure 13.13 presents a schematic diagram of an evaporative cooling tower. The atmospheric air, not saturated, enters the cooling tower at the

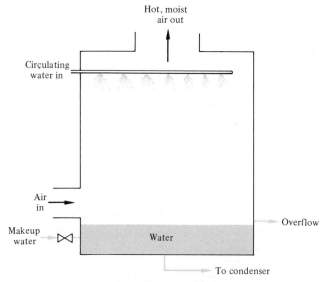

Figure 13.13 A simplified schematic diagram of a cooling tower.

bottom, flowing counter to the direction of the warm circulating water. The circulating water evaporates, saturating the air, and decreasing its own energy and therefore its temperature. By the time it has reached the bottom of the tower, the remaining water is sufficiently cool to return to the condenser. Makeup water must be added to replenish that which evaporated. The temperature and, even more so, the relative humidity of the inlet air are important in determining how much air is required to cool the water. Should the inlet air be saturated, no evaporative cooling could occur.

Example 13.12

An evaporative cooling tower is used by a large power-generating facility located in the Midwest. The circulating water flow rate is 126 kg/s and enters at 38°C. The air inlet conditions are $T_{\text{dry bulb}} = 27°C$ and $T_{\text{wet bulb}} = 15°C$. The exit air temperature is 32°C and 90 percent relative humidity. Determine the air flow rate required to cool the water to 27°C. How much makeup water is required?

An energy balance on cooling tower yields

Energy In = Energy Out

$$\dot{m}_a h_{a_1} + \dot{m}_{fc} h_{f_1} + \dot{m}_{fm} h_{f_2} = \dot{m}_{fc} h_{f_3} + \dot{m}_a h_{a_2} \qquad (13.22)$$

where

$\dot{m}_a$ = air mass flow rate

$\dot{m}_{fc}$ = circulating water flow rate

h_{f_1} = inlet circulating water enthalpy

h_{f_2} = inlet makeup water enthalpy

$\dot{m}_{fm}$ = makeup water flow rate

h_{a_1} = inlet air enthalpy

h_{f_3} = outlet circulating water enthalpy

h_{a_2} = outlet air enthalpy

Assume that $h_{f_2} = h_{f_3}$, which is to say that the makeup water is supplied at the exit temperature of the circulating water. The water enthalpy may be found from the steam tables and the air enthalpy from the psychrometric chart.

The mass of makeup water may be expressed as

$$\dot{m}_{fm} = \dot{m}_a(\omega_2 - \omega_1) \tag{13.23}$$

Substitute Equation (13.23) into Equation (13.22); this leaves one unknown, the air mass, $\dot{m}_a$.

$$\dot{m}_a(42.1) + (126)(159.2) + \dot{m}_a(0.0275 - 0.0056)(113.2)$$
$$= (126)(113.2) + \dot{m}_a(102.5)$$
$$\dot{m}_a = 100 \text{ kg air/s}$$

For makeup water,

$$\dot{m}_{fm} = 100(0.0275-0.0056) = 2.19 \text{ kg water/s}$$

This discussion on cooling towers is by no means definitive; many other factors and effects have to be included. The cooling tower, however, is an increasingly important application of evaporative cooling.

PROBLEMS

1. A gaseous mixture has the following volumetric analysis: O_2, 30%; CO_2, 40%; N_2, 30%. Determine (a) the analysis on a mass basis; (b) the partial pressure of each component if the total pressure is 100 kPa and the temperature is 32°C; (c) the molecular weight of the mixture.

2. A gaseous mixture has the following analysis on a mass basis (gravimetric): CO_2, 30%; SO_2, 30%; He, 20%; N_2, 20%. For a total

pressure and temperature of 101 kPa and 300 K, determine (a) the volumetric (molal) analysis; (b) the component partial pressures; (c) the mixture gas constant; (d) the mixture specific heats.

3. A cubical tank, 1 m on a side, contains a mixture of 1.8 kg of nitrogen and 2.8 kg of an unknown gas. The mixture pressure and temperature are 290 kPa and 340 K. Determine (a) molecular weight and gas constant of the unknown gas; (b) the volumetric analysis.

4. A mixture of ideal gases at 30°C and 200 kPa is composed of 0.20 kg CO_2, 0.75 kg N_2, and 0.05 kg He. Determine the mixture volume.

5. Equal masses of hydrogen and oxygen are mixed. The mixture is maintained at 150 kPa and 25°C. For each component determine the volumetric analysis and its partial pressure.

6. A 3-m^3 drum contains a mixture at 101 kPa and 35°C of 60% methane and 40% oxygen on a volumetric basis. Determine the amount of oxygen that must be added at 35°C to change the volumetric analysis to 50% for each component. Determine also the new mixture pressure.

7. A gaseous mixture of propane, nitrogen, and hydrogen have partial pressures of 83 kPa, 102 kPa, and 55 kPa, respectively. Determine (a) the volumetric analysis; (b) the gravimetric (mass) analysis.

8. A mixture of gases contains 20% N_2, 40% O_2, and 40% CO_2 on a mass basis. The mixture pressure and temperature are 150 kPa and 300 K, respectively. (a) Consider the mixture to be heated in a 20-m^3 tank to 600 K; find the heat required. (b) Consider the mixture to be flowing steadily at 1 kg/s through a heat exchanger until the temperature is doubled; find the heat required.

9. A rigid insulated tank, as shown in Figure 13.3, is divided into two sections by a membrane. One side contains 0.5 kg of nitrogen at 200 kPa and 320 K, and the other side contains 1.0 kg of helium at 300 kPa and 400 K. The membrane is removed; determine (a) the mixture temperature and pressure; (b) the change of entropy for the system; (c) the change of internal energy of the system.

10. A rigid insulated tank, such as shown in Figure 13.3, contains 0.28 m^3 of nitrogen and 0.14 m^3 of hydrogen. The pressure and temperature of each gas is 210 kPa and 93°C, respectively. The membrane separating the gases is removed; determine the entropy of mixing.

11. Ethylene is stored in a 5.6-liter spherical vessel at 260°C and 2750 kPa. To protect against explosion, the vessel is enclosed in another spherical vessel with a volume of 56 liters and filled with nitrogen at 260°C and 10.1 MPa. The entire assembly is maintained at 260°C in a furnace. The inner vessel ruptures; determine (a) the final pressure; (b) the entropy change.

12. An air-conditioning unit receives an air–water vapor mixture at 101 kPa, 35°C, and 80% relative humidity. Determine (a) the dew point;

(b) the humidity ratio; (c) the partial pressure of air; (d) the mass fraction of water vapor.

13. An air–water vapor mixture at 138 kPa, 43°C, and 50% relative humidity is contained in a 1.4 m³. The tank is cooled to 21°C. Determine (a) the mass of water condensed (b) the partial pressure of water vapor initially; (c) the final mixture pressure; (d) the heat transferred.

14. Given for an air–water vapor mixture that $T_{mix} = 60°C$, $p_{mix} = 300$ kPa, and $\phi = 50.1\%$, find the dew point and the humidity ratio.

15. Given for an air–water vapor mixture that $T_{mix} = 70°C$ and $p_{mix} = 200$ kPa and $p_{air} = 180$ kPa, find the dew point, humidity ratio, and relative humidity.

16. Air is to be dehumidified with a silica-gel absorber. The initial air condition is 24°C and a humidity ratio of 0.018 kg water vapor/kg dry air; it is to leave with a humidity ratio of 0.005. The silica gel reduces the humidity ratio to 0.001, so a portion of the initial air by-passes the silica gel and mixes with the air leaving the gel. Determine the mass fraction bypassed.

17. Air enters a drier at 21°C and 25% relative humidity and leaves at 66°C and 50% relative humidity. Determine the air flow rate at 1 atm pressure if 5.44 kg/h of water is evaporated from wet material in the drier.

18. At 10°C and 50% relative humidity, 0.2 m³/s of air is processed, and it leaves at 32°C and 95% relative humidity. The pressure remains constant at 101 kPa. Determine (a) the heat supplied; (b) the water supplied.

19. A continuous dryer is designed to produce, in 24 h, 20 000 kg of product containing 5% water. The product enters the dryer with 35% water. The air used for drying has an inlet temperature of 20°C and a relative humidity of 45% and is preheated to 65°C before reaching the product. The air leaves saturated at 43°C. Determine (a) the air flow rate in m³/s; (b) the heat required in the preheater.

20. A compressor receives an air–water vapor mixture at 96 kPa, 10°C, and with a vapor pressure of 1.0 kPa and compresses it adiabatically to 207 kPa and 65°C. Determine (a) the work required per kilogram; (b) the relative humidity initially and finally.

21. A building has a heat loss of 300 kW. It must be heated by fresh air. The space is to be maintained at 24°C and 45% relative humidity, with an outside temperature of 2°C saturated. Humidification is achieved with 208 kPa saturated steam. Determine (a) the cold air flow rate in m³/s; (b) the steam required in kg/s; (c) the hot air flow.

22. Air is at 37°C dry bulb and 15% relative humidity. The air is to be cooled by evaporative cooling until the relative humidity is 60%. Determine (a) the final temperature; (b) for 0.5 m³/s of air, the water required.

23. In the winter months a building is heated with conditioned air at 25°C dry bulb and 45% relative humidity. Atmospheric air is at 2°C dry bulb and 50% relative humidity. Water is supplied at 65°C. The air flow rate is 5 m³/s. Determine (a) the water added; (b) the heat supplied.

24. An air-conditioning unit removes 35 kW of heat from a building. The unit handles 1.92 m³/s of air, 0.29 m³/s of which is outside air at 35°C dry bulb and 24°C wet bulb, while the remainder is recirculated air from the building. The air in the building is maintained at 25°C dry bulb and 40% relative humidity. The refrigeration coil surface temperature is 10°C. At what condition does the air enter the building? How many kg/s of water are condensed by the air-conditioning unit?

25. A heating system for an office building uses an adiabatic saturation air washer followed by a heating coil. A mixture of 1.2 m³/s of outside air at 5°C dry bulb and 50% relative humidity and 6 m³/s of return air at 20°C dry bulb and 55% relative humidity enter the saturator. The mixture leaves with a temperature 1°C less than saturation. The heating coil heats the mixture to 38°C. For the mixture leaving the heating coil, determine (a) the relative humidity; (b) the specific volume; (c) the heat supplied in the coil.

26. A 5-ton Refrigerant 12 air-conditioning system receives air at 32°C and 75% relative humidity and cools and dehumidifies it to 24°C and 45% relative humidity. The R 12 discharge temperature from the compressor is 43°C and the evaporator temperature is 4°C. Determine (a) the air flow rate; (b) the water removed.

27. Water with a flow rate of 126 kg/s enters a natural-draft cooling tower at 43°C. The water is cooled to 29°C by air entering at 38°C dry bulb and 24°C wet bulb. The air leaves saturated at 40°C. Determine (a) the air supply in m³/s; (b) the makeup water required.

28. A cooling tower is to be installed to supply process cooling water. Water enters the tower at a rate of 190 kg/s at 46°C. Water must leave the tower with a flow rate of 190 kg/s at 29°C. The air enters at 24°C and 50% relative humidity and 1 atm pressure and leaves saturated at 31°C. Determine (a) the air flow rate in m³/s; (b) the makeup water flow rate.

29. An air–water vapor mixture enters an adiabatic device with a pressure of 150 kPa, a temperature of 40°C, and an unknown relative humidity. The air flow rate is 0.2 kg/min. The mixture leaves the device at 30°C, 150 kPa, and 80% relative humidity. Water at 30°C is sprayed into the air for cooling. Determine the water required for one hour of operation.

30. An air conditioner receives 0.23 kg/s of atmospheric air at 28°C and 10% relative humidity. Water at 20°C is sprayed into the air and heat is added and/or removed. The air exits at 16°C and 70% relative

humidity. Determine the heat flow (denote direction) and the water flow rate.

31. At a rate of 15 kg/s, air at 25°C and 30% relative humidity enters a device where heat is added and/or removed and water at 20°C is removed. The exit air dry-bulb temperature is 15°C and the wet-bulb temperature is 5°C. Determine the heat flow (denote direction) and the water flow rate.

Chapter 14
Reactive Systems

This chapter deals with the thermodynamic analysis of the combustion process. The first thing we think of when we hear "combustion" is the burning or oxidation of hydrocarbons. The change of chemical energy into thermal energy is important in most power-producing devices, such as the automotive engine. Nevertheless, the fundamentals that we will develop are also applicable to some chemical reactions in the oxidation of food in living organisms. The chapter will primarily discuss the combustion of hydrocarbon fuels. The first and second laws of thermodynamics and the conservation of mass are the tools of analysis. It is important to remember that no matter how complicated the system under consideration, these laws are applied.

14.1 HYDROCARBON FUELS

One of the basic constituents in the combustion reaction is the fuel; we will be dealing with hydrocarbons—solid, liquid, or vapor. The most important form of solid hydrocarbon is coal, which is mined in several grades, ranging from anthracite (hard) to bituminous (soft). Coal is a mixture of carbon, hydrogen, oxygen, nitrogen, sulfur, water, and non-combustible solid material, ash.

The liquid hydrocarbons, such as gasoline, kerosene, and fuel oil, are obtained by the distillation of petroleum. They have several advantages over the solid fuel: cleanliness and ease of handling and storage. The chemical form of the liquid hydrocarbons is C_xH_y, the values of subscripts x and y depending on the hydrocarbon family. Any fuel, such as gasoline, is actually a mixture of many hydrocarbons. Except in sophisticated analysis, the predominant hydrocarbon is assumed to be the only one present, or an average of the several constituents hydro-

Table 14.1   Volumetric Analysis of Some Fuel–Gas Mixtures (numbers are percent)

	CO	H_2	CH_4	C_2H_6	C_4H_8	O_2	CO_2	N_2
Coal gas	9	53.6	25	—	3	0.4	3	6
Producer gas	29	12	2.6	0.4	—	—	4	52
Blast furnace gas	27	2	—	—	—	—	11	60
Natural gas I	1	—	93	3	—	—	—	3
Natural gas II	—	—	80	18	—	—	—	2
Natural gas III	—	1	93	3.5	—	—	2	0.5

carbons is taken. Alcohols are sometimes used as fuels in internal-combustion engines. In alcohol, one of the hydrogen atoms in the hydrocarbon is replaced by the OH radical; the resulting hydrocarbon is called a "carbohydrate" and written as C_xH_zOH. In this text, unless specifically noted, we will consider gasoline to be a single hydrocarbon, C_8H_{18} (octane), and diesel fuel to be a single hydrocarbon, $C_{12}H_{26}$ (dodecane).

Gaseous hydrocarbon fuels also are a mixture of various constituent hydrocarbons. They have nearly complete combustion and are very clean. The products of their combustion do not have sulfur components, which have adverse environmental effects. There are great differences between natural gas, hydrocarbon fuel found underground, and manufactured gas. These differences lie in the proportion of the constituents found in each, as illustrated in Table 14.1.

14.2 COMBUSTION PROCESS

The combustion process is the oxidation of the fuel constituents, and we may write an equation describing the reaction. During the combustion process the total mass remains the same, so in balancing reaction equations, we are applying the law of conservation of mass. Actually there is a slight reduction in mass from Einstein's equation $E = mc^2$, but this is extremely small, on the order of 10^{-10} kg/kg of fuel, is neglected.

We will first consider the complete oxidation of carbon and in so doing, define the terms commonly used in combustion-reaction analysis.

Reactants Products

$$C + O_2 \rightarrow CO_2 \tag{14.1}$$

In this reaction, carbon and oxygen are the initial substances, the *reactants*; they undergo a chemical reaction, yielding carbon dioxide, the final substance, or *product*. Furthermore, we see that

1 mole C + 1 mole $O_2 \rightarrow$ 1 mole CO_2

and since there are 12 kg/kgmol for carbon, 32 kg/kgmol for oxygen, and 44 kg/kgmol for carbon dioxide, then

$$12 \text{ kg C} + 32 \text{ kg O}_2 \rightarrow 44 \text{ kg CO}_2$$

When a hydrocarbon fuel is completely oxidized, the resulting products are carbon dioxide in water. Let us consider methane as the fuel:

$$CH_4 + 2O_2 \rightarrow CO_2 + 2H_2O \tag{14.2}$$

The water may exist as a solid, liquid, or vapor, depending on the final product temperature and pressure, and this is denoted in parentheses as (s), (1), or (v), respectively. This is an important consideration when we perform energy balances later in the chapter. In the oxidation process, many reactions occur before the final products in Equation (14.2) are formed. These intermediate reactions constitute an important area of investigation, but we will be concerned with only the initial and final products.

Combustion with Air

Most combustion processes depend on air, not pure oxygen. Air contains many constituents, particularly oxygen, nitrogen, argon, and other vapors and inert gases; its volumetric, or molal, composition is approximately 21 percent oxygen, 78 percent nitrogen, and 1 percent argon. Since neither nitrogen nor argon enters into the chemical reaction, we will assume that the volumetric air proportions are 21 percent oxygen and 79 percent nitrogen; and that for 100 moles of air, there are 21 moles of oxygen and 79 moles of nitrogen, or

$$\frac{79}{21} = 3.76 \text{ moles nitrogen/mole oxygen}$$

To account for the argon, which we include as nitrogen, we use 28.16 as the equivalent molecular weight of nitrogen; this is the molecular weight of what is called "atmospheric nitrogen." Pure nitrogen has a molecular weight of 28.016. In analyses in which great accuracy is desired, this distinction should be made; we, however, will consider the nitrogen in the air to be pure.

14.3 THEORETICAL AIR

The combustion of methane in air is

$$CH_4 + 2O_2 + 2(3.76)N_2 \rightarrow CO_2 + 2H_2O + 7.52N_2 \tag{14.3}$$

The nitrogen does not enter into the reaction, but it must be accounted for. There are 3.76 moles nitrogen per mole of oxygen, and since 2 moles of oxygen are necessary for the oxidation of methane, (2)(3.76) moles of nitrogen are present. In Equation (14.3) only a minimum amount of oxygen is included in the equation. The minimum amount of air required to oxidize the reactants is called the *theoretical air*. When combustion is achieved with theoretical air, no oxygen is found in the products. In practice, this is not possible. More oxygen than is theoretically necessary is required to achieve complete combustion of the reactants. The excess air is needed because the fuel is of finite size, and each droplet must be surrounded by more than the necessary number of oxygen molecules to assure oxidation of all the hydrocarbon molecules. This excess air is usually expressed as a percentage of the theoretical air. Thus if 25 percent more air than is theoretically required is used, this is expressed as 125 percent theoretical air or 25 percent excess air. There is 1.25 times as much air actually used than is theoretically required. The combustion of methane in 125 percent theoretical air is

$$CH_4 + (1.25)(2)O_2 + (1.25)(2)(3.76)N_2 \rightarrow$$

$$CO_2 + 2H_2O + 9.4N_2 + 0.5O_2 \qquad (14.4)$$

Equation (14.4) is balanced by first balancing the oxidation equation for theoretical air and then multiplying the theoretical air by 1.25 to account for the 125 percent theoretical air. The amount of nitrogen and oxygen appearing in the products is determined by a mass balance on each term.

If the excess air is insufficient to provide complete combustion, then not all the carbon will be oxidized to carbon dioxide; some will be oxidized to carbon monoxide. When there is considerably less theoretical air, unburned hydrocarbons will be present in the products. This is the soot or black smoke that sometimes pours from chimneys and smoke-stacks. This occurs when one or more of the following three conditions for complete combustion have not been met:

1. The air–fuel mixture must be at the ignition temperature.
2. There must be sufficient oxygen to assure complete combustion.
3. The oxygen must be in intimate contact with the fuel.

Smoky products of combustion during start-up operations usually result from a failure to satisfy requirements (1) and (3). To balance the combustion equation when incomplete combustion occurs, we need information about the products. For instance, assume that, with theoretical air, the oxidation of carbon is 90 percent complete in the combustion of methane; then

$$CH_4 + 2O_2 + 2(3.76)N_2 \rightarrow$$

$$0.9CO_2 + 0.1CO + 2H_2O + 7.52N_2 + 0.05O_2 \qquad (14.5)$$

14.4 AIR–FUEL RATIO

Two important terms in the combustion process give us the proportion of air to fuel; these are the *air–fuel ratio*, $r_{a/f}$, and its reciprocal, the *fuel–air ratio*, $r_{f/a}$. Both may be expressed in terms of the moles or the mass of the fuel and air present.

Example 14.1

Calculate the theoretical air–fuel and fuel–air ratios for the complete combustion of ethane, C_2H_6.

The combustion equation is

$$C_2H_6 + 3.5O_2 + 3.5(3.76)N_2 \rightarrow 2CO_2 + 3H_2O + 13.17N_2$$

$$r_{a/f} = \text{moles air/moles fuel} = \frac{3.5 + 13.7}{1}$$

$$= 16.67 \text{ moles air/mole fuel}$$

$$r_{f/a} = \frac{1}{r_{a/f}} = 0.06 \text{ moles fuel/mole air}$$

Very often in engineering calculations it is desirable to have the air–fuel ratio on a mass basis. The molecular weight of air is 28.97 kg/kgmol and the molecular weight of ethane is 30.068 kg/kgmol, so

$$r_{a/f} = \frac{(28.97)(16.67)}{(30.068)(1)} = 16.05 \text{ kg air/kg fuel}$$

$$r_{f/a} = \frac{1}{r_{a/f}} = 0.0623 \text{ kg fuel/kg air}$$

It is important to specify the units on the air–fuel ratio.

Example 14.2

A fuel oil is burned with 50 percent excess air, and the combustion characteristics of the fuel oil are similar to $C_{12}H_{26}$. Determine the volumetric (molal) analysis of the products of combustion, and determine the dew point for the products if the pressure is 101 kPa.

$$C_{12}H_{26} + (1.5)(18.5)O_2 + (1.5)(18.5)(3.76)N_2 \rightarrow$$

$$12CO_2 + 13H_2O + 104.3N_2 + 9.25O_2$$

The total moles of the products are

$12 + 13 + 104.2 + 9.25 = 138.55$

Molal analysis of the products of combustion is

CO_2: $\dfrac{12}{138.55} = 0.0866$

H_2O: $\dfrac{13}{138.55} = 0.0938$

N_2: $\dfrac{104.2}{138.55} = 0.7528$

O_2: $\dfrac{9.25}{138.55} = 0.0668$

The partial pressure of water is $(0.0938)(101) = 9.47$ kPa. The saturation temperature corresponding to this pressure is 45°C, which is also the dew-point temperature. If the temperature drops below this point, precipitation of water will occur. Should precipitation occur, the liquid water will contain dissolved gases, forming a corrosive substance. To prevent this, the temperature of the products of combustion is kept well above the dew point in smokestacks and exhaust piping.

Coal has played an important role as an energy source in the past and will become increasingly important in the future as oil and natural gas become more expensive. Coal has a variable chemical composition, since it originally was vegetation. It is essentially a hydrocarbon fuel, with impurities such as entrapped water and ash that represent the inorganic matter that does not oxidize. The analysis of coal is on a mass basis, and the ultimate analysis of coal includes all the constituent mass fractions. The proximate analysis is often used to define the percent carbon in the coal; that is,

$\%C = 100\% - \%$ ash $- \%$ moisture $- \%$ volatiles

The volatiles are those compounds that evaporate at low temperature when coal is heated. The following example will illustrate these concepts.

Example 14.3

An ultimate analysis of coal yields the following composition: 74 percent C, 5 percent H_2, 6 percent O_2, 1 percent S, 1.2 percent N_2, 3.8 percent water, 9.0 percent ash. Determine the theoretical air–fuel ratio.

First, determine the mole fractions of the coal's constituents on an ashless basis.

	x_i	M_i	x_i/M_i	y_i
C	0.813	12.0	0.0677	0.674
H_2	0.055	2.0	0.0275	0.274
O_2	0.066	32.0	0.0021	0.021
S	0.011	32.0	0.0003	0.003
N_2	0.013	28.0	0.0005	0.005
H_2O	0.042	18.0	0.0023	0.023
	1.000		0.1004	1.000

Then write the combustion equation for one mole of ashless coal.

$$\overbrace{0.674C + 0.274H_2 + 0.021O_2 + 0.003S + 0.005N_2 + 0.023H_2O}^{\text{fuel}}$$

$$\overbrace{+ aO_2 + bN_2}^{\text{air}} \rightarrow$$

$$cCO_2 + dH_2O + eSO_2 + fN_2$$

carbon balance: $\qquad\qquad\qquad\qquad c = 0.674$

hydrogen balance: $\qquad\qquad\qquad 2d = 2(0.274 + 0.023);$

$$d = 0.297$$

sulfur balance: $\qquad\qquad\qquad\qquad e = 0.003$

oxygen balance: $\quad 0.021 + 0.0115 + a = 0.674 + 0.148 + 0.003$

$$a = 0.792$$

nitrogen balance: $\quad (0.005) + (3.76)(0.792) = f; \qquad f = 2.983$

$$(r_{a/f})_{molar} = \frac{(0.792)(4.76)}{(1.0)} = 3.77 \text{ kgmol air/kgmol fuel}$$

$$M_{fuel} = \frac{1}{0.1004} = 9.96 \text{ kg/kg mol}$$

$$(r_{a/f})_{mass} = \frac{(0.792)(4.76)(28.97)}{(1)(9.96)} = 10.96 \text{ kg air/kg fuel}$$

14.5 PRODUCTS OF COMBUSTION

In power plants and other facilities using large amounts of fuel, it is important that the burning be as efficient as possible. Tiny increases in efficiency (even a fraction of a percent) can save thousands of dollars. One important factor affecting the efficiency is the amount of excess air.

If not enough air is used, combustion will be incomplete and not all the chemical energy of the fuel will be utilized. If too much air is used, the heat released by combustion is wasted in heating this excess air. The object is to oxidize the carbon completely with the smallest amount of air. This will yield the greatest release of energy per mass of air. How may this be determined?

Orsat Analysis

An analysis of the products of combustion tells us how much of each product was formed. The Orsat apparatus performs this task by measuring carbon dioxide, carbon monoxide, and oxygen volumetrically. The combustion gas is passed through various chemicals, which absorb carbon dioxide, carbon monoxide, and oxygen. The volumetric decrease of the combustion gas is noted at each step, and the decrease in volume divided by the initial volume gives the percentage of each combustion product. The remaining volume is assumed to be nitrogen. The Orsat gives the volumetric proportions on a dry basis; the amount of water vapor cannot be determined. Since the analysis usually occurs at room temperature and pressure, which is below the dew point of most hydrocarbon products of combustion, the error involved is quite small.

The Orsat analysis cannot measure partially combusted hydrocarbons, nor can it measure carbon. These quantities are important in dealing with internal-combustion engines, where the combustion is not complete and we want to analyze reactions occurring before the final products are formed. There also may be nitric oxides formed at high temperatures. There are measuring devices, gas chromatographs, that can determine these compounds, but these problems are beyond the scope of this book. They involve reaction times and partial reactions; we will consider only complete combustion and will not consider time-dependent reactions at all.

The principles used in an Orsat analysis will be demonstrated by example problems. We can determine the air–fuel ratio as well as balance the reaction equation.

Example 14.4

Fuel oil, $C_{12}H_{26}$, is burned in air at atmospheric pressure. The Orsat analysis of the products of combustion yields

CO_2:	12.8%
O_2:	3.5%
CO:	0.2%
N_2:	83.5%
	100.0%

Determine the air–fuel ratio, the percent theoretical air, and the combustion equation.

We will write the combustion equation for 100 moles of dry products, allowing us to write the percentage as the number of moles. Then the unknown coefficients on the reactant side may be determined by applying the conservation of mass to each reactant.

$$aC_{12}H_{26} + bO_2 + cN_2 \rightarrow$$
$$12.8CO_2 + 0.2CO + 3.5O_2 + 83.5N_2 + dH_2O$$

C balance: $12a = 12.8 + 0.2 \therefore a = \dfrac{13}{12}$

N_2 balance: $c = 83.5$

O_2–N_2 ratio: $\dfrac{c}{b} = 3.76 \qquad \therefore b = 22.2$

H_2 balance: $2d = \left(\dfrac{13}{12}\right)(26) \qquad \therefore d = 14.08$

Divide the equation by a to determine the combustion equation for 1 mole of fuel.

$$C_{12}H_{26} + 20.5O_2 + 77.08N_2 \rightarrow 11.8CO_2 + 0.18CO + 3.23O_2$$
$$+ 77.08N_2 + 13H_2O$$
$$r_{a/f} = \frac{(20.5)(32) + 77.08(28)}{1(144 + 26)}$$
$$= 16.5 \text{ kg air/kg fuel}$$

The balanced equation for 100 percent theoretical air is

$$C_{12}H_{26} + 18.5O_2 + 18.5(3.76)N_2 \rightarrow 12CO_2 + 13H_2O + 69.56N_2$$
$$r_{a/f} = \frac{(18.5)(32) + (69.56)(28)}{1(144 + 26)}$$
$$= 14.93 \text{ kg/air/kg fuel}$$
$$\text{theoretical air} = \frac{16.5}{14.93} \times 100 = 110.5 \text{ percent}$$

Example 14.5

An unknown fuel has the following Orsat analysis:

CO_2:   12.5%
CO: 0.3%
O_2:   3.1%
N_2:   84.1%

Determine the air–fuel ratio, the fuel composition on a mass basis, and the percent theoretical air. Write the equation for 100 moles of dry products.

$$C_aH_b + cO_2 + dN_2 \rightarrow$$

$$12.5CO_2 + 0.3CO + 3.1O_2 + 84.1N_2 + eH_2O$$

C balance: $\quad a = 12.5 + 0.3 = 12.8$

N$_2$ balance: $\quad d = 84.1; \quad \dfrac{d}{c} = 3.76, \quad \therefore c = 22.36$

O$_2$ balance: $\quad 22.36 = 12.5 + \dfrac{0.3}{2} + 3.1 + \dfrac{e}{2}$

$$e = 13.2$$

H$_2$ balance: $\quad b = 2e = 26.4$

$$C_{12.8}H_{26.4} + 22.35O_2 + 84.1N_2 \rightarrow 12.5CO_2 + 0.3CO + 3.1O_2$$
$$+ 84.1N_2 + 13.2H_2O$$

$$r_{a/f} = \frac{(22.36 + 84.1)(28.97)}{1[(12)(12.8) + (1)(26.4)]}$$

$$= 17.13 \text{ kg air/kg fuel}$$

$$C_{12.8}H_{26.4} + 19.4O_2 + 73N_2 \rightarrow 12.8CO_2 + 13.2H_2O + 73N_2$$

$$r_{a/f} = \frac{(92.4)(28.97)}{(1)(180)}$$

$$= 14.87 \text{ kg air/kg fuel}$$

$$\text{theoretical air} = \frac{17.13}{14.87} \times 100 = 115 \text{ percent}$$

Fuel Composition

C: $\quad \dfrac{(12.8)(12)}{180} = 0.853 \text{ or } 85.3 \text{ percent}$

H: $\quad \dfrac{(26.4)(1)}{180} = 0.147 \text{ or } 14.7 \text{ percent}$

14.6 ENTHALPY OF FORMATION

In the previous chapters the working substance has always been homogeneous and never changed its chemical composition during any process. Tables were developed describing the properties of these substances—for example, the steam tables—and in these tables there was always an arbitrary reference base. The enthalpy of saturated liquid water at 0°C is zero; this is an arbitrary base, but it did not matter as we always dealt with changes in the enthalpy as with changes in any other property. In chemical reactions, however, the substance in the system changes during the course of the process, so the capricious use of arbitrary standards for each substance would make the energy analysis of the process impossible.

To overcome this difficulty, the enthalpy of all elements is assumed to be zero at an arbitrary reference state of 25°C and 1 atm pressure. The enthalpy of formation of a compound is its enthalpy at this temperature and pressure.

Consider a steady-state combustion process in which 1 mole of carbon and 1 mole of oxygen at the reference state of 25°C and 1 atm pressure combine to produce 1 mole of carbon dioxide. Heat is transferred, so the carbon dioxide finally exists at the reference state. Figure 14.1 demonstrates this process.

The reaction equation is

$$C + O_2 \rightarrow CO_2$$

Let H_R be the total enthalpy of all the reactants and H_P be the total enthalpy of all the products. The first law for this process is

$$Q + H_R = H_P \tag{14.6}$$

or

$$Q + \sum_R n_i \bar{h}_i = \sum_P n_j \bar{h}_j \tag{14.7}$$

where the summations are over all the reactants and all the products, respectively.

Since the enthalpy of all the reactants is zero (they are all elements), we find that

$$Q = H_P - 393\ 757 \text{ kJ} \tag{14.8}$$

where the heat transferred has been carefully measured and found to be $-393\ 757$ kJ. The sign is negative because heat is flowing from the control volume, opposite to the assigned direction in Equation (14.6). The enthalpy of CO_2 at the reference state, 25°C and 1 atm pressure, is called the "enthalpy of formation" and designed by the symbol $\bar{h}_f^\circ$.

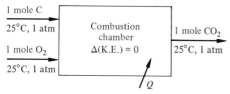

Figure 14.1 *A steady-state combustion process with heat transfer.*

Therefore

$$\left(\bar{h}_f^{\circ}\right)_{CO_2} = -393\ 757\ \text{kJ/kgmol} \tag{14.9}$$

The negative sign for the enthalpy of formation is due to the reaction's being exothermic; heat is released from the combustion of carbon and oxygen. Since the energy must be conserved in the reaction and 393 757 kJ left the control volume, the energy of the carbon dioxide must be less than the energy of the reactants by an amount equal to the heat transferred. The enthalpy of the reactants is zero at 25°C and 1 atm; therefore the enthalpy of carbon dioxide at 25°C and 1 atm must be negative.

Table C.1 lists the enthalpies of formation for several different substances.

14.7 FIRST-LAW ANALYSIS FOR STEADY-STATE REACTING SYSTEMS

In general, the steady-state combustion process will transfer heat and produce work, as illustrated in Figure 14.2. An energy balance would yield

$$Q + \sum_R n_i \bar{h}_i = W + \sum_P n_j \bar{h}_j \tag{14.10}$$

The enthalpies of formation may be readily used in this analysis, since they are all relative to the same reference base. Usually the analysis is done per mole of fuel. The following examples illustrate important concepts that may be used in combustion energy analysis.

Example 14.6

Propane, C_3H_8, undergoes a steady-state, steady-flow reaction with atmospheric air. Determine the heat transfer per mole of fuel entering the combustion chamber. The reactants and products are

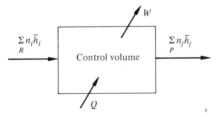

Figure 14.2   A steady-flow system where chemical reaction produces heat and work.

at 25°C and 1 atm pressure.

$$C_3H_8 + 5O_2 + 5(3.76)N_2 \rightarrow 3CO_2 + 4H_2O_{(l)} + 18.8N_2$$

The first law, with no work being done in the control volume is

$$Q + \sum_R n_i \bar{h}_i = \sum_P n_j \bar{h}_j$$

Since the enthalpy of all the elements at 25°C and 1 atm is zero, the summation terms contain only the enthalpies of formation for the compounds, which are found in Table C.1.

$$\sum_R n_i \bar{h}_i = \left(\bar{h}_f^{\circ} \right)_{C_3H_8} = -103\ 909 \text{ kJ/kgmol}$$

$$\sum_P n_j \bar{h}_j = 3 \left(\bar{h}_f^{\circ} \right)_{CO_2} + 4 \left(\bar{h}_f^{\circ} \right)_{H_2O(l)} = -2\ 325\ 311$$

$$Q = -2\ 221\ 402 \text{ kJ/kgmol fuel}$$

In most cases, neither the reactants nor the products are at the reference condition of 25°C and 1 atm pressure. In these cases we must account for the property change between the reference state and the actual state. Appendix Table C.2 tabulates the change in enthalpy between the reference state and the actual state, $(\bar{h}^{\circ} - \bar{h}_{298}^{\circ})$ in kJ/kgmol. The temperature 298 K is the temperature of the reference state in absolute notation. The superscript ° denotes that the pressure is 1 atm. If these tabulated values are not available, then we have to use the ideal gas law, $\bar{h} = \bar{c}_p T$, determining the specific heat by the best means available. For example, if the specific heat varied as a function of temperature and this functional relationship were known, this would be used. If tables of property values existed, such as the gas tables or steam tables, then these, too, could be used.

All problems in this text are done at 1 atm pressure. This is the standard pressure and the one found in most combustion reactions. When it is necessary to include the pressure effects, as when calculating entropy changes, the reactants and products must be expanded to the reference state and the change in doing so included in the calculation. This is a very small effect and need not be included in most engineering applications.

In general, the first law for a steady-state, steady-flow reaction may now be written as

$$Q + \sum_R n_i \left[\bar{h}_f^{\circ} + \left(\bar{h}^{\circ} - \bar{h}_{298}^{\circ} \right) \right]_i = W + \sum_P n_j \left[\bar{h}_f^{\circ} + \left(\bar{h}^{\circ} - \bar{h}_{298}^{\circ} \right) \right]_j$$

$$(14.11)$$

The following examples illustrate some important concepts.

Example 14.7

A diesel engine used dodecane, $C_{12}H_{26}(v)$, for fuel. The fuel and the air enter the engine at 25°C. The products of combustion leave at 600 K and 200 percent theoretical air is used; the heat loss from the engine is 232 000 kJ/kgmol fuel. Determine the work for a fuel-rate of 1 kgmol/h.

Combustion Equation

$$C_{12}H_{26}(v) + 2(18.5)O_2 + 2(18.5)(3.76)N_2 \rightarrow$$
$$12CO_2 + 13H_2O + 18.5O_2 + 139.12N_2$$

First Law, Open System

$$Q + \sum_R n_i \left[\bar{h}_f^\circ + \left(\bar{h}^\circ - \bar{h}_{298}^\circ \right) \right]_i = W + \sum_P n_j \left[\bar{h}_f^\circ + \left(\bar{h}^\circ - \bar{h}_{298}^\circ \right) \right]_j$$

$$Q = -232\ 000 \text{ kJ/kgmol fuel}$$

$$\sum_R n_i \left[\bar{h}_f^\circ + \left(\bar{h}^\circ - \bar{h}_{298}^\circ \right) \right]_i = 1 \left(\bar{h}_f^\circ \right)_{C_{12}H_{26}(v)}$$

$$= -290\ 971 \text{ kJ/kgmol}$$

$$\sum_P n_j \left[\bar{h}_f^\circ + \left(\bar{h}^\circ - \bar{h}_{298}^\circ \right) \right]_j = 12(-393\ 757 + 12\ 916)$$

$$+ 13(-241\ 971 + 10\ 498)$$
$$+ 139.12(8891) + 18.5(9247)$$
$$= -6\ 171\ 255 \text{ kJ/kgmol fuel}$$

$$Q + H_R = W + H_P$$
$$-232\ 000 - 290\ 971 = W - 6\ 171\ 255$$
$$W = 5\ 648\ 284 \text{ kJ/kgmol fuel}$$
$$\dot{W} = \dot{n}_f W = 5\ 648\ 284 \text{ kJ/h}$$
$$\dot{W} = 1568.9 \text{ kW}$$

Example 14.8

A gas-turbine generating unit produces 600 kW and uses $C_8H_{18}(l)$ as a fuel at 25°C, 400 percent theoretical air is used, and the air enters at 310 K. The products of combustion leave at 700 K. The heat transfer is 146.5 kW. Determine the fuel consumption for complete combustion.

The combustion equation is

$$C_8H_{18}(l) + 4(12.5)O_2 + 4(12.5)(3.76)N_2 \rightarrow$$
$$8CO_2 + 9H_2O + 37.5O_2 + 188N_2$$

Determine H_R and H_P for 1 mole of fuel (as combustion equation); then utilize a first-law analysis to determine the fuel flow rate,

$$H_R = \sum_R n_i \left[\bar{h}_f^\circ + \left(\bar{h}^\circ - \bar{h}_{298}^\circ \right) \right]_i$$

$$= 1(-250\ 102) + 50(384) + 188(377)$$

$$H_R = -160\ 026 \text{ kJ/kgmol fuel}$$

$$H_P = \sum_P n_j \left[\bar{h}_f^\circ + \left(\bar{h}^\circ - \bar{h}_{298}^\circ \right) \right]_j$$

$$= 8(-393\ 757 + 17\ 761) + 9(-241\ 971 + 14\ 184)$$

$$+ 37.5(12\ 502) + 188(11\ 937)$$

$$H_P = -2\ 345\ 070 \text{ kJ/kgmol fuel}$$

$$H_P - H_R = -2\ 185\ 044 \text{ kJ/kgmol fuel}$$

First Law, Open System

$$-146.5 + \dot{n}_f H_R = 600 + \dot{n}_f H_P$$

$$\dot{n}_f = 0.000\ 341 \text{ kgmol fuel/second}$$

$$M_f = 114.23 \text{ kg/kgmol}$$

$$\dot{m}_f = \dot{n}_f M_f = 0.039 \text{ kg fuel/s}$$

Heretofore all the problems have been for open systems. The following example illustrates what happens if the system is closed.

Example 14.9

A mixture of methane and oxygen, in the proper ratio for complete combustion and at 25°C and 1 atm, react in a constant-volume calorimeter bomb. Heat is transferred until the products of combustion are at 400 K. Determine the heat transfer per mole of methane.

Combustion Equation

$$CH_4 + 2O_2 \rightarrow CO_2 + 2H_2O$$

First Law, Closed System, $V = C$

$$Q = U_2 - U_1 = \sum_P n_j \bar{u}_j - \sum_R n_i \bar{u}_i$$

$$\bar{u} = \bar{h} - \bar{R}T$$

$$\sum_R n_i \bar{u}_i = \sum_R n_i \left[\bar{h}_f^\circ + \left(\bar{h}^\circ - \bar{h}_{298}^\circ \right) - \bar{R}T \right]_i$$

$$U_R = 1[-74\,917 + 0 - (8.3143)(298)]$$
$$+ 2[0 + 0 - (8.3143)(298)]$$
$$U_R = -82\,350 \text{ kJ/kgmol fuel}$$

$$\sum_P n_j \bar{u}_j = U_P = \sum_P n_j \left[\bar{h}_f^\circ + \left(\bar{h}^\circ - \bar{h}_{298}^\circ \right) - \bar{R}T \right]_j$$

$$U_P = 1[-393\,757 + 4008 - (8.3143)(400)]$$
$$+ 2[-241\,971 + 3452 - (8.3143)(400)]$$
$$U_P = -876\,764 \text{ kJ/kgmol fuel}$$

$$Q = U_2 - U_1 = U_P - U_R = -794\,414 \text{ kJ/kgmol fuel}$$

14.8 ADIABATIC FLAME TEMPERATURE

So far we have assumed that heat transfer and work occur during a combustion process. If no work, no heat transfer, or no change in kinetic or potential energy should occur, then all the thermal energy would go into raising the temperature of the products of combustion. When the combustion is complete under these circumstances, the maximum amount of chemical energy has been converted into thermal energy and the temperature of the products is at its maximum. This temperature is called the "adiabatic flame temperature."

Should the combustion be incomplete or excess air be used, the temperature of the mixture would be less than the maximum adiabatic flame temperature. Excess air is utilized in engine design to keep the temperature within metallurgical limits. If combustion is incomplete, not all the chemical energy is converted into thermal energy; hence the temperature will be lower than the maximum that is possible. When excess air is used, the thermal energy must be used to raise the temperature of a greater mass; hence the temperature rise for the fixed amount of thermal energy will not be as great as the maximum. A third factor that reduces temperature is dissociation of the combustion products. The dissociation reaction is endothermic; it uses some of the available thermal energy to proceed. We will discuss this reaction later.

The following example will illustrate the adiabatic flame temperature. It is a trial-and-error solution process.

Example 14.10

Gaseous propane is burned with 100 and 400 percent theoretical air. Determine the adiabatic flame temperature in each case.

The combustion equation for 100 percent theoretical air is

$$C_3H_8 + 5O_2 + 5(3.76)N_2 \rightarrow 3CO_2 + 4H_2O + 18.8N_2$$

The first law with the absence of heat and work reduces to

$$H_R = H_P$$

$$H_R = \sum_R n_i \left[\bar{h}_f^\circ + \left(\bar{h}^\circ - \bar{h}_{298}^\circ \right) \right]_i$$

$$= 1(-103\ 909) = -103\ 909 \text{ kJ/kgmol fuel}$$

$$H_P = \sum_P n_j \left[\bar{h}_f^\circ + \left(\bar{h}^\circ - \bar{h}_{298}^\circ \right) \right]_j = 3 \left[-393\ 757 + \left(\bar{h}^\circ - \bar{h}_{298}^\circ \right) \right]$$

$$+ 4 \left[-241\ 971 + \left(\bar{h}^\circ - \bar{h}_{298}^\circ \right) \right] + 18.8 \left[\left(\bar{h}^\circ - \bar{h}_{298}^\circ \right) \right]$$

Guess that $T = 2500$ K for the products; then

$$H_P = +10\ 845$$

Try another temperature, $T = 2300$ K:

$$H_P = -206\ 068$$

Plot $(H_P - H_R)$ versus T and where the line connecting the two points intersects the $(H_P - H_R) = 0$ axis gives the linear approximation of the temperature that will satisfy the first-law equation. The temperature that will balance the equation is $T = 2394$ K the adiabatic flame temperature for 100 percent theoretical air.

The balanced equation for 400 percent theoretical air is

$$C_3H_8 + (4)(5)O_2 + (4)(5)(3.76)N_2 \rightarrow$$
$$3CO_2 + 4H_2O + 15O_2 + 75.2N_2$$

$$H_R = H_P$$

$$H_R = -103\ 909 \text{ kJ/kgmol fuel}$$

Guess that $T = 1000$ K:

$$H_P = +10\ 612$$

Guess that $T = 900$ K:

$$H_P = -317\ 290$$

Plot $(H_P - H_R)$ versus T, yielding $T = 942$ K for the adiabatic flame temperature with 400 percent theoretical air.

Notice the significant temperature decrease brought about by using excess air. This can be an important engineering tool in designing optimum energy transfer in equipment, such as steam generators and automotive engines.

14.9 ENTHALPY OF COMBUSTION, HEATING VALUE

The enthalpy of combustion, $\bar{h}_{RP}$, is the difference between the enthalpies of the products and reactants at the same temperature, T, and pressure. Thus

$$\bar{h}_{RP} = H_P - H_R$$

$$\bar{h}_{RP} = \sum_P n_j \left[\bar{h}_f^\circ + \left(\bar{h}_T^\circ - \bar{h}_{298}^\circ \right) \right]_j$$

$$- \sum_R n_i \left[\bar{h}_f^\circ + \left(\bar{h}_T^\circ - \bar{h}_{298}^\circ \right) \right]_i \qquad (14.12)$$

The enthalpy of combustion usually is expressed in the units of kJ/kg of fuel. This is also called the "heating value" of the fuel at constant pressure, because the heat is transferred at constant pressure in an open system and is equal to the same enthalpy difference. The enthalpy of combustion of various fuels is given in Table C.3, where the values are given at the reference state.

The internal energy of combustion, $\bar{u}_{RP}$, is the difference between the internal energies of the products and reactants and may be written as

$$\bar{u}_{RP} = U_P - U_R$$

$$\bar{u}_{RP} = \sum_P n_j \left[\bar{h}_f^\circ + \left(\bar{h}_T^\circ - \bar{h}_{298}^\circ \right) - \bar{R}T \right]_j$$

$$- \sum_R n_i \left[\bar{h}_f^\circ + \left(\bar{h}_T^\circ - \bar{h}_{298}^\circ \right) - \bar{R}T \right]_i \qquad (14.13)$$

where all the reactants and products have been considered as gases. Should this not be the case, then $p\bar{v}$ must be used in place of $\bar{R}T$ in determining the internal energy. Since this is equal to the heat transferred at constant volume, the constant-volume heating value of the fuel is also equal to $\bar{u}_{RP}$. Notice that Equation (14.13) may be written more compactly as

$$\bar{u}_{RP} = \bar{h}_{RP} - \bar{R}T \left(\sum_P n_j - \sum_R n_i \right) \qquad (14.14)$$

We have seen that it is important to know whether the water in the products of combustion is in the liquid or vapor phase. The term *higher heating value* of a fuel indicates that the water in the products of combustion is liquid; thus the latent heat of water is included in determining the heat transferred. The term *lower heating value* means that the water in the products of combustion exists as a vapor, hence the latent heat will not be included in determining $\bar{h}_{RP}$ and it will have a lower numerical value than the *higher heating value*.

Example 14.11

Calculate the enthalpy of combustion of liquid propane at 25°C for the following conditions. The enthalpy of evaporation is 370 kJ/kg. (a) The water is a liquid. (b) The water is a vapor.

Nitrogen is not needed in the combustion equation since both reactants and products are at 25°C.

$$C_3H_8(l) + 5O_2 \rightarrow 3CO_2 + 4H_2O(l)$$

$$\bar{h}_{RP} = H_P - H_R$$

$$H_R = \left(\bar{h}_f^\circ\right)_{C_3H_8(l)} = \left(\bar{h}_f^\circ\right)_{C_3H_8(v)} - Mh_{fg}$$

$$H_R = -103\ 909 - 44.1(370)$$

$$= -120\ 226 \text{ kJ/kgmol fuel}$$

$$H_P = 3\left(\bar{h}_f^\circ\right)_{CO_2} + 4\left(\bar{h}_f^\circ\right)_{H_2O(l)}$$

$$H_P = 3(-393\ 757) + 4(-286\ 010)$$

$$= -2\ 325\ 311 \text{ kJ/kgmol fuel}$$

$$\bar{h}_{RP} = \frac{H_P - H_R}{M} = -\frac{2\ 205\ 085}{44.1} = -50\ 002 \text{ kJ/kg}$$

Thus the higher heating value of propane is 50 002 kJ/kg.

For the case in which water is a vapor, the enthalpy of the reactants is the same, but the enthalpy of the products is

$$H_P = 3\left(\bar{h}_f^\circ\right)_{CO_2} + 4\left(\bar{h}_f^\circ\right)_{H_2O(v)} = 3(-393\ 757) + 4(-241\ 971)$$

$$H_P = -2\ 149\ 155$$

$$\bar{h}_{RP} = \frac{H_P - H_R}{M} = \frac{-2\ 028\ 929}{44.1} = -46\ 007 \text{ kJ/kg}$$

Thus the lower heating value of liquid propane is 46 007 kJ/kg.

Example 14.12

Calculate the enthalpy of combustion of ethane, C_2H_6, at 500 K. The average value of the specific heat at constant pressure of ethane may be taken as 2.22 kJ/kg · K.

The combustion equation is

$$C_2H_6 + 3.5O_2 \rightarrow 2CO_2 + 3H_2O(v)$$

$$\bar{h}_{RP} = H_P - H_R$$

$$H_P = 2(-393\ 757 + 8314) + 3(-241\ 971 + 6920)$$

$$H_P = -1\ 476\ 039 \text{ kJ/kgmol fuel}$$

$$H_R = 1\big[-84\ 718 + (2.22)(30.07)(500 - 298)\big] + 3.5(6088)$$

$$H_R = -49\ 925 \text{ kJ/kgmol fuel}$$

$$\bar{h}_{RP} = \frac{H_P - H_R}{M} = -\frac{1\ 426\ 113}{30.07} = -47\ 426 \text{ kJ/kg}$$

14.10 SECOND-LAW ANALYSIS

The combustion process and most chemical reactions proceed very rapidly. As we have seen previously, processes that proceed rapidly do so irreversibly. The entropy change for such a reaction indicates how irreversibly the reaction occurred and, knowing this, we can determine the available portion of the thermal energy released by the combustion process.

As with enthalpy, we must be able to refer all the substance entropies to a common reference state or base. In this manner entropies of different substances may be added and subtracted, since they are common to the same base. The third law of thermodynamics states that the entropy of all pure substances is zero at zero degrees absolute. This gives us a reference base, and entropy measured from this base is called "absolute entropy." In Tables C.1 and C.2 the values of absolute entropy at 1 atm pressure, $\bar{s}°$, are given for various temperatures for several substances. In these tables and by customary practice, 1 atm is essentially 100 kPa.

To account for the reaction's not occurring at standard conditions, the absolute entropy at some state 2 is

$$\bar{s}_2 = \bar{s}° + \Delta\bar{s}_{0-2}$$

$$\Delta\bar{s}_{0-2} = \bar{\phi} - \bar{\phi}_0 - \bar{R}\ln\left(\frac{p}{p_0}\right)$$

but

$$p_0 = 1 \text{ atm}$$

and

$$\phi_0 = \bar{s}°$$

thus

$$\bar{s}_2 = \bar{\phi} - \bar{R}\ln(p) \qquad (14.15)$$

If values of $\bar{\phi}$ are not available in the gas tables or elsewhere, then the $\int c_p \, dT/T$ must be evaluated.

In Chapter 9 we studied available energy and introduced the concept of availability, $\mathcal{C}$. We mentioned that availability was particularly important when dealing with chemical reactions. Furthermore, we showed that the change of availability was equal to the maximum work possible for a process between two states, and that if this process took place at constant temperature and pressure, the change in availability was equal to the change in the Gibbs function,

$$d\mathcal{C} = (dG)_{T_0, p_0} \tag{14.16}$$

If we integrate this expression between initial and final conditions, then

$$W_{max} = \mathcal{C}_1 - \mathcal{C}_2 = \sum_R n_i \bar{g}_i - \sum_P n_j \bar{g}_j \tag{14.17}$$

where the maximum work is written as a positive quantity. The availability decreases between the initial and final conditions.

Since $g = h - Ts$ and $T = T_0$, then for the reactive systems under consideration

$$g = h - T_0 s$$

Just as enthalpy of formation was necessary when analyzing compounds in first-law analysis of reactive systems, so is the Gibbs function of formation, $\bar{g}_f^\circ$, necessary in second-law analysis. The Gibbs function of formation is given in Table C.1 for several compounds at reference conditions of 25°C and 1 atm pressure. The Gibbs function of all elements is zero at reference conditions. The following example illustrates this.

Example 14.13

Determine the Gibbs function of $H_2O(l)$. The reaction for H_2O may be written as

$$2H_2 + O_2 \rightarrow 2H_2O(l)$$

where both the reactants and products are measured at 25°C and 1 atm and all the water is assumed to be a liquid.

$$G_P - G_R = (H_P - H_R) - T_0(S_P - S_R)$$

$$G_P - G_R = \sum_P n_j \left(\bar{h}_f^\circ\right)_j - \sum_R n_i \left(\bar{h}_f^\circ\right)_i$$

$$- T_0 \left[\sum_P n_j \left(\bar{s}_{298}^\circ\right)_j - \sum_R n_i \left(\bar{s}_{298}^\circ\right)_i \right]$$

Since the enthalpy and enthalpy of formation of elements are zero at reference conditions, this reduces to

$$G_P - G_R = 2(\bar{h}_f^\circ)_{H_2O(l)} - 298[2(\bar{s}_{298}^\circ)_{H_2O} - 2(\bar{s}^\circ)_{H_2} - (\bar{s}^\circ)_{O_2}]$$

$$G_P - G_R = 2(-286\ 010) - 298[2(69.980) - 2(130.684) - 205.142]$$

The Gibbs function of formation of all elements is zero; this reduces to

$$G_P = -474\ 708 \text{ for 2 moles } H_2O$$

$$G_P = 2(\bar{g}_f^\circ)_{H_2O(l)} \qquad \therefore (\bar{g}_f^\circ)_{H_2O(l)} = -237\ 354 \text{ kJ/kgmol}$$

The following example illustrates the concept of maximum possible work from a combustion process.

Example 14.14

Propane at 25°C and 1 atm is burned with 400 percent theoretical air at 25°C and 1 atm. The reaction takes place adiabatically, and all the products leave at 1 atm. The temperature of the surroundings is 25°C. Compute the entropy change and the maximum work and compare to an isothermal reaction. Figure 14.3 illustrates the example.

At what temperature do the products of reaction leave the combustion chamber? The adiabatic flame temperature. To find this,

$$C_3H_8 + 4(5)O_2 + (4)(5)(3.76)N_2 \rightarrow$$
$$3CO_2 + 4H_2O + 15O_2 + 75.2N_2$$

$$H_R = H_P$$

From Example 14.10,

$$T = 942 \text{ K}$$

The entropy may now be computed:

$$S_R = \sum_R n_i(\bar{s}_i^\circ)_{298} = (\bar{s}_{C_3H_8}^\circ + 20\bar{s}_{O_2}^\circ + 75.2\bar{s}_{N_2}^\circ)_{298}$$

$$S_R = 18\ 782 \text{ kJ/kgmol} \cdot \text{K}$$

$$S_P = \sum_P n_j(\bar{s}_j^\circ)_{942} = (3\bar{s}_{CO_2}^\circ + 4\bar{s}_{H_2O}^\circ + 15\bar{s}_{O_2}^\circ + 75.2\bar{s}_{N_2}^\circ)_{942}$$

$$S_P = 22\ 351 \text{ kJ/kgmol} \cdot \text{K}$$

$$\text{irreversibility} = T_0(S_P - S_R) = 298(22\ 351 - 18\ 782)$$

$$\text{irreversibility} = 1\ 063\ 562 \text{ kJ/kgmol fuel}$$

The maximum work for this case may be found by finding the change in availability of the reactants and products. The Gibbs

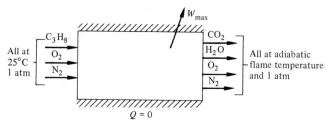

Figure 14.3   *The schematic diagram for Example 14.14.*

function is tabulated only for reference conditions, so we cannot use it in this case.

$$W_{max} = Q_1 - Q_2 = G_R - G_P = (H_R - T_0 S_R) - (H_P - T_0 S_P))$$

$$H_R = H_P$$

$$W_{max} = T_0(S_P - S_R) = 1\ 063\ 562 \text{ kJ/kgmol fuel}$$

$$W_{max} = \frac{1\ 063\ 562}{44.099} = 24\ 117 \text{ kJ/kg fuel}$$

This represents the maximum possible work an engine could perform if the products of combustion were at 942 K. For an isothermal reaction, we may find the maximum work by finding the change in the Gibbs function between the reactants and products at 25°C and 1 atm pressure.

$$W_{max} = G_R - G_P = \sum_R n_i g_i^{\circ} - \sum_P n_j \bar{g}_j^{\circ}$$

$$G_R = 1(-23\ 502) + 0 + 0 = -23\ 502 \text{ kJ/kgmol fuel}$$

$$G_P = 3(-394\ 631) + 4(-237\ 327) + 0 + 0$$

$$= -2\ 133\ 201 \text{ kJ/kgmol fuel}$$

$$W_{max} = 47\ 840 \text{ kJ/kg fuel}$$

The isothermal maximum work represents the ideal energy available from the combustion reaction to do work. If we find the actual work that an engine does using the combustion process as an energy source (such as an internal-combustion engine) and compare the two, we can determine the efficiency of the engine. Actually the heating value of the fuel, not the change in the Gibbs function, is used in current engineering practice.

The previous example assumed all the products left at 1 atm pressure. Actually this is not the case, as the total pressure is constant and is the sum of the partial pressures of all the constituents. The following example illustrates this more realistic situation.

Example 14.15

Propane reacts with 400 percent theoretical air in an isothermal reaction where the initial and final conditions are at 25°C and 1 atm pressure. The surrounding temperature is also 25°C. Calculate the maximum work; Figure 14.4 illustrates the reaction.

The combustion equation is

$$C_3H_8 + 20O_2 + 75.2N_2 \rightarrow 3CO_2 + 4H_2O + 15O_2 + 75.2N_2$$

The values of entropy must be calculated for the partial pressure at which they exist.

$$\bar{s}_i^\circ = \bar{s}^\circ - \bar{R}\ln\left(\frac{p_i}{p_0}\right)$$

The reaction is isothermal, so the temperature term in the entropy equation adds out. The partial pressure of the ith component is

$$p_i = y_i p_0$$

where y_i is the mole fraction of the i component. Thus

$$\bar{s}_i^\circ = \bar{s}^\circ + \bar{R}\ln\left(\frac{1}{y_i}\right)$$

Note that $1/y_i = n/n_i$ and $\bar{R} = 8.3143$ kJ/kgmol · K. For the reactants,

	n	$1/y_i$	$\bar{R}\ln(1/y_i)$	$\bar{s}^\circ$	$\bar{s}_i$
C_3H_8	1	96.2	37.967	270.065	308.032
O_2	20	4.81	13.059	205.142	218.201
N_2	75.2	1.279	2.046	191.611	193.657
	96.2				

For the products,

	n	$1/y_i$	$\bar{R}\ln(1/y_i)$	s°	$\bar{s}_i$
CO_2	3	32.4	28.918	213.795	242.713
H_2O	4	24.3	26.526	188.833	215.359
O_2	15	6.48	15.537	205.142	220.679
N_2	75.2	1.292	2.130	191.611	193.741
	97.2				

The maximum work is equal to the change in the Gibbs function or expressed in terms of enthalpy and entropy as

$$W_{max} = H_R - H_P - T_0(S_R - S_P)$$

$$W_{max} = 47\ 963 \text{ kJ/kg fuel}$$

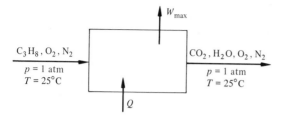

Figure 14.4 The schematic diagram for Example 14.15.

The value for the work including the pressure is essentially the same as when the pressure effect was neglected in the previous example. Thus unless the desired accuracy of the problem demands it, we will not include the pressure effect. This sort of effect would be included in precise scientific measurements and calculations.

14.11 CHEMICAL EQUILIBRIUM AND DISSOCIATION

Thus far in considering chemical reactions we have considered complete reactions, but this is not the whole picture. In considering thermodynamic equilibrium thus far we have concentrated on thermal and mechanical equilibrium. Likewise in studying chemical reactions we must study chemical equilibrium. The following discussion is by no means as extensive as would be found in a textbook on chemical engineering thermodynamics.

As a reaction proceeds, some of the products dissociate into the original products; when chemical equilibrium is reached, the reaction is proceeding in both directions, so there is no net change in either the reactants or the products. This reaction may be written as

$$\nu_a A + \nu_b B + \cdots \rightleftharpoons \nu_x X + \nu_y Y + \cdots \qquad (14.18)$$

where ν_i is the stoichiometric coefficient for the balanced reaction equation. A and B are the reactants, and X and Y are the products. There is a difference between n_i, the moles of i, and ν_i, its stoichiometric coefficient, in that n_i refers to the moles available, where ν_i denotes the equation requirements.

How can we determine when chemical equilibrium has been achieved? When a system is in equilibrium, it can no longer produce work. The Gibbs function can be used to determine the maximum work of a system at constant temperature and pressure. Thus by determining the Gibbs function of the reactants and of the products, we can determine whether or not a chemical reaction can occur. When the Gibbs function of the reactants and products is equal, then no work may be done and chemical equilibrium is achieved, or

$$(dG)_{T,p} = 0 \tag{14.19}$$

for chemical equilibrium.

It is possible to show that the change in the Gibbs function at constant temperature and pressure is the criterion for chemical equilibrium without use of the maximum-work concept.

Heretofore the Gibbs function has been written for a single species, however, for systems of many species we may write that

$$G = G(T, p, n_1, n_2, \ldots, n_i) \tag{14.20}$$

where n_i represents the moles of species i at any instant. The change in the total Gibbs function is

$$dG = \left(\frac{\partial G}{\partial p}\right)_{T,n_i} dp + \left(\frac{\partial G}{\partial T}\right)_{p,n_i} dT + \sum_i \left(\frac{\partial G}{\partial n_i}\right)_{T,p,n_j} dn_i \tag{14.21}$$

The partial derivative in the last term in Equation (14.21) evaluates the change in G with respect to n_i with T, p, and all other species, n_j, held constant. The partial derivatives in Equation (14.21) are defined as

$$\left(\frac{\partial G}{\partial p}\right)_{T,n_i} = V \qquad \left(\frac{\partial G}{\partial T}\right)_{p,n_i} = -S \qquad \left(\frac{\partial G}{\partial n_i}\right) = \mu_i \tag{14.22}$$

In Equation (14.22) μ_i is the chemical potential of the ith component. The chemical potential of species i is the driving force in changing the moles of species i.

Let us apply the change of the Gibbs function to the reaction Equation (14.18), occurring at constant temperature and pressure.

$$(dG)_{T,p} = \sum_i \mu_i \, dn_i$$

$$(dG)_{T,p} = \mu_a \, dn_a + \mu_b \, dn_b + \mu_x \, dn_x + \mu_y \, dn_y \tag{14.23}$$

The change in mole numbers are dependent upon one another as the products come from the reactants. Assume a forward reaction, the left-hand side is being depleted, and let us further assume an infinitesimal change, so a proportionality constant may be used to describe the change in the mole numbers. Thus

$$dn_a = -k\nu_a \qquad dn_b = -k\nu_b \qquad dn_x = +k\nu_x \qquad dn_y = +k\nu_y \tag{14.24}$$

The negative and positive values of k reflect the depletion and gain of specie. Equation (14.23) becomes

$$(dG)_{T,p} = (-\mu_a\nu_a - \mu_b\nu_b + \mu_x\nu_x + \mu_y\nu_y)k \tag{14.25}$$

Since we are interested in chemical equilibrium, assume that the small change occurs at the point of equilibrium where $(dG)_{T,p} = 0$. Thus,

$$\mu_a \nu_a + \mu_b \nu_b = \mu_x \nu_x + \mu_y \nu_y \tag{14.26}$$

for chemical equilibrium.

The concept of work may be related by Equation (14.23) by recalling Equation (3.21), where work is defined as an intensive property acting on the change of its related extensive property.

Consider a system containing methane and oxygen, existing at the same temperature and pressure as the surroundings. The system is in mechanical and thermal equilibrium. Is it in chemical equilibrium? No, because if we computed the Gibbs function for the reaction, we would find that $G_R > G_P$, so a reaction could occur. Nothing will happen, however, until a spark is introduced into the mixture. Then a reaction will occur until $G_R = G_P$. Thus, we find that the Gibbs function has two purposes:

1. It tells us whether or not a reaction can occur. If a system is in chemical equilibrium, $(dG)_{T,p} = 0$, then no reaction can occur.
2. The Gibbs function is the thermodynamic potential for the constant temperature and constant pressure reaction that causes the reaction to occur. It does not tell us how fast the reaction will occur.

Consider the reaction

$$n_a A + n_b B \rightleftharpoons n_x X + n_y Y \tag{14.27}$$

and let the reactants A and B flow slowly into a box so their relative amounts remain constant. A reaction takes place, and products X and Y are formed. We will prescribe that the amounts of the products formed are proportional to the amounts of A and B, and that the ratio of the moles of the products to the moles of the reactants is a constant. For this to occur, products X and Y must be removed from the box in such a manner that the mass entering the box equals the mass leaving. The reactants and products leave at the same temperature and pressure and the reaction also occurs at this temperature. This is called a "Van't Hoff equilibrium box." Figure 14.5 illustrates this process.

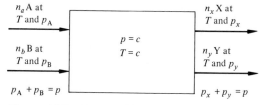

Figure 14.5   Van't Hoff equilibrium box.

Since the Gibbs function is used in determining chemical equilibrium, let us find the Gibbs function for each of the substances in the box.

$$dG = V\,dp - S\,dT$$

For $T = C$,

$$dG = V\,dp$$

For an ideal gas, $V = n\bar{R}T/p$

$$dG = n\bar{R}T\frac{dp}{p}$$

Integrating,

$$(G_2 - G_1) = n\bar{R}T\ln\left(\frac{p_2}{p_1}\right) \tag{14.28}$$

We then apply Equation (14.6) to the conditions in the box. We let $p_2 = p_1$, the partial pressure of the given component in the box. Thus for component A

$$\left(G_{A_2} - G_{A_1}\right)_T = n_a\bar{R}T\ln\left(\frac{p_a}{p}\right) \tag{14.29a}$$

where p is the total pressure.

Similarly, expressions may be written for the other substances. For substance B,

$$\left(G_{B_2} - G_{B_1}\right)_T = n_b\bar{R}T\ln\left(\frac{p_b}{p}\right) \tag{14.29b}$$

For substance X,

$$\left(G_{X_2} - G_{X_1}\right)_T = n_x\bar{R}T\ln\left(\frac{p}{p_x}\right) \tag{14.29c}$$

For substance Y

$$\left(G_{Y_2} - G_{Y_1}\right)_T = n_y\bar{R}T\ln\left(\frac{p}{p_y}\right) \tag{14.29d}$$

Adding Equations (14.29a) to (14.29d) yields

$$\left(G_{A_2} + G_{B_2}\right)_T - \left(G_{A_1} - G_{B_1}\right)_T + \left(G_{X_2} - G_{Y_2}\right)_T - \left(G_{X_1} + G_{Y_1}\right)_T$$

$$= n_a\bar{R}T\ln\left(\frac{p_a}{p}\right) + n_b\bar{R}T\ln\left(\frac{p_b}{p}\right) + n_x\bar{R}T\ln\left(\frac{p}{p_x}\right)$$

$$+ n_y\bar{R}T\ln\left(\frac{p}{p_y}\right) \tag{14.30}$$

Since equilibrium conditions exist within the box, then the sum of the Gibbs function of the reactants must equal the sum of the Gibbs functions of the products in the box. Hence

$$\left(G_{A_2} + G_{B_2}\right)_T - \left(G_{X_1} + G_{Y_1}\right)_T = 0 \tag{14.31}$$

Combining Equations (14.31) and (14.30),

$$\left(G_{X_2} + G_{Y_2}\right) - \left(G_{A_1} + G_{B_1}\right) = \bar{R}T\ln\left(\frac{p_a}{p}\right)^{n_a} + \bar{R}T\ln\left(\frac{p_b}{p}\right)^{n_b}$$

$$+ \bar{R}T\ln\left(\frac{p}{p_x}\right)^{n_x} + \bar{R}T\ln\left(\frac{p}{p_y}\right)^{n_y}$$

We let the total pressure, p, equal 1 atm, and express the partial pressures in atmospheres. Simplifications result if the pressures are so expressed.

$$(G_P - G_R)_T = \bar{R}T\ln(p_a)^{n_a} + \bar{R}T\ln(p_b)^{n_b} - \bar{R}T\ln(p_x)^{n_x}$$

$$- \bar{R}T\ln(p_y)^{n_y}$$

$$(G_P - G_R)_T = \bar{R}T\ln\left(\frac{p_a^{n_a} p_b^{n_b}}{p_x^{n_x} p_y^{n_y}}\right) \tag{14.32}$$

Taking the antilog of Equation (14.32),

$$\frac{p_a^{n_a} p_b^{n_b}}{p_x^{n_x} p_y^{n_y}} = \ln^{-1}\left[\frac{G_P - G_R}{\bar{R}T}\right]_T = K_p \tag{14.33}$$

where K_p is called the "equilibrium constant." Thus K_p is not really a constant but is only constant for a given temperature for an ideal gas. Values for the natural log of the equilibrium constant, K_p, are given in Table C.4.

When the temperature of a gas is increased, there is a tendency for the gas to dissociate. So at any temperature there is an equilibrium mixture of the gas and its dissociated products. The dissociation process is endothermic and, as such, tends to reduce the total energy of a system to a minimum value for a given temperature. The affect of dissociation reduces the adiabatic flame temperature, as some of the chemical energy is used in the dissociation process instead of raising the thermal energy. The following examples will illustrate the effect of dissociation.

Example 14.16

Determine the percent dissociation of carbon dioxide into carbon monoxide and oxygen at 3800 K and 1 atm pressure.

From Table C.4 the $\ln(K_p) = 1.170$ for the reaction $CO + \frac{1}{2}O_2 \rightarrow CO_2$. Let x denote the dissociation of 1 mole of CO_2. The reaction equation is

$$x CO + \frac{x}{2}O_2 \rightleftharpoons (1 - x)CO_2$$

The total number of moles present at equilibrium is $(2 + x)/2$. The partial pressure for each component, p_i, may be expressed in terms of the mole fraction and the total pressure, p.

$$p_{CO} = \frac{x}{(2 + x)/2}p = \frac{2x}{2 + x}p$$

$$p_{O_2} = \frac{x}{2 + x}p$$

$$p_{CO_2} = \frac{2(1 - x)}{2 + x}p$$

Since the total pressure is 1 atm, we may solve Equation (14.33) for the equilibrium constant K_p.

$$K_p = \frac{p_{CO_2}^1}{\left(p_{CO}^1\right)\left(p_{O_2}^{0.5}\right)} = \frac{(2 + x)^{1/2}(1 - x)}{x^{3/2}}$$

Solve this equation for x by a trial and error process, knowing that $\ln(K_p) = 1.170$.

$x = 0.3$	$K_p = 6.46$	$\ln(K_p) = 1.865$
$x = 0.4$	$K_p = 3.673$	$\ln(K_p) = 1.301$
$x = 0.425$	$K_p = 3.231$	$\ln(K_p) = 1.173$

Thus 42.5 percent of the carbon dioxide has dissociated at 3800 K.

As mentioned previously, the dissociation reaction uses some of the chemical energy released by combustion and thus reduces the adiabatic flame temperature. In determining the adiabatic flame temperature, we must find the amount of dissociation before calculating the enthalpy.

Example 14.17

Hydrogen is steadily burned with 100 percent theoretical air at 1 atm pressure and standard temperature. Determine the adiabatic flame temperature and the maximum temperature, taking dissociation into account.

$$H_2 + 0.5O_2 + 1.88N_2 \rightarrow H_2O + 1.88N_2$$

$$H_R = H_P$$

$$H_R = 0$$

Determine the adiabatic flame temperature by trial-and-error solution to be 2525 K.

The combustion equation accounting for dissociation of water in the products of combustion is

$$H_2 + 0.5O_2 + 1.88N_2 \rightarrow (1-x)H_2O + xH_2 + \frac{x}{2}O_2 + 1.88N_2$$

where x is the fraction of H_2O that dissociates.

Procedure

1. first-law analysis:

$$H_R = H_P$$
$$H_R = 0$$

2. guess a temperature: (a) look up $\ln(K_p)$
3. guess a value for x: (a) calculate K_p—correct value of x will have same value of $\ln(K_p)$ as (2a).
4. using the value of x found in (3), calculate H_P for first-law analysis.
5. repeat until $H_P = H_R$.

guess $T = 2200$ K:

$$\ln(K_p) = -6.774$$

$$K_p = \frac{(p_{O_2})^{1/2}(p_{H_2})^1}{(p_{H_2O})^1}$$

Moles of products at equilibrium $= 2.88 + 0.5x$

Partial Pressures

$$p_{H_2O} = \left[\frac{(1-x)}{(2.88+0.5x)}\right]^{(p)}$$

$$p_{H_2} = \left[\frac{x}{(2.88+0.5x)}\right]^{(p)}$$

$$p_{O_2} = \left[\frac{0.5x}{(2.88+0.5x)}\right]^{(p)}$$

$$p = 1 \text{ atm}$$

$$K_p = \frac{0.7071x^{3/2}}{(1-x)(2.88+0.5x)^{1/2}}$$

x	K_p	$\ln(K_p)$
0.1	0.0145	-4.23
0.05	0.00488	-5.32
0.02	0.00120	-6.72

Assume $x = 0.02$ for the initial calculation; the products side of the combustion equation is

$$0.98H_2 + 0.02H_2 + 0.01O_2 + 1.88N_2$$

guess $T = 2400$ K:

$$\ln(K_p) = -5.625$$

A value of $x = 0.04$ balances $\ln(K_p)$ and yields $H_P = -5302$ kJ/kgmol fuel. Plot $(H_P - H_R)$ versus T, the line intersects $(H_P - H_R) = 0$ at $T = 2433$ K. The value of $x = 0.045$ essentially balances the first-law equation.

Thus 4.5 percent of the water dissociated, lowering the maximum temperature by 92 K. In hydrocarbon combustion the dissociation of the hydrocarbon should be considered, as well as the reaction of nitrogen and oxygen to form nitric oxide and the dissociation of other products. All of these reactions are endothermic, tending to reduce the products' temperature.

14.12 STEAM GENERATOR EFFICIENCY

The first-law efficiency of any device is output divided by input. This is no exception for steam generators. The only question to be answered is what we define as input and output.

$$\eta_{\text{stm gen}} = \frac{\text{output}}{\text{input}} = \frac{\text{energy to water/steam across steam generator}}{\text{energy supplied by fuel}}$$

$$(14.34)$$

The energy supplied to the water is the change of enthalpy of the water from the feedwater entering to the superheated steam leaving. All flows must be considered if some of the steam is extracted before entering the superheater. The energy supplied by the fuel is the fuel flow rate times the fuel's higher heating value.

A power plant's efficiency is also denoted by its specific fuel consumption and heat rate. The specific fuel consumption, sfc, is

$$\text{sfc} = \frac{\dot{m}_{\text{fuel}}}{\dot{W}_{\text{net}}} = \frac{\text{kg}}{\text{kW} \cdot \text{s}}$$

$$(14.35)$$

The difficulty with using Equation (14.35) is that it does not account for the fuel's heating value. In comparing plant and/or steam generator designs the heating value of the fuel should be known. The heat rate eliminates this problem, and modern convention recommends its use over specific fuel consumption; that is,

$$\text{heat rate} = \frac{(\dot{m}_{\text{fuel}})(h_{\text{RP}})}{\dot{W}_{\text{net}}}$$

$$(14.36)$$

14.13 FUEL CELLS

A fuel cell transforms chemical energy into electrical energy through a series of catalyst-aided oxygen-reduction reactions. Unlike many energy conversion systems such as the steam power plant and the internal combustion engine, the fuel cell generates electricity in a continuous and direct process. In this manner, the excessive losses that occur in multistep energy conversion systems may be avoided using a single-step process. Thus fuel cells tend to have higher theoretical energy conversion efficiencies.

The fuel cell is composed of an electrolyte sandwiched between electrodes known as the anode and cathode, as shown in Figure 14.6. Unlike the conventional battery, however, the fuel cell consumes externally supplied fuels. The fuel (normally hydrogen) and the oxidizer (normally oxygen in the form of air) are supplied to the anodic and cathodic sides of the cell, respectively. The electrochemical reaction between these two results in the transfer of electrons and the production

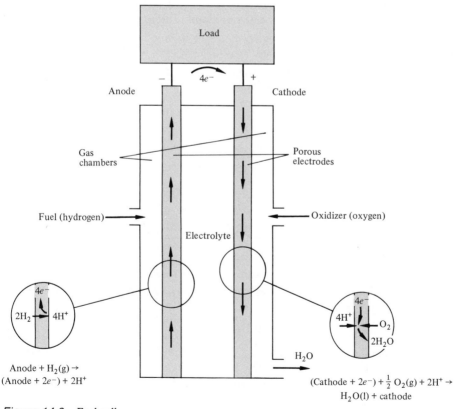

Figure 14.6   Fuel cell.

of a voltage between the two electrodes. When a load is connected in series to these electrodes, a current, calculable by Ohm's law, will result. The theoretical voltages that arise from these fuel cell reactions normally range between 1.0 and 1.3 volts dc at 1–2 kW/m² of electrode.

As an example, let us analyze the most common fuel cell, the hydrogen–oxygen fuel cell. Fuel, as hydrogen enters the fuel cell and loses an electron, giving the anodic plate a negative charge. The half-cell reaction for the anode is

$$\text{anode} + H_2(g) \rightarrow (\text{anode} + 2e^-) + 2H^+$$

At the cathode, the oxygen molecules supplied by the air pick up electrons, giving that plate a positive charge. The half-cell reaction for the cathode is

$$(\text{cathode} + 2e^-) + \tfrac{1}{2}O_2(g) + 2H^+ \rightarrow H_2O(l) + \text{cathode}$$

When a load is connected in series with the electrodes, a current develops as a function of the induced voltage. The excess electrons in the anode travel through the load to the cathode providing power. The H^+ ions formed at the anode according to the half-cell reaction migrate through the electrolyte solution to the cathode. The oxygen then combines with the H^+ ions to form water.

The fuel cell efficiency is

$$\eta_{\text{fc}} = \frac{\text{maximum work}}{\text{change of overall enthalpy}} = \frac{\Delta G}{\Delta H} \qquad (14.37)$$

where ΔG equals the change in the Gibbs function. The Gibbs function is defined as

$$G = H - TS$$

Therefore, at standard temperature and pressure (25°C, 1 atm),

$$\Delta G^\circ = G_p - G_R = \left(H_p - H_R \right) - T_0 \left(S_p - S_R \right)$$

$$\Delta G^\circ = \sum_p n_j (\bar{h})_j - \sum_R n_i (\bar{h})_i - T_0 \left[\sum_p n_j \bar{s}_j^\circ - \sum_R n_i \bar{s}_i^\circ \right]$$

The values of h and s for substances are found in Tables C.1 and C.2. To simplify calculations, the hydrogen–oxygen fuel cell reaction may be written as

$$H_2(g) + \tfrac{1}{2}O_2(g) \rightarrow H_2O(l)$$

Calculate the change in the Gibbs function at standard temperature and pressure, substituting in the tabulated values. The Gibbs function of an element is zero at standard conditions; thus

$$\Delta G^\circ = G_p = \left(\bar{g}_f^\circ \right)_{H_2O(l)} = -237\,327 \text{ kJ/kgmol}$$

The change in enthalpy for the overall fuel cell reaction is

$$\Delta H = H_p = \left(h_f^\circ \right)_{H_2O(l)} = -286\,010 \text{ kJ/kgmol}$$

The ideal fuel cell efficiency is

$$\eta_{fc} = 0.83 \text{ or } 83 \text{ percent}$$

The fuel cell voltage is found from

$$\Delta G = -n\mathscr{F}V \tag{14.38}$$

where

$\Delta G = $ change in Gibbs function per mole of fuel

$n = $ number of electrons transferred per molecule of fuel

$\mathscr{F} = $ Faraday's constant, 96 500 kJ/V · kgmol

$V = $ fuel cell voltage

Hence

$$V = \frac{\Delta G}{n\mathscr{F}} = \frac{+237\,327}{(2)(96\,500)} = 1.229 \text{ volts}$$

A number of other fuels may be used in fuel cells, some of which are listed in Table 14.2. Each of these fuels react and may be analyzed in a similar fashion. The fuel cell reaction for carbon monoxide fuel is

$$CO(g) + \tfrac{1}{2}O_2(g) \rightarrow CO_2(g)$$

To calculate

$$\Delta G^\circ = (H_P - H_R) - T_0(S_P - S_R)$$

Table 14.2   Theoretical Fuel Cell Performance

Fuel Cell Reaction	$-\Delta H$, kJ/kgmol	$-\Delta G$, kJ/kgmol	η_{FC}	V, volts
$H_{2(g)} + \tfrac{1}{2}O_{2(g)} \rightarrow H_2O_{(l)}$				
at 298 K	286 010	237 327	0.830	1.229
at 1000 K	—	—	—	—
$H_{2(g)} + \tfrac{1}{2}O_{2(g)} \rightarrow H_2O_{(g)}$				
at 298 K	241 971	228 729	0.945	1.184
at 1000 K	247 856	175 940	0.777	0.998
$CO_{(g)} + \tfrac{1}{2}O_{2(g)} \rightarrow CO_{2(g)}$				
at 298 K	283 161	257 405	0.909	1.333
at 1000 K	282 796	195 797	0.692	1.014
C (graphite) $+ O_{2(g)} \rightarrow CO_{2(g)}$				
at 298 K	393 757	394 631	1.002	1.002
at 1000 K	394 828	396 042	1.003	1.026
$C_3H_{8(g)} + 5O_{2(g)} \rightarrow$ $3CO_{2(g)} + 4H_2O_{(g)}$				
at 298 K	2 044 884	2 075 023	1.015	1.075
at 1000 K	2 046 558	2 149 953	1.051	1.124

where H_R does not equal zero because $(\bar{h}_f^\circ)_{CO_{(g)}}$ must be included in it. Similarly, G_R does not equal zero. Substituting, we find that

$$\Delta G^\circ = G_P - G_R = (-393\ 757) - (-110\ 596)$$

$$-298\left[213.795 - 197.653 - \tfrac{1}{2}(205.142)\right]$$

$$\Delta G^\circ = -257\ 405 \text{ kJ/kgmol}$$

The change of enthalpy for the fuel cell reaction is

$$\Delta H = H_P - H_R = (-393\ 757) - (-110\ 596)$$

$$= -283\ 161 \text{ kJ/kgmol}$$

The fuel cell efficiency is

$$\eta_{fc} = 0.909 \text{ or } 90.9 \text{ percent}$$

and the fuel cell voltage is

$$V = \Delta G/n\mathfrak{F} = 1.333 \text{ volts}$$

The fuel cell efficiency and voltage may be calculated at other than standard conditions. Consider the carbon monoxide fuel cell at 1000 K and 1 atmosphere pressure

$$\Delta G = \left[\sum_P n_j\left(\bar{h}_f^\circ + (\bar{h} - \bar{h}_{298})\right)_j - \sum_R n_i\left(\bar{h}_f^\circ + (\bar{h} - \bar{h}_{298})\right)_i\right]$$

$$-T\left[\sum_P n_j(\bar{s}^\circ)_j - \sum_R n_i(\bar{s}^\circ)_i\right]$$

$$\Delta G = \left[(-393\ 757 + 33\ 405) - (-110\ 596 + 21\ 686)\right]$$

$$-(0.5)(22\ 707) - (1000)\left[269.325 - 234.531 - 0.5(243.585)\right]$$

$$\Delta G = -195\ 797 \text{ kJ/kgmol}$$

The change in the fuel cell enthalpy is

$$\Delta H = -282\ 796 \text{ kJ/kgmol}$$

The fuel cell efficiency is

$$\eta_{fc} = 0.692 \text{ or } 69.2 \text{ percent}$$

and the fuel cell voltage is

$$V = 1.014 \text{ volts}$$

As may be seen in Table 14.2, some ideal fuel cell efficiencies may exceed unity. These cases correspond to processes in which the fuel cell absorbs heat from the surroundings.

In actual fuel cell operation, losses occurring during electron transfer at the electrode, during mass transport through concentration gradients, and during electron transport through the electrolyte result in energy losses that may exceed 25 percent of the ideal energy generated.

PROBLEMS

1. A fuel mixture of 50% C_7H_{16} and 50% C_8H_{18} is oxidized with 20% excess air. Determine (a) mass of air required for 50 kg of fuel; (b) volumetric analysis of products of combustion.
2. A gas-turbine power plant receives an unknown type of hydrocarbon fuel. Some of the fuel is burned with air yielding the following Orsat analysis of the products of combustion:

CO_2	10.5%
O_2	5.3%
N_2	84.2%

Determine (a) the percent, by mass, of carbon and hydrogen in the fuel; (b) the percent theoretical air.
3. What mass of liquid oxygen is required to completely burn 1000 kg of liquid butane, C_4H_{10}, on a rocket ship?
4. One kilogram of sugar, $C_{12}H_{22}O_{11}$, is completely oxidized with theoretical air. Determine (a) the volumetric product analysis; (b) the mass of air required at standard temperature and pressure.
5. With 110% theoretical air, 1 kgmol of methane is completely oxidized. The products of combustion are cooled and completely dried at atmosphere pressure. Determine (a) the partial pressure of oxygen in the products; (b) the kilograms of water removed.
6. An unknown hydrocarbon had the following Orsat analysis when burned with air:

CO_2	11.94%
O_2	2.26%
CO	0.41%
N_2	85.39%

Determine (a) the air–fuel ratio on a mass basis; (b) the percent of carbon and hydrogen in the fuel on a mass basis.
7. Write the combustion equation for gaseous dodecane and theoretical air. Determine (a) the fuel–air ratio on the mass basis; (b) the fuel–air ratio on the mole basis; (c) the mass of fuel–mass of water formed; (d) the molecular weight of the reactants; (e) the molecular weight of the products; (f) the ratio of moles of reactants to moles of products.
8. A coal sample has the following ultimate analysis on a dry basis: 81% C, 2.5% H_2, 0.6% S, 3.0% O_2, 1.0% N_2, 11.9% ash. Determine the reaction equation for 100% theoretical air.
9. The ultimate analysis of a coal sample is: 77% C, 3.5% H_2, 1.8% N_2, 4.5% O_2, 0.7% S, 6.5% ash, 6.0% H_2O. Determine the reaction equation for 120% theoretical air.

10. How many cubic meters of air at 20°C will be required to completely oxidize 1 m³ of gas with the following volumetric analysis? The pressure is atmospheric.

CH_4	46%
CO	38%
O_2	5%
N_2	11%

11. Given the following volumetric analysis on the dry basis of a coal sample:

C	80%
H	4.5%
O_2	4.5%
N_2	1.0%
S	1.0%
Ash	9.0

and that the heat of combustion of C, H_2, and S are 33 700, 141 875, 9300 kJ/kg, respectively, determine the higher heating value of the coal in kJ/kg.

12. A gaseous mixture containing 60% CH_4, 30% C_2H_6, and 10% CO is burned with 12% excess air. The combustion air is supplied to the furnace by a forced draft fan that increases the pressure by 76 mm of water. Determine for complete combustion (a) the molal air–fuel ratio; (b) the fan power when 0.8 m³/s of fuel is burned at standard temperature and pressure.

13. A cigarette lighter burns butane with 200% theoretical air. The fuel and air are at 25°C, and the products are at 127°C. The pressure is 100 kPa. Determine (a) the balanced reaction equation; (b) the fuel air ratio on a mass basis; (c) the heat released during combustion; (d) the dew point of the products.

14. Octane is burned with 150% theoretical air. The air is at 25°C and has 50% relative humidity. Determine (a) the balanced reaction equation; (b) the dew point of the products; (c) the dew point of the products if dry air were used.

15. A diesel engine uses $C_{12}H_{26}$ in a ratio of 1 to 30 (fuel to air by mass). The engine exhausts at 327°C into a heating system where the room temperature is 25°C. Determine the percent of fuel available for heating. Assume complete combustion of the fuel with air and fuel entering at 25°C.

16. An internal-combustion engine uses liquid octane for fuel and 150% theoretical air at 25°C and 100 kPa. The products of combustion leave the engine at 260°C. The heat loss is equal to 20% of the work. Determine (a) the work/kgmol; (b) the dew point; (c) the kg/s of fuel required to produce 400 kW.

17. Equal moles of hydrogen and carbon monoxide are mixed with theoretical air in an insulated rigid vessel at standard temperature and pressure. The mixture is ignited by a spark. Complete oxidation occurs. Determine (a) the maximum temperature; (b) the maximum pressure.

18. A steam turboelectric generator plant has the following test data:

Orsat analysis of stack gas: 12.1% CO_2; 3.71% O_2
Stack gas temperature: 232°C
Feedwater in at 288°C
Steam leaves at 16 MPa, 570°C
Steam flow rate: 453 kg/s
Fuel oil flow rate: 23.3 kg/s
Fuel: $C_{12}H_{26}$
Air and fuel enter at 25°C

Determine (a) the excess air in kg/s; (b) the work required to move excess air (consider constant pressure at 100 kPa; (c) the steam-generator thermal efficiency.

19. A furnace burns natural gas that has the following volumetric analysis:

CH_4 90%
C_2H_6 7%
C_3H_8 3%

The gas flow is 0.02 m^3/s; and 25% excess air is required for complete combustion. The natural gas and air enter at 25°C and 1 atm pressure. The exhaust gas has a temperature of 1060°C. Determine (a) the volumetric analysis of products of combustion; (b) the dew point of the products; (c) the thermal energy utilized in furnace; (d) the stack has a 1-m diameter. Determine the exit gas velocity.

20. A coal-fired steam boiler had the following test data:

Coal consumption: 1.26 kg/s
Coal heating value: 27 900 kJ/kg

Orsat Analysis of Products

CO_2 11%
O_2 8%
CO   1%
N_2 80%

The stack temperature is 227°C and the ambient temperature is 25°C. The refuse removed from the ash pit is 0.095 kg/s. The coal initially contains 5% ash and 80% carbon. Determine (a) the thermal energy loss to the products of combustion; (b) the heat loss in kilowatts due to incomplete combustion; (c) the loss of energy release due to unburned carbon in the refuse.

21. A coal-fired steam generating plant was operated for a year with an average flue gas analysis of 13% CO_2, 0% CO, 6.25% O_2, and combustible matter to the ash pit was 10%. An attempt to improve efficiency was made and the second-year average was 15% CO_2, 0.1% CO, 3.9% O_2, and 16 percent combustible matter to the ash pit. Coal with a 7% ash content and a heating value of 33 000 kJ/kg, dry, was used. At the end of the second year it was found that the efficiency had remained the same, but the cost of operation had increased. Why?

22. A test of an oil-fired steam generator indicated that 12.77 kg of water evaporated per kg of oil burned. The boiler pressure was 1.4 MPa, superheat temperature was 245°C, feedwater temperature was 34°C, and the boiler efficiency was 82.8%. Determine the fuel's heating value.

23. A steam generator generates 401 430 kg of steam in a 4-hour period. The steam pressure is 2750 kPa and 370°C. The temperature of the water supplied to the steam generator is 138°C. If the steam generator efficiency is 82.5% and the coal has a heating value of 32 200 kJ/kg, find the average amount of coal burned per hour.

24. A steam generator uses coal containing 75% carbon and 8% ash and is completely burned, leaving no unburned residue. The air enters the furnace at 32°C and 80% relative humidity. The stack temperature is 360°C. The average flue gas analysis is: 12.6% CO_2, 6.2% O_2, 1% CO. Determine (a) the percent excess air; (b) the complete analysis of the fuel (include percent hydrogen); (c) the cubic meters of flue gas per kg of coal burned; (d) the cubic meters of air per kg of coal burned.

25. Determine the higher and lower heating value of butane for (a) constant pressure; (b) constant volume.

26. Compute the adiabatic flame temperature of gaseous methane, ethane, and octane for steady combustion. Compare the resultant temperatures.

27. Calculate the Gibbs function of formation for carbon dioxide.

28. Octane is burned with 200% theoretical air in a steady-flow process. The total pressure is 1 atm, and the reactants enter at standard temperature. The products leave at the adiabatic flame temperature. Determine (a) the entropy increase during combustion; (b) the maximum work; (c) the maximum work for isothermal combustion.

29. Methane reacts with 200% theoretical air at standard conditions. The surrounding temperature is also at 25°C. Calculate the maximum work for an isothermal reaction considering the partial pressure of the reactants and products.

30. Calculate the equilibrium constant for the complete oxidation of methane. The total pressure is 1 atm.
31. Calculate the percent dissociation of oxygen, $O_2 \rightarrow 20$ at 4000 K and 1 atm pressure.
32. Calculate the percent dissociation of oxygen, $O_2 \rightarrow 20$, at 4000 K and 3 atm pressure.
33. Methane is steadily burned with 100% theoretical air at standard conditions. Determine the maximum temperature considering the dissociation of carbon dioxide.
34. Determine the ideal-cell voltage and efficiency for a hydrogen–oxygen fuel cell at 1000 K and 1 atm.
35. Determine the ideal-cell voltage and efficiency for a methane–oxygen fuel cell at 298 K and 1000 K and 1 atm.
36. For a hydrogen–oxygen fuel cell compute the flow rates necessary for the cell to produce 25 kW at 298 K and 1 atm.
37. For the methane–oxygen fuel cell operating at 1000 K, determine the fuel flow rate to produce 50 kW.

Chapter 15
Gas Compressors

This chapter is concerned with the energy analysis of gas compressors. Gas compressors are devices in which mechanical work is done on the gas, raising its pressure. Energy analysis is just one of many aspects to be considered in designing a gas compressor. We will not concern ourselves with the mechanics of construction or with the various systems in which compressors are used.

Compressed gas is found in a variety of manufacturing processes: It is used to transport solid material, to provide control air for pneumatic systems, to drive tools in construction industries; and the list goes on. The compressor is a vital component of refrigeration and gas-turbine systems, and the laws developed and applied in this chapter will be applied later to compressors in those systems.

There are two general types of compressors: reciprocating and rotative. For high pressures and low-volume flow rates, the reciprocating compressor is preferred; for lower pressures and high-volume flow rates, the rotative compressor is used. There is not a distinct pressure that separates the two compressor types as modern rotative compressors can develop rather high pressures. We will analyze the reciprocating compressor first.

Figure 15.1 illustrates a two-stage reciprocating compressor. The discharge from the first stage passes through an intercooler and into the second stage. Notice that the valves are flexible disks; they deflect to open and close, relying on a pressure differential.

15.1 COMPRESSORS WITHOUT CLEARANCE

All reciprocating compressors have a clearance volume between the top of the piston and the top of the cylinder, where the exhaust and intake

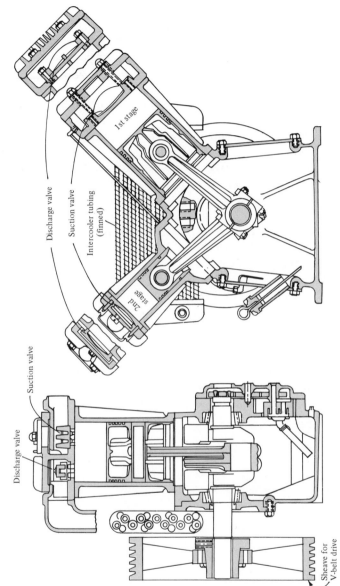

Figure 15.1 A two-stage air compressor.

valves are located. Many compressors are double acting, which means they compress in both stroke directions. We will consider the compressor to be single acting; if double acting is desired, multiply the result by two. We will also consider the clearance volume to be zero, which means that all the gas in the cylinder is pushed out when the piston is at the top of its stroke. From these considerations, the analytical development of compressors with clearance will be made.

Figure 15.2 illustrates a p–V diagram for this cycle. From 0 to 1, gas intake occurs at constant pressure until the piston reaches bottom dead center at state 1; the gas is compressed polytropically from 1 to 2 until the pressure is that of the gas in the discharge line; the exhaust valve opens, and the gas is discharged at constant pressure from state 2 to state 3. Since there is no gas left, the pressure is undefined. As soon as the piston moves an infinitesimal amount, the intake valve opens and gas is drawn in again from 0 to 1. Notice that there will be a difference between the work necessary to compress the gas from states 1 to 2 and the total work of the cycle. The line from 1 to 2′ illustrates the path of isothermal compression. The enclosed area is less, so the cycle work is less for isothermal compression.

Let us calculate the cycle work using the ideal gas laws for each process and add the terms together, which yields

$$W_{\text{cycle}} = \frac{n}{n-1}(p_1V_1 - p_2V_2)$$

(15.1)

The cycle work equation may be further arranged to eliminate V_2 and have an expression for the work in terms of p_1, V_1, and p_2.

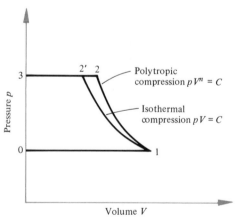

Figure 15.2   A p–V diagram for a single-acting reciprocating compressor without clearance.

Since the process from 1 to 2 is polytropic,

$$\frac{V_2}{V_1} = \left(\frac{p_2}{p_1}\right)^{-1/n}$$

and

$$\frac{p_2 V_2}{p_1 V_1} = \left(\frac{p_2}{p_1}\right)^{(n-1)/n} \tag{15.2}$$

Substituting Equation (15.2) into Equation (15.1) yields

$$W_{\text{cycle}} = \frac{n}{n-1} p_1 V_1 \left[1 - \left(\frac{p_2}{p_1}\right)^{(n-1)/n}\right] \tag{15.3}$$

Equation (15.3) is valid for compressors without clearance. We calculate the cycle work for the isothermal case. The result is given in Equation (15.4),

$$W_{\text{cycle}} = -p_1 V_1 \ln\left(\frac{p_2}{p_1}\right) \tag{15.4}$$

which is valid for compressors without clearance when the compression is isothermal.

15.2 RECIPROCATING COMPRESSORS WITH CLEARANCE

Let us extend our knowledge of reciprocating compressors without clearance so it can be applied to actual compressors, compressors with clearance. In these, the piston does not move to the top of the cylinder, and some space is left around the valves, called the "clearance volume." This volume is usually expressed as a percentage of the total displacement volume; it is called the "percent clearance," c, and defined as

$$c = \frac{\text{clearance volume}}{\text{displacement volume}} = \frac{V_3}{V_{\text{PD}}} \tag{15.5}$$

Typically, the value of c falls between 3 and 10 percent.

Figure 15.3 illustrates a p–V diagram for a compressor with clearance. Starting at state 1 on Figure 15.3, the gas is compressed polytropically to state 2; at state 2 the exhaust valve opens, and from state 2 to state 3 the gas is discharged at constant pressure; at state 3 the piston is at the top of its stroke; as it moves down, the exhaust valve closes and the trapped gas expands doing work on the piston until state 4 is reached; at state 4 the pressure in the cylinder is low enough for gas to be drawn in through the intake valve until state 1 is reached and the cycle is complete.

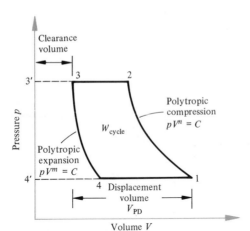

Figure 15.3   *The p–V diagram for a reciprocating compressor with clearance.*

To calculate the cycle work, we note that the area 1234 is equal to the cycle work and

$$\text{area}_{1234} = \text{area}_{123'4'} - \text{area}_{433'4'}$$

where areas 123'4' and 433'4' may be calculated as if they were cycle works for a compressor without clearance, Equation (15.3). The cycle work becomes, on substitution of the work expressions,

$$
W_{\substack{\text{cycle} \\ 1234}} = \frac{n}{n-1} p_1 V_1 \left[1 - \left(\frac{p_2}{p_1} \right)^{(n-1)/n} \right]
$$

$$
- \frac{m}{m-1} p_4 V_4 \left[1 - \left(\frac{p_3}{p_4} \right)^{(m-1)/m} \right] \tag{15.6}
$$

For this development, $p_3 = p_2$, and $p_4 = p_1$. Since the expansion work is small compared with the compression work, the error involved in setting $m = n$ is very small. With these assumptions and the pressure equalities, Equation (15.6) becomes

$$
W_{\text{cycle}} = \frac{n}{n-1} p_1 (V_1 - V_4) \left[1 - \left(\frac{p_2}{p_1} \right)^{(n-1)/n} \right] \tag{15.7}
$$

Equation (15.7) is the cycle work for a gas compressor with clearance. The difference between Equation (15.7) and Equation (15.3) is the volume term, $(V_1 - V_4)$. This term represents the amount of gas drawn into the cylinder at T_1 and p_1. As can be seen from Figure 15.3, the smaller the clearance volume, the greater the volume of gas that can be drawn into the compressor.

Example 15.1

An ideal compressor has a displacement volume of 14 liters and a clearance volume of 0.7 liter. It receives air at 100 kPa and discharges it at 500 kPa. The compression is polytropic with $n = 1.3$, and the expansion is isentropic with $m = k$. Determine the net cycle work and calculate the error involved if $m = n$. Figure 15.3 illustrates the cycle.

$$V_1 - V_3 = V_{PD} = 0.014 \text{ m}^3$$

$$V_3 = 0.0007 \text{ m}^3$$

$$V_1 = 0.0147 \text{ m}^3$$

$$V_4 = V_3 \left(\frac{p_3}{p_4} \right)^{1/m} = 0.0007(5)^{1/1.4} = 0.0022 \text{ m}^3$$

From Equation (15.6),

$$W_{cycle} = \frac{1.3}{0.3}(100)(0.0147)\left[1 - (5)^{0.23}\right]$$

$$- \frac{1.4}{0.4}(100)(0.0022)\left[1 - (5)^{0.286}\right]$$

$$W_{cycle} = -2.40 \text{ kJ}$$

If $m = n$, then

$$V_4 = (0.0007)(5)^{0.769} = 0.0024$$

$$V_1 - V_4 = 0.0123$$

$$W_{cycle} = \frac{1.3}{0.3}(100)(0.0123)\left[1 - (5)^{0.23}\right]$$

$$W_{cycle} = -2.39 \text{ kJ}$$

The percent error is

$$\text{error} = \frac{(0.01)(100)}{2.40} = 0.41 \text{ percent}$$

This illustrates that the assumption of $m = n$ is a good one and that the error involved is very small.

Example 15.2

A double-acting compressor, with a piston displacement of 0.05 m³ per stroke, operates at 500 rev/min. The clearance is 5 percent and it receives air at 100 kPa and discharges it at 600 kPa. The compression is polytropic, $pV^{1.35} = C$. Determine the power required and the air discharged in cubic meters per second.

Single Acting

$$W_{cycle} = \frac{n}{n-1} p_1(V_1 - V_4)\left[1 - \left(\frac{p_2}{p_1}\right)^{(n-1)/n}\right]$$

We need to determine V_1, V_4:

$$V_1 = V_{PD} + V_3 = V_{PD} + cV_{PD} = 0.05 + (0.05)(0.05)$$

$$= 0.0525 \text{ m}^3$$

$$V_4 = V_3\left(\frac{p_3}{p_4}\right)^{1/n} = 0.0025\left(\frac{600}{100}\right)^{1/1.35} = 0.0094 \text{ m}^3$$

$$V_1 - V_4 = 0.0431 \text{ m}^3$$

$$W_{cycle} = \frac{1.35}{0.35}(100)(0.0431)\left[1 - (6)^{(0.35/1.35)}\right] = -9.829 \text{ kJ}$$

Double Acting

$$W_{cycle} = (2)(-9.829) = -19.658 \text{ kJ}$$

$$\dot{W} = (W_{cycle})(\text{cycles/s}) = (-19.658)\left(\frac{500}{60}\right)$$

$$\dot{W} = -163.81 \text{ kW}$$

$$\dot{V}_{discharged} = (V_2 - V_3)(2)\left(\frac{500}{60}\right)$$

$$V_2 = V_1\left(\frac{p_1}{p_2}\right)^{1/n} = (0.0525)\left(\frac{1}{6}\right)^{1/1.35} = 0.0139 \text{ m}^3$$

$$V_2 - V_3 = 0.0114 \text{ m}^3$$

$$\dot{V}_{discharged} = (0.0114)(2)\left(\frac{500}{60}\right) = 0.19 \text{ m}^3/\text{s}$$

Notice that the mass, but not the volume, is conserved. If the inlet-volume flow rate is calculated, the result is $0.718 \text{ m}^3/\text{s}$. Also, it is easiest to calculate the work for a single-acting compressor and then correct for double acting.

15.3 VOLUMETRIC EFFICIENCY

The compressor's function is to take gas in and raise its pressure to a higher level. The volume of gas drawn in (during one stroke) is a function of the piston displacement. The term *volumetric efficiency* is used to described how efficiently gas is drawn into a compressor. The ideal volumetric efficiency, η_v, is the ratio of the volume (mass) of gas drawn in divided by the maximum possible amount of gas that could be drawn in, the displacement volume (mass). Either mass of volume may be used in defining volumetric efficiency.

$$\eta_v = \frac{\text{volume actual}}{\text{displacement volume}} = \frac{V_1 - V_4}{V_{PD}} \tag{15.8}$$

Referring to Figure 15.4, we note that $V_3 = cV_{PD}$, and the ideal volumetric efficiency may be reduced as follows:

$$V_1 = V_{PD} + cV_{PD}$$

Substituting in Equation (15.8) yields

$$\eta_v = 1 + c - c\left(\frac{p_2}{p_1}\right)^{1/n} \tag{15.9}$$

In analyzing Equation (15.9), we should note that the volumetric efficiency decreases as the clearance increases and as the discharge pressure increases. An increase of either effect will cause the mass of gas entering to be less because of a greater mass of trapped gas at top dead center.

In an actual compressor, ideal processes do not occur. Figure 15.5 illustrates an actual p–V diagram for a reciprocating compressor. The gas surrounding the compressor intake must be greater than the pressure in the cylinder, or the gas would not flow into the cylinder. There are frictional effects to overcome in flowing around the intake valves, as well as flow irreversibilities in the cylinder itself. Furthermore, the cylinder walls of the compressor are warm, and this raises the temperatures of the incoming gas. These combined effects serve to reduce the mass, and hence the volume, of surrounding gas that can be drawn into the compressor. To account for these effects, the ideal volumetric efficiency is reduced by the ratio of pressure inside the cylinder at state 1 to the pressure of the surrounding gas, p_0. The heating-effect term is the ratio of

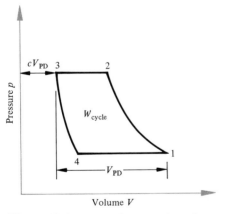

Figure 15.4 *A p–V diagram where the expansion and compression processes follow $pV^n = C$.*

the temperature of the surrounding gas, T_0, to the temperature of the gas at state 1.

$$\eta_{v(\text{actual})} = \eta_v \left(\frac{p_1}{p_0} \right) \left(\frac{T_0}{T_1} \right) \tag{15.10}$$

The work of the compressor will be increased since it must compress from less than the surrounding pressure to more than the discharge pressure. Why more? For the gas to flow from the cylinder, past the exhaust valves to the discharge line, there must be a pressure differential. This higher pressure must be used in calculating the compressor work. Figure 15.5 also illustrates the higher pressure effect. The work to overcome valve friction is the crosshatched area on Figure 15.5.

Once the volumetric efficiency of a compressor is known, then the capacity may be readily determined by multiplying the volumetric efficiency by the piston displacement. Another problem with air-compressor standards is that the density, hence the specific volume, of air varies with altitude. A compressor at sea level will be able to deliver a greater mass of air than a compressor at several thousand meters of elevation. In this case the compressor ratings are determined for "free air" or a standard condition. Corrections for temperature and pressure may be made, as in Equation (15.10), but the surrounding air would be "free air" at 1 atm pressure and 20°C.

Example 15.3

An air compressor receives air from the surroundings at 100 kPa and 21°C. There is a 2.0 kPa drop through the intake valves, and

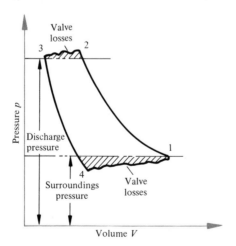

Figure 15.5 *The p–V diagram of a reciprocating compressor with valve losses denoted.*

the temperature at the end of intake is 38°C. The discharge pressure is 480 kPa and there is a 20 kPa pressure drop through the discharge valves. Determine (a) and volumetric efficiency, ideal and actual; (b) the power if the piston displacement is 14.0 liters and the ideal intake volume is 11.2 liters. The compressor is operating at 200 rev/min and $n = 1.35$.

$$p_1 = 100 - 2.0 = 98 \text{ kPa} \qquad p_0 = 100 \text{ kPa}$$

$$p_2 = 480 + 20 = 500 \text{ kPa} \qquad T_0 = 294 \text{ K} \qquad T_i = 311 \text{ K}$$

Refer to Figure 15.5 to verify that $p_2 >$ discharge pressure.

$$\eta_v = \frac{11.2}{14.0} = 0.80 \qquad (\eta_v)_{actual} = (0.80)\left(\frac{98}{100}\right)\left(\frac{294}{311}\right) = 0.741$$

$$W_{cycle} = \frac{n}{n-1}p_1(V_1 - V_4)\left[1 - \left(\frac{p_2}{p_1}\right)^{(n-1)/n}\right]$$

$$= \frac{n}{n-1}p_1(\eta_v)_{actual}V_{PD}\left[1 - \left(\frac{p_2}{p_1}\right)^{(n-1)/n}\right]$$

$$W_{cycle} = \frac{1.35}{0.35}(100)(0.741)(0.014)\left[1 - \left(\frac{500}{98}\right)^{0.35/1.35}\right]$$

$$= -2.10 \text{ kJ/cycle}$$

$$\dot{W} = \frac{(-2.10)(200)}{60} = 7 \text{ kW}$$

We have seen that the isothermal compressor uses the least work. To approach this ideal, some compressors have water-jacketed cylinders and those that do not have finned cylinders. A greater amount of heat is transferred when fins are used than when the cylinder walls are smooth. The greater the heat transfer, the lower will be the temperature at state 2, and the nearer the compression process will be to isothermal compression.

15.4 MULTISTAGE COMPRESSION

When pressures of 300 kPa and above are desired, it takes less work to use two or more stages for compression. The exact trade-off would be made on the basis of cost, as virtually all trade-offs are made. Multiple-stage compression is more efficient in that the gas may be cooled between the compression stages. This is also necessary to prevent vaporization of the lubricating oil and to prevent its ignition should the temperature become too high. This could easily happen in single-stage compression to a high pressure.

Figure 15.6 illustrates a two-stage compressor with an intercooler between the first and second stages. Ideally the intercooler will bring the temperature of the gas leaving the intercooler down to the ambient temperature. Figures 15.7(a) and 15.7(b) illustrate the p–V and T–s diagrams for the compressor. To accomplish this temperature drop, the intercooler may be water-jacketed. For two-stage compressors, the intercooler may consist of a parallel set of finned pipes connecting the low-pressure discharge header to the high-pressure intake header. The air from the fluted vanes on the compressor flywheel blows over the tubes, cooling the compressed gas within the tubes.

The work for the first- and second-stage cylinders may be calculated from Equation (15.7):

$$W_{1st} = \frac{n}{n-1} p_1 (V_1 - V_8) \left[1 - \left(\frac{p_2}{p_1} \right)^{(n-1)/n} \right] \tag{15.11}$$

The work for the second stage is

$$W_{2nd} = \frac{m}{m-1} p_2 (V_3 - V_6) \left[1 - \left(\frac{p_4}{p_2} \right)^{(m-1)/m} \right] \tag{15.12}$$

Experience with gas compressors has shown that $m = n$. The total work is the sum of the work for the two stages:

$$W_{total} = \frac{n}{n-1} p_1 (V_1 - V_8) \left[1 - \left(\frac{p_2}{p_1} \right)^{(n-1)/n} \right]$$

$$+ \frac{n}{n-1} p_2 (V_3 - V_6) \left[1 - \left(\frac{p_4}{p_2} \right)^{(n-1)/n} \right] \tag{15.13}$$

For steady flow through the compressor, the mass entering the first stage enters the second stage,

$$\dot{W}_{total} = \frac{n}{n-1} \dot{m} R T_1 \left[1 - \left(\frac{p_2}{p_1} \right)^{(n-1)/n} \right]$$

$$+ \frac{n}{n-1} \dot{m} R T_3 \left[1 - \left(\frac{p_4}{p_2} \right)^{(n-1)/n} \right] \tag{15.14}$$

For an ideal compressor, $T_3 = T_1$. Let us now find the value of p_2 that will minimize the total work. Since $\dot{W}_{total}$ will be a minimum, the first

Figure 15.6   A schematic diagram of multistage compression with intercooling.

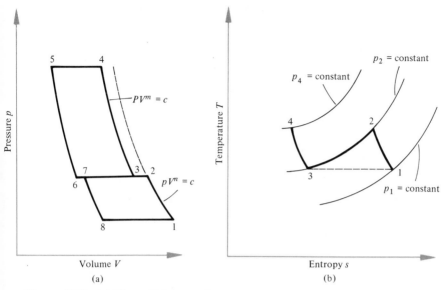

Figure 15.7 *(a) The p–V diagram for a two-stage reciprocating compressor. (b) The T–s diagram for two-stage compression with ideal intercooling.*

derivative of this with respect to the variable p_2 is zero.

$$p_2 = \sqrt{p_1 p_4} \qquad (15.15)$$

When the value of the pressure for the intercooler is determined as in Equation (15.15), then the work is equal in all stages and the total work is a minimum. The second derivative of the total work, expressed as the negative of Equation (15.14), is positive; thus the work is a minimum.

For a three-stage compressor, the low-pressure intercooler, p_2, may be found in a similar manner to be

$$p_2 = \sqrt[3]{(p_1)^2 p_4} \qquad (15.16)$$

and the pressure of the high-pressure intercooler, $p_{2'}$, is

$$p_{2'} = \sqrt[3]{(p_1)(p_4)^2} \qquad (15.17)$$

where p_1 is the intake pressure and p_4 is the final discharge pressure.

Example 15.4

A two-stage air compressor receives 0.238 m³/s of air at 100 kPa and 27°C and discharges it at 1000 kPa. The value of n for the compression is 1.35. Determine (a) the minimum power necessary for compression; (b) the power for one-stage compression to the same pressure; (c) the maximum temperature for (a) and (b); (d) the heat removed in the intercooler.

Find the optimum pressure; calculate the work for one stage of compression and double the result for two stages.

(a) $p_2 = \sqrt{(100)(1000)} = 316$ kPa

$$\dot{W}_{1st} = \frac{1.35}{0.35}[(100)(0.238)]\left[1 - \left(\frac{316}{100}\right)^{0.35/1.35}\right]$$

$$= -31.9 \text{ kW}$$

$\dot{W}_{total} = -63.8$ kW

(b) For a single-stage of compression $p_2 = 1000$ kPa,

$$\dot{W} = \frac{1.35}{0.35}[(100)(0.238)]\left[1 - \left(\frac{1000}{100}\right)^{0.35/1.35}\right] = -74.9 \text{ kW}$$

This is a 17.4 percent increase in the required power.

(c) $T_{max} = T_1\left(\frac{p_2}{p_1}\right)^{(n-1)/n} = 300\left(\frac{316}{100}\right)^{0.35/1.35} = 404.2$ K
   (a)

$T_{max} = 300(10)^{0.35/1.35} = 544.9$ K
   (b)

(d) For the first-law, open-system analysis of the intercooler.

$q = (h_2 - h_3) = c_p(T_2 - T_3) \qquad T_3 = T_1$

$q = (1.0047)(404.2 - 300) = 104.7 \text{ kJ/kg}$

$\dot{Q} = \dot{m}q \qquad \dot{m} = \frac{p_1\dot{V}_1}{RT_1} = 0.276 \text{ kg/s}$

$\dot{Q} = (0.276)(104.7) = 28.9$ kW

15.5 COMPRESSOR PERFORMANCE FACTORS

Thus far we have been able to compute the work for a compressor in which the processes were reversible. In an actual compressor irreversibilities occur, and the ideal work is not equal to the actual work that must be supplied to the compressor. To account for these irreversibilities, we introduce several compressor performance factors.

The *compression efficiency*, η_{cn}, is an indication of how closely the actual compression process approaches the ideal process:

$$\eta_{cn} = \frac{\text{theoretical work}}{\text{indicated work}} \text{(isentropic or isothermal)} \qquad (15.18)$$

Should the theoretical be based on isentropic compression, then the compression efficiency is called the *isentropic* or *adiabatic compression efficiency.* If the theoretical process is isothermal, then it is the *isothermal compression efficiency.* The *indicated work* is the work that the gas indicates was done on it. This is not the work supplied by the motor. There are mechanical losses that diminish the amount of the work the motor delivers. To account for the loss, the compressor has a mechanical efficiency, η_m, which is defined as

$$\eta_m = \frac{\text{indicated work}}{\text{shaft work}} \tag{15.19}$$

where the shaft work is the work delivered by the power source. If the two efficiencies are multiplied, the overall efficiency, or *compressor efficiency*, η_c, is obtained:

$$\eta_c = \eta_{cn}\eta_m = \frac{\text{theoretical work}}{\text{shaft work}} \tag{15.20}$$

The compressor efficiency then shows how well the compressor utilizes the energy supplied. Note that the theoretical work is the minimum energy that must be supplied, whereas the shaft work is the available energy that must actually be supplied.

15.6 ROTATIVE COMPRESSORS

This is the other major class of compressors. The principles of operation of these two compressor types—rotative and centrifugal—are entirely different. Figure 15.8 illustrates a Roots blower type of compressor. This is often used for supercharging diesel engines. The air is trapped by the lobes and casing of the blower and pushed up to the pressure of the discharge line. The clearance between the lobes and the casing and between the lobes themselves is very close to minimize leakage. Note that the lobes are rotating in opposite directions and that a steady supply of air or gas is pushed through the blower.

The centrifugal compressor, illustrated in Figure 15.9, uses a different principle of operation. Gas enters the rotating impeller axially (into

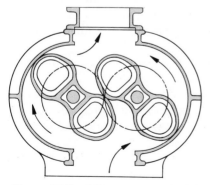

Figure 15.8 A Roots-type compressor.

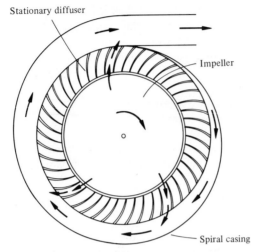

Stationary diffuser

Impeller

Spiral casing

Figure 15.9   The diffuser for a centrifugal compressor.

the paper) and has a component of velocity changed to the radial direction because of centrifugal force, while it develops an angular velocity approaching the speed of the impeller. As the gas moves outward, more gas flows into the impeller, creating a continuous gas flow. As the gas flows radially through the stationary diffuser, which has an increasing area with radial distance, the velocity head is changed to a pressure head. From the conservation of energy viewpoint, the kinetic energy decreases because of a decrease in velocity, and energy is conserved by an increase in enthalpy. Even if the process were isothermal, the pv component of enthalpy would increase.

The axial-flow compressor is similar in appearance to the reaction steam turbine, where the blades are inserted in reverse order. The larger blades, to handle the low-density gas, occur first and the smaller blades later, as the density of the gas increases. The flow of gas is essentially axial, with pressure rises occurring in both the moving and fixed blades. A pair of moving and fixed blades is considered to be one stage of compression. Figure 15.10 illustrates the blade arrangements and the pressure rise in an axial-flow compressor.

This type of compressor is used in gas-turbine units. It has a high-volume flow rate capacity with a small pressure rise. The efficiency of the unit is higher than that of comparable centrifugal compressors at design conditions. A problem with the compressor is its tendency to stall at off-design conditions when the revolutions per minute, and thus the flow, is lower. Stalling is due to backflow in the later compression stages, which blocks the flow of gas through the compressor. Typically, a bypass around the later stages is used until the design conditions are reached.

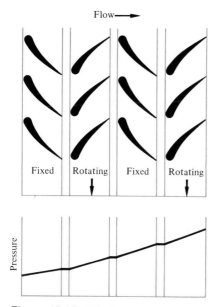

Flow⟶

Fixed | Rotating | Fixed | Rotating

Pressure

Figure 15.10   *The blading and pressure diagrams for two stages of a symmetric axial compressor.*

The efficiency of the centrifugal compressor is flat over a wide range of operating conditions, so a trade-off exists between higher efficiency and a small operating range or lower efficiency and a wider operating range when comparing axial and centrifugal compressors.

There is a great deal yet to be learned about compressor design, particularly in regard to fluid motion through the compressor as well as the mechanical design. A text in turbomachinery will provide information regarding these areas.

A final note in comparing compressors. The reciprocating compressors may use more than 11 200 kW and produce discharge pressures of 200 MPa. Also, a two-stage reciprocating compressor may be able to handle 2.4 m^3/s at inlet conditions. Rotative compressors are able to produce discharge pressures of several MPa and have flow rates of several hundred cubic meters per second. Obviously there is a wide range of overlap in operating conditions for these two basic types, and other factors, such as maintenance and frequency of operation, must be considered before selecting one type or the other.

Energy Analysis

To find the work of rotative compressors, or of reciprocating compressors using steady-flow assumptions, we can write an energy balance across the compressor. The change in kinetic energy is essentially zero as the gas

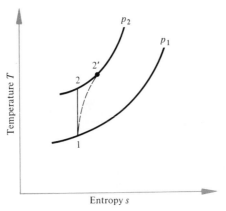

Figure 15.11 The T–s diagram for a compressor showing isentropic and actual states.

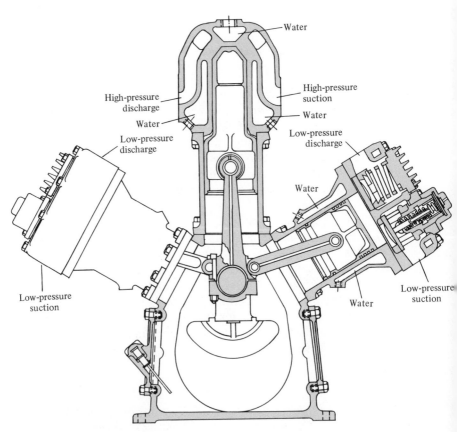

Figure 15.12 A water-cooled two-stage air compressor.

leaves with nearly the same velocity with which it entered. Intake and discharge piping are of different diameters to accommodate the changes in pressure and specific volume. The heat loss per kilogram of gas is very small considering the large flow rate through the compressor; potential-energy changes are also negligible. The energy analysis then yields

$$\dot{W} = \dot{m}(h_{2'} - h_1) \tag{15.21}$$

where $h_{2'}$ is the actual enthalpy of the gas leaving the compressor and h_1 is the inlet enthalpy. Figure 15.11 shows a T–s diagram for the compressor. The solid line from 1 to 2 is the isentropic compression line, and the dashed line denotes the irreversible process between the inlet and actual exit conditions. The ideal work is

$$\dot{W} = \dot{m}(h_2 - h_1)$$

and the compression efficiency, η_{cn}, Equation (15.18), becomes

$$\eta_{cn} = \frac{(h_2 - h_1)_s}{(h_{2'} - h_1)} \tag{15.22}$$

Example 15.5

Air is compressed in a centrifugal compressor from 110 kPa and 300 K to 330 kPa. The compression efficiency is 0.90. Determine the work per unit mass of air using the ideal gas law.

To solve the problem, we must find $h_{2'}$, the actual compressor discharge enthalpy, and knowing that and the inlet enthalpy, the work may be calculated from Equation (15.21).

$$T_2 = T_1 \left(\frac{p_2}{p_1}\right)^{(k-1)/k} = 300 \left(\frac{330}{110}\right)^{0.4/1.4} = 410.62 \text{ K}$$

$$\eta_{cn} = 0.90 = \frac{h_2 - h_1}{h_{2'} - h_1} = \frac{T_2 - T_1}{T_{2'} - T_1} = \frac{410.62 - 300}{T_{2'} - 300}$$

$$T_{2'} = 422.91 \text{ K}$$

$$w = (h_{2'} - h_1) = c_p(T_{2'} - T_1) = 123.5 \text{ kJ/kg}$$

Example 15.6

Consider a two-stage reciprocating gas compressor with water cooling as an open system. Figure 15.12 illustrates this type of compressor. Water enters at 21°C and leaves at 38°C with a flow rate of 0.038 kg/s. The air, 0.227 kg/s, enters at 300 K and 100 kPa and leaves at 1000 kPa and 450 K. Determine the power supplied.

First Law, Open System

Energy In = Energy Out

$$\dot{m}_a h_1 + (-\dot{Q}) = \dot{m}_a h_{4'} - \dot{W}_1 - \dot{W}_2$$

$$\dot{W} = \dot{W}_1 + \dot{W}_2$$

$$\dot{Q} = \dot{m}_{H_2O}(h_{out} - h_{in}) = (0.038)(159.21 - 88.14)$$

$$= 2.7 \text{ kW}$$

$$h_{4'} = 451.83 \text{ kJ/kg}$$

$$h_1 = 300.19 \text{ kJ/kg}$$

$$\dot{W} = (0.227)(451.83 - 300.19) + 2.7$$

$$\dot{W} = 37.1 \text{ kW}$$

PROBLEMS

1. A two-stage compressor receives 0.151 kg/s of helium at 140 kPa and 300 K and delivers it at 7.0 MPa. The compression is polytropic with $n = 1.5$. The intercooler is ideal. Determine (a) the power required; (b) the intercooler pressure; (c) the maximum temperature; (d) the temperature for one stage of compression; (e) the heat transferred in the intercooler.

2. A three-stage compressor operates under the conditions described in Problem 1. Determine (a) the power required; (b) the intercooler pressures; (c) the maximum temperature; (d) the heat transferred in the intercoolers.

3. Determine the expression, in terms of percent clearance and the inlet and discharge pressure, for the mean effective pressure of an ideal, single-stage compressor with isothermal compression and expansion.

4. An air compressor is tested and it is found that the electric motor used 37.3 kW when the compressor handled 0.189 m³/s of air at 101.4 kPa and 300 K and discharged it at 377.1 kPa. Determine (a) the overall adiabatic efficiency; (b) the overall isothermal efficiency.

5. Calculate the volumetric efficiency of a single-cylinder, double-acting compressor with a bore and stroke of 0.45 × 0.45 m. The compressor is tested at 150 rev/min and found to deliver a gas from 101.3 kPa and 300 K to 675 kPa at a rate of 0.166 m³/s when $n = 1.33$ for expansion and compression processes.

6. For Problem 5 calculate the percent clearance and estimate the clearance volume in cubic meters.

7. A reciprocating compressor with a 3% clearance receives air at 100 kPa and 300 K and discharges it at 1.0 MPa. The expansion and compression are polytropic with $n = 1.25$. There is a 5% pressure

drop through the inlet and discharge valves. The air is warmed to 38°C by the cylinder walls by the end of the intake stroke. Determine (a) the theoretical and actual volumetric efficiency; (b) the work per kilogram; (c) the percentage of the work needed to overcome the throttling losses.

8. A reciprocating compressor has a 5% clearance with a bore and stroke of 25 × 30 cm. The compressor operates at 500 rev/min. The air enters the cylinder at 27°C and 95 kPa and discharges at 2000 kPa. If $n = 1.3$ for compression and expansion processes determine (a) the volumetric efficiency; (b) the volume of air handled at inlet conditions in m^3/s; (c) the power required; (d) the mass of air discharged in kg/s; (e) the mass of air left at top dead center; and (f) sketch the p–V diagram.

9. A reciprocating double-acting, single-cylinder air compressor operates at 220 rev/min with a piston speed of 200 m/s. The air is compressed isentropically from 96.5 kPa and 289 K to 655 kPa. The compressor clearance is 5.4% and the air flow rate is 0.4545 kg/s. Determine for $n = 1.35$ (a) the volumetric efficiency; (b) the piston displacement; (c) the power; (d) the bore and stroke if $L = D$.

10. A double-acting, single-cylinder reciprocating air compressor has a piston displacement of 0.015 m^3 per revolution, operates at 500 rev/min, and has a 4% clearance. The air is received at 96 kPa and delivered at 584 kPa. The compression and expansion processes obey $pV^{1.35} = C$. Determine (a) the power required; (b) the free air in m^3/s; (c) the volumetric efficiency; (d) the heat transfer during compression if $T_1 = 289$ K.

11. Derive expressions for the optimum intercooler pressures for a three-stage compressor with two stages of intercooling.

12. A two-stage, double-acting air compressor operates at 150 rev/min. The conditions of the air at the beginning of compression are $p_1 = 97.9$ kPa and $T_1 = 27°C$. The low-pressure cylinder with a bore and stroke of 35.5 × 38.1 cm discharges the air at 379 kPa into the intercooler. The air in the intercooler suffers a 17.2-kPa pressure drop and enters the high-pressure cylinder at 29°C. The discharge pressure from the compressor is 2000 kPa. Compression and expansion processes in both cylinders are $pV^{1.3} = C$. The surroundings are at 100 kPa and 20°C. The percent clearance of each cylinder is 5%. Determine (a) the "free air" capacity in m^3/s; (b) the heat loss in the intercooler; (c) the total power required; (d) the optimum interstage pressure; (e) the diameter of the high-pressure piston if the stroke is the same as the low-pressure piston; (f) the heat loss in the low-pressure and high-pressure compression processes.

13. A natural-gas compressor handles 100 m^3/s of gas at 101 kPa and 280 K. The discharge pressure is 500 kPa. The compression is

polytropic with $n = 1.45$. Determine (a) the power required; (b) the discharge temperature; (c) the isothermal power required.

14. A compressor receives 0.143 m³/s of oxygen at 110 kPa and delivers it at 550 kPa. The volumetric efficiency is 85%, the isothermal compression efficiency is 70%, and the mechanical efficiency is 91%. Determine (a) the shaft power; (b) the indicated power.

15. A Roots-type blower is used on an internal-combustion engine. The air enters at 300 K and 98 kPa and leaves at 122 kPa. The flow rate is 1.5 m³/s. Determine the power required. Note: The compression is constant total volume.

16. A six-stage axial-flow compressor handles 4.75 m³/s of air at 100 kPa and 21°C and discharges it at 400 kPa (gage) and 235°C. Determine (a) the isentropic compression efficiency; (b) the power required; (c) the second law efficiency.

17. A turbine-driven compressor handles 10 kg/s of air from 100 kPa to 600 kPa with an inlet temperature of 300 K and a discharge temperature of 530 K. The inlet tubing has a 0.5-m inside diameter and the discharge piping, a 0.2-m diameter. The compression is adiabatic. Determine (a) the air inlet and exit velocities; (b) the isentropic compression efficiency; (c) the power required.

18. A large centrifugal compressor handles 9.1 kg/s of air and compresses it from 100 kPa and 15°C and with an initial velocity of 110 m/s to discharge pressure. The velocity of the high-pressure air stream is 90 m/s. The pressure ratio is 4:1. The compressor has an isentropic compression efficiency of 80%. Determine (a) the exit pressure; (b) the exit temperature; (c) the power required; (d) the second law efficiency.

19. A two-stage reciprocating air compressor is required to deliver 0.70 kg/s of air from 98.6 kPa and 305 K to 1276 kPa. The compressor operates at 205 rev/min, compression and expansion processes follow $pV^{1.25} = C$, and both cylinders have a 3.5% clearance. There is a 20-kPa pressure drop in the intercooler. The low-pressure cylinder discharges at the optimum pressure into the intercooler. The air enters the high-pressure cylinder at 310 K. The intercooler is water cooled, with the water entering at 295 K and leaving at 305 K. Determine (a) the "free air" in m³/s; (b) the low-pressure and high-pressure discharge temperature; (c) the optimum interstage pressure; (d) the cooling water required for the intercooler in kg/s; (e) the theoretical power required; (f) the low-pressure cylinder dimensions if $L/D = 0.70$; (g) the motor requirements in kW, if the adiabatic compression efficiencies are 83% and the mechanical efficiency is 85%.

20. A small freezer is to have a cooling load of 1.1 tons at $-25°C$. The compressor and motor must be selected. The following conditions

may be assumed:

1. Refrigerant, R12
2. Reciprocating compressor operating at 600 rev/min
3. Discharge pressure, 800 kPa
4. Liquid receiver temperature, 20°C
5. $\eta_{cn} = 0.80$ and $\eta_m = 0.80$

Determine (a) the compressor displacement in liters; (b) the power required.

21. A water-jacketed air compressor handles 0.143 m³/s of air entering at 96.5 kPa and 21°C and leaving at 480 kPa and 132°C; 10.9 kg/h of cooling water enters the jacket at 15°C and leaves at 21°C. Determine the compressor power.

22. A single-stage, single-acting compressor has a 20-cm bore and a 25-cm stroke and rotates at 200 rev/min. The compressor intake receives dry saturated ammonia vapor at −18°C and compresses it adiabatically to 1000 kPa. The actual shaft work of the compressor is 20% more than the reversible, isentropic, compression work. The volumetric efficiency is 88%. After leaving the compressor and the condenser, the liquid ammonia is at 20°C. Determine (a) the ammonia flow rate; (b) the power input to the compressor; (c) the heat removed in the condenser.

23. A fan whose efficiency is 40% has a capacity of 0.5 m³/s at 15°C and a pressure of 1 atm. The fan provides a static pressure of 5 cm of water at full load. What size motor is required to drive the fan?

24. A single-acting, twin-cylinder 30-×30-cm compressor receives saturated ammonia vapor at 226 kPa and discharges it at 1200 kPa. Saturated liquid ammonia enters the expansion valve. Ice is to be manufactured at −9°C and the water is available at 26°C. The compressor runs at 150 rev/min and the volumetric efficiency is 80%. Assuming at specific heat of ice to be 2.1 kJ/kg · K, determine (a) the ammonia flow rate; (b) the mass of ice manufactured; (c) the compressor power.

25. A single-stage, double-acting reciprocating air compressor is guaranteed to deliver 0.24 m³/s of free air with a clearance of 3% and inlet conditions of 100 kPa and 21°C and a discharge pressure of 725 kPa. When tested under these conditions the compression and expansion curves followed $pV^n = C$, where $n = 1.34$. Determine (a) the piston displacement in cubic meters per second; (b) the capacity and discharge pressure if the percent clearance is held constant and the compressor is operated at an altitude of 1800 m, where the barometric pressure is 604 mm Hg absolute and the temperature is 21°C.

Chapter 16
Internal-Combustion Engines

We all know that one of the most important power-producing devices is the internal-combustion engine, of which the automobile engine is one. Why is this called an internal-combustion engine? Because the heat addition, the combustion process, occurs within the engine, not outside it as with the Rankine cycle. The chemical energy of the fuel is converted to thermal energy (heat), which in turn is converted into mechanical energy, or work. The work per unit time is the power propelling the car. The fuel used in internal-combustion engines is a hydrocarbon mixture, such as gasoline, diesel fuel, alcohol, or gas.

Attempts have been made since the mid-1800s to develop an internal-combustion engine. Attempts were made to operate an engine on gunpowder. The exploding gunpowder would force a piston upward. The downward motion of the piston would engage a rachet and turn a shaft, which was connected to a load. This engine was not a success, and other avenues were explored, In 1862 Beau de Rochas developed the theoretical steps necessary to have an efficient engine. However, it remained for Nicholas A. Otto, who developed the theory independently, to implement theory into practice by constructing an operable engine. The first successful internal-combustion engine was built by Otto in 1876.

Another example of the internal-combustion engine is the diesel engine, developed by Rudolf Diesel, who wished to operate an engine on powdered coal; it exploded. Later designs, however, in which the engine operated on liquid fuel, were successful.

16.1 AIR-STANDARD CYCLES

Both Otto's and Diesel's engines operate on open cycles. The products of combustion leaving the engine cannot be continuously reused in a closed system. Fresh air must be drawn in. It is, however, theoretically conveni-ent to have these engines operating on a thermodynamic cycle. The theoretical engines are called *air-standard* engines—air is the working fluid in the engine. Instead of fuels being burned, heat is added from an external source; and instead of exhausting the products of combustion, a heat sink is used to remove heat from the air and return the air to its original state.

Air-Standard Otto Cycle

The air-standard Otto cycle has the following processes.

1. Starting with the piston at bottom dead center, compression proceeds isentropically from 1 to 2.
2. Heat is added at constant volume from 2 to 3.
3. Expansion occurs isentropically from 3 to 4.
4. Heat is rejected at constant volume from 4 to 1.

Figures 16.1(a) and 16.1(b) illustrate the $T–S$ and $p–V$ diagrams for the cycle. The mass of air remains constant throughout the cycle, and the cycle is a thermodynamic cycle. This is different from an actual internal-combustion engine cycle, which will be discussed later in the chapter.

Let us calculate the thermal efficiency, η_{th}, for the Otto cycle. The thermal efficiency is defined as the work produced (the desired effect)

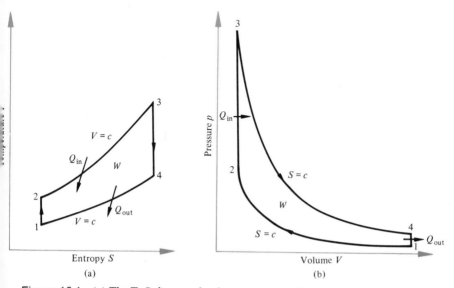

Figure 16.1 (a) The $T–S$ diagram for the air-standard Otto cycle. (b) The $p–V$ diagram for the air-standard Otto cycle.

divided by the heat added (what it costs to achieve that effect).

$$\eta_{th} = \frac{W_{net}}{Q_{in}} = \frac{\Sigma Q}{Q_{in}} \tag{16.1}$$

The heat is added at constant volume from state 2 to state 3. Since the system is a closed one, because the mass is constant, the first law tells us that

$$Q_{2-3} = U_3 - U_2 = mc_v(T_3 - T_2) \tag{16.2}$$

Heat is rejected at constant volume from state 4 to state 1. The first law again states

$$Q_{4-1} = U_1 - U_4 = -mc_v(T_4 - T_1) \tag{16.3}$$

Substituting Equations (16.2) and (16.3) into Equation (16.1) yields

$$\eta_{th} = 1 - \frac{T_4 - T_1}{T_3 - T_1} \tag{16.4}$$

The compression ratio, r, is defined as

$$r = \frac{\text{volume at bottom dead center}}{\text{volume at top dead center}} \tag{16.5}$$

and for the Otto cycle this becomes

$$r = \frac{V_1}{V_2} = \frac{V_4}{V_3} \tag{16.6}$$

Also, the temperature is related by an isentropic process between state 1 and state 2, so

$$\frac{T_2}{T_1} = (r)^{k-1} \tag{16.7}$$

and similarly

$$\frac{T_3}{T_4} = (r)^{k-1}$$

so

$$\frac{T_3}{T_4} = \frac{T_2}{T_1} \tag{16.8}$$

Equation (16.4) may be further simplified by eliminating T_3 and T_4, yielding

$$\eta_{th} = 1 - \frac{1}{(r)^{k-1}} \tag{16.9}$$

Thus the thermal efficiency of the Otto cycle is a function of the compression ratio only. As the compression ratio increases, the efficiency increases. In an actual engine the compression is limited by the temperature at state 2. If this temperature is too great, the gasoline–air mixture will ignite spontaneously and at the incorrect time.

The efficiency of the Otto cycle is much greater than that of an actual engine. Figure 16.2 shows the efficiency versus compression ratio graph for two values of k and indicates where the actual engine efficiency may lie. The value of the thermal efficiency for $k = 1.4$ is called the "cold-air standard efficiency," and for $k = 1.3$, the hot-air standard efficiency. As we will find out, the temperature of the air throughout most of the cycle is quite high, 1000 K to 2000 K, and the value of k at these temperatures is less than at lower temperature levels. When we analyze the engine as an open system considering the combustion process, the results will be more accurate. Finally, the effects of dissociation and nonideal considerations must be made before the engine may be correctly modeled.

Example 16.1

An engine operates on the air-standard Otto cycle. The conditions at the start of compression are 27 C and 100 kPa. The heat added is 1840 kJ/kg. The compression ratio is 8. Determine the temperature and pressure at each point in the cycle, the thermal

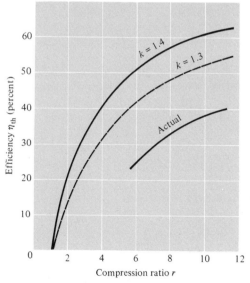

Figure 16.2 *The thermal efficiency of the air-standard and actual engines as a function of compression ratio.*

efficiency, and the mean effective pressure.

$$T_1 = 300 \text{ K} \qquad p_1 = 100 \text{ kPa}$$

Process 1–2 isentropic,

$$pV^k = C \qquad \text{and} \qquad V_1/V_2 = r = 8$$

$$\frac{T_2}{T_1} = \left(\frac{V_1}{V_2}\right)^{k-1} \qquad T_2 = 300(8)^{0.4} = 689.2 \text{ K}$$

$$\frac{p_2}{p_1} = \left(\frac{V_1}{V_2}\right)^{k} \qquad p_2 = 100(8)^{1.4} = 1837.9 \text{ kPa}$$

Process 2–3 constant volume,

$$q = u_3 - u_2 = c_v(T_3 - T_2)$$

$$1840 = (0.7176)(T_3 - 689.2) \qquad T_3 = 3253.5 \text{ K}$$

$$p_3 = p_2\left(\frac{T_3}{T_2}\right) = 1837.9\left(\frac{3253.5}{689.2}\right) = 8676.1 \text{ kPa}$$

Process 3–4 isentropic,

$$pV^k = C \qquad \text{and} \qquad V_3/V_4 = 1/r$$

$$\frac{T_4}{T_3} = \left(\frac{V_3}{V_4}\right)^{k-1} \qquad T_4 = 3253.5\left(\frac{1}{8}\right)^{0.4} = 1416.2 \text{ K}$$

$$\frac{p_4}{p_3} = \left(\frac{V_3}{V_4}\right)^{k} \qquad p_4 = 8676.1\left(\frac{1}{8}\right)^{1.4} = 472.1 \text{ kPa}$$

$$\eta_{th} = 1 - \frac{1}{(r)^{k-1}} = 1 - \frac{1}{(8)^{0.4}} = 0.565 \text{ or } 56.5 \text{ percent}$$

$$p_m = \frac{w_{net}}{\text{displacement volume}} = \frac{w_{net}}{v_1 - v_2}$$

$$v_1 = \frac{RT_1}{p_1} = \frac{(0.287)(300)}{100} = 0.861 \text{ m}^3/\text{kg}$$

$$v_2 = \frac{v_1}{r} = 0.1076 \text{ m}^3/\text{kg}$$

$$p_m = \frac{1039.6}{0.861 - 0.1076} = 1379.9 \text{ kPa}$$

From this example, we find that the thermal efficiency of the Otto cycle is high and also that the mean effective pressure is a useful measure in comparing engines operating on different cycles; the greater the mean effective pressure, the smaller the engine may be for a given work output. The Carnot cycle mean effective pressure is quite low, seldom exceeding 138 kPa.

The maximum temperature and pressures were also very high in the air-standard cycle. This is because specific heat variation is neglected, there is no dissociation, and heat is added at constant volume. In an actual engine none of the preceding assumptions are true. Why, then, bother with the air-standard analysis in the first place? The reason is that the model is simple and gives a sense of direction to changes affecting an actual engine. Thus we can determine that increasing the compression ratio improves the thermal efficiency without running experiments to verify it.

The following example will demonstrate the use of the air tables in an air-standard Otto cycle problem.

Example 16.2

An engine operating on the air-standard Otto cycle has a 15 percent clearance volume, a total displacement volume of 2.8 liters, and operates at 2500 rev/min. The heat added is 1400 kJ/kg. Determine the maximum temperature and pressure, the thermal efficiency, power, and the available portion of the heat rejected. The inlet conditions are 27°C and 100 kPa, and $T_0 = 300$ K. Use the air table, Table A.2.

$$V_2 = V_3 = cV_{PD} = 0.000\ 42\ m^3$$

$$V_4 = V_1 = V_{PD} + V_2 = 0.003\ 22\ m^3$$

Find T_3, p_3 for maximum T, p. Find state 1 to 4 to evaluate thermal efficiency and determine power and available energy. Process 1–2, isentropic, use p_r and v_r where applicable.

$$\frac{v_{r_2}}{v_{r_1}} = \frac{V_2}{V_1} \qquad v_{r_2} = 144.32 \left(\frac{0.000\ 42}{0.003\ 22} \right) = 18.82$$

From Table A.2,

$$p_{r_2} = 23.49 \qquad T_2 = 662.7\ K \qquad u_2 = 483.2\ kJ/kg$$

$$\frac{p_2}{p_1} = \frac{p_{r_2}}{p_{r_1}} \qquad p_2 = 100 \left(\frac{23.49}{1.3860} \right) = 1695\ kPa$$

Process 2–3, constant volume,

$$q = u_3 - u_2$$

$$1400 = u_3 - 483.2 \qquad u_3 = 1883.2 \text{ kJ/kg}$$

$$p_{r_3} = 3210 \qquad v_{r_3} = 1.9793 \qquad T_3 = T_{max} = 2211 \text{ K}$$

$$p_3 = p_2 \left(\frac{T_3}{T_2} \right) = 1695 \left(\frac{2211}{662.7} \right) = 5.65 \text{ MPa} = p_{max}$$

Process 3–4, isentropic, use p_r and v_r where applicable.

$$\frac{v_{r_4}}{v_{r_3}} = \frac{V_4}{V_3} = \frac{V_1}{V_2} \qquad v_{r_4} = 1.9793 \left(\frac{0.003\ 22}{0.000\ 42} \right) = 15.174$$

$$u_4 = 917.1 \text{ kJ/kg} \qquad p_{r_4} = 223.6$$

$$p_4 = p_3 \left(\frac{p_{r_4}}{p_{r_3}} \right) = 5.65 \left(\frac{223.6}{3210} \right) = 393.5 \text{ kPa}$$

Process 4–1, constant volume,

$$q = u_1 - u_4 = 214.07 - 917.1 = -703 \text{ kJ/kg}$$

$$w_{net} = \Sigma q = 1400 - 703 = 697 \text{ kJ/kg}$$

$$\eta_{th} = \frac{w_{net}}{q_{in}} = \frac{697}{1400} = 0.498 \text{ or } 49.8 \text{ percent}$$

Note that the thermal efficiency cannot be evaluated by Equation (16.9) because the assumption of constant specific heat is no longer invoked when using the air tables.

Power:

$$\dot{W} = m w_{net} n$$

where

m = mass of air in cycle

n = cycles per minute

$$m = \frac{p_1 V_1}{R T_1} = 0.003\ 74 \text{ kg}$$

$$\dot{W} = (0.003\ 74)(697) \left(\frac{2500}{60} \right) = 108.6 \text{ kW}$$

Available portion of q_{out},

$$(A.E.)_{4-1} = q_{4-1} - T_0(s_1 - s_4)$$

$$s_1 - s_4 = \phi_1 - \phi_4 - R\ln\left(\frac{p_1}{p_4}\right)$$

$$s_1 - s_4 = 2.5153 - 3.4112 - 0.287\ln\left(\frac{100}{393.5}\right)$$

$$= -0.5027 \text{ kJ/kg} \cdot \text{K}$$

$$(A.E.)_{4-1} = -703 - (300)(-0.5027)$$

$$(A.E.)_{4-1} = -552.2 \text{ kJ/kg}$$

$$(A.E.)_{4-1} = \frac{(552.2)(100)}{(703)} = 78.5 \text{ percent of } q_{out}$$

The high temperature at state 4 indicates that there is a substantial amount of available energy not being used.

The Air-Standard Diesel Cycle

The diesel cycle, developed by Rudolf Diesel, is characterized by constant-pressure heat addition, constant-volume heat rejection, and isentropic compression and expansion processes. This engine is a compression-ignition type; the air is compressed to a high temperature, fuel is injected into the air, ignition is due to the high air temperature, and combustion occurs at constant pressure. The piston expands isentropically to bottom dead center, where heat is rejected at constant volume. Figures 16.3(a) and 16.3(b) illustrate the air-standard diesel cycle on the p–V and T–S diagrams. In the air standard, cycle heat, not fuel, is added.

The processes in the air-standard diesel cycle are as follows:

1. Starting with the piston at bottom dead center, compression occurs isentropically from 1 to 2.
2. Heat is added at constant pressure from state 2 to state 3.
3. Expansion occurs isentropically from state 3 to state 4.
4. Heat rejection occurs at constant volume from state 4 to state 1.

The thermodynamic cycle is complete. Let us calculate the thermal efficiency for the diesel cycle. To accomplish this, we must find the expressions for the heat supplied and heat rejected and substitute these into Equation (16.1). The heat supplied is at constant pressure, and for a closed system the first law tells us that

$$Q_{2-3} = H_3 - H_2 = mc_p(T_3 - T_2) \tag{16.10}$$

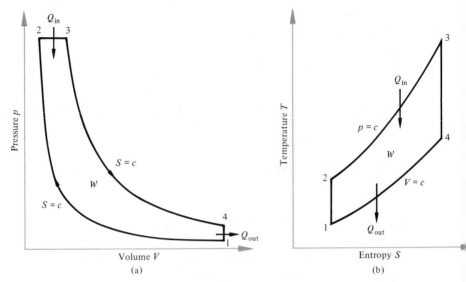

Figure 16.3 *(a) The p–V diagram for the air-standard Diesel cycle. (b) The T–S diagram for the air-standard Diesel cycle.*

The heat rejection occurs at constant volume, and the first law for a closed system tells us that

$$Q_{4-1} = U_1 - U_4 = -mc_v(T_4 - T_1) \qquad (16.11)$$

Substituting in Equation (16.1) yields

$$\eta_{th} = 1 - \frac{1}{k}\left(\frac{T_4 - T_1}{T_3 - T_2}\right) \qquad (16.12)$$

Because the cycle is not symmetrical, the expression for the thermal efficiency cannot be further reduced as can that for the Otto cycle. We note that $V_1/V_2 \neq V_4/V_3$.

The thermal efficiency of the diesel cycle is slightly less than that of the Otto cycle for the same heat addition. This is because part of the expansion process is occurring while heat is being added in the diesel cycle, whereas the expansion of air in the Otto cycle occurs after all the heat is added. There is an important factor favoring diesels, however. The compression ratio can be much greater in the diesel cycle than in the Otto cycle because only air, not a fuel–air mixture, is being compressed. Thus for actual engines, the diesel-cycle engine is more efficient than the Otto-cycle engine because the possible compression ratio is greater. The maximum allowable temperature after combustion is about the same for both engines, so the advantage of high peak temperatures occurring in the Otto cycle does not accrue to the actual engine.

Example 16.3

An engine operates on the air-standard diesel cycle. The conditions
at the start of compression are 27°C and 100 kPa. The heat supplied
is 1840 kJ/kg. The compression ratio is 16. Determine the maximum
temperature and pressure, the thermal efficiency, and the mean
effective pressure.

Process 1–2, isentropic,

$$pV^k = C \qquad r = \frac{V_1}{V_2} = 16$$

$$\frac{T_2}{T_1} = \left(\frac{V_1}{V_2}\right)^{k-1} \qquad T_2 = (300)(16)^{0.4} = 909.4 \text{ K}$$

$$\frac{p_2}{p_1} = \left(\frac{V_1}{V_2}\right)^k \qquad p_2 = (100)(16)^{1.4} = 4850.3 \text{ kPa}$$

Process 2–3, constant pressure,

$$q = (h_3 - h_2) = c_p(T_3 - T_2)$$
$$1840 = (1.0047)(T_3 - 909.4) \qquad T_3 = T_{max} = 2740.8 \text{ K}$$
$$p_{max} = p_3 = p_2 = 4850.3 \text{ kPa}$$
$$v_3 = \frac{RT_3}{p_3} = \frac{(0.287)(2740.8)}{4850.3} = 0.1622 \text{ m}^3/\text{kg}$$

Process 3–4, isentropic,

$$pV^k = C \qquad v_4 = v_1$$
$$v_1 = \frac{RT_1}{p_1} = \frac{(0.287)(300)}{100} = 0.861 \text{ m}^3/\text{kg}$$

$$\frac{T_4}{T_3} = \left(\frac{v_3}{v_4}\right)^{k-1} \qquad T_4 = 2740.8\left(\frac{0.1622}{0.861}\right)^{0.4} = 1405.7 \text{ K}$$

$$\frac{p_4}{p_3} = \left(\frac{v_3}{v_4}\right)^k \qquad p_4 = 4850.3\left(\frac{0.1622}{0.861}\right)^{1.4} = 468.6 \text{ kPa}$$

Process 4–1, constant volume,

$$q = (u_1 - u_4) = c_v(T_1 - T_4)$$
$$q_{out} = (0.7176)(300 - 1405.7) = -793.4 \text{ kJ/kg}$$
$$w_{net} = \Sigma q = 1046.6 \text{ kJ/kg}$$

$$\eta_{th} = \frac{w_{net}}{q_{in}} = \frac{1046.6}{1840} = 0.569 \text{ or } 56.9 \text{ percent}$$

The mean effective pressure, p_m, is

$$p_m = \frac{w_{net}}{v_1 - v_2} = \frac{1046.6}{(0.861 - 0.0538)} = 1296.6 \text{ kPa}$$

Example 16.4

An engine operates on the air-standard diesel cycle with inlet air conditions of 300 K and 100 kPa. The compression ratio is 16, and the heat added is 1400 kJ/kg. Determine the maximum temperature and pressure and the thermal efficiency for the engine. Use the air tables.

Process 1–2, isentropic, p_r and v_r relationships

$$\frac{v_{r_2}}{v_{r_1}} = \frac{v_2}{v_1} = \frac{1}{r} \qquad v_{r_2} = \frac{621.2}{16} = 38.82$$

$$p_{r_2} = 63.8$$

$$h_2 = 890.9 \text{ kJ/kg}$$

$$T_2 = 662.4 \text{ K}$$

$$\frac{p_{r_2}}{p_{r_1}} = \frac{p_2}{p_1} \qquad p_2 = 100\left(\frac{63.8}{1.386}\right) = 4603 \text{ kPa}$$

Process 2–3, constant pressure,

$$q = h_3 - h_2$$

$$1400 = h_3 - h_2 \qquad h_3 = 2290.9 \text{ kJ/kg}$$

$$T_3 = 2031 \text{ K}$$

$$p_3 = p_{max} = p_2 = 4603 \text{ kPa}$$

$$p_{r_3} = 2214$$

$$v_{r_3} = 2.639$$

$$T_{max} = T_3$$

$$v_3 = \frac{RT_3}{p_3} = \frac{(0.287)(2031)}{4603} = 0.1266 \text{ m}^3/\text{kg}$$

Process 3–4, isentropic p_r and v_r relationships:

$$v_4 = v_1 = \frac{RT_1}{p_1} = \frac{(0.287)(300)}{100} = 0.861 \text{ m}^3/\text{kg}$$

$$\frac{v_{r_4}}{v_{r_3}} = \frac{v_4}{v_3} \qquad v_{r_4} = 2.639\left(\frac{0.861}{0.1226}\right) = 17.947$$

$$T_4 = 1118.8 \text{ K}$$

$$u_4 = 861.7 \text{ kJ/kg}$$

$$p_{r_4} = 178.9$$

$$\frac{p_4}{p_3} = \frac{p_{r_4}}{p_{r_3}} \qquad p_4 = 4603\left(\frac{178.9}{2214}\right) = 371.9 \text{ kPa}$$

Process 4–1, constant volume,

$$q_{out} = u_1 - u_4 = 214.07 - 861.7 = -647.6 \text{ kJ/kg}$$

$$w_{net} = \sum q = +752.4 \text{ kJ/kg}$$

$$\eta_{th} = \frac{w_{net}}{q_{in}} = \frac{752.4}{1400} = 0.537 \text{ or } 53.7 \text{ percent}$$

Cutoff Ratio

Another ratio is used in describing diesel engine performance and that is the cutoff ratio, r_c. It is defined as

$$r_c = \frac{\text{volume at end of heat addition}}{\text{volume at start of heat addition}} = \frac{V_3}{V_2} \qquad (16.13)$$

In an actual engine the cutoff ratio refers to the volume ratio at the start of injection to that at the end of fuel injection. It is also sometimes defined as the volume change during injection (heat addition) divided by the displacement volume. Using Equation (16.13) the thermal efficiency for an air standard diesel engine, Equation (16.12), may be reduced to

$$\eta = 1 - \frac{1}{(r)^{k-1}}\left[\frac{r_c^k - 1}{k(r_c - 1)}\right] \qquad (16.14)$$

Figure 16.4 illustrates the effect of cutoff and compression ratios on the diesel cycle efficiency. As the cutoff ratio increases, there is more power. This means there is a longer period of heat addition or, in an actual engine, a longer period of fuel injection and hence more energy input. However, there is a limit to r_c. If the cutoff ratio is more than 10 percent of the stroke, smoking tends to occur in an actual engine because there is not sufficient time for the combustion process to be completed

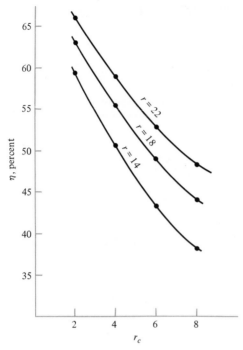

Figure 16.4 Cycle efficiency as a function of cutoff and compression ratios.

before the exhaust valve opens. Thus, the products of combustion leave the cylinder before all the chemical energy has been converted into thermal energy and unburned fuel is present. There is a lower limit on r_c; there must be some heat added, or fuel consumption, to overcome the friction of the various moving parts, as well as the piston and cylinder, and the friction associated with bearings in the engine.

Air-Standard Stirling and Ericsson Cycles

The Stirling and Ericsson cycles are two less-known thermodynamic cycles. A regenerator is the key element in both cycles; heat must be stored in the regenerator during one part of the cycle and reused in another part. The development of such a heat exchanger has proved to be a very limiting factor in the development of these cycles into machinery.

The Stirling cycle, illustrated in Figures 16.5(a) and 16.5(b), is characterized by the following processes. From state 1 to state 2, heat is added at constant temperature, causing the volume to increase. The air is then forced through a regenerator from state 2 to state 3 at constant

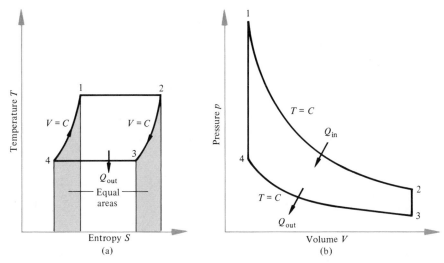

Figure 16.5 (a) The T–S diagram for the air-standard Stirling cycle. (b) The p–V diagram for the air-standard Stirling cycle.

volume. In the regenerator the air is progressively cooled, and the thermal energy, transferred from the air, is stored in the regenerator. From state 3 to state 4, heat is rejected at constant temperature and the volume decreases. The air is now forced back into the regenerator from state 4 to state 1. The entering air is cool, and heat is ideally transferred from the regenerator to the air. All the heat stored is returned to the air and the air exits the regenerator at state 1. Thus on the T–S diagram, the areas denoting the heat transferred are equal. Obviously this can happen only in an ideal regenerator; otherwise a temperature difference must exist between two systems for there to be heat transfer. Two problems have existed for the Stirling hot-air engine: the regenerator design and constant-volume regeneration.

To overcome the constant-volume regeneration, Ericsson developed a constant-pressure regenerative cycle, illustrated in Figures 16.6(a) and 16.6(b).

Why were these cycles devised? The thermal efficiencies of both cycles match that of the Carnot cycle. Furthermore, for engines operating between the same temperatures reservoirs and with equal changes in specific volume, the work of the Stirling cycle is greater than that of the Carnot and Ericsson cycles. The work of the Ericsson cycle lies between that of the Stirling and Carnot cycles. This is significant when applied to practical engines, where the work per cycle is very important. The greater the work per cycle, the smaller the engine.

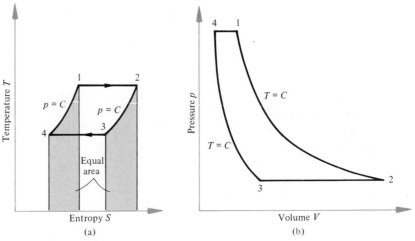

Figure 16.6 *(a) The T–S diagram for the air-standard Ericsson cycle. (b) The p–V diagram for the air-standard Ericsson cycle.*

Stirling Cycle Thermal Efficiency

Let us calculate the thermal efficiency of the Stirling cycle; the same method may be used for the Ericsson cycle, yielding the identical thermal efficiency.

$$\eta_{th} = \frac{W_{net}}{Q_{in}} = \frac{\Sigma Q}{Q_{in}}$$

The first law for a closed system:

$$Q = \Delta U + W$$

$$\Delta U = 0 \qquad \text{for } T = C$$

$$Q_{in} = W = mRT_1 \ln\left(\frac{V_2}{V_1}\right)$$

Similarly,

$$Q_{out} = -mRT_3 \ln\left(\frac{V_3}{V_4}\right)$$

However, $V_2 = V_3$ and $V_4 = V_1$, so

$$\eta_{th} = \frac{T_1 - T_3}{T_1} = 1 - \frac{T_L}{T_H} \tag{16.15}$$

where T_L is the low temperature and T_H is the high temperature. The thermal efficiency is the same as the Carnot cycle. This is not so surprising when we realize the heat is transferred by the same process in both engines.

The Stirling engine is gaining in interest today in both power-producing and power-consuming cycles. The reversed Stirling cycle is used for gas liquefaction and cryogenic work because advances in heat transfer make better regenerator designs possible. Because of the high thermal efficiency potential, actual engine designs have been developed, but they do not adhere to constant-volume regeneration. The major problem for the engine is that the air is heated from an external source, with resulting inefficiencies. The engine has also been used in solar-power systems, with the solar energy acting as the heat source. This chapter, however, is concerned more with the internal-combustion engine, particularly the diesel and Otto cycles.

Air-Standard Dual Cycle

Neither the air-standard Otto cycle nor the air-standard diesel cycle approximates the cycles of the actual engines. An air-standard approximation, the dual cycle, was developed to compensate for nonideal behavior in both engine types. In this cycle, heat is added at constant volume and at constant pressure. Figures 16.7(a) and 16.7(b) illustrate the p–V and T–S diagrams, respectively. The heat addition simulates the behavior of either engine as both engines experience pressure and volume changes during the combustion process. The equations for work, thermal

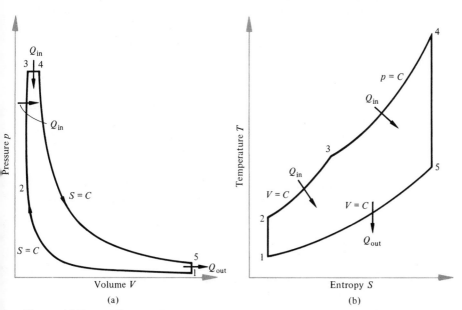

Figure 16.7   (a) The p–V diagram for the air-standard dual cycle. (b) The T–S diagram for the air-standard dual cycle.

efficiency, and heat supplied and rejected may be calculated in a similar manner to that illustrated for the Otto and diesel cycles.

16.2 OPEN-CYCLE ANALYSIS

In considering an engine as an open system, which indeed it is, greater accuracy may be achieved in the determination of engine work and efficiency. Let us consider the internal-combustion engines as an open system shown in Figure 16.8, where $\dot{m}_a$ is the air flow rate, h_a is the enthalpy of the air, $\dot{m}_f$ is the fuel flow rate, h_f is the total enthalpy of the fuel, $\dot{Q}$ is the heat lost to cooling water surroundings, $\dot{W}$ is the time rate of work produced, and h_e is the enthalpy of the products of combustion leaving the engine.

The difference between the diesel and Otto cycles lies in how the combustion process occurs. Figure 16.9(a) shows a T–S diagram for the Otto cycle and Figure 16.9(b) shows a T–S diagram for the diesel cycle.

In order to calculate the exit enthalpy, something must be known of the processes leading to it. If the exit temperature and pressure are known, then the enthalpy may be determined directly from tabulated property values. If not, we must proceed around the cycle to determine state 4.

The enthalpy of the fuel, h_f, has two components, the enthalpy of combustion and the enthalpy due to a temperature other than that of the reference state. The latter enthalpy is very small compared with the enthalpy of combustion, 150 versus 45 000 kJ/kg. Also, the products leave the engine at a fairly high temperature, so the water remains a vapor and the lower heating value will be used. Thus

$$h_f \approx h_{RP} \qquad\qquad (16.16)$$

The following two examples will illustrate the open-system energy analysis for Otto and diesel cycles.

Example 16.5

An engine operates on the Otto cycle and has a compression ratio of 7.5. The air–fuel ratio is 30 : 1, and the engine burns gaseous

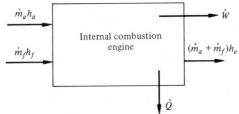

Figure 16.8   A schematic diagram for an open-system internal-combustion engine.

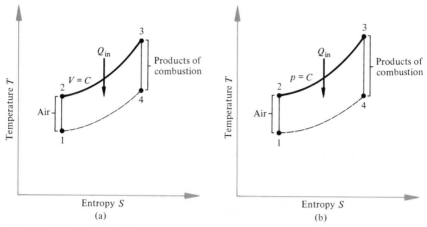

Figure 16.9 (a) The T–S diagram for the open Otto-cycle analysis. (b) The T–S diagram for the open Diesel-cycle analysis.

octane with a lower heating value of 44 232 kJ/kg. The air enters the engine at 27°C and 100 kPa. Determine the maximum possible work per kilogram of air. Use the air tables.

The maximum possible work for the engine occurs under adiabatic conditions, $\dot{Q} = 0$.
From Table A.2,

$$h_a = 300.19 \text{ kJ/kg} \qquad v_{r_1} = 144.32$$

For an isentropic process from 1 to 2,

$$v_{r_2} = v_{r_1}\left(\frac{v_2}{v_1}\right) = \frac{144.32}{7.5} = 19.24$$

$$T_2 = 657 \text{ K}$$

$$u_2 = 479.0 \text{ kJ/kg}$$

For the Otto cycle, combustion is a constant-volume process; the energy released is u_{RP}, where $u_{RP} = h_{RP} - RT$. The temperature is taken at the reference datum, 25°C, for u_{RP} because we do not have a method to calculate the variation of h_{RP} with temperature. This results in an error of about 1 percent, which is less than the error involved in assuming an isentropic compression and expansion.

$$u_{RP} = 44\ 232 - \frac{(8.3143)(298)}{114.23} = 44\ 210.0 \text{ kJ/kg}$$

Thus from state 2 to state 3, the first law yields

$$u_2 + r_{f/a} u_{RP} = u_3 \left(1 + r_{f/a}\right)$$

$$r_{f/a} = \frac{1}{30} = 0.0333$$

$$u_3 = 1888.3 \text{ kJ/kg} \qquad T_3 = 2215 \text{ K}$$

$$p_{r_3} = 3554 \qquad v_{r_3} = 0.4972$$

From 3–4 the process is isentropic, but we cannot say that $v_4 = v_1$. What is true is that $p_4 = p_1$, so we must find state 4 from the p_r relationships, knowing the pressure at state 3.

$$p_2 = p_1 \left(\frac{p_{r_2}}{p_{r_1}}\right) = 100 \left(\frac{22.75}{1.3860}\right) = 1641 \text{ kPa}$$

$$p_3 = p_2 \left(\frac{T_3}{T_2}\right) = 1641 \left(\frac{2215}{657}\right) = 5532 \text{ kPa}$$

$$p_{r_4} = p_{r_3} \left(\frac{p_4}{p_3}\right) = 3554 \left(\frac{100}{5532}\right) = 64.24$$

$$T_4 = 864 \text{ K} \qquad h_4 = h_e = 892.7 \text{ kJ/kg}$$

A first-law analysis of the entire engine yields

$$h_a + r_{f/a} h_{RP} = \left(1 + r_{f/a}\right) h_e + w$$

$$w = 300.19 + (0.0333)(44\ 232) - (1.0333)(892.7)$$

$$w = 850.7 \text{ kJ/kg}$$

$$\eta_{th} = \frac{w}{q_{in}} = \frac{w}{r_{f/a} h_{RP}} = \frac{850.7}{1472.9} = 0.577 \text{ or } 57.7 \text{ percent}$$

Notice that the entire enthalpy of formation is chargeable against the engine, although it did not use it in the combustion process. This is used in determining the thermal efficiency of the engine.

In an actual engine the cylinder walls, and on larger engines the cylinder heads and pistons, are cooled by water. This means that the assumption of isentropic compression and expansion is not valid, but more information must be known about the heat removed from the engine. If this quantity is given, we may assume the exhaust gas of the adiabatic engine is cooled by this amount of heat, to give a better approximation of the actual engine exit temperature. The heat released to the cooling water, as a percent of total energy varies with engine load and type, but an indicative value for large diesel engines is 25 percent.

Example 16.6

A diesel engine receives air at 27°C and 100 kPa and fuel, liquid dodecane, at the same temperature. The air–fuel ratio is 30 : 1, and the compression ratio is 15. The lower heating value of dodecane is 44 460 kJ/kg. Calculate the work if the heat loss to cooling is 25 percent of the energy supplied to the engine. The products of combustion are similar to 200 percent theoretical air.

To perform a first-law analysis on the engine, we must find the enthalpy of the products of combustion leaving the engine. We must find the adiabatic exit temperature first.

From 1–2, isentropic, use p_r and v_r relationships

$$\frac{v_{r_2}}{v_{r_1}} = \frac{v_2}{v_1} = \frac{1}{r} \qquad v_{r_2} = \frac{144.32}{15} = 9.621$$

$$h_2 = 869.6 \text{ kJ/kg} \qquad p_{r_2} = 58.45 \qquad T_2 = 843.1 \text{ K}$$

$$\frac{p_2}{p_1} = \frac{p_{r_2}}{p_{r_1}} \qquad p_2 = 100\left(\frac{58.45}{1.3860}\right) = 4217 \text{ kPa}$$

From 2–3, constant pressure, and the first law for the combustion process is

$$h_2 + r_{f/a} h_{RP} = \left(1 + r_{f/a}\right) h_3$$

$$869.6 + (0.0333)(44\ 460) = (1.0333) h_3$$

$$h_3 = 2274.4 \text{ kJ/kg}$$

$$p_{r_3} = 2494 \qquad T_3 = 1929 \text{ K}$$

From 3–4, isentropic, use p_r and v_r relationships

$$\frac{p_{r_4}}{p_{r_3}} = \frac{p_4}{p_3} \qquad p_{r_4} = 2494\left(\frac{100}{4217}\right) = 59.14$$

$$h_4 = 869.7 \text{ kJ/kg} \qquad T_4 = 817.6 \text{ K}$$

At this point we know the adiabatic exhaust temperature is 817.6 K. The actual exhaust temperature will be less by the heat removed in cooling. Thus,

$$q_{out} = (0.25)(0.0333)(44\ 460) = 370.1 \text{ kJ/kg}$$

$$370.1 = \left(1 + r_{f/a}\right)\left(h_4 - h_e\right) = (1.0333)(869.7 - h_e)$$

$$h_e = 511.5 \text{ kJ/kg} \qquad T_e = 496 \text{ K}$$

The first-law analysis of the entire engine is

$$-q + h_a + r_{f/a}h_{RP} = (1 + r_{f/a})h_e + w$$

$$- 370.1 + 300.19 + (0.0333)(44\ 460) = (1.0333)(511.5) + w$$

$$w = 882.1\ \text{kJ/kg}$$

$$\eta_{th} = \frac{w}{q_{in}} = \frac{882.1}{(0.0333)(44\ 460)}$$

$$= 0.569 \text{ or } 59.6 \text{ percent}$$

16.3 ACTUAL DIESEL AND OTTO CYCLES

One of the more difficult concepts in engine analysis is that of the cycle. The thermodynamic cycle, used for analyzing the air-standard cycle, is not an actual engine cycle. The mass of air and products of combustion are actually continually undergoing change. As we know, it is impossible to recombust air, so a fresh air supply must continually be drawn into the engine and the products of combustion removed.

Four-Stroke Cycle

Let us consider the four-stroke mechanical cycle, such as that of the automotive engine. In the spark-ignition engine, the air–fuel mixture is compressed, the spark plug discharges, and a spark ignites the fuel mixture. The combustion process is very rapid and occurs over a small volume change in the cylinder; thus the ideal process is a constant-volume heat addition. (If the engine is compression ignition—a diesel cycle, for example—then at the top of the compression stroke a selected amount of fuel is injected into the cylinder and ignites because of the high air temperature. The combustion of the fuel continues as the piston expands on the power stroke, and the pressure remains essentially constant during the combustion process.) Because there is only one power stroke for every two revolutions, it is necessary to know the operating cycle when calculating engine power. The four strokes may be classified as follows.

1. *Intake stroke:* The intake valve is open, the exhaust valve is closed, and the piston moves down bringing a fresh air–fuel spark-ignition (engine) or air compression-ignition (engine) mixture into the cylinder.
2. *Compression stroke:* Both intake and exhaust valves are closed and the air–fuel spark-ignition and air compression-ignition mixture is compressed by the upward movement of the piston.
3. *Power stroke:* Both intake and exhaust valves are closed and combustion occurs with the resultant pressure increase forcing the piston downward.

4. *Exhaust stroke:* The exhaust valve is open, the intake valve is closed, and the upward movement of the piston forces the products of combustion from the engine.

Since not all the products of combustion are removed from the cylinder on the exhaust stroke, there is a dilution of the incoming air charge by the remaining products. The greater the clearance volume, the greater the dilution.

Actual p–V Diagram

The p–V diagram for a spark-ignition engine is illustrated in Figure 16.10, with the lines for the Otto cycle superimposed. The compression process from 1 to 2 is not adiabatic, so the actual pressure is less than the ideal. Ignition occurs before top dead center, allowing time for the combustion to develop. The combustion process does not occur at constant volume; there is energy loss to the cylinder walls and piston head and the piston is moving downward; thus the peak pressure is less than that in the Otto cycle. The expansion process is nonadiabatic, hence the lower pressures from state 3 to state 4. We notice that the exhaust valve opening occurs before bottom dead center, and this reduces the power produced (it reduces area I). Why is this done? Area II represents the work the engine must do in pushing out the exhaust products, state 4 to state 5, and pulling in the fresh air charge, state 5 to state 1. A trade-off must be made that will allow the net area, I–II, to be a maximum. If the exhaust valve opens too early, then the area I reduction

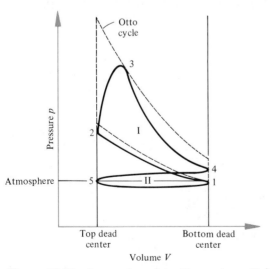

Figure 16.10   An actual spark-ignition engine p–V diagram with the Otto cycle superimposed.

cannot compensate for the resulting decrease in area II. For this condition the exhaust pressure at state 4 is lower, and the line from 4 to 5 is more horizontal. If the exhaust valve is opened too late, then the work to remove the exhaust gases is greater because of a higher pressure at state 4. This is indicated by an increase in area II. Each engine is unique and is a function of the operating conditions as well as the design. In the design we would want the resistance in the intake and exhaust systems to be as small as possible. The $p-V$ diagram for a compression-ignition engine is very similar to that shown in Figure 16.10.

Two-Stroke Cycle

If there were only one cycle that the spark-ignition and compression-ignition engines used, then it would not be necessary to designate the strokes per cycle. There is, however, another cycle that both engines can use, the two-stroke cycle, in which all four events occur in two strokes, or one revolution of the engine. The two-stroke cycle is found on compression-ignition engines, in large and small power ranges, and on small spark ignition engines such as those in lawn mowers, chain saws, and motorcycles. Figure 16.11 illustrates the $p-V$ diagram for a two-stroke cycle diesel engine that is not supercharged. To assist the exhaust process in the two-stroke cycle engine, the engines are equipped with scavenging blowers, which raise the inlet air pressure to 13–35 kPa above atmospheric pressure. Thus the intake air pushes the exhaust gas out. This is not the same as supercharging, which raises the inlet pressure much higher. The work required in operating the scavenging air pump or

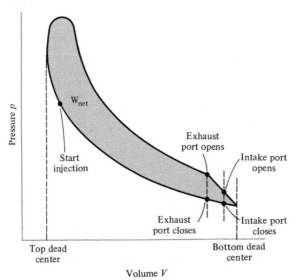

Figure 16.11   *A p–V diagram for two-stroke cycle (nonsupercharged).*

compressor is charged against the engine, so the net work of the two-stroke cycle engine is reduced by this amount. In addition to this loss of work, the combustion process often does not go as far toward completion as it does in the four-stroke cycle engine, so that the work of a two-stroke cycle engine is not twice, but only about $1\frac{1}{2}$ times, that of a four-stroke cycle engine.

16.4 CYCLE COMPARISONS

Of all heat engines, the internal-combustion engine has the highest thermal efficiency. This is because the maximum temperature in the cylinder may reach as high as 2400°C during the combustion process. The metal parts of the engine do not come in contact with this temperature as it occurs only in the gas mixture. The economic trade-offs between the Otto, or spark-ignition engine, and the diesel, or compression-ignition engine, are such that at high speeds, 4000 to 6000 rev/min, and reasonably low power, 150 to 225 kW, the spark-ignition engine is more advantageous, as well as lighter. In the middle range (several hundred kilowatts) the diesel and Otto engines overlap, and at higher powers the diesel engine dominates. Diesel engines are commonly found in trucks, buses, auxiliary or emergency power generators, and as the main propulsion engine on ships. The marine diesel engine sizes are currently around 20 000 kW, with a bore and stroke of 0.9 m by 1.5 m, and these engines may have ratings up to 30 000 kW. The diesel engine uses a less expensive fuel than the Otto engine, but both engines require precise timing for the combustion process. The combustion process is not continuous and must occur cyclically and be completed in times of 10^{-3} s.

In the spark-ignition engine the fuel mixture is ignited by an electric spark from the spark plug. The temperature of the fuel–air mixture in the vicinity of the spark plug is raised above the ignition temperature and combustion occurs. When the fuel–air mixture is correct, the flame of this localized mixture spreads throughout the cylinder burning the mixture. The fuel–air mixture must be $0.055 \leqslant r_{f/a} \leqslant 0.10$ or combustion will not occur. The propagation of the flame throughout the cylinder occurs very quickly, but the crank travel may be 20 to 30 degrees.

16.5 ENGINE PERFORMANCE ANALYSIS

An increase in thermal efficiency occurs if the combustion time is decreased; a greater temperature can be achieved before expansion and more work accomplished for the same energy supplied. One method is to increase the flame velocity. As the flame velocity increases, however, the engine begins to run rough, because of unbalanced pressure waves moving across the piston top. The rapidly burning fuel in one region causes a localized pressure increase; the unbalanced pressure causes a pressure wave to move across the cylinder trying to achieve equilibrium. If localized self-ignition of the mixture occurs, then severe pressure rises will be created. The pressure waves produced may be supersonic and move across the cylinder very quickly. This is accompanied by a knocking noise; knock is the noise caused by autoignition of the mixture. The combined effects of autoignition, plus the noise, is called "detonation." This obviously is not good for the engine surfaces. Pitting may occur on the metal surfaces, and cracks may develop from high-speed, cyclic knocking. The formation and movement of shock waves is an irreversible process; we know that energy thus spent cannot be used for work and has a deleterious effect on engine efficiency. There are fuel additives that resist detonation. The octane rating of fuel is a measure of its resistance to detonation. The greater the octane number, the more resistant the fuel is to detonation.

In compression-ignition engines a spark plug is not needed to ignite the fuel; the temperature of the compressed air is sufficiently high to accomplish this. Fuel is sprayed into the cylinder at the proper instant, and the hot air causes the fuel oil droplets to vaporize; then as ignition temperature is reached, combustion occurs. There is a delay between the injection and the combustion, and this is called *ignition delay*. The fuel droplets are a cool liquid when they enter and must be broken into a fine spray so vaporization and combustion can occur. As in the spark-ignition engine, there can be knocking due to a discontinuous combustion process. One means of preventing autoignition is use of a fuel with a low ignition temperature. The fuel will ignite as injected into the cylinder, and the flame will sustain the combustion as more fuel is injected. The ignitability of the fuel is rated by its cetane number—the greater the ignitability, the higher the rating. Considerable excess air must also be provided in the diesel engine to ensure combustion of the liquid fuel droplets. If the fuel is heated to a high temperature in the absence of oxygen, then cracking occurs, and carbon is formed. This is manifested as a sooty exhaust.

At reduced power the compression-ignition engine is more efficient than the spark-ignition engine. This is due, in part, to the combustion process. At a reduced load less fuel is needed. In the spark-ignition engine the air flow must be restricted to maintain the correct fuel–air ratio. The throttling loss is irreversible, and the loss of available energy reduces the thermal efficiency of the engine. In the compression-ignition engine, on the other hand, the fuel and air systems are separate and the fuel can be decreased independently of the air with no resulting throttling losses. The fuel–air ratio may be smaller than 0.01 and, as the ratio decreases, the engine efficiency approaches that of the air-standard diesel engine.

16.6 WANKEL ENGINE

The Wankel engine operates by transferring rotary motion from the engine's rotary piston to the car's driveshaft. In a piston–cylinder engine the vertical motion of the piston is transferred to the rotary motion of the crankshaft by the connecting rod. The Wankel engine is an internal-combustion engine and must have the same four processes: intake, compression, power, and exhaust. Figure 16.12 illustrates the continuous cycle. On the three sides of the rotary piston different processes are continuously occurring. There are few moving parts in the engine; there are, for instance, ports instead of exhaust and inlet valves. An air and fuel mixture is continuously drawn in from 1 to 4. At state 4, the inlet port is closed by one of the lobes. Compression continues from 5 to 8. At state 9, ignition of the mixture occurs, and the power from the burning mixture, demonstrated by increased pressure and temperature, forces the rotor lobe from 10 to 12. The exhaust port is uncovered and the burned mixture exhausts from 13 to 18. The process now repeats itself. The

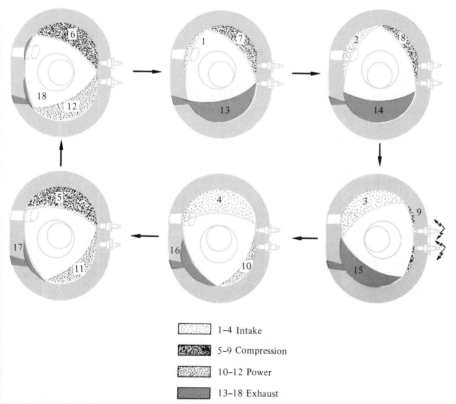

Figure 16.12   The processes occurring in a Wankel engine.

reader will notice that each revolution has three power strokes. Therefore the power may be produced in a more compact volume than in a piston–cylinder engine.

One of the problems with the engine is in sealing the space between the lobes and the rotor housing. Also, the combustion process is incomplete, typical also of the two-stroke internal-combustion engine, and thus the exhaust contains comparatively large amounts of unburned hydrocarbons. This results in a decrease in overall efficiency compared with the piston–cylinder type. However, the high power produced per unit engine volume makes the Wankel engine attractive for certain applications.

16.7 ENGINE EFFICIENCIES

There are basic measurements and indications of an engine's performance. From the measurements we get an indication of how well the engine is producing power. As a fluid passes through an engine, it performs work on the engine, for example, it moves a piston, turns a turbine wheel. This is called *indicated work*, or W_i, the work the fluid indicates it has accomplished. This is usually expressed in the power, the time rate of doing work, as the indicated power $\dot{W}_i$, expressed in kilowatts. Not all the work produced by the fluid is delivered to the engine shaft, however; some of it must overcome the friction in the engine, and we have to consider the "pumping losses" in the exhaust and intake processes. This is the frictional work, and the power is the friction power $\dot{W}_f$. The remaining portion of the indicated work can be used to drive an automobile or ship or to accomplish whatever purpose the engine may have. The work is called the "shaft" or "brake" work W_b and the associated power is the brake power $\dot{W}_b$.

indicated work − friction work = brake work

$$W_i \qquad - \qquad W_f \qquad = \qquad W_b$$

There are several efficiencies that may be determined for an engine. There are two engine efficiencies, the brake engine efficiency, η_b, and the indicated engine efficiency, η_i. All efficiencies are the desired effect divided by the expense of producing the effect.

$$\eta_{\text{engine}} = \frac{\text{actual work of a system}}{\text{work of corresponding ideal system}} \qquad (16.17)$$

Thus

$$\eta_b = \frac{\text{brake work}}{\text{theoretical work}} = \frac{W_b}{W} \qquad (16.18)$$

$$\eta_i = \frac{\text{indicated work}}{\text{theoretical work}} = \frac{W_i}{W} \qquad (16.19)$$

The mechanical efficiency of an engine, η_m, is an indication of how well the engine could convert the indicated work into brake work, or

$$\eta_m = \frac{\text{brake work}}{\text{indicated work}} = \frac{W_b}{W_i} \qquad (16.20)$$

There are thermal efficiencies for the engine also. We know that thermal efficiencies are

$$\eta_{\text{th}} = \frac{\text{work output}}{\text{energy input}} \qquad (16.21)$$

and since we are discussing two work terms, there are two thermal efficiencies. The brake thermal efficiency, η_{tb},

$$\eta_{tb} = \frac{\text{brake work}}{\text{energy in}} = \frac{\dot{W}_b}{\dot{m}_f h_{RP}} \qquad (16.22)$$

and the indicated thermal efficiency, η_{ti},

$$\eta_{ti} = \frac{\text{indicated work}}{\text{energy in}} = \frac{\dot{W}_i}{\dot{m}_f h_{RP}} \qquad (16.23)$$

The higher, not the lower, heating value of the fuel is typically used. There are many ways to relate the various efficiencies by algebraic manipulation. That is best left for the individual problem at hand.

Example 16.7

A spark-ignition engine produces 224 kW while using 0.0169 kg/s of fuel. The fuel has a higher heating value of 44 186 kJ/kg and the engine has a compression ratio of 8 : 1. The friction power is found to be 22.4 kW. Determine: $\eta_{tb}, \eta_{ti}, \eta_m, \eta_b, \eta_i$.

$$\eta_{tb} = \frac{\dot{W}_b}{\dot{m}_f h_{RP}} = \frac{(224)}{(0.0169)(44\ 186)} = 0.30$$

$$\dot{W}_i = \dot{W}_b + \dot{W}_f = 246.4 \text{ kW}$$

$$\eta_{ti} = \frac{\dot{W}_i}{\dot{m}_f h_{RP}} = \frac{(246.4)}{(0.0169)(44\ 186)} = 0.33$$

$$\eta_m = \frac{\dot{W}_b}{\dot{W}_i} = \frac{224}{246.4} = 0.909$$

$$\eta_b = \frac{\dot{W}_b}{\dot{W}}$$

However, $\dot{W}$ is not yet known; $\dot{W}$ is the power of the Otto cycle.

$$\eta_{\text{Otto}} = 1 - \frac{1}{(r)^{k-1}} = 1 - 0.435 = 0.565$$

$$\eta_{\text{Otto}} = \frac{\dot{W}}{\dot{Q}_{\text{in}}} = 0.565 = \frac{\dot{W}}{(0.0169)(44\ 186)}$$

$$\dot{W} = 421.9 \text{ kW}$$

$$\eta_b = \frac{224}{421.9} = 0.53$$

$$\eta_i = \frac{246.4}{421.9} = 0.58$$

The engine efficiency is, therefore, a function of how the ideal work is calculated. If this is not otherwise noted, we will use the air-standard cycle for calculating the theoretical work.

16.8 POWER MEASUREMENT

We have discussed the various work forms and the engine efficiencies. How are these work terms measured experimentally? The brake work may be measured by experimentally determining the horsepower generated by the rotating engine shaft. For low-speed, low-horsepower engines, this may be accomplished with the use of a prony brake (Figure 16.13). For high-speed engines, a dynamometer is used and the output measured electrically. Let us consider the prony brake, as this is the simplest one.

The brake work is dissipated as friction on a water-cooled friction band, similar to a brake shoe on an automobile. Work is defined as a force acting through a distance. For one revolution of the engine, a point on the perimeter of the flywheel will move a distance $2\pi r$. There is a uniform friction force, f, acting against the flywheel throughout the

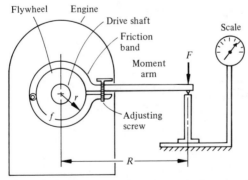

Figure 16.13 *A schematic diagram of a prony brake.*

revolution; thus the work dissipated as friction during the revolution is

work during one revolution = force × distance = $2\pi f$

The turning moment, rf, produced by the driveshaft is balanced by an equal and opposite moment, which is the product of the length of the moment arm, R, and the force on the scale, F. Thus

$$rf = RF$$

work during one revolution = $2\pi RF$

If the number of revolutions per unit time are known, then the time rate of doing work, or power, may be calculated.

power = $2\pi RFn$

where n is the revolutions per minute. To determine the power

$$\dot{W}_b\,(W) = 2\pi R\,(m/rev) \cdot F\,(N) \cdot n\,(rev/s)$$

$$\dot{W}_b\,(W) = 2\pi RFn \tag{16.24}$$

where $\dot{W}_b\,(W)$ is the brake power in watts.

The product of the moment arm, R, and the force, F, is called the *torque*, T, of the engine. The torque is the turning moment exerted by a tangential force acting at a distance from the axis of rotation. Thus Equation (16.24) could be written as

$$\dot{W}_b\,(W) = 2\pi Tn \tag{16.25}$$

Since the torque of an engine is frequently referred to and is sometimes confused with power, the following example should help in keeping the two distinct. Torque is the capacity to do work, whereas power is the rate at which work may be done. Let us consider a truck pulling a load. The torque will determine whether or not the truck is able to pull the load; the power will determine the rate, that is, how fast the load may be pulled.

If a test is being run on the prony brake, the moment arm will have an initial weight. This must be subtracted from the scale reading to find the force exerted by the engine. This is called the *tare*.

The indicated work is found by determining the p–V diagram of the engine. An indicator card of the cycle is taken either mechanically or with an oscilloscope. The work is the net area enclosed by the cycle processes. This area divided by the length of stroke will give the mean effective pressure. It may be helpful to review the section in Chapter 7 in which the concept of mean effective pressure was developed. This is called the "indicated mean effective pressure," p_{mi}, as it is determined by knowing the indicated work.

$$p_{mi} = \frac{W_i}{\text{piston displacement}} \tag{16.26}$$

The brake mean effective pressure, p_{mb}, is

$$p_{mb} = \frac{W_b}{\text{piston displacement}} \qquad (16.27)$$

The power may be determined in terms of the mean effective pressure.

$$\text{piston displacement} = LA$$

where L is the length of the stroke, and A is the area of the piston.

$$p_{mi} LA = \frac{\text{indicated work}}{\text{revolution}}$$

$$p_{mi} LAn = \text{power}$$

$$\dot{W}_i = p_{mi} LAn \text{ W} \qquad (16.28)$$

$$\dot{W}_b = p_{mb} LAn \text{ W} \qquad (16.29)$$

The important concept to remember about mean effective pressure is that it is an artificial pressure based on the work of the cycle.

Example 16.8

A six-cylinder automotive engine with a 9-×9-cm bore and stroke has a fuel consumption of 8.5×10^{-5} kg/s at 3000 rev/min; $\dot{W}_b = 86$ kW, $\dot{W}_i = 105$ kW. The thermal efficiency of the ideal cycle is 47 percent, and the fuel has a heating value of 44 186 kJ/kg. Compute (a) the mechanical efficiency, the brake and indicated thermal efficiencies; (b) the brake and indicated engine efficiencies; and (c) the brake and indicated mean effective pressures.

(a)

$$\eta_m = \frac{\dot{W}_b}{\dot{W}_i} = \frac{86}{105} = 0.819 \text{ or } 81.9 \text{ percent}$$

$$\eta_{tb} = \frac{\dot{W}_b}{\dot{m}_f h_{RP}} \qquad \dot{m}_f = (8.5 \times 10^{-5})(86)$$

$$\dot{m}_f = 7.31 \times 10^{-3} \text{ kg fuel/s}$$

$$\eta_{tb} = \frac{86}{(7.3 \times 10^{-3})(44\ 186)} = 0.266 \text{ or } 26.6 \text{ percent}$$

$$\eta_{ti} = \frac{105}{(7.31 \times 10^{-3})(44\ 186)} = 0.325 \text{ or } 32.5 \text{ percent}$$

(b)

$$\eta_b = \frac{\dot{W}_b}{\dot{W}} \qquad \dot{W} = (0.47)(\dot{Q}_{in})$$

$$\dot{W} = (0.47)(7.31 \times 10^{-3})(44\ 186)$$

$$\dot{W} = 151.8\ \text{kW}$$

$$\eta_b = \frac{86}{151.8} = 0.566 \text{ or } 56.6 \text{ percent}$$

$$\eta_i = \frac{\dot{W}_i}{\dot{W}} = \frac{105}{151.8} = 0.692 \text{ or } 69.2 \text{ percent}$$

(c) From Equations (16.26) and (16.27),

$$p_{mb} = 3004\ \text{kPa}$$

$$p_{mi} = 3668\ \text{kPa}$$

PROBLEMS

1. An air-standard Otto cycle has a compression ratio of 8.0 and has air conditions at the beginning of compression of 100 kPa and 25°C. The heat added is 1400 kJ/kg. Determine (a) the four cycle state points; (b) the thermal efficiency; (c) the mean effective pressure.

2. An engine operates on an air-standard Otto cycle with a compression ratio of 9 to 1. The pressure and temperature at the end of the compression stroke are 800 kPa and 700°C, respectively. Determine the net cycle work per kilogram if the pressure at the end of heat addition is 3.0 MPa.

3. An engine operates on the air standard Otto cycle. The cycle work is 900 kJ/kg, the maximum cycle temperature is 3000°C and the temperature at the end of isentropic compression is 600°C. Determine the engine's compression ratio.

4. An engine operates on the air standard Otto cycle. The pressure and temperature at the beginning of isentropic compression are 120 kPa and 35°C, respectively. The peak pressure and temperature are 4.8 MPa and 2500°C. Determine (a) the net cycle work in kJ/kg; (b) the cycle efficiency.

5. A supercharged spark-ignition engine operates on the air standard Otto cycle. The inlet conditions are 210 kPa and 100°C and the maximum temperature and pressure are 2250°C and 7.5 MPa. Determine (a) the net work; (b) the cycle efficiency.

6. An air-standard Otto cycle uses 0.1 kg of air and has a 17% clearance. The intake conditions are 98 kPa and 37°C, and the

energy release during combustion is 1600 kJ/kg. Using the air tables, determine (a) the compression ratio; (b) the pressure and temperature at the four cycle state points; (c) the displacement volume; (d) the thermal efficiency; (e) the work; (f) the cycle second-law efficiency.

7. An air-standard Otto cycle has the following cycle states, where state 1 is at the beginning of the isentropic compression: $p_1 = 101$ kPa, $T_1 = 333$ K, $V_1 = 0.28$ m³, $T_3 = 2000$ K, $r = 5$. Determine (a) the remaining cycle state points; (b) the thermal efficiency; (c) the heat added; (d) the heat rejected; and (e) if $T_1 = T_0$, find the available portion of the heat rejected.

8. An air-standard diesel cycle receives 28.5 kJ/cycle of heat while operating at 300 rev/min. At the beginning of compression, $p_1 = 100$ kPa, $T_1 = 305$ K, $V_1 = 0.0425$ m³. At the beginning of heat addition, the pressure is 3450 kPa. Determine (a) p, V, T at each cycle state point; (b) the work; (c) the power; (d) the mean effective pressure.

9. The heat addition process in an air standard diesel cycle adds 800 kJ/kg. The cycle minimum temperature and pressure are 20°C and 100 kPa, and the maximum temperature is 1000°C. Determine the cycle thermal efficiency.

10. An engine operates on the air standard diesel cycle with a compression ratio of 18. The pressure and temperature at the beginning of compression are 120 kPa and 43°C. The maximum temperature is 1992°K and the heat added is 1274 kJ/kg. Determine (a) the maximum pressure; (b) the temperature at the beginning of heat addition.

11. An air-standard diesel cycle has a compression ratio of 20 and a cutoff ratio of 3. Inlet pressure and temperature are 100 kPa and 27°C. Determine: (a) the heat added per kilogram; (b) the net work per kilogram.

12. An air-standard diesel cycle engine operates as follows: at the end of expansion the pressure is 240 kPa and the temperature is 550°C; at the end of compression the pressure is 4.2 MPa and the temperature is 700°C. Determine (a) the compression ratio; (b) the cutoff ratio; (c) the heat added per kilogram of air; (d) the cycle efficiency; (e) the cycle second-law efficiency.

13. An air-standard diesel cycle engine operates as follows: inlet temperature, 30°C; temperature after compression, 700°C; net cycle work, 590.1 kJ/kg; and heat added per cycle, 925 kJ/kg. Determine (a) the compression ratio; (b) the maximum cycle temperature; (c) the cutoff ratio.

14. In the air-standard diesel cycle, the compression ratio is 17 : 1. The cutoff ratio, the ratio of the volume after heat addition to that before heat addition (V_3/V_2), is 2.5 : 1. The air conditions at the beginning

of compression are 101 kPa and 300 K. Determine (a) the thermal efficiency; (b) the heat added per kilogram of air; (c) the mean effective pressure.

15. In the air-standard diesel cycle, the air is compressed isentropically from 26°C and 105 kPa to 3.7 MPa. The entropy change during heat rejection is -0.6939 kJ/kg · K. Determine (a) the heat added per kilogram of air; (b) the thermal efficiency; (c) the maximum temperature; (d) the temperature at the start of heat rejection.

16. In the air-standard dual cycle, the isentropic compression starts at 100 kPa and 300 K. The compression ratio is 13 : 1; the maximum temperature is 2750 K; the maximum pressure is 6894 kPa. Determine (a) the cycle work per kilogram; (b) the heat added per kilogram; (c) the mean effective pressure.

17. An engine operates on an air-standard Stirling cycle with regeneration. At the beginning of isothermal compression the air pressure is 150 kPa. At the end of isothermal compression the pressure is 300 kPa. The peak cycle pressure and temperature are 1.5 MPa and 850°C, respectively. Determine (a) the heat added per kilogram; (b) the heat rejected per kilogram; (c) the thermal efficiency; (d) the cycle second-law efficiency.

18. An engine operates on the air-standard Stirling cycle. The pressure and temperature at the beginning of isothermal compression are 3.5 MPa and 150°C, respectively. The engine thermal efficiency is 40%. Determine the heat transferred to the regenerator per kilogram.

19. An engine operates on an air-standard Stirling cycle. The pressure and temperature at the beginning of isothermal compression are 700 kPa and 100°C, respectively. The engine has a compression ratio of 3 and a mean effective pressure of 1.0 MPa. Determine the heat transferred to the regenerator per kilogram.

20. An automobile engine, operating on the open Otto cycle, receives air at 100 kPa and 27°C and fuel (octane) at the same temperature. The engine has a compression ratio of 8 : 1. The engine has an air–fuel ratio of 25 : 1. The heat loss from the engine is equal to 30% of the work produced. Determine (a) the maximum temperature and pressure; (b) the work produced per cycle; (c) the heat loss; (d) the available portion of the products if $T_0 = 300$ K.

21. A test is run on a high-speed, four-cycle, spark-ignition engine operating at 5000 rev/min. Air enters at 100 kPa and 27°C, and fuel, with a heating value of 43 000 kJ/kg, enters at the same temperature. The maximum temperature is 2100 K. Cooling water enters the engine at 21°C and leaves at 43°C with a flow of 27 kg water/kg fuel. Use the open Otto cycle for determining necessary state points. For a six-cylinder engine with a bore and stroke of 9 × 9 cm and a compression ratio of 7.5 determine the following: (a) the fuel–air

ratio; (b) the fuel flow rate; (c) the power produced; (d) the water flow rate; (e) the exhaust temperature.

22. A stationary internal-combustion engine operates on the Otto cycle and develops 335 kW at 2200 rev/min with full throttle at standard atmospheric pressure and temperature. The mechanical efficiency is 80% and the brake specific fuel consumption is 7.6×10^{-5} kg/s · kW. The engine is moved to a higher elevation and tests indicate that only 261 kW will be produced at full throttle. The thermal efficiency is assumed to be constant; determine (a) the percent change in the volumetric efficiency at the new elevation; (b) the elevation; (c) the fuel tank required for two days' fuel, if the fuel density is 672 kg/m³.

23. A six-cylinder spark-ignition engine has a bore and stroke of 10.9×10.5 cm. The engine requires 0.00315 kg/s of $C_8H_{18(l)}$ when operating at half load with a speed of 3000 rev/min. The reduction of engine speed to axle speed is 3.78 : 1. The tires have an effective radius of 35.5 cm. (a) Determine the car speed in kilometers per hour and the fuel consumption in kilometers per liter (the specific gravity may be assumed to be 0.85). (b) The air–fuel ratio on the mass basis is 15.3 : 1; the products of combustion leave the engine at 900 K with air and fuel inlet temperature of 25°C. Determine the percentage of the heat release lost to the products of combustion.

24. An automotive engine is rated at 150 kW at 4500 rev/min. The fuel has a heating value of 46 500 kJ/kg. The overall thermal efficiency is 27%. Find the brake specific fuel consumption.

25. In an air-standard Ericsson cycle the maximum pressure is 4.1 MPa and the minimum pressure is 210 kPa. The heat supplied is 581 kJ/kg and the minimum temperature is 21°C. Determine (a) the cycle work; (b) the heat rejected; (c) the heat stored in the regenerator; (d) the entropy change during heat addition.

26. An engine operating on the air-standard Stirling cycle is examined and found to have the following conditions: $p_1 = 725$ kPa, $T_1 = 590$ K, and $V_1 = 0.0567$ m³ at the beginning of isothermal expansion; $V_2/V_1 = 1.5$; $T_3 = 300$ K. For one cycle, determine (a) the work; (b) the thermal efficiency; (c) the mean effective pressure; (d) the heat rejected; (e) the heat added.

27. An eight-cylinder diesel engine with a 10-$\times$10-cm bore and stroke operates at 2000 rev/min. Dodecane $(C_{12}H_{26})_1$ fuel is used with 80% excess air. The air enters the engine at 100 kPa and 37°C and is compressed to 3.0 MPa. The heat loss from the engine is one-third of the work produced. Use the open-system diesel cycle to calculate state points. Determine (a) the compression ratio; (b) the fuel consumption; (c) the thermal efficiency; (d) the power produced; (e) the engine-cooling water required if the water enters at 21°C and leaves at 49°C.

28. A diesel engine with a compression ratio of 14.5 : 1 starts the compression stroke with air at 101 kPa and 312 K. Fuel with a heating value of 43 260 kJ/kg is used with a fuel–air ratio of 0.0333 kg fuel/kg air. The air flow to the engine is measured and found to be 0.10 m^3/s. Use the ideal gas laws for air and determine (a) the maximum temperature; (b) the mean effective pressure; (c) the maximum power.

29. A four-cylinder, four-cycle diesel engine has a bore and stroke of 30 × 53 cm and operates at 250 rev/min. The fuel is natural gas with a heating value of 37 200 kJ/m^3, measured at 25°C and 1 atm pressure. The theoretical air–fuel ratio is 10.3 m^3 air per cubic meter of gas. At full load, the engine requires 25% excess air. The indicated thermal efficiency is 25 percent, and the mechanical efficiency is 75%. Determine (a) the full-load power; (b) the full-load fuel consumption in m^3/s; (c) the brake mean effective pressure at full load.

30. Find the mean effective pressure of a nine-cylinder diesel aircraft engine rated at 186 kW. The engine operates on a four-stroke cycle, runs at 2000 rev/min, and has a 12.4-cm bore and a 15.2-cm stroke.

31. A four-stroke cycle diesel engine with a compression ratio of 16 : 1 drives a 500-kW generator at 1200 rev/min. The generator efficiency is 90%. At this condition the engine's air–fuel ratio is 20 : 1 and the brake specific fuel consumption is 6.75×10^{-5} kg/s · kW. The inlet air conditions are 90 kPa and 35°C. The engine volumetric efficiency is 85%. The stroke is 1.2 times the bore. Determine for a six-cylinder engine (a) the bore and stroke per cylinder; (b) the clearance volume per cylinder; (c) the total fuel consumption in kg/s; (d) the thermal efficiency if the fuel is $C_{12}H_{26}$.

32. A diesel engine develops 750 kW at 200 rev/min when the ambient pressure is 100 kPa and the temperature is 17°C. The air–fuel ratio is 23 kg air/kg fuel and 7.6×10^{-5} kg/s of fuel are consumed per brake kilowatt developed. Determine for $h_{RP} = 43\ 200$ kJ/kg (a) the thermal efficiency; (b) the fuel consumption for 52 kW if the thermal efficiency is constant; (c) the second-law cycle efficiency.

33. A prony brake is used to analyze the performance of a small, double-acting, reciprocating steam engine. The engine operates at 200 rev/min, has an indicated mean effective pressure of 516 kPa, a stroke of 17.7 cm, a bore of 15 cm, a moment arm of 58 cm, and a scale reading of 667 N. The tare reading is 44 N. Determine (a) the brake mean effective pressure; (b) the indicated power; (c) the friction power; (d) the mechanical efficiency.

34. An engine has 14 cylinders with a 13.6-cm bore and a 15.2-cm stroke and develops 2850 kW at 250 rev/min. The clearance volume of each cylinder is 380 cm^3. Determine (a) the compression ratio; (b) the brake mean effective power.

35. A test on a one-cylinder Otto cycle engine yields the following data: 950-N · m torque; 758-kPa mean effective pressure; 28-cm bore; 30.5-cm stroke; 300 rev/min; 0.003-kg/s fuel consumption with a heating value of 41 860 kJ/kg. Determine (a) the engine thermal efficiency; (b) the engine mechanical efficiency; (c) the fuel cost per hour if fuel costs 50 cents per liter. The specific gravity is 0.82.

36. Calculate the bore and stroke of a six-cylinder engine that delivers 22.4 kW at 1800 rev/min, with a ratio of bore to stroke of 0.71. Assume the mean effective pressure in the cylinder is 620 kPa and the mechanical efficiency is 85%.

Chapter 17
Gas Turbines

Various mechanical divides have been used to produce power for industry and society needs. We have analyzed the steam power plant, where heat was added to the water and the water vapor expanded through a steam turbine, producing work. The thermal efficiency of 500-MW plants is around 40 percent. One case of the inefficiency is that an intermediate fluid, water, is used to transfer the energy of the hot combustion gases to the steam turbine. Gas-turbine units overcome this by using the combustion gases directly in the turbine. Nevertheless, the basic gas turbine does not achieve a higher thermal efficiency, for reasons that we will explore. A very important factor in gas-turbine selection is that gas-turbine power plants are very compact and lightweight. The conventional steam power plant must occupy a far greater area and is much heavier.

17.1 FUNDAMENTAL GAS-TURBINE CYCLE

For the gas turbine to produce any work, the hot gases must expand from a high pressure to a low pressure. Therefore the gases must first be compressed. If after the compression the fluid were expanded through the turbine, the power produced would equal that used by the compressor, provided that both the turbine and compressor functioned ideally. If heat were added to the fluid before it reached the turbine, raising its temperature, then an increase in the power output could be achieved. Figure 17.1 illustrates this. If more and more thermal energy could be added to the fluid, then more and more power output could be produced. Unfortunately this cannot occur; the turbine blades have a metallurgical thermal limit. If the gases continuously enter higher than this temperature, the combined thermal and material stresses in the blade will cause it

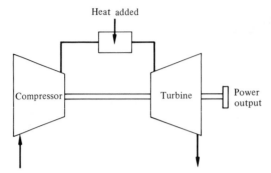

Figure 17.1   A simple open gas-turbine cycle.

to fail. Typically, inlet temperatures of 1300 K may be found on industrial turbines, and inlet temperatures on experimental models are up to 1500 K.

17.2 CYCLE ANALYSIS

The gas-turbine cycle may be either closed or open. The more common cycle is the open, in which atmospheric air is continuously drawn into the compressor, heat is added to the air by the combustion of fuel, and the fluid expands through the turbine and exhausts to the atmosphere. This is illustrated in Figure 17.1. In the closed cycle, the heat must be added to the fluid in a heat exchanger from an external source, such as a nuclear power plant, and the fluid must be cooled after it leaves the turbine and before it enters the compressor. Figure 17.2 illustrates this.

The air-standard Brayton cycle is the ideal closed-system gas-turbine cycle. It is characterized by constant-pressure heat addition and heat rejection and isentropic compression and expansion processes. Air is the working fluid and may be considered an ideal gas. Figure 17.2 illustrates the schematic for this cycle, and Figures 17.3(a) and 17.3(b) illustrate the $p-V$ and $T-S$ diagrams, respectively, for the cycle.

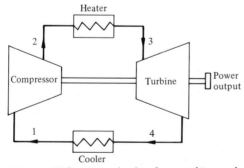

Figure 17.2   A simple closed gas-turbine cycle.

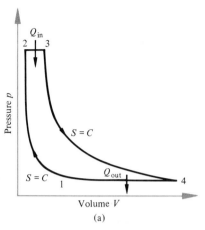
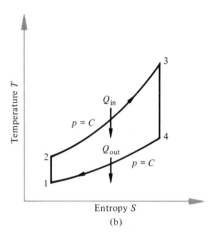

Figure 17.3 (a) The p–V diagram for the air-standard Brayton cycle. (b) The T–S diagram for the air-standard Brayton cycle.

The thermal efficiency, η_{th}, of the Brayton cycle may be found as follows:

$$\eta_{th} = \frac{W_{net}}{Q_{in}} = \frac{\Sigma Q}{Q_{in}} = 1 - \left(\frac{T_4 - T_1}{T_3 - T_2}\right) \tag{17.1}$$

The pressure ratio, r_p, is defined as

$$r_p = \frac{p_2}{p_1} \tag{17.2}$$

and from isentropic expansion and compression processes, we find that

$$\frac{T_2}{T_1} = \frac{T_3}{T_4} \tag{17.3}$$

We eliminate T_4 from Equation (17.1) by means of Equation (17.3).

$$\eta_{th} = 1 - \frac{1}{(r_p)^{(k-1)/k}} \tag{17.4}$$

Thus for the Brayton cycle the thermal efficiency is a function of the pressure ratio, r_p. Figure 17.4 illustrates this for two values of k, one of air, $k = 1.44$, and the other of helium, $k = 1.66$.

The maximum temperature does have an effect on the optimum performance. The previous derivation does not account for a fixed temperature T_3. If T_3 and T_1 are fixed, then there will be an optimum pressure ratio to produce a maximum amount of work, W_{net}. The variable temperature is T_2, the temperature of the fluid leaving the compressor.

$$W_{net} = mc_p(T_3 - T_4) - mc_p(T_2 - T_1) \tag{17.5}$$

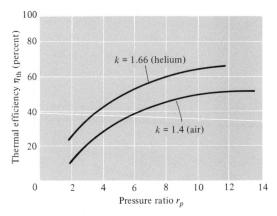

Figure 17.4 *The variation of thermal efficiency with pressure ratio for the air-standard Brayton cycle.*

but

$$T_4 = \frac{(T_3)(T_1)}{T_2}$$

$$W_{net} = mc_p \left(T_3 - \frac{T_3 T_1}{T_2} - T_2 + T_1 \right)$$

For W_{net} to be a maximum, then $dW_{net}/dT_2 = 0$, yielding

$$T_2 = \sqrt{T_3 T_1}$$

Also,

$$r_p = \frac{p_2}{p_1} = \left(\frac{T_2}{T_1} \right)^{k/(k-1)} = \left(\frac{T_3}{T_1} \right)^{k/2(k-1)}$$

$$(r_p)^{(k-1)/k} = \frac{T_2}{T_1} = \frac{T_3}{T_4} \qquad \therefore (r_p)^{2(k-1)/k} = \frac{T_3}{T_1}$$

hence

$$T_2 = T_4 \tag{17.6}$$

Thus when Equation (17.6) is substituted in Equation (17.5), an expression for the optimum net work in terms of the maximum temperature and the inlet temperature is obtained. Thus the optimum net work, $W_{net} = mc_p(T_3 + T_1 - 2\sqrt{T_3 T_1})$ occurs when the air temperature at the compressor exit is equal to the air temperature at the turbine exit. For industrial use $(T_3/T_1) = 3.5 \rightarrow 4$, whereas for aircraft engines, $(T_3/T_1) = 5.0 \rightarrow 5.5$, with the use of air-cooled blades. The pressure ratio may be calculated as in the previous derivation, knowing the optimum temperature.

17.3 EFFICIENCIES

One of the greatest problems gas-turbine manufacturers had to overcome was the poor compressor efficiencies. During the 1930s and 1940s, compressor efficiencies of 60 percent were common. Turbine efficiencies were greater, but the combination meant that virtually all the turbine work went to drive the compressor. Only by the gas dynamic analysis of the fluid passing through compressor and turbine blading could design changes be made—changes that now result in compressor efficiencies between 75 and 90 percent and turbine efficiencies of a few percent better. Thus for a compressor with an efficiency of 85 percent for a given pressure ratio, the turbine would have an efficiency of 87 percent.

The compressor may use more than 50 percent of the total work the turbine produces. The compressor efficiency, η_c, and the turbine efficiency, η_t, are defined as

$$\eta_c = \frac{W_{ideal}}{W_{actual}} = \frac{(h_2 - h_1)_s}{(h_{2'} - h_1)} \qquad (17.7)$$

$$\eta_t = \frac{W_{actual}}{W_{ideal}} = \frac{(h_3 - h_{4'})}{(h_3 - h_4)_s} \qquad (17.8)$$

where Figure 17.5 illustrates the states on a h–s diagram.

The following examples will illustrate the effects of component efficiencies on the overall cycle efficiency.

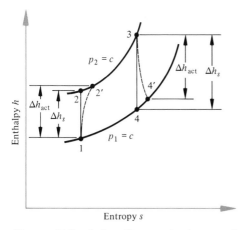

Figure 17.5   *An h–s diagram showing actual and ideal compressor and turbine states.*

Example 17.1

An air-standard Brayton cycle has air enter the compressor at 27°C and 100 kPa. The pressure ratio is 10 and the maximum allowable temperature in the cycle is 1350 K. Determine (a) the pressure and temperature at each state in the cycle; (b) the compressor work, turbine work, and the cycle efficiency per kilogram of air.

(a) $p_1 = 100$ kPa

   $T_1 = 300$ K

   $r_p = \dfrac{p_2}{p_1} = 10$

   $p_2 = 1000$ kPa

   $T_2 = T_1(r_p)^{(k-1)/k} = 300(10)^{0.286} = 579.6$ K

   $p_3 = p_2 = 1000$ kPa $T_3 = T_{max} = 1350$ K

   $T_4 = T_3\left(\dfrac{p_4}{p_3}\right)^{(k-1)/k} = (1350)(0.10)^{0.286} = 698.8$ K

   $p_4 = p_1 = 100$ kPa

(b) From a first-law analysis for each component:

   $w_c = -(h_2 - h_1) = -c_p(T_2 - T_1) = -(1.0047)(579.6 - 300)$

   $\quad = -280.9$ kJ/kg

   $w_t = (h_3 - h_4) = c_p(T_3 - T_4) = (1.0047)(1350 - 698.8)$

   $\quad = 654.3$ kJ/kg

   $w_{net} = \Sigma w = 373.4$ kJ/kg

   $q_{in} = (h_3 - h_2) = c_p(T_3 - T_2) = (1.0047)(1350 - 579.6)$

   $\quad = 774.0$ kJ/kg

   $\eta_{th} = \dfrac{w_{net}}{q_{in}} = 0.482$ or 48.2 percent

Check:

   $\eta_{th} = 1 - \dfrac{1}{(r_p)^{(k-1)/k}} = 0.482$

Example 17.2

An air-standard Brayton cycle has air enter the compressor at 27°C and 100 kPa. The pressure ratio is 10 and the maximum allowable temperature is 1350 K. The compressor and turbine efficiencies are

85 percent and there is a 27 kPa drop between the compressor discharge and the turbine inlet. Determine (a) the pressure and temperature at each state in the cycle; (b) the compressor work, turbine work, and thermal efficiency of the cycle per kilogram of air; (c) compare the results of (b) to those of the previous example.

(a) $p_1 = 100$ kPa

$T_1 = 300$ K

$T_2 = 579.6$

$\eta_c = 0.85 = \dfrac{579.6 - 300}{T_{2'} - 300}$ $T_{2'} = 628.9$ K

$p_{2'} = p_2 = 1000$ kPa

$T_3 = T_{max} = 1350$ K $p_3 = 973$ kPa

$T_4 = T_3 \left(\dfrac{p_4}{p_3}\right)^{(k-1)/k} = 1350\left(\dfrac{100}{973}\right)^{0.286} = 704.3$ K

$\eta_t = 0.85 = \dfrac{1350 - T_{4'}}{1350 - 704.1}$ $T_{4'} = 801.0$ K

(b) A first-law analysis of each component yields

$w_c = -(h_{2'} - h_1) = -c_p(T_{2'} - T_1) = -(1.0047)(628.9 - 300)$

$\quad = -330.4$ kJ/kg

$w_t = (h_3 - h_{4'}) = c_p(T_3 - T_{4'}) = (1.0047)(1350 - 801)$

$\quad = 551.6$ kJ/kg

$q_{in} = (h_3 - h_{2'}) = c_p(T_3 - T_{2'}) = (1.0047)(1350 - 628.9)$

$\quad = 724.5$ kJ/kg

$w_{net} = \Sigma w = 221.2$ kJ/kg

$\eta_{th} = \dfrac{w_{net}}{q_{in}} = 0.305$ or 30.5 percent

(c) change in $w_c = \dfrac{(49.5)(100)}{280.9} = 17.6$ percent increase

change in $w_t = \dfrac{(102.7)(100)}{654.3} = 15.7$ percent decrease

change in $q_{in} = \dfrac{(49.5)(100)}{774.0} = 6.4$ percent decrease

change in $\eta_{th} = \dfrac{(0.177)(100)}{(0.482)} = 36.7$ percent decrease

17.4 OPEN-CYCLE ANALYSIS

In the open gas-turbine cycle, the air entering the compressor is increased in pressure; it then enters the combustion chamber, where fuel is added, and the combustion process raises the temperature of the products of combustion, which enter the turbine. The products of combustion leave the turbine and exhaust to the atmosphere. Figure 17.6 illustrates this cycle schematically, and Figure 17.7 shows the $T–s$ diagram for the cycle. Note that air is the working fluid up to the combustion chamber and that a different fluid, the products of combustion, is the working substance after the combustion chamber. In analyzing this power cycle, we should use the properties of each fluid. In using air, we may use the air tables, or if none are available, the ideal gas law. In using the products of combustion, we may use the 400 percent theoretical air tables, or the ideal gas law. The specific heats of the air and the products of combustion will be slightly different with the products of combustion specific heat being greater than that of air.

Let us look at the combustion chamber to find out why 400 percent theoretical air is a good assumption when evaluating the products of combustion. Figure 17.8 illustrates the combustion chamber, or combustor, for a gas turbine. The temperature of the hot gases leaving the combustion chamber has metallurgical limitations. We noted when we studied adiabatic flame temperatures in Chapter 14 that combustion temperatures may easily exceed this limit. To prevent the hot gas temperature from being greater than the allowable exit temperature from the combustion chamber, excess air is used. For a gas turbine, this may typically have properties similar to those of 400 percent theoretical air. Not all the excess air is used in the combustion process, as that would cool the air-fuel mixture and cause incomplete combustion. Some of the air passes around the sides of the combustion chamber, cooling the metal

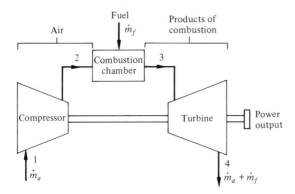

Figure 17.6   Schematic diagram of an open gas-turbine cycle.

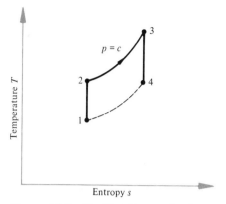

Figure 17.7 *The T–s diagram for the gas turbine in Figure 17.6.*

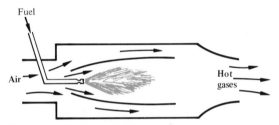

Figure 17.8 *A combustion chamber.*

walls and mixing with the products of combustion in the latter part of the combustion chamber.

The following example will illustrate the analysis of an open-system gas-turbine unit where the dissimilarity in the working substances and the combustion chamber analysis is included.

Example 17.3

A gas-turbine unit produces 600 kW while operating under the following conditions: the inlet air pressure and temperature are 100 kPa and 300 K, respectively; the pressure ratio is 10; the fuel is similar to $C_{12}H_{26}$ and has a fuel–air ratio of 0.015 kg fuel/kg air; the products of combustion are similar to 400 percent theoretical air. Calculate the air flow rate, the total turbine work, the compressor work, and the thermal efficiency.

From Table A.2 we find the properties of air:

$$h_1 = 300.19 \text{ kJ/kg} \qquad p_{r_1} = 1.3860$$

In analyzing the compressor, it is an open system, so the first law is

Energy In = Energy Out

$$\dot{m}_a h_1 = \dot{m}_a h_2 + \dot{W}_c$$

$$\dot{W}_c = -\dot{m}_a(h_2 - h_1)$$

Since no internal efficiencies are given, $h_2 = (h_2)_s$. For an isentropic process,

$$p_{r_2} = p_{r_1}(r_p) = (1.3860)(10) = 13.86$$

$$h_2 = 579.8 \text{ kJ/kg}$$

The adiabatic combustion chamber is the next element in the cycle. It too is an open system, and so the first law applied around it is

Energy In = Energy Out

$$\dot{m}_a h_2 + \dot{m}_f h_{RP} = (\dot{m}_a + \dot{m}_f)h_3$$

where $\bar{h}_{RP}$ is the enthalpy of combustion. From Table C.3, $h_{RP} = 44\ 102 \text{ kJ/kg}$

$$h_2 + r_{f/a}h_{RP} = (1 + r_{f/a})h_3$$

Thus $h_3 = 1241.3 \text{ kJ/kg}$, but the substance leaving the combustion chamber is 400 percent theoretical air, found in Table A.3. Thus

$$h_3 = 1241.3 \text{ kJ/kg}$$

$$p_{r_3} = 220.1$$

The flow through the turbine may be considered isentropic, so

$$p_{r_4} = p_{r_3}\left(\frac{1}{r_p}\right) = 22.01$$

$$h_4 = 662.5 \text{ kJ/kg}$$

An open-system energy balance on the turbine yields

Energy In = Energy Out

$$\dot{m}_a(1 + r_{f/a})h_3 = \dot{W}_c + \dot{W}_{net} + \dot{m}_a(1 + r_{f/a})h_4$$

$$\dot{W}_c = -\dot{m}_a(h_2 - h_1)$$

$$\dot{W}_{net} = 600 \text{ kW}$$

Solve for $\dot{m}_a$, yielding

$$\dot{m}_a = 1.949 \text{ kg/s}$$

$$\dot{W}_t = \dot{m}_a(1 + r_{f/a})(h_3 - h_4) = 1145.0 \text{ kW}$$

$$\dot{W}_c = -\dot{m}_a(h_2 - h_1) = -545.0 \text{ kW}$$

$$\eta_{th} = \frac{\dot{W}_{net}}{\dot{Q}_{in}} = \frac{600}{(1.949)(0.015)(44\ 102)} = 0.465$$

17.5 COMBUSTION EFFICIENCY

The combustion process may not proceed to completion, so not all the chemical energy is converted into thermal energy. The temperature of the products of combustion will not be as high as ideally expected. To account for this, we introduce a combustion efficiency, η_{cc}, which tells us the percentage of energy released from the fuel. Therefore the thermal energy generated in a combustion chamber per kilogram of air is $r_{f/a}\eta_{cc}h_{RP}$, not $r_{f/a}h_{RP}$.

17.6 REGENERATION

The thermal efficiency of the gas turbine unit is quite low. We observe that the exhaust temperature of the turbine is quite high, indicating that a large portion of available energy is not being used. One way to put this high-temperature energy to use is to preheat the combustion air before it enters the combustor. This increases the overall efficiency by decreasing the fuel, hence heat, added. The net work from the turbine does not increase however, in an actual unit, will decrease because of pressure effects, which will be shown later. The optimum place for reheating is after the air has been compressed; Figure 17.9(a) illustrates a schematic for the system and Figure 17.9(b) illustrates a typical diagram for the unit. Figure 17.10 denotes the T–s diagram for the regenerative cycle. If the heat exchanger were 100 percent effective, then the temperature at state x would be equal to the temperature at state 4. However, the heat exchanger is not 100 percent effective. To denote how effectively the heat exchanger operates, we have a coefficient called the "regenerator effectiveness," $\mathcal{E}_{reg}$.

$$\mathcal{E}_{reg} = \frac{\text{actual heat transferred}}{\text{maximum possible heat transferred}}$$

$$\mathcal{E}_{reg} = \frac{h_x - h_2}{(1 + r_{f/a})(h_4 - h_c)} \tag{17.9}$$

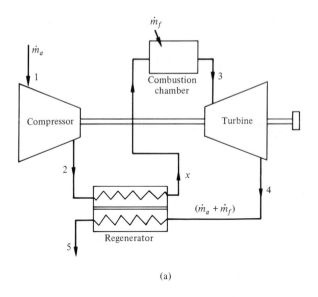

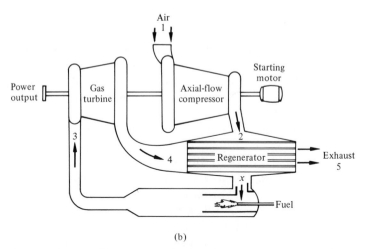

Figure 17.9 (a) The schematic diagram for a regenerative gas-turbine cycle. (b) An open-cycle gas turbine with regenerative heating.

where h_c is the enthalpy of the products at temperature T_2. The actual heat transferred is equal to the heat picked up by the air. The maximum possible heat transfer would occur when the products of combustion are cooled down to the temperature of the air entering the regenerator.

The following examples will illustrate the effect of a regenerator on the gas-turbine cycle.

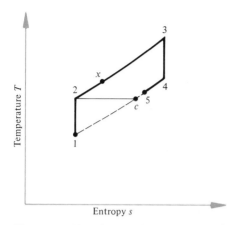

Figure 17.10   The T–s diagram for the regenerative gas-turbine cycle.

Example 17.4

A gas-turbine unit receives air at 100 kPa and 300 K and compresses it adiabatically to 620 kPa with $\eta_c = 88$ percent. The fuel has a heating value of 44 186 kJ/kg, and the fuel–air ratio is 0.017 kg fuel/kg air. The turbine internal efficiency is 90 percent; calculate the turbine work, compressor work, and thermal efficiency.

Calculate the state points, referring to Figure 17.6 for numbering sequence.

$$h_1 = 300.19 \text{ kJ/kg} \qquad p_{r_1} = 1.3860$$

Process 1–2, isentropic,

$$p_{r_2} = 1.3860 \left(\frac{620}{100} \right) = 8.593$$

$$h_2 = 506.04 \text{ kJ/kg}$$

$$\eta_c = 0.88 = \frac{h_2 - h_1}{h_{2'} - h_1} = \frac{506.04 - 300}{h_{2'} - 300}$$

$$h_{2'} = 534.11 \text{ kJ/kg}$$

Process 2–3, constant pressure, heat added,

$$h_{2'} + r_{f/a} h_{RP} = (1 + r_{f/a}) h_3$$

$$534.11 + (0.017)(44\ 186) = (1.017) h_3$$

$$h_3 = 1263.8 \text{ kJ/kg}$$

$$T_3 = 1166 \text{ K} \qquad p_{r_3} = 236.0$$

Process 3–4, isentropic,

$$p_{r_4} = (236)\left(\frac{100}{620}\right) = 38.06$$

$$h_4 = 771.3 \text{ kJ/kg}$$

$$\eta_t = 0.90 = \frac{h_3 - h_{4'}}{h_3 - h_4} \qquad h_{4'} = 820.6 \text{ kJ/kg}$$

A first-law analysis of each component yields

$$w_t = (1 + r_{f/a})(h_3 - h_{4'}) = (1.017)(1263.8 - 820.6)$$
$$= 450.7 \text{ kJ/kg}$$

$$w_c = -(h_{2'} - h_1) = -(534.11 - 300.19) = -233.92$$

$$w_{net} = w_t + w_c = 216.78 \text{ kJ/kg}$$

$$q_{in} = r_{f/a}h_{RP} = 751.2 \text{ kJ/kg}$$

$$\eta_{th} = \frac{w_{net}}{q_{in}} = 0.288 \text{ or } 28.8 \text{ percent}$$

Example 17.5

A regenerator with an effectiveness of 60 percent is added to the gas-turbine unit of Example 17.4. Calculate the new fuel–air ratio for the same value of T_3, and calculate the overall thermal efficiency. Consider the case in which pressure losses in the regenerator must be accounted for. There is a 20-kPa drop on the air side and a 3.4-kPa drop on the exhaust side of the regenerator. Calculate the new net work and thermal efficiency. Refer to Figure 17.9(a) for numbering sequence.

$$h_1 = 300.19 \text{ kJ/kg} \qquad h_{2'} = 534.11 \text{ kJ/kg} \qquad w_c = -233.92 \text{ kJ/kg}$$

$$p_3 = 600 \text{ kPa} \qquad p_4 = 103.4 \text{ kPa} \qquad T_3 = 1166 \text{ K}$$

$$h_3 = 1263.8 \text{ kJ/kg} \qquad p_{r_3} = 236.0$$

Process 3–4, isentropic,

$$p_{r_4} = (236.0)\left(\frac{103.4}{600}\right) = 40.67$$

$$h_4 = 785.5 \text{ kJ/kg}$$

$$\eta_t = 0.90 = \frac{h_3 - h_{4'}}{h_3 - h_4} = \frac{1263.8 - h_{4'}}{1263.8 - 785.5} \qquad h_{4'} = 833.3 \text{ kJ/kg}$$

From the definition for regenerator effectiveness,

$$0.60 = \frac{h_x - h_{2'}}{(1 + r_{f/a})(h_{4'} - h_c)}$$

where $h_c = 541$ kJ/kg at $T_c = T_{2'} = 530.1$ K; and from the first-law analysis of the combustion chamber,

$$h_x + r_{f/a}h_{RP} = (1 + r_{f/a})h_3$$

the two unknowns, $r_{f/a}$ and h_x, may be determined. Solving these equations yields

$$r_{f/a} = 0.01286 \text{ kg fuel/kg air}$$

A first-law analysis of each component yields

$$w_t = (1 + r_{f/a})(h_3 - h_{4'}) = (1.01286)(1263.8 - 833.3)$$

$$w_t = 436.0 \text{ kJ/kg}$$

$$w_{net} = w_t + w_c = 202.08 \text{ kJ/kg}$$

$$q_{in} = r_{f/a}h_{RP} = (0.01286)(44\ 186) = 568.2 \text{ kJ/kg}$$

$$\eta_{th} = \frac{w_{net}}{q_{in}} = 0.355 \text{ or } 35.5 \text{ percent}$$

Closed Regenerative Cycle

The closed air-standard Brayton cycle may also use a regenerator. The same losses may occur, but the closed system allows us the opportunity to investigate analytically the effect the regenerator has on the cycle efficiency. Figure 17.11 illustrates the schematic diagram, and Figure 17.12 illustrates the T–s diagram for the cycle.

The thermal efficiency may be written as

$$\eta_{th(reg)} = \frac{W_{net}}{Q_{in}} = 1 - \frac{Q_{out}}{Q_{in}} = 1 - \frac{T_5 - T_1}{T_3 - T_x} \tag{17.10}$$

For an ideal counterflow heat exchanger,

$$T_x = T_4 \qquad \text{and} \qquad T_5 = T_2$$

and the thermal efficiency is

$$\eta_{th(reg)} = 1 - \frac{T_2 - T_1}{T_3 - T_4} \tag{17.11}$$

From Equation (17.6),

$$\frac{T_2}{T_1} = \frac{T_3}{T_4} = (r_p)^{(k-1)/k}$$

so Equation (17.11) becomes

$$\eta_{th(reg)} = 1 - \left(\frac{T_1}{T_3}\right)(r_p)^{(k-1)/k} \tag{17.12}$$

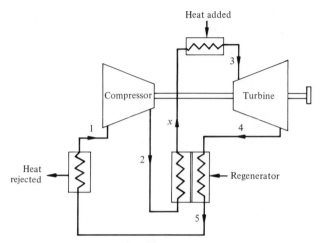

Figure 17.11   The closed regenerative Brayton cycle.

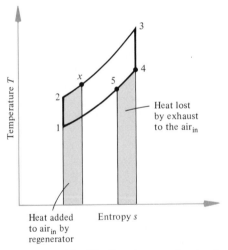

Figure 17.12   The T–s diagram for the closed regenerative Brayton cycle.

Thus in the ideal cycle with regeneration, the overall thermal efficiency is a function of the pressure ratio and the ratio of minimum to maximum temperature occurring in the cycle. A plot of Equations (17.12) and (17.4) is given in Figure 17.13, where $T_3/T_1 = t$.

In actual cycles the temperature and pressure ratios are important. The closed-system analysis does permit us to find that adding a regenerator may or may not increase the cycle efficiency. When pressure drops due to the regenerator are considered, the desirability of adding a regenerator must be reevaluated. The ultimate analysis is an economic

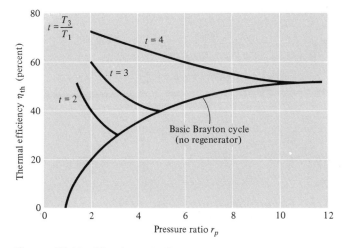

Figure 17.13 *The thermal efficiency of the regenerative Brayton cycle for different values of t and r_p.*

one; will adding the regenerator save money, balancing the increased capital cost and maintenance against the fuel savings? All engineering analyses must eventually undergo the same economic projection: Is it worthwhile? As fuel costs increase, the desirability of efficiency-producing additions to the basic cycle increases. Thus what has been accepted practice may no longer be valid when the economic constraints change.

17.7 REHEATING, INTERCOOLING

Other methods for improving the cycle efficiency are available. Although they all may not be useful on small gas-turbine units, large units usually profit by incorporating them. As we determined when analyzing the Rankine cycle, improvement in the cycle efficiency can be made by reheating the fluid after it has performed some work. In the gas-turbine unit this is accomplished by passing the products of combustion through another combustion chamber. Since there is more than sufficient air for combustion, we can inject some more fuel. The reheated products of combustion return to the turbine. Figure 17.14(a) illustrates the schematic for a reheat-regenerative gas-turbine cycle, and Figure 17.14(b) illustrates the T–s diagram. The products of combustion reentering the turbine at state 5 are usually at the same temperature as those entering the turbine at state 3.

Another way to improve the cycle efficiency is to reduce the work of the compressor. This may be accomplished by compressing in stages and, with an intercooler, cooling the air as it passes from one stage to another.

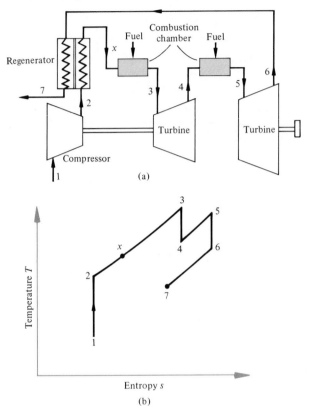

Figure 17.14 (a) The schematic diagram for a regenerative-reheat gas-turbine cycle. (b) The T–s diagram for the regenerative-reheat gas-turbine cycle.

The following example illustrates the combined effects of reheating and intercooling on a gas-turbine unit. Figure 17.15(a) illustrates the schematic arrangement of the equipment, and Figure 17.15(b) denotes the T–s diagram.

Example 17.6

A gas-turbine unit operates on a regenerative-reheat cycle with compressor interstage cooling. Air enters the compressor at 100 kPa and 290 K and is compressed to 410 kPa; it is cooled at constant pressure until the temperature drops by 13°C and is finally compressed to 750 kPa. The regenerator has an effectiveness of 70 percent. The products of combustion enter the turbine at 1350 K and expand to 410 kPa, where they are reheated to 1350 K. The exhaust pressure is 100 kPa. The fuel has a heating value of 44 186 kJ/kg. Determine the fuel–air ratio in each combustion chamber, the compressor work, the total turbine work, and the overall thermal

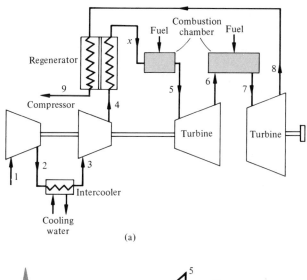

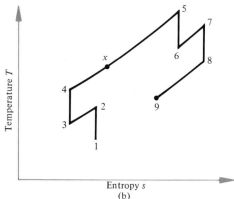

Figure 17.15 (a) *The schematic diagram for a regenerative-reheat gas-turbine cycle with intercooling.* (b) *The T–s diagram for a regenerative-reheat gas-turbine cycle with intercooling.*

efficiency where all expansion and compressor processes are isentropic.

Find all the state points, then solve for energy inputs and outputs.

$h_1 = 290.17 \text{ kJ/kg}$ $p_1 = 100 \text{ kPa}$

$T_1 = 290 \text{ K}$ $p_{r_1} = 1.2311$

Process 1–2, isentropic,

$$p_{r_2} = (1.2311)\left(\frac{410}{100}\right) = 5.048$$

$h_2 = 434.68 \text{ kJ/kg}$ $T_2 = 433 \text{ K}$

Process 2–3, constant pressure,

$T_3 = 420$ K

$h_3 = 421.26$ kJ/kg $p_{r_3} = 4.522$

Process 3–4, isentropic,

$$p_{r_4} = 4.522\left(\frac{750}{410}\right) = 8.272$$

$T_4 = 498$ K $h_4 = 500.58$ kJ/kg

$T_5 = 1350$ K $h_5 = 1487.8$ kJ/kg (400 percent theoretical air)

Process 5–6, isentropic,

$$p_{r_6} = (438)\left(\frac{410}{750}\right) = 239.44$$

$h_6 = 1268.5$ kJ/kg $T_6 = 1170$ K

$T_7 = 1350$ K $h_7 = 1487.8$ kJ/kg

$p_{r_7} = 438.0$

Process 7–8, isentropic,

$$p_{r_8} = (438)\left(\frac{100}{410}\right) = 106.83$$

$T_8 = 963$ K $h_8 = 1022.8$ kJ/kg

Find h_x, r_{f/a_1}, r_{f/a_2} by first-law analysis:

(a) $\left(1 + r_{f/a_1}\right)h_6 + r_{f/a_2}h_{RP} = \left(1 + r_{f/a_1} + r_{f/a_2}\right)h_7$

(b) $h_x + r_{f/a_1}h_{RP} = \left(1 + r_{f/a_1}\right)h_5$

and by the definition of regenerator effectiveness,

(c) $0.70 = \dfrac{h_x - h_4}{\left(1 + r_{f/a_1} + r_{f/a_2}\right)(h_8 - h_c)}$

   Solve equations (a), (b), and (c) simultaneously for h_x, r_{f/a_1}, and r_{f/a_2} where r_{f/a_1} is the fuel–air ratio in the first combustion chamber and r_{f/a_2} is the fuel–air ratio in the second combustion chamber. The following are the results:

$r_{f/a_1} = 0.01448$ kg fuel/kg air $r_{f/a_2} = 0.00523$ kg fuel/kg air

$h_x = 868.7$ kJ/kg

From a first-law analysis of each component

$$w_c = -(h_2 - h_1) - (h_4 - h_3) = -223.8 \text{ kJ/kg}$$

$$w_t = \left(1 + r_{f/a_1}\right)\left(h_5 - h_6\right) + \left(1 + r_{f/a_1} + r_{f/a_2}\right)\left(h_7 - h_8\right)$$

$$w_t = 696.6 \text{ kJ/kg}$$

$$w_{net} = w_t + w_c = 472.8 \text{ kJ/kg}$$

$$q_{in} = \left(r_{f/a_1} + r_{f/a_2}\right)h_{RP} = 870.9 \text{ kJ/kg}$$

$$\eta_{th} = \frac{w_{net}}{q_{in}} = 0.543 \text{ or } 54.3 \text{ percent}$$

The reheating and intercooling improve the effectiveness of the regenerator. The temperatures of the fluids are further apart when they enter the regenerator, so more heat may be transferred for a given tube surface or regenerator size. The intercooling may be accomplished by either air or water. The cost of intercooling is quite low, so it offers a substantial advantage in overall efficiency.

17.8 COMBINED CYCLE

One of the power plant cycles being used by utilities and other industries needing large amounts of power is the combined gas-turbine/Rankine-cycle power plant. This is called the *gas-turbine combined cycle*. The exhaust from the gas turbine is hot enough to generate steam in a waste-heat boiler. The steam generated in the boiler can power another turbine, thus increasing the total work produced.

One of the inherent problems with the gas-turbine plant is that a large portion of the available energy added is not used for work. The available energy of the exhaust gases, at or near atmospheric pressure, may be utilized to generate steam at 5.5 MPa to 6.0 MPa. Many of the features used in both the gas-turbine and Rankine-cycle analyses are included in the combined cycle, but these will not be included in the following example, which illustrates the combined cycle benefits. The combined cycle has a thermal efficiency greater than either cycle may have by itself; the power plant occupies less area, and the energy (fuel) requirements are necessarily less. Figure 17.16(a) illustrates the schematic diagram for the combined cycle plant in the next problem, and Figure 17.16(b) denotes the *T-s* diagram for the cycle.

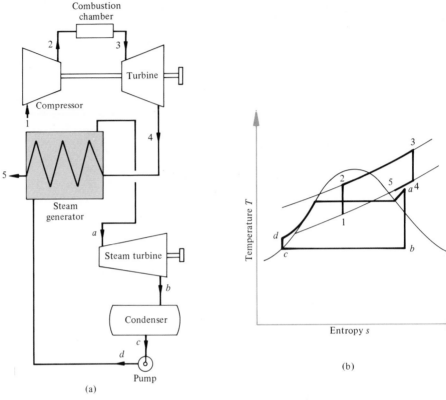

Figure 17.16 (a) A combined gas-turbine steam power cycle. (b) The superimposed T–s diagrams for the combined cycle in Figure 17.16(a).

Example 17.7

A combined cycle gas-turbine/steam power plant produces a total of 37 300 kW. The exhaust from the gas turbine leaves the steam generator at 147°C. Steam leaves the generator at 6.0 MPa and 400°C. Air enters the gas-turbine unit at 290 K and 100 kPa and leaves the compressor that operates with a pressure ratio of 10. The entrance temperature to the gas turbine is 1400 K. Calculate the flow rates required in each cycle, the overall efficiency, and the available energy utilization compared with a conventional gas-turbine unit. The steam power plant operates with a condenser pressure of 13 kPa. Assume air is the working fluid in the gas-turbine system. Calculate the cycle state points.

$$h_1 = 290.17 \text{ kJ/kg} \qquad p_{r_1} = 1.2311$$

Process 1–2, isentropic,

$$p_{r_2} = p_{r_1}\left(\frac{p_2}{p_1}\right) = 12.31 \qquad h_2 = 560.6 \text{ kJ/kg}$$

$$h_3 = 1515.41 \text{ kJ/kg} \qquad p_{r_3} = 450.5$$

Process 3–4, isentropic,

$$p_{r_4} = p_{r_3}\left(\frac{p_4}{p_3}\right) = 45.05$$

$$h_4 = 808.65 \text{ kJ/kg} \qquad h_5 = 421.26 \text{ kJ/kg}$$

For the steam cycle,

(a) $h = 3177.2$ kJ/kg steam $\qquad s = 6.5408$ kJ/kg · K

Process (a)–(b), isentropic,

$$s_b = s_a$$

(b) $x = 0.7932 \qquad h = 2101.8$ kJ/kg steam

(c) $h = 213.67$ kJ/kg steam

(d) $h = h_c + \int v \, dp = 219.73$ kJ/kg steam

From a first-law analysis of the steam cycle,

$$w_t = h_a - h_b = (3177.2 - 2101.8) = 1075.4 \text{ kJ/kg steam}$$

$$w_p = -(h_d - h_c) = -(219.73 - 213.67) = -6.06 \text{ kJ/kg steam}$$

$$w_{net} = w_t + w_p = 1069.34 \text{ kJ/kg steam}$$

From a first-law analysis of the gas-turbine cycle,

$$w_t = h_3 - h_4 = (1515.41 - 808.65) = 706.76 \text{ kJ/kg air}$$

$$w_c = -(h_2 - h_1) = -(560.6 - 290.17) = -270.43 \text{ kJ/kg air}$$

$$w_{net} = w_t + w_c = 436.33 \text{ kJ/kg air}$$

$$q_{in} = h_3 - h_2 = 1515.41 - 560.6 = 954.81 \text{ kJ/kg air}$$

Determine the ratio of the flow of steam to the flow of air by a first-law analysis of the steam generator:

$$\dot{m}_a(h_4 - h_5) = \dot{m}_s(h_a - h_d)$$

$$\frac{m_s}{\dot{m}_a} = 0.1310 \text{ kg steam/kg air}$$

The total power is the sum of the gas turbine and steam power, thus

$$\dot{W}_{net} = \dot{m}_a\left(w_{net\,a} + \frac{\dot{m}_s}{\dot{m}_a}w_{net\,s}\right)$$

$$\dot{m}_a = \frac{37\,300}{436.33 + (0.131)(1075.4)} = 64.62 \text{ kg air/s}$$

$$\dot{m}_s = 8.46 \text{ kg steam/s}$$

The overall thermal efficiency is

$$\eta_{th} = \frac{\dot{W}_{net}}{\dot{Q}_{in}} = \frac{37\,300}{(64.62)(954.81)} = 0.604$$

The available portion of the heat added is

$$(\text{A.E.})_{2-3} = q_{2-3} - T_0(s_3 - s_2)$$
$$(\text{A.E.})_{2-3} = (h_3 - h_2) - T_0(\phi_3 - \phi_2)$$
$$(\text{A.E.})_{2-3} = (1515.41 - 560.6) - 290(4.1753 - 3.1420)$$
$$(\text{A.E.})_{2-3} = 655.15 \text{ kJ/kg air}$$

The percent of available energy used in the gas-turbine unit is

$$(\text{A.E.})_{gt} = \frac{w_{net\,a}}{\text{A.E.}_{2-3}} = \frac{436.33}{655.15} \times 100 = 66.67 \text{ percent}$$

The percent of available energy used in the combined unit is

$$(\text{A.E.})_{combined} = \frac{w_{net\,a} + \dfrac{\dot{m}_s}{\dot{m}_a}w_{net\,s}}{\text{A.E.}_{2-3}}$$

$$(\text{A.E.})_{combined} = 88.0 \text{ percent}$$

The available energy utilization in the combined plant is very high, which accounts for the significant increase in overall thermal efficiency.

17.9 AIRCRAFT GAS TURBINES

The aircraft industry has pioneered many of the advances made in gas-turbine design. The space and weight limitations of the airplane engine make the gas turbine an ideal unit because it can provide large amounts of power in a small volume with a very low weight. The stationary gas turbine and the jet engine differ greatly in the manner by which work is produced. The jet engine gas turbine provides work to drive the compressor (and, on a turboprop, the propeller); the exhaust

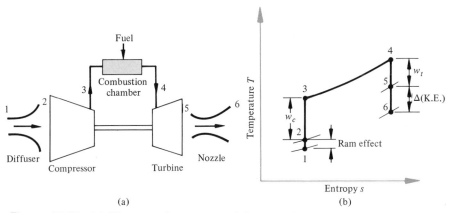

Figure 17.17 (a) The gas-turbine unit used for aircraft engines; (b) The corresponding T–s diagram.

gases from the turbine are expanded through a nozzle, increasing the velocity. The change in velocity of the gas entering and leaving the gas-turbine unit causes a force to be exerted on the aircraft. Figure 17.17(a) illustrates the gas-turbine unit schematically, and Figure 17.17(b) illustrates the *T–s* diagram. Located before the compressor is the *diffuser*, a nozzle operating for the purpose of decreasing velocity and increasing pressure. This is called the *ram effect*. Figure 17.17(b) is drawn for isentropic expansion and compression processes. As we realize, no actual process is isentropic, and so we deal in terms of efficiencies for each component—diffuser, compressor, turbine, and nozzle.

Energy Analysis of Jet Engines

Since the unit produces no net work, as we have used the term in previous sections (all the work of the turbine drives the compressor), let us find out how the aircraft moves. The force that moves the airplane is caused by the gas change of momentum through the engine. The analysis may be made thermodynamically, on an energy basis, or on the impulse-

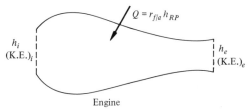

Figure 17.18 The energy flows for a turbojet engine with the observer on the plane.

momentum principle. We will use both approaches, the thermodynamic method first.

Consider a turbojet engine driving a plane at a velocity v_p. The relative velocities of the gas entering and leaving the engine depend on where the observer is located. The observer on the plane finds that the air enters the engine at state i. He makes an energy balance on the engine, Figure 17.18, and finds that, from the first law of thermodynamics for an open system,

$$h_i + (\text{K.E.})_i + r_{f/a}h_{RP} = (1 + r_{f/a})h_e + (1 + r_{f/a})(\text{K.E.})_e$$
$$h_i - (1 + r_{f/a})h_e = (1 + r_{f/a})(\text{K.E.})_e - (\text{K.E.})_i - r_{f/a}h_{RP}$$

$$(17.13)$$

where

$$(\text{K.E.})_i = \frac{v_i^2}{2} \qquad \text{and} \qquad v_i = v_p$$

$$(\text{K.E.})_e = \frac{v_e^2}{2}$$

where v_e is the relative exit velocity.

The observer on the ground sees the same engine and makes an energy analysis, Figure 17.19. In this case energy is added to the system as above, but work is also observed, as a force is moving through a distance. The fuel enters with a kinetic energy since it is located on the plane moving at velocity v_p. Thus an energy balance yields

$$h_1 - (1 + r_{f/a})h_2 = (1 + r_{f/a})(\text{K.E.})_2 - (\text{K.E.})_1 - r_{f/a}h_{RP}$$
$$- r_{f/a}(\text{K.E.})_f + w \qquad (17.14)$$

where $(\text{K.E.})_2 = v_2^2/2$ and $v_2 = v_e - v_p$ and where $-v_p$ is the speed of the plane in the opposite direction to gas flow.

$$(\text{K.E.})_1 = \frac{v_1^2}{2} \qquad \text{but} \qquad v_1 = 0 \text{ (still air)}$$

$$(\text{K.E.})_f = \frac{v_p^2}{2}$$

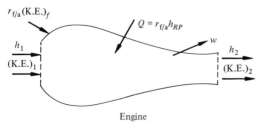

Engine

Figure 17.19   The energy flows when the observer is on the ground.

The temperature is not a function of where the observer is located, so

$$h_i = h_1 \quad \text{and} \quad h_e = h_2 \tag{17.15}$$

Thus Equations (17.13) and (17.14) are equal, and the right-hand sides may be equated.

$$w = \left(1 + r_{f/a}\right)\left[(\text{K.E.})_e - (\text{K.E.})_2\right] + r_{f/a}(\text{K.E.})_f - (\text{K.E.})_i \tag{17.16}$$

Substituting for the kinetic energy, yields

$$w = v_e v_p\left[1 + r_{f/a} - \frac{v_p}{v_e}\right] \text{ J/kg} \tag{17.17}$$

where the work in Equation (17.17) is the work the fluid performs in passing through the engine.

The same result may be obtained by conventional impulse-momentum analyses of the engine. The propulsive force in a jet engine is brought about by a continuous change in the momentum of the gas as it passes through the engine. From basic mechanics, $F + d(mv)/d\tau$, where (mv) is the momentum. Consider the engine to be moving at velocity v_p (observer on the plane), the momentum entering the engine is $\dot{m}_a v_i = \dot{m}_a v_p$ where $\dot{m}_a$ is the air flow rate. An amount of fuel, $\dot{m}r_{f/a}$, is added to the engine, and the momentum leaving the engine is $\dot{m}_a(1 + r_{f/a})$ where v_e is the exit velocity relative to the engine. The force generated by the change in momentum is

$$F = \dot{m}_a v_e\left[1 + r_{f/a} - \frac{v_p}{v_e}\right] \tag{17.18}$$

where the force generated per kilogram of air, denoted in Equation (17.18), is called the "specific thrust."

Work is a force through a distance, and the plane travels a distance v_p during 1 s; thus

$$\dot{W} = Fv_p = \dot{m}_a v_e v_p\left[1 + r_{f/a} - \frac{v_p}{v_e}\right] \text{ W} \tag{17.19}$$

There is no difference between Equations (17.19) and (17.17), other than that (17.19) is derived for an air flow rate of $\dot{m}_a$. This is what we would have expected, as the engine will not perform more or less work as a function of the analysis.

Maximum Work and Propulsive Efficiency

There are two more terms to define: the *maximum work* developed as a function of plane speed and the *propulsive efficiency*. The maximum work developed may be found by finding the derivative of work with respect to plane velocity from Equation (17.19).

$$\frac{d\dot{W}}{dv_p} = 0 = \left(1 + r_{f/a} - \frac{2v_p}{v_e}\right)$$

$$v_p = \frac{v_e(1 + r_{f/a})}{2} \tag{17.20}$$

Thus the maximum work will be developed when the plane speed is about one-half the gas exit velocity.

The propulsive efficiency, η_p, is defined as the ratio of the work of the propulsive force divided by the sum of this same work plus the unused kinetic energy of the engine relative to the ground:

$$\eta_p = \frac{w}{w + (\text{K.E.})_{\text{unused}}}$$

$$(\text{K.E.})_{\text{unused}} = \frac{(1 + r_{f/a})(v_e - v_p)^2}{2} \tag{17.21}$$

$$w = v_e v_p\left[1 + r_{f/a} - \frac{v_p}{v_e}\right]$$

After considerable algebraic operations and with the assumption that $r_{f/a}$ is much less than 1, so it may be neglected, we finally determine that the propulsive efficiency is

$$\eta_p = \frac{2}{1 + v_e/v_p} \tag{17.22}$$

This is often called the "Froude efficiency." From an analysis of Equations (17.22) and (17.18), we find that

When $v_p = 0$, the specific thrust is a maximum, but $\eta_p = 0$. This occurs at takeoff, where maximum thrust is needed.

When $v_p = v_e$, $\eta_p = 100$ percent, but the specific thrust is zero. Thus v_e should be greater than v_p but not substantially greater.

There is a great deal more information on jet engines and gas-turbine units in general. We have only scratched the surface, presenting concepts from which more detailed analyses are developed. A thorough understanding of gas dynamics is very important for advanced work in this area. The following example illustrates some of the concepts developed in the last section.

Example 17.8

A jet engine in which all expansions and compressions are isentropic is operating at an altitude where the entrance pressure is 83 kPa and the entrance temperature is 230 K. The aircraft is moving at a velocity of 305 m/s. Fuel is added so the maximum temperature is 1400 K and the compressor discharge pressure is 480 kPa. Neglect the mass of the fuel, assuming air as the substance throughout. Determine (a) the pressure to which the turbine expands; (b) the exit velocity of the combustion gas from the engine; (c) the specific thrust; (d) the work; (e) the propulsive efficiency; (f) the thermal efficiency.

Calculate the enthalpies through the engine:

$$h_1 = 230.01 \text{ kJ/kg} \qquad p_{r_1} = 0.5477$$

$$h_2 = h_1 + \frac{v_1^2}{2(1000)} = 230.01 + 46.51 = 276.52 \text{ kJ/kg}$$

$$p_{r_2} = 1.0420$$

$$p_2 = p_1\left(\frac{p_{r_2}}{p_{r_1}}\right) = 83\left(\frac{1.0420}{0.5477}\right) = 157.9 \text{ kPa}$$

$$p_3 = 480 \text{ kPa}$$

Process 2–3, isentropic,

$$p_{r_3} = p_{r_2}\left(\frac{p_3}{p_2}\right) = 3.168$$

$$h_3 = 380.48 \text{ kJ/kg}$$

From a first-law analysis of the compressor

$$w_c = (h_3 - h_2) = (380.48 - 276.52) = 103.96 \text{ kJ/kg}$$

$$h_4 = 1515.41 \text{ kJ/kg}$$

$$p_{r_4} = 450.5$$

Since all the turbine work provides for the compressor work,

$$w_t = w_c = (h_4 - h_5) \qquad \therefore h_5 = 1411.45 \text{ kJ/kg}$$

$$p_{r_5} = 344.9$$

(a) $$p_5 = p_4\left(\frac{p_{r_5}}{p_{r_4}}\right) = 480\left(\frac{344.9}{450.5}\right) = 367.5 \text{ kPa}$$

$$p_6 = 83 \text{ kPa}$$

Process 5–6, isentropic,

$$p_{r_6} = p_{r_5}\left(\frac{p_6}{p_5}\right) = 77.90$$

$$h_6 = 941.8 \text{ kJ/kg}$$

From a first-law analysis of the engine, Equation (17.13),

$$h_1 + (K.E.)_1 + q_{in} = h_6 + (K.E.)_6$$

$$230.01 + \frac{(305)^2}{2(1000)} + (1515.41 - 380.48) = 941.8 + \frac{v_6^2}{2(1000)}$$

(b) $v_6 = v_e = 969.2$ m/s
The specific thrust is

(c) $\dfrac{F}{\dot{m}_a} = v_e\left[1 - \dfrac{v_p}{v_e}\right] = 969.2\left[1 - \dfrac{305}{969.2}\right] = 664.2$ kN/kg/s

The work is

(d) $w = v_e v_p\left[1 - \dfrac{v_p}{v_e}\right] = \dfrac{(305)(969.2)}{1000}\left[1 - \dfrac{305}{969.2}\right]$

$= 202.58$ kJ/kg
The propulsive efficiency is

(e) $\eta_p = \dfrac{2}{1 + v_e/v_p} = \dfrac{2}{1 + 969.2/305} = 0.48$

The overall thermal efficiency is

$$\eta_{th} = \frac{w}{q_{in}} = \frac{202.58}{1134.93} = 0.18 \text{ or } 18 \text{ percent}$$

PROBLEMS

When performing an open-system analysis of a gas turbine in the following problems, use the 400% gas tables, unless otherwise instructed, for the products of combustion.

1. An air-standard Brayton cycle has a pressure ratio of 8. The air properties at the start of compression are 100 kPa and 25°C. The maximum allowable temperature is 1100°C. Determine (a) the thermal efficiency; (b) the net work; (c) the heat added.

2. A Brayton cycle uses helium as the working substance. The pressure ratio is 4, but the low pressure is 500 kPa. The temperature at the start of compression is 300 K. The heat added is 600 kJ/kg. Determine (a) the cycle work per kilogram; (b) the maximum temperature; (c) the displacement volume per kilogram.

3. The same as Problem 2 except that the low pressure is 105 kPa. Compare the results of parts (a) and (c).

4. An air-standard Brayton cycle has a net power output of 100 kW. The working substance is air, entering the compressor at 30°C, leaving the high-temperature heat exchanger at 750°C and leaving the turbine at 300°C. Determine (a) the compressor pressure ratio;

(b) the compressor work in kJ/kg; (c) the mass flow rate of air; (d) the thermal efficiency; (e) the second-law efficiency.

5. The compressor for an actual gas turbine requires 300 kJ/kg of work to quadruple the inlet pressure. The inlet air temperature is 100°C. Determine (a) the compressor exit air temperature; (b) the compressor efficiency.

6. An ideal gas turbine operates with a pressure ratio of 8:1 and temperature limits of 20°C and 1000°C. The energy input in the high temperature heat exchanger is 200 kW. Determine (a) the air flow rate; (b) the percentage of turbine work used by the compressor.

7. A furnace needs hot pressurized gas at 200 kPa. This gas is to be provided by the exhaust from a gas turbine operating on the Brayton cycle. The turbine will produce no power beyond that required by the compressor. The compressor inlet conditions are 100 kPa and 290 K. The turbine inlet temperature is 815°C. Determine the compressor pressure ratio.

8. In an air-standard Brayton cycle the compressor inlet conditions are 100 kPa and 280 K. The turbine inlet conditions are 1000 kPa and 1167 K. The turbine produces 11 190 kW. Determine (a) the air flow rate; (b) the compressor power; (c) the cycle efficiency; (d) the pressure ratio for maximum work; (e) the second-law efficiency.

9. A 75-kW air-standard Brayton cycle is designed for maximum work. The compressor inlet conditions are 100 kPa and 27°C and the pressure ratio is 5.5. Determine (a) the turbine inlet temperature; (b) the cycle efficiency; (c) the air flow rate.

10. An air-standard Brayton cycle operates at 1000 cycles/min from a low temperature and pressure of 290 K and 100 kPa to a high pressure of 820 kPa. The heat added is 700 kJ/kg. Use Table A.2 and determine (a) the maximum temperature; (b) the thermal efficiency; (c) the power for 1 kg of air; (d) the available portion of the heat rejected if $T_0 = 290$ K.

11. The same as Problem 1, except that the compressor and turbine efficiencies are equal to 90%.

12. A Brayton cycle uses argon as the working substance. At the beginning of compression, the temperature is 335 K and the pressure is 480 kPa. The compression process is adiabatic with discharge conditions of 645 K and 1930 kPa. The argon is heated and enters the turbine at 1390 K and 1930 kPa and expands adiabatically to 890 K and 480 kPa. Determine (a) the compressor efficiency; (b) the turbine efficiency; (c) the thermal efficiency.

13. A gas turbine receives 4.72 m³/s of air at 27°C and 100 kPa. The compression is adiabatic with discharge at 517 kPa and 220°C. Dodecane is used with a fuel–air ratio of 0.015 kg fuel/kg air. The turbine exit pressure and temperature are 105 kPa and 454°C.

Determine (a) the thermal efficiency; (b) the turbine efficiency; (c) the compressor efficiency; (d) the maximum temperature; (e) the total turbine power; (f) the compressor power.

14. A gas turbine unit is to drive a natural-gas compressor located on a pipeline. The natural gas (assume it is methane) has inlet conditions to the compressor of 140 kPa and 295 K and an exit pressure of 690 kPa. The isentropic compressor efficiency is 85 percent, and the gas flow is 9.5 m^3/s at inlet conditions. The gas-turbine unit receives air at 101 kPa and 300 K, the isentropic compressor discharges it at 550 kPa. The fuel–air ratio is 0.0165 kg fuel/kg air and $h_{RP} = 44\ 000$ kJ/kg. The products of combustion expand isentropically through the turbine to 101 kPa. Determine (a) the gas-turbine unit thermal efficiency; (b) the power required to drive the natural-gas compressor; (c) the air flow rate required to drive the compressor.

15. A gas-turbine unit receives 4.72 m^3/s of air at 27°C and 100 kPa and compresses it isentropically to 450 kPa (gage). In the combustion chamber, fuel with a heating value of 43 200 kJ/kg is added so the maximum temperature is 1250 K. The turbine exhausts to atmospheric pressure. Determine (a) the fuel flow rate; (b) the unit thermal efficiency; (c) the turbine exit temperature; (d) the available portion of the energy of the products of combustion leaving the turbine, if $T_0 = 25°C$.

16. An air turbine operates between pressures of 410 kPa and 100 kPa and receives 0.45 kg/s of air at 650°C. For an ideal turbine, what is the power developed? When operating under these conditions an actual turbine develops 111.9 kW and has a discharge temperature of 150°C. The turbine blades are water-cooled, with water entering at 10°C and leaving at 38°C. Determine the water flow rate in kilograms per second.

17. Air enters the combustion chamber of a gas turbine unit at 550 kPa, 227°C, and 43 m/s. The products of combustion leave the combuster at 517 kPa, 1004°C and 140 m/s. Liquid fuel enters with a heating value of 43 000 kJ/kg. The combustor efficiency is 95%. Determine the fuel-air ratio.

18. A regenerator in a gas-turbine unit receives air from the compressor at 400 kPa and 455 K. The air leaves at 395 kPa and 77 K. The products of combustion enter at 105 kPa and 823 K and leave at 102 kPa and 611 K. The specific heat of air is 1.0047 kJ/kg K and of the products of combustion is 1.044 kJ/kg · K. The air flow rate is 22.89 kg/s, and the fuel flow rate is 0.231 kg/s. Determine (a) the fuel–air ratio; (b) the regenerator effectiveness; (c) the net entropy change across the regenerator.

19. A gas-turbine unit is equipped with a regenerator. The compression process is isentropic with air entering the compressor at 290 K and

100 kPa. The pressure ratio is 8 : 1. The maximum allowable temperature entering the turbine is 1400 K. The turbine exhausts to the regenerator, with an effectiveness of 60% at 100 kPa. The expansion process through the turbine is isentropic. Determine, for h_{RP} = 43 000 kJ/kg, (a) the fuel–air ratio; (b) the temperature entering the combustor; (c) the thermal efficiency; (d) the temperature of the products of combustion leaving the regenerator; (e) the thermal efficiency if the regenerator had not been installed.

20. The same as Problem 19, except that the compressor efficiency is 87%, the turbine efficiency is 90%, and there is a 40-kPa pressure drop between the compressor and turbine and a 2.7-kPa pressure drop between the turbine exhaust and atmosphere.

21. In designing a gas turbine for maximum efficiency, a decision is made to use inter-cooling of the compressor. The air is delivered from 100 kPa and 290 K to a final discharge pressure of 950 kPa. There are two stages of compression, with intercooling at the optimum interstage pressure. The intercooling cools the air temperature to 25°C of the inlet temperature. The regenerator has an effectiveness of 65 percent, and the maximum allowable turbine inlet temperature is 1350 K. All expansion and compression processes are isentropic. Determine, for h_{RP} = 43 000 kJ/kg, (a) the thermal efficiency; (b) the fuel–air ratio; (c) the turbine work per kilogram; (d) the compressor work per kilogram; (e) the heat removed in the intercooler; (f) the available energy of the products of combustion leaving the regenerator; (g) the thermal efficiency with no inter-cooling.

22. The same as Problem 21 except that (a) there is 7.0-kPa drop in the intercooler; (b) there is a 3.4-kPa drop between the turbine exhaust and atmosphere; (c) there is a 34-kPa drop between the compressor and the turbine; (d) the turbine and compressor efficiencies are 90%.

23. A closed gas-turbine unit uses a regenerator to improve the cycle efficiency. The working substance in the cycle is helium. The compressor inlet condition is 400 kPa and 320 K and discharge conditions are 1600 kPa and 590 K. The regenerator effectiveness is 70% and the helium enters the turbine at 1400 K and 1550 kPa. The turbine exhausts at 420 kPa and 860 K and enters the regenerator. Determine (a) the compressor and turbine efficiencies; (b) the heat added per kilogram; (c) the thermal efficiency; (d) the heat rejected per kilogram; (e) the helium flow rate for a net power output of 100 MW.

24. A closed gas-turbine unit uses neon as the working fluid. There are two stages of compression, with ideal intercooling, from 480 kPa and 290 K to 2.75 MPa. A regenerator has an effectiveness of 60% and delivers the neon to the high-temperature heat exchanger, where it is

heated to 890 K. There are two turbine stages, the first discharging at 1170 kPa. Following the first turbine stage, a heat exchanger raises the neon temperature to 890 K. The second turbine exhausts at 480 kPa to the regenerator. All expansion and compression processes are isentropic. Determine (a) the compressor work per kilogram; (b) the turbine work per kilogram; (c) the heat added per kilogram; (d) the overall thermal efficiency; (e) the power output for a neon flow rate of 45 kg/s, (f) the available portion of the heat rejected in the low-temperature heat exchanger if $T_0 = 275$ K.

25. The same as Problem 24 except that (a) the compressor and turbine efficiencies are 90%; (b) there is a pressure drop of 1.5% through each heat exchanger.

26. A combined gas-turbine/steam power plant is to be used for the generation of electric power. The combined unit must produce 600 MW. There are two stages for the compressor with ideal intercooling at the optimum interstage pressure and two stages for the turbine with reheating to the same turbine inlet temperature. The compressor unit receives air at 100 kPa and 290 K and operates with a pressure ratio of 9. The turbine inlet temperature is 1220 K with reheating occurring at 340 kPa. The turbine exhausts to the steam generator, and the products of combustion are cooled to 150°C. The steam generator produces steam at 5.5 MPa and 450°C. The steam turbine exhausts at 13 kPa. All expansion and compression processes are isentropic. Determine (a) the net gas turbine work per kilogram of air; (b) the net steam turbine work per kilogram of steam; (c) the overall thermal efficiency; (d) the air flow required; (e) the fuel–air ratio if $h_{RP} = 43\,200$ kJ/kg fuel; (f) the fuel flow rate; (g) the cost in dollars per kilowatthour of electricity produced if the fuel costs $0.45/kg; (h) the second-law efficiency.

27. A plant modification to an existing 20 000-kW gas-turbine unit is proposed. The existing unit has the following test data:

1. Inlet air	300 K, 100 kPa
2. Pressure ratio	8 : 1
3. Compressor efficiency	90 percent
4. Fuel–air ratio	0.0165 kg fuel/kg air
5. Turbine efficiency	90 percent
6. Heating value of fuel	43 950 kJ/kg

The proposal suggests that the addition of a steam generator using the energy of the turbine exhaust will provide significant increased power output and raise the overall plant thermal efficiency. The steam leaves the generator at 4.8 MPa and 370°C and enters the turbine. The expansion process is adiabatic with a turbine efficiency of 90% and an exhaust pressure of 13 kPa. The gas-turbine products of combustion must leave the steam generator at 160°C to prevent condensation on the tube and stack surfaces. Determine (a) the available energy in the gas-turbine exhaust entering the steam generator ($T_0 = 300$ K); (b) the overall unit thermal efficiency, old and proposed; (c) the total power output under proposed conditions; (d) the percent of the available energy in (a) that was used.

28. A jet plane is traveling at 0.309 km/s and has an engine that develops a thrust of 13 344 N. The gas exiting the engine has a relative velocity of 340 m/s, and the fuel/air ratio is 0.02 kg fuel/kg air. Determine (a) the air flow rate; (b) the propulsive efficiency; (c) the fuel flow rate.

29. A jet aircraft is flying with a velocity of 965 km/h at an altitude of 12.2 km, where the pressure is 20 kPa and the temperature is 220 K. The air enters an ideal diffuser and leaves the combustor at 1350 K and 100 kPa. The fuel has a heating value of 43 000 kJ/kg. All expansion and compression processes are isentropic. Determine (a) the compressor work per kilogram; (b) the fuel–air ratio; (c) the pressure entering the nozzle; (d) the specific thrust; (e) the propulsive efficiency; (f) the thermal efficiency; (g) the total thrust for an air flow of 32 kg/s.

30. There is 0.567 m³/min of air available at 6.2 mPa and 135°C. Air is needed at 750 kPa and between 15°C and 38°C. A proposal suggests using an air turbine to expand the air and perform useful work. (a) If the turbine internal efficiency is 80%, what is the exhaust temperature? (b) What heat must be added in part (a) in a heat exchanger to raise the temperature of the air to 15°C? (c) An alternative method is to throttle the air before it enters the turbine. What should be the inlet pressure of the turbine? (d) What power is produced in (a) and (c)?

31. An air turbine operates at a 7600-m altitude and receives 34 kg/s of air at 413 kPa and 400 K. Determine (a) the turbine isentropic exhaust temperature if the expansion is to 300 mm of mercury absolute; (b) the actual exhaust temperature is 240 K; find the turbine internal efficiency; (c) the mechanical losses are 3%, and the actual exit temperature is 240 K; find the shaft power.

32. A gas-turbine turboelectric plant is to provide 30 000 kW to the electric generator. The turbine unit receives air at 100 kPa and 295 K. The maximum temperature is 1200 K and the maximum pressure is 400 kPa. Neglecting the mass of the fuel, determine (a) the overall thermal efficiency; (b) energy added in the combustion chamber; (c) compressor power; (d) total turbine power; (e) volumetric flow rate of air at inlet conditions in cubic meters per second.

Chapter 18
Fluid Flow and Nozzles

The purpose of this chapter is to explore the thermodynamic aspects of fluid flow. This will include property variations of the fluid while undergoing a process. The process may be reversible or irreversible, the fluid compressible or incompressible. Thermodynamics and fluid dynamics overlap a great deal, and the purpose here is not to present a course in fluid dynamics, but rather to analyze those situations that require thermodynamic understanding. All actual fluid flow is irreversible and three dimensional, but we could not proceed very far in an analysis if we held to these restrictions. A large number of flow conditions may be approximated by assuming one-dimensional, steady-state, steady-flow conditions. These are the restrictions we will use in this chapter: The flow is one dimensional, the mass flow rate through a passage is constant with time, and the property variations at a point does not change with time. These may seem overly restrictive, but in many situations the actual variations from these conditions are small and may be neglected.

We will be particularly interested in the flow of fluids through nozzles, a very important aspect of turbine design. The nozzle has two basic functions, to direct the fluid at the correct angle and to convert the fluid thermal energy into kinetic energy. We will discuss the latter function in detail; optimization and analysis of nozzle and blade angle will not be covered.

18.1 CONSERVATION OF MASS

If we are to analyze the fluid, the tools of analysis must be defined. The conversation of mass for steady, one-dimensional flow is

$$\dot{m} = \frac{A\mathrm{v}}{v} \tag{18.1}$$

where $\dot{m}$ is the mass flow rate, v is the specific volume, A is the flow area, and v is the average fluid velocity. For steady flow, the mass flow rate is the same at one plane as at another plane, thus

$$\dot{m} = \frac{A_1 v_1}{v_1} = \frac{A_2 v_2}{v_2} = \cdots \frac{A_i v_i}{v_i} \qquad (18.2)$$

It will prove convenient to express the mass flow rate as a differential equation. Taking the logarithm of Equation (18.1) and then the derivative yields

$$0 = \frac{dA}{A} + \frac{d\mathrm{v}}{\mathrm{v}} - \frac{dv}{v} \qquad (18.3)$$

Equation (18.3) will be useful a little later in the development.

18.2 CONSERVATION OF MOMENTUM

In developing the equation for the change of momentum and force acting on a control volume, let us recall Newton's second law of motion. This states that the rate of change of momentum in a direction is equal to the sum of external forces acting on the body in the same direction.

$$\Sigma \mathbf{F} = \frac{d}{dt}(m\mathbf{v}) \qquad (18.4)$$

There are two general types of forces that $\Sigma \mathbf{F}$ comprises: body and surface. Body forces are functions of the system size (volume or mass) and would include forces from force fields. The greater the system size the greater would be the total force acting on the system as a result of a magnetic field. Should acceleration be important, as in the case in turbo machines, then centrifugal and Coriolis forces must be included. Surface forces are forces exerted on the system surface (control volume surface) by the surroundings. These surface forces may be represented by the viscous pressure tension which has normal and shearing components.

Figure 18.1 illustrates a control volume with an object in it, such as a blade, vane, or strut. Considering only one component (direction) of Equation (18.4) for Cartesian coordinates yields.

$$\Sigma F_x = \frac{d}{dt}(m\mathrm{v}_x) \qquad (18.5)$$

Considering the right-hand side of Equation (18.5), it is desirable to change systems from constant mass to constant volume. To accomplish this consider that at time t,

(a) $(m\mathrm{v}_x)_t = \Sigma \mathrm{v}_x m_{1t} + (\mathrm{v}_x m)_{11t} + \Sigma \mathrm{v}_x m_{111t}$

and at time $t + \Delta t$,

(b) $(m\mathrm{v}_x)_{t+\Delta t} = \Sigma \mathrm{v}_x m_{1t+\Delta t} + (\mathrm{v}_x m)_{11t+\Delta t} + \Sigma \mathrm{v}_{m111t+\Delta t}$

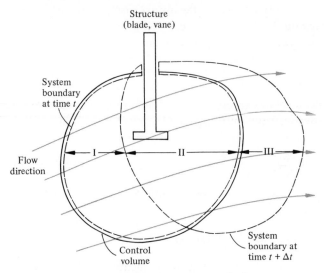

Figure 18.1   *Control volume in a moving fluid.*

Subtract (a) from (b) and divide by Δt:

$$\frac{(mv_x)_{t+\Delta t} - (mv_x)_t}{\Delta t} = \frac{(mv_x)_{IIt+\Delta t} - (mv_x)_{IIt}}{\Delta t}$$

$$+ \sum v_x \frac{(m_{It+\Delta t} - m_{It})}{\Delta t}$$

$$+ \sum v_x \frac{(m_{IIIt\Delta t} - m_{IIIt})}{\Delta t}$$

The reason a summation of velocities $\sum v_x$ is used is that there are a variety of x velocities in the fluid; each stream line would have its own. When the limit of Δt approaching zero is taken, the summations become integrals and mass differences become differentials.

(c) $\dfrac{d(mv_x)}{dt} = \dfrac{d}{dt}(mv_x)_{II} - \int v_x \dfrac{dm_I}{dt} + \int v_x \dfrac{dm_{III}}{dt}$

(d) $\dfrac{d(mv_x)}{dt} = \dfrac{d}{dt}(mv_x)_{CV} - \int v_x \dot{m}_{in} + \int v_x \dot{m}_{out}$

Equation (18.5) becomes

$$\sum F_x = \frac{d}{dt}(mv_x)_{CV} + \int v_x \dot{m}_{out} - \int v_x \dot{m}_{in} \tag{18.6}$$

Equation (18.6) is not in terms of measurable properties; referring to Equations (3.7) and (3.8), Equation (18.6) becomes

$$\sum F_x = \frac{d}{dt}\int_{CV} \rho v_x \, d\mathcal{V} + \int_A (\rho v_x v_n \, dA)_{out} - \int_A (\rho v_x v_n \, dA)_{in} \tag{18.7}$$

In vector notation Equation (18.7) may be written as

$$\sum \mathbf{F} = \frac{d}{dt} \int_{CV} \rho \mathbf{v} \, d\mathcal{V} + \int_{A} \rho(\mathbf{v} \cdot d\mathbf{A})\mathbf{v} \tag{18.8}$$

For one-dimensional, steady flow, Equation (18.7) becomes

$$\sum F_x = \rho_2 A_2 v_{x_2}^2 - \rho_1 A_1 v_{x_1}^2 \tag{18.9}$$

Moment of Momentum

Centrifugal forces are important in the analysis of rotating machinery. In this instance not only are forces of interest, but the moment of these forces—or torque—is important. By looking at Equation (18.8) we may see that the moment of forces is the same as the moment of momentum. Since the expressions for momentum have measurable fluid quantities, this is often used. A torque is produced when a force is acting at distance. When a wrench is used to loosen or tighten a nut, a force is applied (your hand) normal to the wrench and at a distance r (wrench length) from the nut. In vector notation this is the vector product, or

$$\sum \mathbf{F} \times \mathbf{r} = \frac{d}{dt} \int_{CV} \rho \mathbf{v} \times \mathbf{r} \, d\mathcal{V} + \int \rho(\mathbf{v} \cdot d\mathbf{A})\mathbf{v} \times \mathbf{r} \tag{18.10}$$

Consider Figure 18.2, which illustrates a force acting at a distance r in the xy plane. The torque (moment of momentum) is in the z direction (out of the page). Equation (18.10) in Cartesian coordinates for this

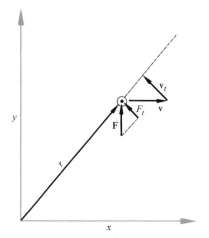

Figure 18.2 Point at a distance r acted by force **F** and fluid velocity **v**.

example is

$$\sum rF_t = \frac{d}{dt}\int_{CV}\rho r v_t\, d\mathcal{V} + \int_A (\rho v_n\, dA\, r v_t)_{out} - \int_A (\rho v_n\, dA\, r v_t)_{in}$$

(18.11)

Example 18.1

Calculate the force on a control volume with an inlet area of 0.1 m^2 and exit area of 0.05 m^2. The density is 3.89 kg/m^3 and constant, and the inlet velocity is 30 m/s.

All the variables are known except v_{x_2}. To determine this, we note that for steady flow

$$\rho_1 A_1 v_{x_1} = \rho_2 A_2 v_{x_2}$$

$$v_{x_2} = 2v_{x_1} = 60 \text{ m/s}$$

$$F_{net} = (3.89)(0.05)(60)^2 - (3.89)(0.1)(30)^2$$

$$F_{net} = 350 \text{ N}$$

18.3 ACOUSTIC VELOCITY

An important velocity in the study of fluid mechanics is the *acoustic* or *sonic velocity*. This is the velocity at which a small pressure wave moves through a fluid. What pressure wave? When we talk, for instance, we are generating pressure waves of varying intensities. These waves move through the air at a certain velocity, the acoustic velocity, commonly called the "speed of sound." This is about 336 m/s at standard atmospheric conditions. The speed of sound is finite; for example, there is the case where a gun is fired at some distance, several hundred yards. The flash or movement of the gun is seen, then the sound of gun fire is heard. The speed of light is 2.998 × 10^8 m/s, exceedingly greater than the acoustic velocity of 336 m/s. For larger pressure waves, such as shock waves, the propagation may be several times the acoustic velocity. The analysis of shock waves is typically included in a course on compressible fluid flow.

Let us consider Figure 18.3, which illustrates a small pressure wave, caused by the piston and moving through a compressible fluid. The pressure wave must be weak enough for the property changes to be differentially small. If the wave is too great, the criterion of differential property change is not met, and the following derivation is not valid. The wave, generated by the piston movement in Figure 18.3, moves at the acoustic velocity, a, into the stationary gas. If we place ourselves on the

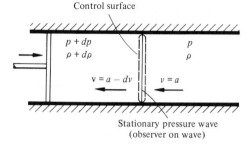

Figure 18.3   *A small pressure wave in the stationary coordinate system.*

pressure wave, then it seems as if the fluid is approaching from the right, the stationary region, at the acoustic velocity and leaving at a different velocity because of the pressure change across the wave front. The mass must be conserved entering and leaving the control surface. The sum of the forces must be equal. The sum of the forces may be written by using Equation (18.9) or by considering the difference in forces acting on the control surface.

$$F_{net} = pA - (p + dp)A$$

or

$$F_{net} = \dot{m}v_{x_2} - \dot{m}v_{x_1}$$

where shear forces, friction, are negligible;

$$\dot{m} = \rho A v = \rho A a$$

Equating the net force across the control surface yields

$$pA - (p + dp)A = Aa[(a - dv) - a]\rho$$

which reduces to

$$dp = \rho a \, dv \qquad\qquad (18.12)$$

The continuity equation across the control surface is

$$\rho a A = (\rho + d\rho)(a - dv)A$$

We neglect second-order terms as being very much less than first-order terms. Thus

$$dv = a\frac{d\rho}{\rho} \qquad\qquad (18.13)$$

Substituting Equation (18.13) into Equation (18.12) yields

$$a = \left(\frac{dp}{d\rho}\right)^{1/2} \qquad\qquad (18.14)$$

For small-pressure waves, the compression process is essentially isentropic. For an ideal gas, the isentropic process is

$$pv^k = C, \quad \text{or} \quad p = C\rho^k$$

We differentiate

$$dp = kpv \, d\rho$$

$$\left(\frac{\partial p}{\partial \rho} \right)_s = kpv = kRT$$

Therefore

$$a = \left(\frac{\partial p}{\partial \rho} \right)_s^{1/2} = (kRT)^{1/2} \tag{18.15}$$

From Chapter 10 we found that physical coefficients were defined in terms of partial derivatives. We defined the coefficient of isothermal compressibility, β_T, in Equation (10.49). Another coefficient, the coefficient of adiabatic compressibility, β_s, is

$$\beta_s = -\frac{1}{v} \left(\frac{\partial v}{\partial p} \right)_s \tag{18.16}$$

where the minus sign allows β_s to be always positive. This coefficient is tabulated for liquids and gases. The acoustic velocity may be defined in terms of β_s as follows:

$$a = \left(\frac{1}{\rho \beta_s} \right)^{1/2} \tag{18.17}$$

where consistent units must be used.

18.4 STAGNATION PROPERTIES

Let us consider a stream of fluid flowing through an insulated pipe. The energy of this fluid at any plane is $h + \text{K.E.}$, the sum of the enthalpy and kinetic energy. If we follow the fluid back to its starting point, where the velocity is zero, we will find that the energy of the fluid is h_0, where the subscript zero stands for zero velocity. Thus since the energy is constant throughout the adiabatic duct,

$$h_0 = h + \frac{v^2}{2} \text{ J/kg} \tag{18.18}$$

Let us work with Equation (18.18) for the time being. We substitute for the enthalpy, $h = c_p T$, thus

$$T_0 = T + \frac{v^2}{2c_p}$$

$$\frac{T_0}{T} = 1 + \frac{v^2}{2c_p T} \tag{18.19}$$

Thus the temperature at zero velocity may be expressed in terms of the fluid temperature and velocity at any plane. This zero velocity condition is called the *stagnation condition*. We may determine the temperature at the stagnation state by specifying that the fluid decelerate adiabatically. If a second condition, that the deceleration be isentropic, is invoked, then the isentropic stagnation pressure may be found. For an isentropic process for an ideal gas

$$\frac{T_2}{T_1} = \left(\frac{p_2}{p_1} \right)^{(k-1)/k}$$

Thus

$$\frac{p_0}{p} = \left(1 + \frac{v^2}{2c_p T} \right)^{k/(k-1)} \tag{18.20}$$

Equation (18.20) gives the isentropic stagnation pressure in terms of the fluid pressure, temperature, and velocity. The stagnation temperature, T_0, is the same for the adiabatic and isentropic deceleration. Figure 18.4 shows a T–s diagram with the various states. The irreversibilities associ-

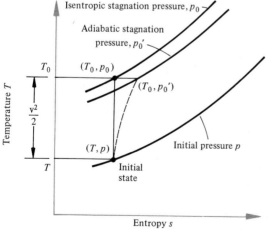

Figure 18.4 *A T–s diagram illustrating stagnation temperature and pressure.*

ated with deceleration in the adiabatic case result in an entropy increase. The irreversibilities are also manifested in the adiabatic stagnation pressure, p_0', being less than the maximum pressure, the isentropic stagnation pressure, p_0.

18.5 MACH NUMBER

The Mach number, M, is defined as the ratio of the actual velocity, v, divided by the local speed of sound, a, in the fluid.

$$M = \frac{v}{a} \qquad (18.21)$$

Equations (18.15) and (18.21) may be substituted into equation (18.19) to yield the temperature in terms of the Mach number.

$$\frac{T_0}{T} = 1 + \frac{k-1}{2} M^2 \qquad (18.22)$$

Example 18.2

A meteorite enters the earth's atmosphere with a velocity of 1200 m/s. The atmospheric pressure is 70 Pa and the temperature is 150 K. Determine the stagnation temperature, the isentropic stagnation pressure and the Mach number of the meteorite.

Find the speed of sound first.

$$a = [kRT]^{1/2} = [(1.4)(287.02)(150)]^{1/2} = 245.5 \text{ m/s}$$

The Mach number is

$$M = \frac{v}{a} = \frac{1200}{245.5} = 4.89$$

$$\frac{T_0}{T} = 1 + \left(\frac{k-1}{2}\right)(4.89)^2 = 5.78$$

$$T_0 = 867 \text{ K}$$

$$\frac{p_0}{p} = \left(\frac{T_0}{T}\right)^{k/(k-1)}$$

$$p = 32.49 \text{ MPa}$$

18.6 FIRST-LAW ANALYSIS

The conservation of energy for steady flow with heat transfer between state 1 and state 2 is

$$\dot{Q} + \dot{m}[(P.E.)_1 + h_1 + (K.E._1)] = \dot{m}[(P.E.)_2 + h_2 + (K.E._2)]$$

We divide by $\dot{m}$, putting the equation on a unit mass basis. This permits us to use the constant mass expression for the heat flow. On the differential basis the equation becomes

$$\delta q = dh + d(\text{K.E.}) + d(\text{P.E.}) \qquad (18.23)$$

and

$$dh = du + p\,dv + v\,dp$$

and

$$\delta q = du + p\,dv$$

also

$$d(\text{K.E.}) = v\,dv$$

$$d(\text{P.E.}) = g\,dz$$

Substituting into Equation (18.23) yields

$$\frac{dp}{\rho} + v\,dv + g\,dz = 0 \qquad (18.24)$$

Equation (18.24) is *Euler's equation for steady flow.* Bernoulli's equation in fluid mechanics is the same, but it is not written on a differential basis. Euler's equation was developed from thermodynamic arguments, whereas in fluid mechanics the equation is typically developed from principles of mechanics. This demonstrates, in part, the close relationship of thermodynamics and fluid mechanics and indicates the breadth of thermodynamic analysis.

18.7 NOZZLES

A nozzle is a device with two purposes: to convert thermal energy into kinetic energy and to direct the mass flow from it at a specified angle. This is accomplished by means of a variable area duct.

Let us talk about an adiabatic, variable area duct as shown in Figure 18.5. For steady flow, the energy at any plane is constant and equal to the sum of the kinetic energy and the enthalpy.

total energy $= h + \text{K.E.}$

However, this is equal to the stagnation enthalpy, h_0.

$$h_0 = h + \frac{v^2}{2} \text{ J/kg}$$

We solve for the velocity,

$$v = \sqrt{2(h_0 - h)} \text{ m/s} \qquad (18.25)$$

Thus the velocity at any plane may be expressed in terms of the enthalpy at that plane and the stagnation enthalpy.

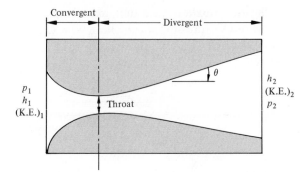

Figure 18.5 An adiabatic convergent-divergent nozzle.

What happens if the flow has an initial velocity, v_1, as illustrated in the figure. The first law for an open system is

Energy In = Energy Out

$$h_1 + (\text{K.E.})_1 = h_2 + (\text{K.E.})_2$$

$$h_1 + \frac{v_1^2}{2} = h_2 + \frac{v_2^2}{2}$$

$$v_2 = \left[2(h_1 - h_2) + v_1^2\right]^{1/2} \text{ m/s} \qquad (18.26)$$

This is far more awkward to use than Equation (18.25). Note that $h_1 + (\text{K.E.})_1 = h_0$, and if h_0 is used, the solution for v_2 is greatly simplified. In most instances it is easier to solve for h_0 initially from the inlet conditions, and use the stagnation enthalpy in the subsequent calculations.

One purpose of a nozzle is to convert thermal energy, enthalpy, into kinetic energy. Figure 18.5 shows a converging (area decreasing) section followed by a diverging (area increasing) section. Since both sections must increase the velocity, let us find out why the area changes as it does.

Combine the first law and the continuity equation to find an expression for the area change in terms of the Mach number.

$$\frac{dA}{A} + \frac{dv}{v} + \frac{d\rho}{\rho} = 0 \qquad \text{continuity} \qquad (18.27)$$

$$\frac{dp}{\rho} + v\,dv = 0 \qquad \text{Euler's equation or first law} \qquad (18.28)$$

The potential-energy term is not included in Equation (18.28). Solve for ρ from Equation (18.28) and substitute this value into Equation (18.27), solving for dA/A,

$$\frac{dA}{A} = \frac{dv}{v}(M^2 - 1) \qquad (18.29)$$

There are three flow regions. When M is less than one, the flow is subsonic; when M equals one, the flow is sonic; and when M is greater than one, the flow is supersonic. The change in velocity, $d\text{v}$, must always be positive, as that is the purpose of a nozzle. Let us find how the area must change in these three flow regions for this to be true. For M less than one, the area change must be negative for $d\text{v}$ to be positive. This is the converging portion of the nozzle. At M equal to one, the area change is zero, or an undefined condition exists. At the throat, the Mach number is one and the area is a minimum; thus the derivative of the area is zero. When the Mach number is greater than one, the area change must be positive for a positive $d\text{v}$; therefore the diverging portion of the nozzle increases the cross-sectional area.

If a liquid, such as water, were flowing through a nozzle, would we design a converging-diverging nozzle? No. Let us examine the continuity equation, Equation (18.27). The liquids are essentially incompressible, hence $d\rho$ is zero. Thus the converging portion of the nozzle is satisfied, as the decreasing area results in increasing velocity. The diverging portion is impossible, however, as both the area and velocity cannot increase simultaneously. In compressible fluids, the density change compensates for the area and velocity changes. For incompressible fluids, the density does not vary and so only a converging nozzle is used. This means that the fluid velocity cannot exceed $M = 1$. This seems a little strange, but the acoustic velocity in an ideal incompressible fluid is infinite, so the possible velocity change is great. For water, the acoustic velocity is around 1525 m/s, about 4.5 times that of air, so in terms of exist velocities, both the compressible and incompressible fluid nozzles permit a wide range of exit velocities.

Critical Pressure Ratio

At the throat conditions for compressible flow, special properties exist if the flow is at $M = 1$. The properties are called the "critical properties" and are denoted as p^* and T^*, the critical pressure and temperature, respectively. From Equation (18.22) for $M = 1$, we find that

$$\frac{T_0}{T^*} = \frac{k + 1}{2} \tag{18.30}$$

and

$$\frac{p_0}{p^*} = \left(\frac{k + 1}{2} \right)^{k/(k-1)} \tag{18.31}$$

If Equation (18.31) is inverted, then the ratio of p^*/p_0 is called the "critical pressure ratio." Thus the critical pressure, p^*, is a function of k

only. The following list is for the values we will typically use.

$$\frac{p^*}{p_0} = \left(\frac{2}{k+1}\right)^{k/(k-1)}$$ (18.32)

$k = 1.2$	$p^* = 0.577p_0$	saturated steam
$k = 1.3$	$p^* = 0.545p_0$	low-pressure superheated steam and supersaturated steam
$k = 1.4$	$p^* = 0.528p_0$	air and diatomic gases at 25°C
$k = 1.67$	$p^* = 0.487p_0$	monatomic gases

When may we use Equations (18.30) to (18.32)? When the flow is isentropic and when the pressure that the throat senses is less than p^*. Fortunately, the flow may be considered isentropic to the throat almost all the time. The irreversibilities in actual nozzle flow occur in the diverging portion of the nozzle. Since the area is a minimum at the throat, the condition at the throat determines the mass flow rate through the nozzle. Once critical conditions are achieved at the throat, the density and velocity are constant, so the flow rate is constant. This is called a "choked flow condition." The nozzle cannot pass a greater flow. Pressure waves travel at the speed of sound in the divergent portion of the nozzle. Any signal to increase the flow will travel upstream at the speed of sound and since the flow is supersonic, the signal to increase the mass flow cannot reach the throat. The following example will illustrate many of the concepts we have been discussing.

Example 18.3

Air, which may be considered an ideal gas, is flowing through a nozzle isentropically. It enters the nozzle with a velocity of 100 m/s, a pressure of 600 kPa and a temperature of 600 K. The exit pressure

Table 18.1   Table for Example 18.3

p (kPa)	v (m³/kg)	V (m/s)	T (K)	A (cm²)
600	0.287	100	600	28.7
550	0.305	199	585.25	15.3
500	0.327	267	569.5	12.2
450	0.352	324	552.6	10.9
400	0.383	377	534.3	10.2
350	0.422	427	514.3	9.9
326	0.444	450	503.9	9.8
300	0.471	476	492.1	9.9
250	0.536	526	467.1	10.2
200	0.629	579	438.2	10.9
150	0.772	636	403.6	12.1

is 150 kPa. Tabulate and plot the area, specific volume, velocity, and temperature for 50 kPa increments. Include the throat conditions. The flow rate is 1 kg/s.

Calculate the stagnation properties and then velocities, areas, and specific volumes.

$$h_0 = h_1 + \frac{v_1^2}{2} = (1.0047)(600) + \frac{100^2}{2(1000)} = 607.82 \text{ kJ/kg}$$

$$T_0 = 604.97 \text{ K}$$

$$\frac{p_0}{p} = \left(\frac{T_0}{T}\right)^{k/(k-1)} = \left(\frac{604.97}{600}\right)^{3.5} \qquad p_0 = 617.6 \text{ kPa}$$

$$p^* = 0.528 p_0 = 326 \text{ kPa}$$

$$v_1 = \frac{RT_1}{p_1} = \frac{(0.287)(600)}{(600)} = 0.287 \text{ m}^3/\text{kg}$$

$$A_1 = \frac{v_1 \dot{m}}{v_1} = \frac{(0.287)(1)}{100} = 28.7 \text{ cm}^2$$

Find T_2. The process 1–2 is isentropic,

$$T_2 = T_1 \left(\frac{p_2}{p_1}\right)^{(k-1)/k} = 600 \left(\frac{550}{600}\right)^{0.286} = 585.25 \text{ K}$$

$$v_2 = \sqrt{(2)(1004.7)(604.97 - 585.25)} = 199 \text{ m/s}$$

$$v_2 = \frac{RT_2}{p_2} = \frac{(0.287)(585.25)}{(550)} = 0.305 \text{ m}^3/\text{kg}$$

$$A_2 = \frac{v_2 \dot{m}}{v_2} = \frac{(0.305)(1)}{199} = 15.3 \text{ cm}^2$$

Continue with other pressures, yielding the results in Table 18.1. The plot of the area, in Figure 18.6, does not indicate how long the nozzle should be to effect these area changes. In Figure 18.5 the angle of divergence is indicated by 2θ. The angle is around 6 to 12 degrees for turbine nozzles. For angles greater than this, flow irreversibilities occur, limiting the efficiency of the nozzle. A trade-off is made in nozzle design between the irreversibilities due to overexpansion, which occurs if 2θ is large, and the increased frictional irreversibilities due to boundary layer thickness at small angles of divergences. As indicated, this results in an angle of 6 to 12 degrees for turbine nozzles.

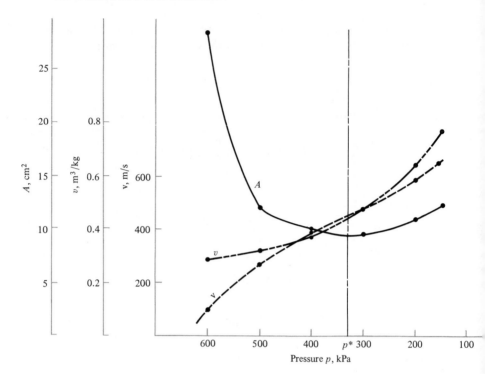

Figure 18.6   *A plot of area, specific volume, and velocity for Example 18.3.*

Nozzle Efficiency

The purpose of the nozzle is to convert enthalpy into kinetic energy. Actually irreversibilities in flow prevent the ideal conversion from occurring. The nozzle efficiency, η_n, relates the actual and ideal conversion:

$$\eta_n = \frac{(\text{K.E.}_2)_{\text{actual}}}{(\text{K.E.}_2)_s} \qquad (18.33)$$

$$\eta_n = \frac{(h_0 - h_{2'})}{(h_0 - h_2)_s} \qquad (18.34)$$

Figure 18.7 illustrates a T–s diagram for a nozzle. The shaded area represents the reheat. This energy irreversibly increases the exit enthalpy to 2′. The reheat is equal to $h_{2'} - h_2$. Typical values of nozzle efficiency are from 94 to 99 percent.

Example 18.4

The expansion of air through a nozzle with an efficiency of 95 percent is 1.5 kg/s. The air enters at 600 K and 500 kPa and exits at 101 kPa. Determine the actual exit area, velocity, and enthalpy.

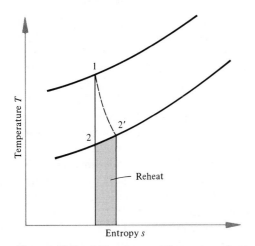

Figure 18.7 *A T–s diagram illustrating reheat caused by irreversibilities in flow through a nozzle.*

Calculate isentropic exit conditions and from this determine actual exit conditions.

$$T_2 = T_1 \left(\frac{p_2}{p_1} \right)^{(k-1)/k} = 600 \left(\frac{101}{500} \right)^{0.286} = 379.9 \text{ K}$$

$$\eta_n = \frac{h_0 - h_{2'}}{(h_0 - h_2)_s} = \frac{T_0 - T_{2'}}{(T_0 - T_2)_s} = 0.95$$

$$T_{2'} = 390.9 \text{ K}$$

$$h_{2'} = c_p T_{2'} = 392.7 \text{ kJ/kg}$$

$$v_{2'} = \sqrt{2(h_0 - h_{2'})} = \sqrt{2(1004.7)(600 - 390.9)}$$

$$v_{2'} = 648.17 \text{ m/s}$$

$$v_{2'} = \frac{RT_{2'}}{p_{2'}} = \frac{(0.287)(390.9)}{101}$$

$$v_{2'} = 1.111 \text{ m}^3/\text{kg}$$

$$A_{2'} = \frac{\dot{m} v_{2'}}{v_{2'}} = \frac{(1.5)(1.111)}{648.17} = 2.57 \times 10^{-3} \text{ m}^2$$

18.8 SUPERSATURATION

In analyzing the flow of steam through a nozzle we have dealt with equilibrium flow. The properties of steam were considered to be continuous, and for equilibrium flow an isentropic path joined the end states. In actual steam flow through a nozzle, the steam may or may not be in equilibrium during its expansion in the nozzle. A criterion for judging whether or not the steam is in metastable equilibrium is to determine whether condensation (the steam path's crossing the saturated vapor line) could have occurred for equilibrium flow. The condensation process takes time. Several molecules of steam must come together to form the nucleus of a droplet. Once the droplet is formed, other steam molecules are readily attracted. The time required is small, but the velocity through a nozzle is high, so the time that the steam is in the nozzle is also small.

In equilibrium flow, the energy released by the condensing molecules provides for increasing kinetic energy of the steam as it passes through the nozzle. If the steam does not condense, then the energy for the kinetic energy increase must come from another source. It comes from the molecular translational energy, which is denoted on the macroscopic level as temperature. There is a reduction in the temperature level compared with equilibrium steam, and steam in this condition is called *super-saturated*. The supersaturated state is a *metastable*, or *nonequilibrium*, state. If it is a nonequilibrium state, then the properties could not be determined from equilibrium tables of properties. This condition refers to steam not lying on the equilibrium surface.

Figure 18.8 illustrates the T–s diagram for steam and denotes the paths for an equilibrium process from 1 to 2 and for the nonequilibrium process 1 to 2′. How may the properties at state 2′ be found? By extending the constant-pressure line until it reaches the same value of entropy as state 1. Condensation has not occurred, so we may relate state 1 and state 2′ by $pv^k = C$. The values of steam at state 1 are known, so

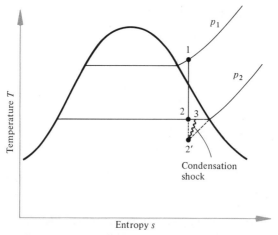

Figure 18.8   *A T–s diagram for steam illustrating equilibrium path 1–2 and nonequilibrium path 1–2′.*

the value of v for state $2'$ may be calculated knowing $p_{2'} = p_2$. If we need to know the velocity, we cannot use the typical change of enthalpy involving the temperature. Although it is permissible to use pv relationships, $pv \neq RT$ in the supersaturated region. The molecular forces of attraction are great, so the ideal gas equation of state is not indicative of the mixture behavior. The change of kinetic energy is still equal to the change of enthalpy, but we must develop an equation to express the enthalpy change in terms of pressure and specific volume. For isentropic flow,

$$dh = v\, dp$$

$$\int_1^{2'} dh = h_{2'} - h_1 = \frac{k}{k-1} p_1 v_1 \left[1 - \left(\frac{p_{2'}}{p_1} \right)^{(k-1)/k} \right]$$

Thus

$$(K.E.)_{2'} - (K.E.)_1 = h_{2'} - h_1$$

$$\frac{v_{2'}^2 - v_1^2}{2} = \frac{k}{k-1} p_1 v_1 \left[1 - \left(\frac{p_{2'}}{p_1} \right)^{(k-1)/k} \right] \qquad (18.35)$$

where the velocity is in meters per second.

The diagram in Figure 18.8 also shows something called "condensation shock." This occurs when the droplet nuclei start to form and then grow very rapidly. This rapid growth appears like a shock, a sudden change, from metastable conditions to equilibrium conditions at state 3. If further expansion occurs after state 3, the expansion may be assumed to be equilibrium. This is because the droplets are present and form a nucleus for the vapor molecules to condense on. It is because no nuclei were present initially that supersaturated conditions existed. In cloud seeding, nuclei are injected into a water–vapor mixture to assist the droplet formation.

When is it necessary to account for supersaturated flow in a nozzle and why? We have seen that the throat conditions determine the mass flow rate through the nozzle. The supersaturated steam has a significantly lower specific volume than the equilibrium value, so if supersaturated flow existed at the throat, the mass flow rate through the nozzle would be greater than the equilibrium flow rate. Studies have shown that the condensation shock does not occur in the converging portion of the nozzle, but rather in the diverging portion. Since the condensation shock is an irreversibility, it must be accounted for. The nozzle efficiency will account for this irreversibility. As a guideline, if supersaturated flow can exist at the throat, we will assume it does exist. The condensation shock occurs in the diverging portion of the nozzle, and equilibrium flow exists at the nozzle exit. The following example compares equilibrium and supersaturated flow of steam in a nozzle.

Example 18.5

Steam expands in a nozzle from an initial condition of 2.8 MPa and 240°C to 140 kPa. The nozzle efficiency is 95 percent. Isentropic flow may be assumed to exist to the throat. The throat area is 5 cm². Determine the mass flow rate for equilibrium and supersaturated flow. Determine the exit area.

$$p_1 = 2.8 \text{ MPa} \qquad T_1 = 240 \text{ C}$$

$$h_1 = 2835.6 \text{ kJ/kg} \qquad v_1 = 0.074\,01 \text{ m}^3/\text{kg}$$

$$s_1 = 6.2761 \text{ kJ/kg} \cdot \text{K} \qquad p_2 = 140 \text{ kPa}$$

$$s_2 = s_1 \qquad\qquad x = 0.8337$$

$$h_2 = 2319.3 \text{ kJ/kg}$$

$$\eta_n = 0.95 = \frac{h_1 - h_{2'}}{(h_1 - h_2)_s};$$

$$h_{2'} = 2345.1 \text{ kJ/kg} \qquad \text{and} \qquad h_{2'} = h_f + x h_{fg}$$

$$x' = 0.8452 \qquad v_{2'} = 1.045 \text{ m}^3/\text{kg}$$

For equilibrium flow at the throat,

$$p^* = 0.545 p_0 = 1526 \text{ kPa} \qquad s^* = s_1$$

$$x = 0.961 \qquad h^* = 2718.8 \text{ kJ/kg}$$

$$v^* = 0.1228 \text{ m}^3/\text{kg}$$

$$v^* = \sqrt{2(h_0 - h^*)(1000)} = 483.3 \text{ m/s}$$

$$\dot{m} = \frac{A^* v^*}{v^*} = 1.97 \text{ kg/s}$$

For supersaturated flow at the throat

$$\frac{v_t^2}{2} = \frac{1.3}{0.3}(2800)(1000)(0.07401)\left[1 - \left(\frac{1526}{2800}\right)^{0.3/1.3}\right];$$

$$v_t = 484.5 \text{ m/s}$$

$$v_t = v_1\left(\frac{p_1}{p_t}\right)^{1/k} = (0.07401)\left(\frac{2800}{1526}\right)^{1/1.3} = 0.1180 \text{ m}^3/\text{kg}$$

$$\dot{m} = \frac{A v_t}{v_t} = \frac{(0.0005)(484.5)}{(0.1180)} = 2.05 \text{ kg/s}$$

$$\text{error for } \dot{m} = \frac{(0.08)(100)}{1.97} = 4.0 \text{ percent}$$

Exit conditions are as follows:

$$v_{2'} = \sqrt{2(1000)(h_0 - h_{2'})} = 990.4 \text{ m/s}$$

$$A_{2'} = \frac{\dot{m}v_{2'}}{v_{2'}} = \frac{(2.05)(1.045)}{(990.4)} = 21.6 \text{ cm}^2$$

The steam vapor tables, the superheated steam tables, do list the properties of supersaturated steam for certain conditions.
These properties are in italics. The property values in boldface type are equilibrium values.

18.9 SHOCK WAVES

We have looked at isentropic flow through a nozzle and have used nozzle efficiency to correct for flow irreversibilities with the nozzle. A nozzle is designed to operate at certain conditions, but often in turbines the operating conditions are other than the design conditions. What happens then? Figure 18.9 illustrates a convergent-divergent nozzle and the pressure distribution in it for various back pressures for steady-flow conditions. Let p_{b_i} be the back pressure for any line i, $i = 1, 2, \ldots, 6$. If the back pressure is equal to the inlet pressure, then no flow occurs and the pressure is uniform through the nozzle, as shown by line 1 on Figure 18.9. If the back pressure is decreased to p_{b_2}, then there is flow through the nozzle. The pressure decreases through the converging portion to the throat; the flow is subsonic, so the diverging portion of the nozzle acts as a diffuser, decelerating the fluid with the resultant rise in pressure. When the back pressure is lowered to p_{b_3}, then the flow at the throat reaches sonic velocity and the divergent portion of the nozzle acts as a diffuser and decelerates the flow. For steam, the value of p_{b_3}/p_0 must be about 0.82 for this condition to be achieved. As the back pressure is lowered further, the pressure decreases through the convergent portion of the nozzle, sonic velocity is achieved at the throat, and the pressure continues to decrease in the divergent portion of the nozzle, as the flow is supersonic. Lines 4 and 5 illustrate this phenomenon. If the process is isentropic, the flow through the entire nozzle is indicated by line 6.

Frequently the back pressure is p_{b_4} or p_{b_5}. The exit pressure wave propagates into the fluid stream, indicating that the pressure for the supersonic flow is too low; it must increase. There is a discontinuity in pressure, and something called a "shock wave" patches the two flow regimes. The shock wave is a compression wave, where the kinetic energy of the fluid irreversibility increases the pressure of the fluid by decelerating in a very, very short distance. This is called a "normal" shock wave, as it is normal to the fluid flow. The flow on one side of the shock wave is supersonic; the flow on the other side is subsonic. The diverging

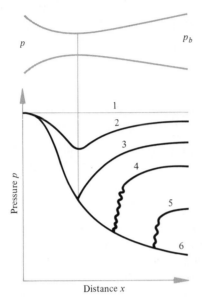

Figure 18.9   *Pressure distribution in a convergent-divergent nozzle for various back pressures.*

portion of the nozzle acts as a diffuser for the subsonic flow in the remaining length of the nozzle.

The disturbance caused by the shock wave cannot be felt at the throat because after the throat, the flow is supersonic, whereas the disturbance propagates at the sonic velocity and cannot reach the throat.

Is the mass flow rate affected by the changes in back pressure? Only until p_{b_3} is reached. After that, any lowering of the back pressure does not affect the properties at the throat, and choked flow is said to exist. The nozzle cannot pass any greater mass flow, so it is in a choked condition. If we want to assure a given mass flow rate to a device under varying pressures, then a nozzle can be used as a metering device as long as the back pressures are less than p_{b_3}.

18.10 DIFFUSER

The diffuser has been mentioned previously and its purpose is opposite to that of the nozzle. It converts kinetic energy into thermal energy; more specifically, it compresses a fluid to a higher pressure by decelerating the flow. The ideal process for this is isentropic. From Equation (18.29), $dA/A = (dv/v)(M^2 - 1)$, we see that for supersonic flow to exist the area must converge to decrease the velocity. For subsonic flow, the area must increase for the velocity to decrease. At $M = 1$, the area and

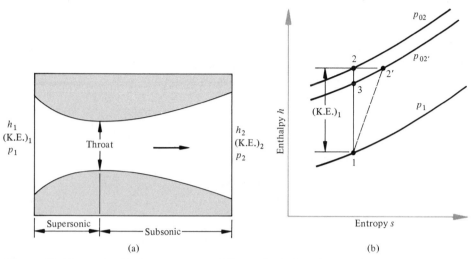

Figure 18.10 (a) *An adiabatic supersonic diffuser.* (b) *An h–s diagram for a diffuser when the exit velocity is negligibly small.*

velocity change is zero, as in the case of the nozzle. Thus a supersonic diffuser would look similar to Figure 18.10(a). Figure 18.10(b) shows an h–s diagram for the diffuser when the exit velocity is negligibly small. For the isentropic compression, $p_{02} = p_{01}$, by the definition of the stagnation state. Actually, the compression is adiabatic, the dotted path, and the stagnation pressure, p_{02}, cannot be achieved. Note that the stagnation enthalpy or temperature is only a function of the process, being adiabatic, so $h_{02} = h_{02'}$.

The same equation apply for a diffuser as for a nozzle. The first law for the diffuser, an energy balance, is

$$(\text{K.E.})_1 - (\text{K.E.})_2 = h_2 - h_1$$

Since we want the diffuser to increase the fluid pressure, the diffuser efficiency, η_d, would relate how this was performed.

$$\eta_d = \frac{\text{change in kinetic energy to achieve } p_{02'} \text{ ideally}}{\text{change in kinetic energy to achieve } p_{02'} \text{ actually}} \qquad (18.36)$$

$$\eta_d = \frac{(\text{K.E.})_1 - (\text{K.E.})_3}{(\text{K.E.})_1 - (\text{K.E.})_{2'}} = \frac{(h_1 - h_3)_s}{(h_1 - h_{2'})} \qquad (18.37)$$

Supersonic diffuser efficiencies tend to be lower than nozzle efficiencies because of shock-wave formation, especially at other than design conditions.

Example 18.6

Air enters a diffuser at 100 kPa and 300 K with Mach number equal to four. It is decelerated in a convergent-divergent diffuser with an efficiency of 80 percent, to negligible exit velocity. Determine the actual exit pressure and specific volume.

Find the inlet velocity from the acoustic velocity:

$$a = (kRT)^{1/2} = [(1.4)(287.02)(300)]^{1/2} = 347.2 \text{ m/s}$$
$$v = Ma = 4a = 1388.8 \text{ m/s}$$

From the first-law analysis

$$\Delta(\text{K.E.})_{\text{actual}} = \Delta h_{\text{actual}}$$

$$\frac{v^2}{2} = \Delta h = \frac{(1388.8)^2}{2} = 964.38 \text{ kJ/kg}$$

$$h_{2'} - h_1 = c_p(T_{2'} - T_1) = 1.0047(T_{2'} - 300) = 964.38$$

$$T_{2'} = 1260 \text{ K}$$

From diffuser efficiency, find state 3, and then use isentropic relationships to find $p_3 = p_{02'}$.

$$\eta_d = \frac{(h_1 - h_3)_s}{h_1 - h_{2'}} = 0.8 \qquad (h_1 - h_3)_s = -771.5 \text{ kJ/kg}$$

$$c_p(T_1 - T_3) = -771.5 \qquad T_3 = 1068 \text{ K}$$

$$\left(\frac{p_3}{p_1}\right) = \left(\frac{T_3}{T_1}\right)^{k/(k-1)} = \left(\frac{1068}{300}\right)^{1.3/0.3} ;$$

$$p_3 = 24\,525 \text{ kPa}$$

$$v_{2'} = \frac{RT_{2'}}{p_{02'}} = \frac{(0.287)(1260)}{24\,525}$$

$$v_{2'} = 0.0147 \text{ m}^3/\text{kg}$$

18.11 FLOW MEASUREMENT

The principles we have covered may be applied to flow-measuring devices, such as the venturi meter, shown in Figure 18.11. The fluid pressure decreases as the fluid accelerates to the throat, the pressure at the throat, p_2, being a minimum. The velocity may be determined by knowing the pressure change and the flow rate, determined from $\dot{m} = \rho A v$. The throat contraction is small, so the pressure change is also small. If the flow is compressible, then the velocity change may be found as follows. If a density change occurs and must be accounted for, then the

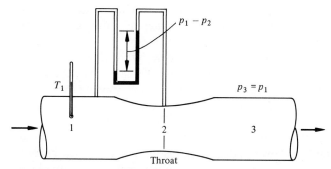

Figure 18.11   A schematic diagram illustrating a venturi flow-measuring device.

equation $pv^k = c$ can be used to determine the functional relationship in the following integral.

The conservation of energy applied between 1 and 2 yields for isentropic flow

$$h_1 + (K.E.)_1 = h_2 + (K.E.)_2$$

$$\frac{v_2^2}{2} = \frac{v_1^2}{2} + \int_1^2 v \, dp$$

For incompressible flow, $v = $ constant,

$$v_2 = \left[v_1^2 + 2v_1(p_2 - p_1) \right]^{1/2} \text{ m/s} \tag{18.38}$$

The velocity, specific volume, and area being known, the mass flow rate may be calculated.

There are other flow-measuring devices, such as orifices, but we will not include these, nor will we discuss the empirical coefficients that must be included in any study of flow-measuring devices. Most fluid dynamics texts cover this area in greater detail.

18.12 WIND POWER

The use of wind power as a power source appears attractive for several reasons. First, wind power is a renewable resource, for which the required technology has already been developed. Second, there is no air, water, or thermal pollution associated with it. Third, weather modification due to wind utilization is negligible. The drawbacks to wind power utilization are relatively low efficiencies, high capital costs, and noise and aesthetic problems.

Wind power is not a reliable energy source, the output being a function of the wind velocity. Since wind velocity fluctuations do not normally coincide with power requirements variations, it is necessary that most wind power systems include energy storage systems.

There are several types of wind-driven machines, operating on several different principles. Although each type has its advantages, the one type that stands out as the most promising is the horizontal-axis, two-bladed propeller wind turbine, illustrated in Figure 18.12.

A wind turbine should have the following characteristics: the ability to maintain optimum alignment with the wind; a low starting torque; the ability to endure high winds; and, if used to drive a generator, a high rotational speed.

The propeller-type windmill is almost always either a two- or three-bladed design; the two-bladed design is more widely used because it is strong, simple, and less expensive. The horizontal axis windmill can be positioned so that the blades lie upwind or downwind of the tower. The downwind design is usually preferred for larger machines, where a tail vane is not practical. On smaller models, a tail vane keeps the blades pointing into the wind. Large windmills are usually steered by a pilot wind vane, coupled to the drive gear, which operates to keep the windmill in constant alignment with the wind. The pilot wind vane is more

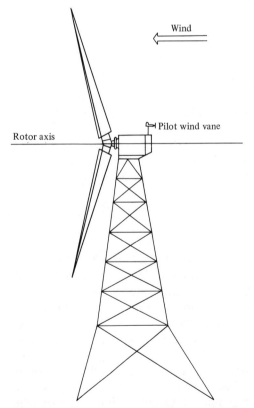

Figure 18.12   A two-bladed horizontal windmill.

sensitive to wind shifts than the large windmill. At the cutout wind speed, the blades are turned edgewise to the wind to protect the machine against high wind damage.

Let us develop the equations governing the windmill. Figure 18.13 illustrates a windmill propeller located in a moving fluid.

When the fluid between planes 1 and 4 is isolated, the only force acting is that exerted by the fluid on the propeller. The force acting on the windmill is equal to the pressure drop across the blades.

$$F = (p_2 - p_3)A \qquad (18.39)$$

The force is also equal to

$$F = \dot{m}(v_1 - v_4)$$

If we let v be the mean velocity across the blades and ρ be the air density,

$$F = \rho v A(v_1 - v_4) \qquad (18.40)$$

Combining Equation (18.39) and (18.40) yields

$$p_2 - p_3 = \rho v(v_1 - v_4) \qquad (18.41)$$

Applying the first law to the flow between planes 1 and 2 yields

$$u_1 + p_1 v_1 + \tfrac{1}{2}v_1^2 = u_2 + p_2 v_2 + \tfrac{1}{2}v_2^2 \qquad (18.42)$$

The temperature is constant; hence $u_1 = u_2$ and

$$p_1 + \tfrac{1}{2}\rho_1 v_1^2 = p_2 + \tfrac{1}{2}\rho_2 v_2^2 \qquad (18.43a)$$

Similarly for planes 3 and 4,

$$p_3 + \tfrac{1}{2}\rho_3 v_3^2 = p_4 + \tfrac{1}{2}\rho_4 v_4^2 \qquad (18.43b)$$

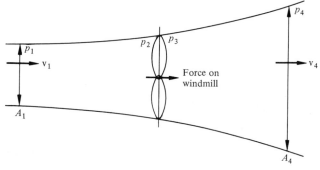

Figure 18.13   Windmill propeller in a flow field.

For a windmill operating in an unconfined fluid the pressures p_1 and p_4 are equal. Since this is so, Equations (18.42) and (18.43) may be combined, on the assumption that the density is constant, as follows:

$$p_2 - p_3 = \tfrac{1}{2}\rho(v_1^2 - v_4^2) \qquad (18.44)$$

Combining Equation (18.41) and (18.44) yields

$$v = \frac{v_1 + v_4}{2} \qquad (18.45)$$

The mean velocity across the propeller blades is equal to the average of the upstream and downstream velocities, measured at some distance from the windmill. Thus, the velocity drop through the propeller is the same ahead as behind.

The windmill efficiency is the ratio of the power output to the total power available in the airstream of area A and velocity v. This was derived in Chapter 3.

$$\text{Power available} = \frac{\dot{m}v_1^2}{2}$$

$$\text{Power output} = \frac{(v_1^2 - v_4)^2 A v \rho}{2}$$

$$\eta = \frac{(v_1 + v_4)(v_1^2 - v_4^2)}{2v_1^3} \qquad (18.46)$$

The maximum efficiency is found by differentiating η with respect to (v_4/v_1) and setting the result equal to zero. Let

$$y = \frac{v_4}{v_1}$$

Then

$$\eta = \frac{(v_1 + yv_1)(v_1^2 - y^2 v_1^2)}{2v_1^3} = \frac{(1 + y)(1 - y^2)}{2} \qquad (18.47)$$

$$\frac{d\eta}{dy} = 0 = 3y^2 + 2y - 1$$

The only physically possible solution is

$$y = \tfrac{1}{3} \qquad (18.48)$$

This results in a value of $v_4/v_1 = \frac{1}{3}$, which when substituted in Equation (18.47) yields a maximum efficiency of 59.3 percent. This is the maximum percentage of energy that can be utilized from the available energy, the inlet wind's kinetic energy.

PROBLEMS

1. Helium is flowing in a pipeline at a velocity of 1000 m/s and with a pressure of 120 kPa and a temperature of 300 K. Determine the stagnation temperature and the isentropic stagnation pressure.

2. Helium is flowing in a duct with a velocity of 350 m/s and pressure and temperature of 100 kPa and 25°C. Determine the Mach number and the stagnation pressure and temperature.

3. Air flows in a 200-cm² duct at a velocity of 180 m/s, a pressure of 119 kPa, and a temperature of 28°C. Determine the temperature, pressure, and area required when the flow becomes sonic.

4. Air in a converging nozzle has the following stagnation conditions: $p_0 = 200$ kPa, $T_0 = 370°$K. If the back pressure is 100 kPa, find the exit Mach number and velocity.

5. Air flows in a converging-diverging nozzle. The throat area is 90 cm² and the exit area is 93.4 cm². The stagnation pressure is 150 kPa and the stagnation temperature is 450 K. The nozzle discharges into air at 100 kPa. Determine (a) the exit velocity and Mach number; (b) whether or not the flow is choked.

6. Air is flowing through a nozzle at a rate of 3 kg/s. Inlet conditions are 1400 kPa, 890 K, and 100 m/s, and the exit pressure is 300 kPa. For increments of 100 kPa, plot the variation of area, specific volume, and temperature through the nozzle. Use the air tables.

7. Air is flowing through a nozzle with inlet conditions of 600 kPa and 1200 K and an exit condition of 150 kPa. The nozzle efficiency is 96%. Determine (a) the exit velocity; (b) the exit Mach number; (c) the flow rate, if the throat area is 6.45×10^{-4} m².

8. At 630 kPa and 1200 K, 27.2 kg/s of helium flows through the inlet nozzles of a gas turbine. The exit pressure is 280 kPa. The throat diameter of each circular nozzle is 1.5 cm. Determine (a) the critical pressure; (b) the minimum number of nozzles required; (c) the force on the nozzles.

9. At a rate of 5 kg/s, carbon dioxide flows through an adiabatic nozzle that has an efficiency of 95%. The inlet velocity is 150 m/s, inlet temperature is 600 K, and inlet pressure is 1.4 MPa. The exit pressure is 350 kPa. Determine (a) the exit temperature; (b) the exit area; (c) the critical temperature; (d) the throat area; (e) the exit Mach number.

10. Steam flows isentropically through a nozzle from $v = 200$ m/s, $T = 370°C$, and $p = 3.5$ MPa. Determine (a) the isentropic stagnation pressure; (b) the exit specific volume; (c) the exit velocity; (d) the critical temperature if the exit pressure is 700 kPa.

11. The high-pressure turbine in a power plant receives 25.0 kg/s of steam at 420°C and 3.5 MPa. The steam exits the first nozzle block at 2.0 MPa and enters the blades. Determine (a) the enthalpy of steam exiting the nozzle; (b) the velocity of steam leaving the nozzle; (c) the total force on the nozzle block; (d) the minimum number of nozzles with a 1-cm exit diameter required.

12. A nozzle receives 10 kg/s of steam at 4.0 MPa and 260°C and discharges it at 1.4 MPa. The nozzle efficiency is 97%. Determine (a) the throat diameter for equilibrium and supersaturated flow; (b) the exit velocity; (c) the exit area.

13. Steam at 3.8 MPa and 260°C enters a nozzle with a throat area of 4.5 cm². The nozzle discharges at 1.0 MPA. Determine (a) the mass flow rate; (b) the exit steam quality; (c) the exit area; (d) the specific volume at the throat.

14. Methane flows through an ideal nozzle with the following inlet conditions: Static pressure is 700 kPa, static temperature is 300 K, and velocity is 125 m/s. The nozzle discharges into a static pressure of 550 kPa. Determine (a) the exit static temperature; (b) the exit specific volume; (c) the exit velocity.

15. Steam enters an adiabatic nozzle at 1.4 MPa, 260°C, and negligible velocity, and expands to 140 kPa and a quality of 97%. Determine (a) the steam exit conditions; (b) the nozzle efficiency.

16. A circular converging-diverging nozzle handles 0.15 kg/s of air from 20°C and 1.0 MPa. The nozzle has an efficiency of 95%. Determine (a) the exit velocity; (b) the exit temperature; (c) the length from the throat for $\theta = 5$ degrees. The discharge pressure is 100 kPa.

17. A mixture (by volume) of 50% hydrogen and 50% carbon dioxide is flowing isentropically through a nozzle with a flow rate of 1 m³/s measured at inlet conditions of 700 kPa and 373 K. There is negligible inlet velocity. Determine (a) the specific volume at inlet in m³/kg; (b) the temperature after expanding to 150 kPa; (c) the exit diameter; (d) the mass flow rate in kg/s.

18. A gaseous mixture of 75% CO_2, 25% He, on a molal basis, flows through a 30-cm I.D. pipe with a velocity of 150 m/s at a pressure of 600 kPa and a temperature of 115°C. The mixture velocity is to be increased to 350 m/s by reducing the pipe diameter. During the process the static temperature remains constant at 115°C by heat transfer. Determine (a) the heat transfer to or from the gas; (b) the static pressure where the velocity is 350 m/s; (c) the diameter where the velocity is 350 m/s.

19. An ideal gas in a rocket has the following conditions:

1.	Nozzle inlet chamber pressure	2800 kPa
2.	Nozzle exit pressure	28 kPa
3.	k for the gas	1.2
4.	Molecular weight	21.0
5.	Nozzle inlet chamber temperature	2500 K

Determine (a) the critical pressure ratio; (b) the velocity at the throat; (c) the exit temperature; (d) the exit velocity; (e) the ratio of exit area to throat area.

20. A Pitot tube is installed in a 76-mm I.D. pipe to determine the water flow rate by measuring the velocity head. The water temperature is 20°C. The manometer reads 38 mm of mercury. Determine the water flow rate in kilograms per second.

21. Dry saturated steam enters a diffuser at 14 kPa and an unknown velocity. It exits at 42 kPa with negligible velocity. Determine (a) the discharge temperature for isentropic flow; (b) the discharge temperature if $\eta_d = 0.80$.

22. Air at 140 kPa and 95°C enters an ideal diffuser with a Mach number of 3.5 and exits at $M = 1.0$. The air flow is 12.6 kg/s. Determine (a) the exit pressure; (b) the exit temperature; (c) the exit area; (d) the minimum area.

23. Carbon monoxide enters a diffuser at 100 kPa and 60°C with a Mach number of 3.0. The diffuser efficiency is 85 percent, the flow rate is 10 kg/s, and the exit velocity is negligible. Determine (a) the exit pressure; (b) the exit temperature; (c) the minimum area.

24. A venturi flow meter is used in a plant to measure water flowing through a 30-cm pipe. The throat diameter of the venturi is 23 cm and the pressure drop is 15 cm of water. Determine the mass flow rate.

25. Air is flowing through a 80-mm I.D. pipe at 100 kPa and 50°C. The venturi diameter is 40 mm and a pressure drop of 280 mm of water is recorded across the venturi. Determine the volume flow rate of air.

26. Air with a velocity of 225 m/s enters an isentropic diffuser with an inlet area of 0.15 m². The inlet static temperature and pressure are 255 K and 63 kPa, respectively. The exit velocity is 100 m/s. Determine (a) the air flow rate; (b) the static and total exit temperature; (c) the static and total exit pressure; (d) the exit area.

27. A wind turbine with a 61-m blade diameter produces 2000 kW when the wind velocity is 40 km/h. The air temperature is 25°C and the pressure is 101 kPa. Determine the wind turbine efficiency.

28. For the turbine in Problem 27 determine the velocity leaving the blades.

29. A wind turbine has a start-up wind speed of 17 km/h and cuts out at a wind velocity of 57 km/h. The wind turbine has a blade diameter of 30 m. Consider the wind turbine to have a constant efficiency of 40%. For a 24-hour period, the following wind speeds, in kilometers per hour, are available. The air density is constant at 1.181 kg/m^3.

Time Period	Average Wind Speed
0000–0400	10
0400–0800	18
0800–1200	25
1200–1600	20
1600–2000	33
2000–2400	30

Determine the power produced.

Chapter 19
Heat Transfer and
Heat Exchangers

The purpose of this chapter is to introduce the concept of heat transfer and apply it to some special cases. The field of heat transfer is quite extensive, and this chapter is not meant to replace study in the field, but rather to introduce you to the various modes of heat transfer. In equilibrium thermodynamics we have used heat flow in the analysis of problems, but we have not discussed how the heat transfer occurs physically. We have dealt with different thermodynamic states and have observed the property changes, and from this deduced the heat transfer. Heat flow is a transient problem, so its analysis will involve more than investigation of equilibrium states. The laws of heat transfer do obey the first and second laws of thermodynamics; energy is conserved, and heat must flow from hot to cold. The purpose of including rudiments of heat-exchanger thermal analysis is that heat exchangers have formed an essential element in many of the systems we have analyzed—power plants, refrigeration units, gas compressors—and by the time this basic course in thermal analysis is finished, you should have an understanding of the selection process for heat exchangers.

19.1 MODES OF HEAT TRANSFER

There are three modes of heat transfer—*conduction, radiation, and convection*. We will define them as follows.

1. Conduction is the heat transfer within a medium. In solids, particularly metals, conduction is due to (a) the drift of free electrons and (b) phonon vibration. At low temperatures, phonon vibration, the vibration of crystalline structure, is the primary mechanism for conduction, and at higher temperatures electron drift is the primary mechanism. Regardless of the mechanism, energy is transferred from one atom or

molecule to another, resulting in a flow of energy within a medium. In gas, the mechanism for conduction is primarily molecular collision. The conduction is dependent on the pressure and temperature, which act in obvious ways to increase the chance of molecular collisions. In liquids, the mechanism for conduction is the combination of electron drift and molecular collision. Conduction in liquids is temperature, not pressure, dependent.

2. Radiation is thermal energy flow, via electromagnetic waves, between two bodies separated by a distance. Electromagnetic waves, which are a function of body-surface temperature, transfer heat and thus constitute thermal radiation.

3. Convection is the heat transfer between a solid surface and a fluid. This is a mixed mode, in that at the solid–fluid interface heat is transferred by conduction, molecular collisions between the solid and fluid molecules. As a result of these collisions a temperature change in the fluid occurs, a density variation is produced, and a bulk fluid motion occurs. There is a mixing of the high- and low-temperature fluid elements, and heat is transferred between the solid and fluid by convection.

19.2 LAWS OF HEAT TRANSFER

There are laws for radiation and conduction heat transfer. These are based on experiments and, as with all laws, cannot be proved; but because nothing contradicts them, they are assumed to be valid. The expression for convective heat flow is not a law, but rather an empirical equation.

Conduction

The law for conduction heat transfer is called "Fourier's law," named for the man who proposed it. It states that the conductive heat flow, q_λ, is a product of (a) the thermal conductivity of the material, λ; (b) the area normal to the heat flow, A; and (c) the temperature gradient, dT/dx, across the area.

The thermal conductivity is a property of the material, like specific heat, and is a measure of how well a material conducts heat. Fourier's law for one-dimensional flow in rectangular coordinates is

$$q_\lambda = -\lambda A \frac{dT}{dx} \quad \text{W} \tag{19.1}$$

The units on each term are

λ: $\text{W/m} \cdot \text{K}$
A: m^2
dT/dx: K/m

The reason for the minus sign on Equation (19.1) is that we want the heat flow to be positive. Since the gradient is negative in the $+x$ direction—hot to cold, hence decreasing—the minus sign corrects for this. The values for the thermal conductivity depend on the molecular structure and are highest for a solid phase, the most compact phase structure, less for the liquid phase, and still less for the vapor phase. Table A.13 lists the properties of many types of materials.

Before Equation (19.1) can be solved, we must know whether the heat flow is constant with time, steady state, or whether it varies with time, transient or unsteady-state heat flow. We will only consider steady-state heat transfer in this chapter. Let us consider a plane wall, shown in Figure 19.1, with constant temperatures on each surface. The wall is a distance L thick. The variables, T and x, may be separated in Equation (19.1)

$$q_\lambda \, dx = -\lambda A \, dT$$

We integrate from $x = 0$ and $T = T_h$, to $x = L$ and $T = T_c$,

$$\int_0^L q_\lambda \, dx = - \int_{T_h}^{T_c} \lambda A \, dT = \int_{T_c}^{T_h} \lambda A \, dT$$

A, λ, and q_λ are constant, so

$$q_\lambda = \lambda A \frac{(T_h - T_c)}{L} = \frac{T_h - T_c}{R_\lambda} \tag{19.2}$$

where

$$R_\lambda = \frac{L}{\lambda A} \tag{19.3}$$

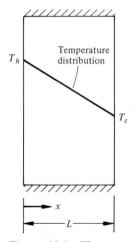

Figure 19.1   *The temperature distribution in a plane wall for steady-state heat transfer.*

is called the *conductive resistance*. This is developed to illustrate the similarity between electric current flow being equal to a voltage potential divided by an electrical resistance, and heat flow being equal to a temperature potential divided by a thermal resistance. We were able to integrate Equation (19.1) because the problem is steady state and thus is constant; it is a plane wall, so A is constant; and the thermal conductivity did not vary with temperature, so it is constant. The thermal conductivity does vary with temperature, as might be suspected, but we will assume it to be constant and compensate for the variation by using values averaged over the temperature range.

Example 19.1

A plane wall is 15 cm thick with an area of 1 m^2 and has a thermal conductivity of 0.5 W/m · K. A temperature difference of 55°C is imposed across it. Determine the heat flow and the conductive resistance.

$$q_\lambda = \lambda A \frac{\Delta T}{L} = \frac{(0.5)(1)(55)}{(0.15)} = 183.3 \text{ W}$$

$$R_\lambda = \frac{L}{\lambda A} = \frac{0.15}{(0.5)(1)} = 0.3 \text{ K/W}$$

Radiation

The law for radiative heat transfer was discovered by two men: J. Stefan determined the law experimentally and L. Boltzmann derived it theoretically from statistical mechanics. The law is that the radiant heat flow, q_r, for a blackbody is proportional to the surface area, A, times the absolute surface temperature to the fourth power. A proportionality constant, the Stefan-Boltzmann constant, σ, makes the proportionality an equation. Thus

$$q_r = \sigma A T^4 \text{ W} \tag{19.4}$$

A blackbody, or black surface, is one that absorbs all the radiation incident upon it. The value of the Stefan-Boltzmann constant is 5.67×10^{-8} W/m^2 · K^4.

Equation (19.4) describes the emission of radiation from a black surface; however, it does not indicate what the net radiative heat transfer will be between two surfaces. Consider that surface 1 at temperature T_1 is completely enclosed by another black surface, surface 2, at temperature T_2. The net radiant heat transfer is

$$q_r = \sigma A_1 \left(T_1^4 - T_2^4 \right) \tag{19.5}$$

where all temperatures are absolute. Unfortunately, real bodies, surfaces, are not perfect absorbers and radiators, but emit, for the same surface temperature, a fraction of the blackbody radiation. This fraction is called the emittance, ϵ:

$$\epsilon = \frac{\text{actual surface radiation at } T}{\text{black surface radiation at } T} \tag{19.6}$$

The actual surfaces are called gray surfaces. Thus, the net rate of heat transfer between a gray surface at temperature T_1 to a surrounding black surface at temperature T_2 is

$$q_r = \sigma A_1 \epsilon_1 \left(T_1^4 - T_2^4 \right) \tag{19.7}$$

Table A.13 lists emittances for various surfaces.

The enclosure being total and a black surface may be modified by the modulus, $\mathscr{F}_{1-2}$, which accounts for the relative geometries of the surfaces (not all radiation leaving 1 reaches 2) and the surface emittances. Thus Equation (19.7) becomes

$$q_r = \sigma A_1 \mathscr{F}_{1-2} \left(T_1^4 - T_2^4 \right) \tag{19.8}$$

Radiant heat transfer frequently occurs with other modes of heat transfer, and the use of a radiative resistance R_r is helpful. Let us define q_r to also be

$$q_r = \frac{T_1 - T_{2'}}{R_r} \tag{19.9}$$

where $T_{2'}$ is an arbitrary reference temperature. The radiative resistance is found by combining Equations (19.9) and (19.8) to obtain

$$R_r = \frac{T_1 - T_{2'}}{\sigma A_1 \mathscr{F}_{1-2} \left(T_1^4 - T_2^4 \right)} \tag{19.10}$$

Example 19.2

A gray body with a constant surface temperature of 700°C and a surface area of 500 cm² radiates in a large room, whose surfaces are black and maintained at 100°C. If the gray surface emittance is 0.6, determine the radiative heat transfer betwen the surface and the room.

$$q_r = (5.67 \times 10^{-8})(0.05)(0.6)(973^4 - 373^4)$$
$$q_r = 1491.6 \text{ W}$$

Convection

Convection heat transfer is a combination of conduction, fluid flow, and mixing. There are two types of convection: free convection, where density changes cause the bulk fluid motion, and forced convection, where a pump or fan provides the motive force for the fluid's motion. The expression for convective heat flow is not a law, but rather an empirical equation. The convective heat flux, q_c, is the product of three terms.

1. The solid–fluid surface area, A
2. The temperature difference between the solid surface, T_s, and the fluid temperature far from the surface, T_∞
3. The average unit convective coefficient, $\bar{h}_c$

$$q_c = \bar{h}_c A (T_s - T_\infty) \text{ W} \tag{19.11}$$

Equation (19.5) rigorously defines $\bar{h}_c$, but this is the equation conventionally used to describe convective heat flow. The units on the terms are

$\bar{h}_c$: $\text{W}/(\text{m}^2 \cdot \text{K})$
A: m^2
T: K

Equation (19.11) may also be expressed in terms of a convective resistance, R_c.

$$q_c = \frac{T_s - T_\infty}{R_c} \tag{19.12}$$

Comparing Equations (19.12) and (19.11) yields an expression for R_c.

$$R_c = \frac{1}{\bar{h}_c A} \tag{19.13}$$

In calculating the heat flux the term that may be difficult to determine is $\bar{h}_c$. This term relates the fluid's physical properties and the fluid velocity over the solid surface. These affect the rate at which thermal energy may enter or leave the fluid. Table 19.1 illustrates the wide range of values $\bar{h}_c$ may have.

The tremendous variation in values of $\bar{h}_c$ makes its selection very important in convective heat transfer analysis.

Figure 19.2 illustrates the forced convection velocity and temperature profiles of a cold fluid moving over a heated plate. There are several

Table 19.1 Typical Unit Convective Coefficient Values

Mode	$\bar{h}_c$, $\text{W}/(\text{m}^2 \cdot \text{K})$
Free convection, air	5–25
Forced convection, air	10–200
Free convection, water	20–100
Forced convection, water	50–10 000

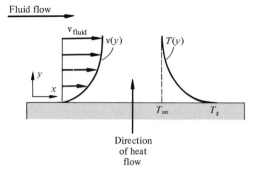

Fluid flow

Figure 19.2   Velocity and temperature profiles in convective heat transfer.

observations that may be made. The velocity decreases as it approaches the solid surface, reaching zero in the fluid layer next to the surface. The heat transfer from the solid to the fluid is by conduction, since the fluid layer has zero velocity. This must equal the heat transfer by convection into the rest of the fluid, or

$$\frac{q_c}{A} = -\lambda_f \frac{\partial T}{\partial y}\bigg|_{y=0} = \bar{h}_c(T_s - T_\infty) \tag{19.14}$$

where λ_f is the thermal conductivity of the fluid. From this expression we see that the heat flux is directly proportional to the thermal conductivity of the fluid and to the temperature gradient at the wall. The temperature gradient is directly proportional to the fluid velocity, with higher velocities allowing higher temperature gradients.

Convection heat transfer involves fluid velocity, fluid properties, and the overall convective coefficient. It is possible to relate these properties, but first dimensionless parameters are needed. The dimensionless number for velocity is the Reynolds number. For flow through a tube or duct the Reynolds number, Re_D, is defined as

$$\text{Re}_D = \frac{v\rho D}{\mu} = \frac{vD}{\nu} \tag{19.15}$$

v = average fluid velocity
ρ = fluid density
D = inside diameter
μ = fluid dynamic viscosity
ν = fluid kinemetic viscosity

Often times the duct is not circular, but rectangular, or even the annulus formed by a tube within a tube. In these cases the characteristic dimension is not the tube diameter, but an equivalent diameter, called the hydraulic diameter, D_H, and defined as

$$D_H = 4\left(\frac{\text{cross-sectional area for fluid flow}}{\text{wetted perimeter}}\right)$$

Thus, for flow through a duct with a width a and height b

$$D_H = 4\left[\frac{a \cdot b}{2(a+b)}\right] = \frac{2ab}{a+b}$$

For flow through an annulus with inside diameter D_1 and outside diameter D_2

$$D_H = 4\left[\frac{\pi(D_2^2 - D_1^2)}{4\pi(D_2 + D_1)}\right] = D_2 - D_1$$

Low Reynolds numbers, up to about 2300, are indicative of laminar flow. From 2300 to 6000 the laminar flow begins a transition to turbulent flow. The flow is usually completely turbulent at 6000. This becomes a matter for concern just a little later when different correlations are used for laminar and turbulent flow. Turbulent flow has a flatter velocity profile than laminar flow, allowing a greater average velocity and a greater temperature at the wall.

The Prandtl number, Pr, is a dimensionless number relating fluid properties.

$$\text{Pr} = \frac{\mu c_p}{\lambda} = \frac{\nu}{\alpha} \tag{19.16}$$

where

$$\alpha = \frac{\lambda}{\rho c_p}$$

The Nusselt number, Nu, is the dimensionless form of the convective coefficient.

$$\text{Nu} = \frac{\bar{h}_c D}{\lambda} \tag{19.17}$$

A great number of tests have been run to correlate these dimensionless numbers. There are many correlations and the following ones are typical. There are two flow regimes, laminar and turbulent. For laminar flow in short tubes, Seider and Tate developed the expression

$$\text{Nu} = 1.86\,(\text{Re})^{1/3}(\text{Pr})^{1/3}\left(\frac{D}{L}\right)^{1/3}\left(\frac{\mu}{\mu_s}\right)^{0.14} \tag{19.18}$$

In this expression L is the pipe length and all fluid properties are evaluated at the average fluid temperature except for μ_s. This term is evaluated at the wall surface temperature. The unit convective coefficient is dependent on the fluid viscosity, which varies with temperature. This additional viscosity term accounts for the temperature variation.

In turbulent flow the following equation may be used when the temperature difference between the tube or duct surface and the average fluid temperature is not greater than 5.5°C for liquids or 55.5°C for gases,

$$\text{Nu} = 0.023(\text{Re})^{0.8}(\text{Pr})^n \tag{19.19}$$

where $n = 0.4$ for heating and $n = 0.3$ for cooling.

In those cases where the previous temperature limits are exceeded or the fluid's viscosity is greater than that of water, use

$$\text{Nu} = 0.027(\text{Re})^{0.8}(\text{Pr})^{1/3}\left(\frac{\mu}{\mu_s}\right)^{0.14} \tag{19.20}$$

In this case all properties are evaluated at the average fluid temperature except μ_s, which is evaluated at the wall temperature.

How are these equations used? Table A.14 list properties of various substances. Equations (19.18)–(19.20) are used to solve for $\bar{h}_c$. This is used in Equation (19.12), where T_∞ is now the average fluid temperature, to calculate the heat flux.

Example 19.3

Water enters a 3-cm diameter tube with a velocity of 50 m/s and a temperature of 20°C and is heated. Calculate the average unit convective coefficient.

Calculate Re to determine the flow regime.

$$\text{Re} = \frac{(50)(0.03)}{1.006 \times 10^{-6}} = 1.491 \times 10^6 \qquad \therefore \text{ turbulent}$$

$$\text{Pr} = 7.0 \qquad \lambda = 0.597$$

Use Equation (19.19) to obtain

$$\text{Nu} = 0.023\,(1.491 \times 10^6)^{0.8}(7.0)^{0.4}$$

$$\text{Nu} = 4350$$

$$\bar{h}_c = \frac{(4350)(0.597)}{(0.03)} = 8.656 \times 10^4 \text{ W/m}^2 \cdot \text{K}$$

19.3 COMBINED MODES OF HEAT TRANSFER

Most engineering applications involve a combination of heat-transfer modes. Figure 19.3 shows a plane wall. Heat is transferred by the fluid at a high temperature to the wall by convection and radiation. Within a medium, the plane wall, heat is transferred by conduction, q_λ. At the outside wall surface heat is again transferred by convection to the surrounding fluid at a low temperature. At low temperatures radiation does not have an appreciable effect. It is possible to express the total heat flow, q, in terms of the extreme temperatures, T_h and T_l, and the resistances to heat that occur between the two temperature extremes.

On the furnace side, the total heat flow, q, is equal to

$$q = q_r + q_{ch} = \frac{T_h - T_{wh}}{R_1} \tag{19.21}$$

where

$$R_1 = \frac{(R_{ch})(R_r)}{R_{ch} + R_r}$$

In the wall, the total heat flow is equal to the conductive heat flow, q_λ; that is,

$$q = q_\lambda = \frac{T_{wh} - T_{wl}}{R_\lambda} \tag{19.22}$$

and at the outside surface

$$q = q_{cl} = \frac{T_{wl} - T_l}{R_{cl}} \tag{19.23}$$

Figure 19.4 shows the thermal circuit for the wall in Figure 19.3.

$$q = \frac{T_h - T_{wh}}{R_1}$$

$$q = \frac{T_{wh} - T_{wl}}{R_2}$$

$$q = \frac{T_{wl} - T_l}{R_3}$$

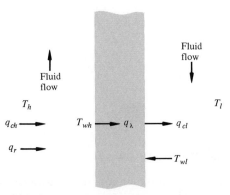

Figure 19.3 *A plane wall with convective and radiative heat transfer at the surface.*

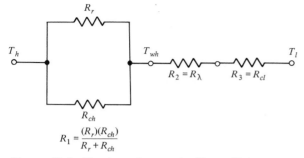

$$R_1 = \frac{(R_r)(R_{ch})}{R_r + R_{ch}}$$

Figure 19.4   The thermal circuit for Figure 19.3.

We multiply each equation by its resistance, add, and solve the resulting equation for q. This results in the following equation:

$$q = \frac{T_h - T_l}{R_1 + R_2 + R_3} = \frac{\Delta T}{\sum\limits_{i=1}^{3} R_i} \qquad (19.24)$$

Another coefficient, called the "overall coefficient of heat transfer," U, is frequently referred to in heat-transfer analyses. The equation for heat flow using the overall coefficient of heat transfer is

$$q = UA \, \Delta T \qquad (19.25)$$

and comparing the Equations (19.25) and (19.24) yields

$$UA = \frac{1}{\sum\limits_{i=1}^{3} R_i} \qquad (19.26)$$

19.4 CONDUCTION THROUGH A COMPOSITE WALL

It is possible to have conduction through more than one material. For instance, an actual furnace wall is composed of different materials in series, and parallel, such as illustrated in Figure 19.5 The material might be chrome ore, followed by three types of fire bricks, followed by the steel casing. For a steady state, the heat flow may be written in terms of the sum of thermal resistances:

$$q = \frac{T_{wh} - T_{wl}}{\sum\limits_{i=1}^{4} R_i} \qquad (19.27)$$

where

$$R_3 = \frac{R_a R_b}{R_a + R_b}$$

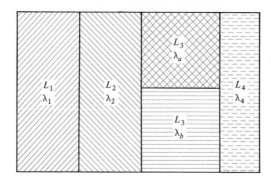

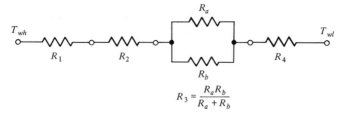

$$R_3 = \frac{R_a R_b}{R_a + R_b}$$

Figure 19.5   Composite wall with thermal circuit.

19.5 CONDUCTION IN CYLINDRICAL COORDINATES

Thus far we have dealt only with plane walls; the area has been constant in the direction perpendicular to the heat flow. In cylindrical coordinates, such as heat transfer through a pipe, the area varies with radial distance, and the conduction equation must account for this variation. Figure 19.6 illustrates a cylinder of length l. We will assume that heat flows only in the radial direction, so Fourier's law yields

$$q_\lambda = -\lambda A \frac{dT}{dr}$$

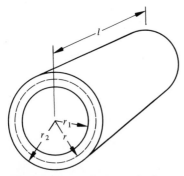

Figure 19.6   A hollow cylinder.

The area is the area normal to the heat flow, $A = 2\pi r l$, hence

$$q_\lambda = -2\pi r l \lambda \frac{dT}{dr} \tag{19.28}$$

For steady state, q_λ is constant; we may separate the variables and integrate from $r = r_1$ and $T = T_1$ to $r = r_2$ and $T = T_2$.

$$\int_{r_1}^{r_2} q_\lambda \frac{dr}{r} = \int_{T_2}^{T_1} 2\pi \lambda l \, dT$$

$$q_\lambda = \frac{2\pi \lambda l (T_1 - T_2)}{\ln(r_2/r_1)} \tag{19.29}$$

Equation (19.29) may be expressed in terms of thermal resistance, R_λ, as follows:

$$q_\lambda = \frac{T_1 - T_2}{R_\lambda}$$

where

$$R_\lambda = \frac{\ln(r_2/r_1)}{2\pi \lambda l}$$

Equation (19.29) was developed for the inside temperatures being greater than the outside temperature, with the heat flow in the positive r direction being positive. Should the outside temperature be greater, then the heat flow would be negative, indicating flow opposite to the direction assumed in the derivation of Equation (19.29). A more common occurrence is when the pipe is insulated. Typically steam pipes are insulated to prevent heat loss and refrigerant pipes are insulated to prevent heat gain. Figure 19.7 illustrates a cross-sectional sketch of an insulated pipe. The heat flow, q, may be determined for each section.

$$q = q_{ci} = h_i 2\pi r_1 l (T_i - T_1) = \frac{T_i - T_1}{R_i} \tag{19.30}$$

where

$$R_i = \frac{1}{(2\pi r_1 l h_i)}$$

$$q = q_{\lambda_1} = -2\pi l \lambda_1 r \frac{dT}{dr}$$

$$q = q_{\lambda_1} = \frac{2\pi l \lambda_1 (T_1 - T_2)}{\ln(r_2/r_1)} = \frac{T_1 - T_2}{R_{\lambda_1}} \tag{19.31}$$

where

$$R_{\lambda_1} = \frac{\ln(r_2/r_1)}{2\pi l \lambda_1}$$

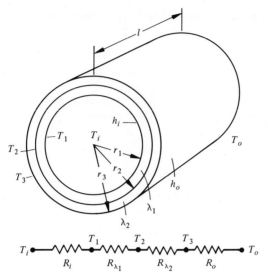

Figure 19.7   *The schematic diagram and thermal circuit for an insulated pipe.*

Similarly,

$$q = q_{\lambda_2} = \frac{2\pi l \lambda_2 (T_2 - T_3)}{\ln(r_3/r_2)} = \frac{T_2 - T_3}{R_{\lambda_2}} \tag{19.32}$$

where

$$R_{\lambda_2} = \frac{\ln(r_3/r_2)}{2\pi l \lambda_2}$$

For the outside surface,

$$q = q_{co} = 2\pi r_3 h_o l (T_3 - T_o) = \frac{T_3 - T_o}{R_o} \tag{19.33}$$

where

$$R_o = \frac{1}{2\pi r_3 h_o l}$$

The heat flow can now be calculated by summing the resistances and knowing the overall temperature difference:

$$q = \frac{T_i - T_o}{\sum\limits_{i=1}^{4} R_i}$$

Example 19.4

Saturated steam at 500 K flows in a 0.2-m I.D., 0.21-m O.D. pipe. The pipe is covered with 0.08 m of insulation with a thermal conductivity of 0.10 W/m · K. The pipe's conductivity is 52 W/m · K. The ambient temperature is 300 K. The unit convective coefficients are $h_i = 18\ 000$ W/m^2 · K and $h_o = 12$ W/m^2 · K. Determine the heat loss from 4 m of pipe. Calculate the overall coefficient of heat transfer based on the outside area.

It is easiest to calculate the heat loss per unit length of pipe, $l = 1$ m, and then correct for the total length asked for in the problem.

$$q = \frac{\Delta T}{\sum\limits_{i=1}^{4} R_i}$$

$$R_1 = \frac{1}{2\pi r_1 h_i} = 0.000\ 088$$

$$R_2 = \frac{\ln(r_2/r_1)}{2\pi\lambda_1} = 0.001\ 49$$

$$R_3 = \frac{\ln(r_3/r_2)}{2\pi\lambda_2} = 0.9014$$

$$R_4 = \frac{1}{2\pi r_3 h_o} = 0.0716$$

$$q = \frac{500 - 300}{0.9733} = 205.2\ \text{W}/m$$

$$Q = lq = 4(205.2) = 820.8\ \text{W}$$

$$q = UA\ \Delta T \qquad UA = \frac{1}{\sum R_i}$$

$$UA = 1.026$$

$$U = \frac{1.026}{\pi dl} = \frac{1.026}{\pi(0.37)(1)} = 0.8826\ \text{W}/m^2 \cdot \text{K}$$

19.6 CRITICAL INSULATION THICKNESS

We might intuitively believe that adding insulation to a pipe would decrease the heat flow from the pipe. While this is usually the case, it is not always true for small-diameter pipes and wires. Consider a cylinder with an inside radius r_i at temperature T_i, an outside radius r_o, and a thermal conductivity λ in a room where the ambient temperature is T_o.

There are two resistances to heat transfer, the conductive resistance from r_i to r_o, and the convective resistance at the outside surface. Therefore

$$q = \frac{2\pi\lambda l(T_i - T_o)}{\ln(r_o/r_i) + \lambda/\bar{h}_o r_o} \tag{19.34}$$

Thus for a fixed inside radius and temperature potential the heat flow will be a function of the two thermal resistances. As r_o increases, one term increases logarithmically, while the other decreases linearly. The sum of these two terms is the total resistance, so a given change in radius may increase or decrease the denominator of Equation (19.34). It is possible to develop an expression for the maximum heat transfer as a function of the outside radius. The outside radius that yields this maximum heat flow is called the "critical radius." For wires, it may be desirable to maximize the heat flow from the wire so as to minimize line voltage drops. This may be accomplished by adding enough insulation to the wire so that the outside radius is the critical radius. This is called the "critical insulation thickness." If the object is to prevent heat loss, then the outside radius must be greater than the critical radius. Economics determines how much insulation will be added in this case. This typically occurs in refrigeration systems.

To find the critical radius, we take the derivative of q with respect to r_o in Equation (19.34) and set it equal to zero. Solve for the value of r_o. This value of r_o is the critical radius, r_{oc}.

$$r_{oc} = \frac{\lambda}{h_o} \tag{19.35}$$

Thus the critical radius, r_{oc}, is a function of the outside convective coefficient and the thermal conductivity of the insulation. The critical radius is usually less than about 2.5 cm, so for pipes or wires with a radius greater than the critical radius, the critical insulation thickness is not a physically obtainable phenomenon.

19.7 HEAT EXCHANGERS

Heat exchangers are commonly used in most thermal systems. They allow the transfer of heat from one fluid to another. There are direct-contact heat exchangers, in which the same substance, at two different states, is mixed. More common is the heat exchanger in which one fluid flows through tubes and the other flows over the tubes. There are a number of tube configurations; we shall analyze a few, but the principles of analysis apply to all.

Why analyze heat exchangers at all? In the thermal system we will need a heat exchanger that will allow the transfer of a given amount of heat. How large a heat exchanger will this be? The thermal analysis will provide the answer. We will be able to determine the heat exchanger size.

There are two subsequent steps, mechanical design and manufacturing design; they, however, are not within the scope of this text. After the thermal analysis, we can use manufacturer's catalogs to select the heat exchanger that satisfies our requirements.

It is desirable to have an equation that will denote the heat transferred in a heat exchanger in terms of the overall coefficient of heat transfer, U, the total tube surface area, A, and some temperature difference between the inlet and outlet states, $\overline{\Delta T}$. Consider a parallel-flow, tube-in-tube heat exchanger, with its temperature distribution, shown in Figure 19.8

Let us examine a differential length of heat exchanger, with differential area, dA. The heat transferred across this differential area can be expressed in three equivalent ways: the heat lost by the hot fluid, the heat gained by the cold fluid, or the heat transferred in the heat exchanger. Thus

$$q = -\dot{m}_h c_{ph}\, dT_h = \dot{m}_c c_{pc}\, dT_c = U\, dA(T_h - T_c) \qquad (19.36)$$

In Equation (19.36) the negative sign allows the heat loss by the hot fluid to be positive in accordance with the other terms denoting a positive quantity; ΔT at any dA is the difference between the hot and cold fluid temperatures at that differential area. Let

$$(19.37)$$

$$(19.38)$$

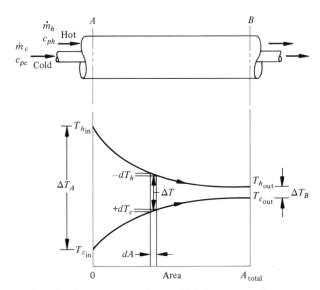

Figure 19.8   A tube-in-tube parallel-flow heat exchanger with temperature-area diagram.

Rearrange Equation (19.36) as follows:

$$\frac{UdA}{C_h} = \frac{-dT_h}{T_h - T_c}$$

$$\frac{UdA}{C_c} = \frac{dT_c}{T_h - T_c}$$

Adding these two equations, we obtain

$$\frac{UdA}{C_h} + \frac{UdA}{C_c} = \frac{dT_c - dT_h}{T_h - T_c}$$

Let

$$\theta = T_h - T_c$$

Then

$$\frac{UdA}{C_h} + \frac{UdA}{C_c} = \frac{-d\theta}{\theta} \tag{19.39}$$

Let us assume that U is constant, or an effective value, across the entire heat exchanger and integrate Equation (19.39) across the heat exchanger, from the A end to the B end.

$$\left(\frac{U}{C_h} + \frac{U}{C_c}\right)\int_A^B dA = -\int_A^B \frac{d\theta}{\theta}$$

$$A_A = 0 \qquad \theta_A = \Delta T_A$$

$$A_B = A \qquad \theta_B = \Delta T_B$$

$$\left(\frac{UA}{C_h} + \frac{UA}{C_c}\right) = \ln\left(\frac{\Delta T_A}{\Delta T_B}\right) \tag{19.40}$$

$$q = UA\,\Delta T = C_h\left(T_{h_{in}} - T_{h_{out}}\right) = C_c\left(T_{c_{out}} - T_{c_{in}}\right)$$

$$UA = \frac{C_h\left(T_{h_{in}} - T_{h_{out}}\right)}{\overline{\Delta T}} = \frac{C_c\left(T_{c_{out}} - T_{c_{in}}\right)}{\overline{\Delta T}}$$

We substitute into Equation (19.40) and solve for $\overline{\Delta T}$,

$$\overline{\Delta T} = \text{LMTD} = \frac{\Delta T_A - \Delta T_B}{\ln(\Delta T_A/\Delta T_B)} \tag{19.41}$$

LMTD is the log mean temperature difference, and thus the total heat transferred in a tube-in-tube heat exchanger for parallel flow is

$$q = UA\,\text{LMTD} \tag{19.42}$$

Equation (19.42) is usually used to calculate the total surface area, hence the tube length, of the heat exchanger. The heat transferred may be calculated from the energy gain or loss of one of the fluids. Before we apply these tools of analysis, let us examine two other heat exchanger types, the counterflow and the shell and tube.

Figure 19.9 illustrates a counterflow heat exchanger with the fluid temperature distributions. Figure 19.10 illustrates a shell and tube heat exchanger with the fluid temperature distributions. In the shell and tube and parallel-flow heat exchangers, it is not possible to have $\Delta T_A = \Delta T_B$ unless there is no heat transfer. However, in the counterflow heat exchanger it is possible to have heat exchange with $\Delta T_A = \Delta T_B$. This means the log mean temperature difference is undefined. How to find the heat exchanged? Physically, $\Delta T_A = \Delta T_B$ means that the temperature difference throughout the heat exchanger is constant, or

$$q = UA\,\overline{\Delta T} = UA\,\Delta T_A \tag{19.43}$$

Let us examine the counterflow heat exchanger in a little more detail. Note that as the area becomes greater, the difference between the hot and cold fluid temperature distribution becomes smaller and, in the limit of infinite area, the lines are coincident. This means that the hot fluid is cooled to the cold-fluid inlet temperature and the cold fluid is heated to the hot-fluid inlet temperature. This complete exchange of heat is possible only in the counterflow heat exchanger. What thermodynamic principle makes this possible? The heat exchanger that has the more

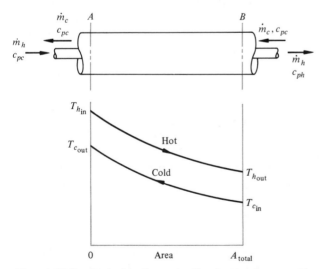

Figure 19.9 *A tube-in-tube counterflow heat exchanger with a temperature-area diagram.*

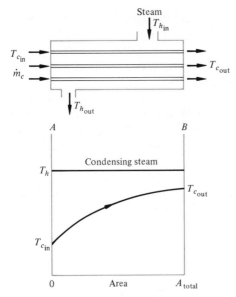

Figure 19.10 A shell and tube heat exchanger with temperature-area diagram for the case of condensing steam on the shell side.

reversible heat exchange is more efficient and transfers more heat for a given area of tube surface. Constant-temperature heat exchange is the reversible case, so the heat exchanger that can, in the limit, achieve this is thermodynamically more efficient. In the limit the counterflow exchangers transfer heat at constant temperature. No other heat exchanger approaches this state, so the counterflow is more efficient. An example illustrates this point also.

Example 19.5

It is desired to cool 0.63 kg/s of oil, $c_p = 3.35$ kJ/kg · K from 125°C to 65°C. Water, $c_p = 4.18$ kJ/kg · K, is available with a flow rate of 0.57 kg/s at a temperature of 10°C. The overall coefficient of heat transfer is 85 W/m² · K. Determine the length of 3-cm I.D. tubing required for counterflow and for parallel-flow heat exchange.

$$q = UA\overline{\Delta T}$$

$\overline{\Delta T} =$ LMTD for tube-in-tube counterflow heat exchangers. Determine the water outlet temperature by a first-law analysis. From this we can determine ΔT_A and ΔT_B.

$$q_{H_2O} = q_{oil}$$

$$\dot{m}c_p(\Delta T)_{H_2O} = \dot{m}_o c_{po}(\Delta T)_{oil}$$

$$(\Delta T)_{H_2O} = \frac{(0.63)(3.35)(60)}{(0.57)(4.18)} = 53.1$$

$$\left(T_{H_2O}\right)_{out} = 63.1°C$$

$$\Delta T_B = 65 - 10 = 55$$

$$\Delta T_A = 125 - 63.1 = 61.9$$

$$\text{LMTD} = \frac{\Delta T_A - \Delta T_B}{\ln(\Delta T_A/\Delta T_B)} = \frac{61.9 - 55}{\ln(61.9/55)} = 58.4$$

$$q = \dot{m}c_p(\Delta T)_{oil} = (0.63)(3.35)(60) = 126.63 \text{ kW}$$

$$A = \frac{q}{(U)(\text{LMTD})} = \frac{126\,630}{(85)(58.4)} = 25.5 \text{ m}^2$$

$$A = \pi\,dL \qquad L = \frac{25.5}{\pi(0.03)} = 270.5 \text{ m}$$

For parallel flow,

$$\Delta T_A = 125 - 10 = 115$$

$$\Delta T_B = 65 - 63.1 = 1.9$$

$$\text{LMTD} = \frac{115 - 1.9}{\ln(115/1.9)} = 27.5$$

$$A = \frac{q}{(U)(\text{LMTD})} = \frac{126\,630}{(85)(27.5)} = 54.1 \text{ m}^2$$

$$L = \frac{A}{\pi d} = \frac{(54.1)}{(\pi)(0.03)} = 574.0 \text{ m}$$

The parallel-flow configuration requires more than twice the area to achieve the same heat transfer.

Correction Factor

Thus far the heat exchangers have had a simple geometry. When the construction becomes more complex, then $\overline{\Delta T} \neq \text{LMTD}$, and the log mean temperature difference must be modified by a correction factor F,

$$\overline{\Delta T} = (\text{LMTD})(F) \tag{19.44}$$

$$q = UA\overline{\Delta T} \tag{19.45}$$

All that remains is to find a means for calculating the correction factor. Figures 9.11 and 19.13 show the correction factor for more complex shell and tube constructions illustrated schematically by Figures 19.12 and 19.14. By evaluating the two dimensionless numbers, P and Z, the correction factor F may be found from the appropriate chart. There are other charts for other exchangers, but we will consider only the shell and tube type, as that is the most common, or one of the most common, types of heat exchangers.

The first dimensionless parameter, P, is

$$P = \frac{T_{t_{\text{out}}} - T_{t_{\text{in}}}}{T_{s_{\text{in}}} - T_{t_{\text{in}}}} \tag{19.46}$$

where the subscripts t and s denote the shell and tube conditions. This is a measure of the transfer effectiveness: The numerator is an indication of the actual heat transferred, while the denominator indicates the maximum heat transfer possible.

The second dimensionless parameter, Z, is

$$Z = \frac{\dot{m}_t c_{pt}}{\dot{m}_s c_{ps}} = \frac{T_{s_{\text{in}}} - T_{s_{\text{out}}}}{T_{t_{\text{out}}} - T_{t_{\text{in}}}} \tag{19.47}$$

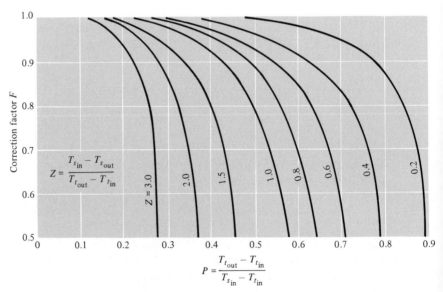

Figure 19.11 *Correction factor for heat exchanger with one shell pass and two, or a multiple of two, tube passes.* (Source: R. A. Bowman, A. C. Mueller, and W. M. Nagle, "Mean Temperature Differences in Design," *Trans. ASME* 62 (1940), 283–294. Reproduced with permission.)

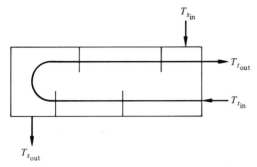

Figure 19.12 *Heat exchanger with one shell pass and two, or multiple of two, tube passes.*

and is a measure of the relative heat capacity of the fluids. If the terms in Equation (19.47) are cross-multiplied, the result is the conservation of energy applied to the two fluids in the heat exchanger.

Example 19.6

A two-shell pass, four-tube pass, counterflow heat exchanger is used for cooling oil with a specific heat of 3.55 kJ/kg · K and a flow rate of 2.52 kg/s. The oil enters the tube side at 125°C and leaves at

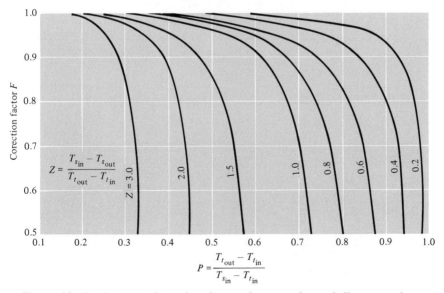

Figure 19.13 *Correction factor for a heat exchanger with two shell passes and two, or a multiple of two, tube passes.* (Source: R. A. Bowman, A. C. Mueller, and W. M. Nagle, "Mean Temperature Differences in Design," *Trans. ASME* 62 (1940), 283–294. Reproduced with permission.)

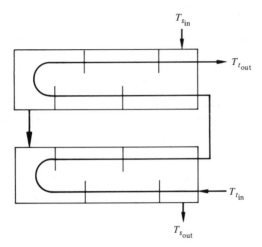

Figure 19.14   Heat exchanger with two shell passes and two tube passes.

50°C. Water enters the shell side at 20°C with a flow rate of 3.15 kg/s. The overall coefficient of heat transfer is 116 W/m² · K Determine (a) the heat transferred, (b) LMTD; (c) F; (d) surface area.

(a) q = heat loss by oil = $\dot{m}c_p(\Delta T)_{oil}$

$q = (2.52)(3.55)(125 - 50) = 670.95 \text{ kW}$

(b) We need to determine water outlet temperature before the log mean temperature difference may be calculated. From the conservation of energy, the heat loss by the oil is gained by the water, or

$$670.95 = \dot{m}_{H_2O}c_p(\Delta T)_{H_2O}$$

$$(\Delta T)_{H_2O} = 50.9$$

$$\left(T_{H_2O}\right)_{out} = 70.9°C$$

$$\Delta T_B = 50 - 20 = 30 \qquad \Delta T_A = 125 - 70.9 = 54.1$$

$$\text{LMTD} = \frac{54.1 - 30}{\ln(54.1/30)} = 40.87$$

(c) $P = \dfrac{50 - 125}{20 - 125} = 0.714$

$Z = \dfrac{20 - 70.9}{50 - 125} = 0.678$

From Figure 19.13,

$F = 0.882$

(d) $\overline{\Delta T} = (F)(\text{LMTD}) = 36.04$

$$A = \frac{q}{U\overline{\Delta T}} = \frac{670\,950}{(116)(36.04)} = 160.5 \text{ m}^2$$

Heat Exchanger Effectiveness

With the use of the correction factors we are able to compensate for complex constructions and calculate the heat transferred for a given flow condition. What if the conditions change or we wish to use the exchanger for another purpose? Will we be able to determine, from what we know about the existing performance, future performance? No. However, this is soon to be remedied.

Let us define heat exchanger effectiveness, $\mathcal{E}$.

$$\mathcal{E} = \frac{\text{actual heat transfer}}{\text{maximum possible heat transfer}} \qquad (19.48)$$

$$q_{act} = C_h\left(T_{h_{in}} - T_{h_{out}}\right) = C_c\left(T_{c_{out}} - T_{c_{in}}\right)$$

$$q_{max} = C_{min}\left(T_{h_{in}} - T_{c_{in}}\right)$$

where C_{min} is the smaller value of C_h and C_c. The maximum possible heat exchange would occur in an infinite area counterflow heat exchanger. Thus

$$\mathcal{E} = \frac{C_h\left(T_{h_{in}} - T_{h_{out}}\right)}{C_{min}\left(T_{h_{in}} - T_{c_{in}}\right)} \qquad (19.49)$$

or

$$\mathcal{E} = \frac{C_c\left(T_{c_{out}} - T_{c_{in}}\right)}{C_{min}\left(T_{h_{in}} - T_{c_{in}}\right)} \qquad (19.50)$$

Equations (19.49) and (19.50) are equal, and the choice of which to use is up to you. Generally, you use the one where the C's cancel.

$$\mathcal{E}C_{min}\left(T_{h_{in}} - T_{c_{in}}\right) = C_h\left(T_{h_{in}} - T_{h_{out}}\right) = C_c\left(T_{c_{out}} - T_{c_{in}}\right) \qquad (19.51)$$

Once the effectiveness and the inlet conditions are known, the outlet conditions may be calculated from Equation (19.51). Figures 19.15 and 19.16 are charts used to determine the heat exchanger effectiveness for shell and tube heat exchangers.

There are two dimensionless quantities that must be determined before the effectiveness may be found. The ratio, C_{min}/C_{max}, is one quantity; it is the ratio of the flow specific heats. The second quantity, NTU_{max}, is

$$NTU_{max} = \frac{AU}{C_{min}} \qquad (19.52)$$

where AU is the value for the new condition. The value $(AU)_{old}$ may be determined from $q = UA\overline{\Delta T}$ at the original operating conditions. Generally U will change with a change in operating conditions because of variation in the convective coefficient, and this change must be determined before the effectiveness may be found. In this text we will

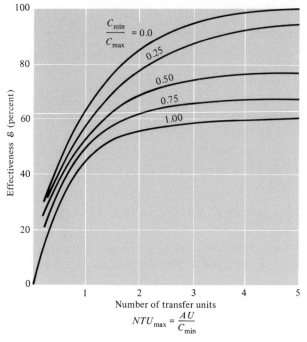

Figure 19.15 *Effectiveness for heat exchanger with one shell pass and a multiple of two tube passes.* (Source: Adapted by permission from W. M. Kays and A. L. London, *Compact Heat Exchangers* (New York: McGraw-Hill, 1958).)

simplify this step, unless directed otherwise, we will assume that UA is a constant for the heat exchanger.

Example 19.7

A single-shell pass, four-tube pass, counterflow heat exchanger operates under the following conditions. Flue gases enter the exchanger at 260°C and leave at 150°C with a specific heat, $c_p = 1.05$ kJ/kg · K, and a flow rate of 0.51 kg/s. Water enters the exchanger at 125°C with a flow rate of 0.38 kg/s. A change in operating conditions occurs; a feed heater must be bypassed, so the water enters at 65°C with a flow rate of 0.31 kg/s. The specific heat of the water is 4.18 kJ/kg · K. Determine the new outlet water conditions.

Assume that water is in the tubes, with gases on the shell side.

$$C_{gas} = \dot{m}c_p = (0.51)(1.05) = 0.5355$$

$$C_{H_2O} = \dot{m}c_p = (0.38)(4.18) = 1.5884$$

$$\frac{C_{min}}{C_{max}} = 0.337$$

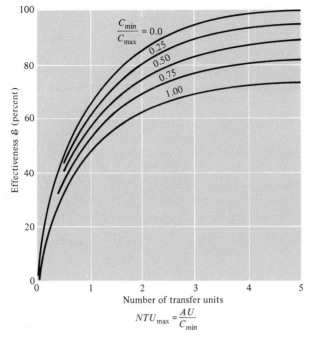

Figure 19.16   Heat exchanger effectiveness for two shell passes and a multiple of two tube passes. (Source: Adapted by permission from W. M. Kays and A. L. London, *Compact Heat Exchangers* (New York: McGraw-Hill, 1958).)

From a first-law analysis find the water outlet temperature,

$$(\dot{m}c_p \Delta T)_{H_2O} = (\dot{m}c_p \Delta T)_{gas}$$

$$(T_{H_2O})_{out} = 162.1°C$$

$$Z = \frac{260 - 150}{162.1 - 125} = 2.96$$

$$P = \frac{162.1 - 125}{260 - 125} = 0.275$$

From Figure 19.11

$$F = 0.75$$

$$\text{LMTD} = 53.4°C$$

$$\overline{\Delta T} = (\text{LMTD})(F) = 40.05°C$$

$$UA = \frac{q}{\overline{\Delta T}} = \frac{(0.51)(1.05)(110)}{40.05}$$

$$UA = 1.470 \text{ kW/K}$$

$$\frac{UA}{C_{min}} = 2.74$$

Determine the effectiveness from Figure 19.15.

$$\mathscr{E} = 81 \text{ percent}$$

$$\mathscr{E}C_{min}\left(T_{h_{in}} - T_{c_{in}}\right) = C_c\left(T_{c_{out}} - T_{c_{in}}\right)$$

$$(0.81)(0.5355)(260-65) = (0.31)(4.18)(\Delta T)_c$$

$$(\Delta T)_c = 65.2°C$$

$$T_{c_{out}} = 130.2°C$$

There remains a great deal of information about heat transfer and heat exchangers to be learned in a separate course. It is hoped, though, that you have gained an appreciation of, and an interest in, heat-transfer phenomena. The methods of analysis presented in this chapter are obviously simplified or special cases, but they are often enough to get one started on the first step in problem analysis.

PROBLEMS

1. An experiment is undertaken to determine the thermal conductivity of an unknown material. The material is 5 cm thick and has a diameter of 15 cm. It is placed on a hot plate of equal diameter where the surface temperature is maintained constant at 60°C. The outer surface temperature is 24°C and the power used by the hot plate is 45 W. Determine the thermal conductivity of the material.

2. A wall with a thermal conductivity of 0.30 W/m · K is maintained at 40°C. The heat flow through the wall is 250 W. The wall surface area is 1.5 m² and its thickness is 1 cm. Determine the temperature of the other surface.

3. The sum of the resistances for the outside wall of a house is 8.0 K · m²/W. For a temperature difference of 25°C across the wall and a total surface area of 150 m², determine the heat flow through the wall.

4. A dry ice storage chest is a wooden box lined with glass fiber insulation 5 cm thick. The wooden box is 2 cm thick, and is cubical, 60 cm on an edge. The inside surface temperature is −76°C and the outside surface temperature is 18°C. Determine (a) the wood–insulation interface temperature; (b) the heat gain per day.

5. The surface of a furnace wall is at a temperature of 1200°C. The outside wall temperature is 38°C. The furnace wall construction has 15 cm of refractory material, $\lambda = 1.73$ W/m · K, and the outside wall is 1 cm steel, $\lambda = 44$ W/m · K. What thickness of refractory brick must be used between the refractory material and the wall if the heat loss is not to exceed 0.7 kW/m²? The thermal conductivity of the brick is 0.34 W/m · K.

6. A furnace is constructed with 20 cm of firebrick, $\lambda = 1.36$ W/m · K, 10 cm of insulating brick, $\lambda = 0.26$ W/m · K, and 20 cm of building brick, $\lambda = 0.69$ W/m · K. The inside surface temperature is 650°C and the outside air temperature is 32°C. The heat loss from the furnace wall is 0.56 kW/m². Determine (a) unit convective coefficient for the air; (b) the temperature 25 cm in from the outside surface.

7. A composite furnace wall has an inside wall temperature of 1100°C and an outside wall temperature of 38°C. Three types of bricks are available as follows:

Brick	λ (W/m · K)	Thickness (cm)	Maximum Allowable Temperature (°C)
(a)	1.55	11.5	—
(b)	0.21	7.5	950
(c)	0.70	10.0	150

The heat loss must not exceed 0.73 kW/m². Determine (a) the minimum wall thickness; (b) the actual heat loss for (a).

8. The oven of a stove must have sufficient insulation that the outside surface temperature of the stove is not greater than 50°C. To accomplish this insulation, $\lambda = 0.11$ W/m · K, is used between the inside and outside metal surfaces. Neglecting the resistance of the metal, what is the minimum thickness of insulation required, if the inside temperature reaches 315°C? The room temperature is 20°C, and the outside unit convective coefficient is 8 W/m² · K.

9. For the composite wall illustrated in Figure 19.5, the following values apply:

$L_1 = 20$ cm $\lambda_1 = 75$ W/m · K $L_2 = 25$ cm

$\lambda_2 = 60$ W/m · K $L_3 = 30$ cm $\lambda_a = 20$ W/m · K

$\lambda_b = 60$ W/m · K $L_4 = 50$ cm $\lambda_4 = 50$ W/m · K

One surface is maintained at 400°C while the other is maintained at 100°C. Determine the heat flow and the temperature at the L_3/L_4 interface.

10. Referring to Figure 19.3, the fluid on the left is air at 800°C, the fluid on the right is water at 80°C and the wall is 30 cm thick with a thermal conductivity of 0.50 W/m · K. The left-hand side convective coefficient is 100 W/m² · K and the right-hand side convective coefficient is 1000 W/m² · K. The gas radiates as a blackbody. Determine: (a) the overall circuit resistance; (b) the heat flow; (c) the wall temperatures.

11. The filament of a 75-W light bulb may be considered a blackbody radiating into a black enclosure at 70°C. The filament diameter is 0.10 mm and the length is 5 cm. Considering only radiation, determine the filament temperature.

12. Water with a flow rate of 0.4 kg/s and a temperature of 10°C enters a 2.5-cm tube. A constant wall temperature of 50°C is maintained. Determine the convective heat transfer coefficient and, for a tube 5.0 m long, the water exit temperature.

13. Air enters a heating duct with dimensions of 7.5 × 15 cm. The air enters the 5-m duct with a temperature of 0°C and the duct surface is maintained at 67°C. If the air exit temperature is to be 20°C, what is the air flow rate?

14. Engine oil flows at 0.3 m/s through a 2.5-cm tube. The oil temperature is 160°C and the tube surface temperature is 150°C. Determine the unit convective coefficient for tube lengths of 2 m and 4 m.

15. Water with a flow rate of 2.6 kg/s is heated from 10°C to 24°C as it passes through a 5-cm pipe. The inside pipe surface temperature is 95°C. Determine the pipe length required.

16. A dry saturated vapor enters 40 m of 11.5-cm O.D., 10-cm I.D. pipe with a flow rate of 0.12 kg/s. The pipe is covered with 5 cm of 85% magnesia insulation. The surface temperature of the insulation is constant at 38°C and the surrounding air temperature is 27°C. The unit outside convective coefficient is 10 W/m^2 · K. The latent heat of the vapor is 140 kJ/kg. Determine (a) the total heat loss in 40 m; (b) the quality of the vapor exiting the pipe.

17. A 15-cm steam line carries saturated steam at 6 MPa and is located in a tunnel with stagnant air with a temperature of 50°C. The outside unit convective coefficient is 8.5 W/m^2 · K. The steam line is covered with 3 cm of 85% magnesia insulation. It is decided to reduce the heat loss by one-half. How much 85% magnesia insulation is required? Neglect thermal resistance of pipe and inside surface.

18. A 5-cm O.D. steel tube is used to transport saturated steam at 480 kPa. The tube is insulated with a 5-cm layer of 85% magnesia insulation followed by 3.5 cm of cork. The cork surface temperature is 24°C. Find the heat loss per meter of pipe, neglecting pipe resistance.

19. Exhaust gases, cooling from 370°C to 175°C are used to heat 18.9 kg/s of oil from 10°C to 100°C in a parallel-flow heat exchanger. The gas side area is 2000 m^2. The average specific heat of the gases is 1.09 kJ/kg · K and the average specific heat for the oil is 2.5 kJ/kg · K. Determine (a) the overall coefficient of heat transfer based on the gas side area; (b) the log mean temperature difference; (c) the heat transferred.

20. The stack gas from a chemical operation contains noxious vapors that must be condensed by lowering their temperature from 315°C to 38°C. The gas volume flow rate is 0.7 m^3/s. Water is available at 10°C at 1.26 kg/s. A shell and tube exchanger with two shell passes and four tube passes will be used with water flowing through the tubes. The gas has a specific heat of 1.11 kJ/kg · K and a gas constant of 0.26 kJ/kg · K. Determine (a) the required surface area if $U = 60$ W/m^2 · K; (b) the log mean temperature difference.

21. A 22.7-kg/s flow of air enters a preheater at 28°C and leaves at 150°C, 23.7 kg/s of exhaust gases, $c_p = 1.09$ kJ/kg · K, enters at 315°C. The overall coefficient of heat transfer is 710 W/m^2 · K. Determine (a) the exit exhaust gas temperature; (b) the surface area for parallel flow; (c) the surface area for counterflow; (d) the log mean temperature difference.

22. Crude oil, $c_p = 1.92$ kJ/kg · K, flows at the rate of 0.315 kg/s through the inside of a concentric, double-pipe heat exchanger and is heated from 32°C to 96°C. Another hydrocarbon, $c_p = 2.5$ kJ/kg · K, enters at 240°C. The overall coefficient of heat transfer is found to be 4400 W/m^2 · K. Determine for a minimum temperature difference of 17°C between the fluids (a) the log mean temperature difference for parallel and counterflow flow; (b) the surface area for parallel and counterflow flow.

23. A shell and tube heat exchanger must transfer 205 kW to a solution with a specific heat of 3.26 kJ/kg · K and change its temperature from 65°C to 93°C. Steam is available at 250 kPa. The inside unit convective coefficient is 3400 W/m^2 · K and for the outside it is 7300 W/m^2 · K. The thermal conductivity for the tubes is 111 W/m · K and the tubes have 4.0 cm O.D., 3.0 cm. I.D., and are 3 m long. Determine the number of tubes required.

24. Exhaust steam with a flow of 63 kg/s at 7 kPa enters a single-pass surface condenser with an outside surface area of 7500 m^2. The steam quality is 90% and the outlet condensate temperature is subcooled 5°C. The cooling-water flow rate is 4413 kg/s entering at 20°C. Find the overall coefficient of heat transfer.

25. A one-shell-pass, six-tube-pass heat exchanger is used as an economizer on a steam generator. The flue gas, $c_p = 1.09$ kJ/kg · K, enters at 350°C and leaves at 205°C with a flow rate of 58.0 kg/s. The feedwater enters at 175°C with a flow rate of 63 kg/s. A change of operating conditions occurs where the water flow is now 25.2 kg/s, entering at 138°C. The new gas flow rate is 23.8 kg/s, but the gas temperature remains the same. Determine (a) the old surface area required if $U = 170$ W/m^2 · K; (b) the effectiveness; (c) the new water outlet temperature.

26. A two-shell-pass, four-pass heat exchanger is used as a feed heater and has an effectiveness of 80%. Water with a flow rate of 12.6 kg/s enters at 115°C and air enters at 290°C with a flow rate of 25.2 kg/s. Determine (a) the air exit temperature; (b) the water exit temperature.

27. Liquid R 12 at a temperature of 15°C leaves the condenser through a 2.5-cm O.D. copper tube and goes to the throttling valve in a refrigeration system. The question arises: should the line be insulated? There are three possible solutions: (a) insulate with 9.5 mm of foam rubber, $\lambda = 0.16$ W/m · K; (b) insulate with 15 mm of foam rubber, $\lambda = 0.16$ W/m · K; (c) do not insulate. The outside convective coefficient is estimated to be 7.0 W/m² · K, and the ambient temperature is 29°C. What is your recommendation and why?

28. Water is flowing through a 20-mm. I.D., 25-mm O.D. brass tube, at a rate of 0.38 kg/s. The water enters at 10°C and leaves at 65°C. Heat is supplied by saturated steam condensing at 200 kPa on the outside tube surface. The outside unit convective coefficient, h_0, is 5600 W/m²·K; the inside unit convective coefficient, h_i, is 1700 W/m²·K. Determine (a) the overall coefficient of heat transfer, U, based on the outside tube area; (b) the length of tubing required.

29. A new plant process requires 50.4 kg/s of air to be heated from 4°C to 32°C. Saturated water at 280 kPa is available for heating the air and has a supply capacity of 8.82 kg/s. An old heat exchanger is suggested for use in the new process. Records show the following: (1) dry, saturated steam flow with no subcooling, 1.94 kg/s at 2000 kPa; (2) air flow, 96.5 kg/s, exiting at 65°C. Determine (a) the original log mean temperature difference; (b) the effectiveness; (c) whether the heat exchanger can be used in the new process.

30. A counterflow shell and tube heat exchanger cools 1.6 liters per second of oil ($c_p = 2.5$ kJ/kg · K $\rho = 720$ kg/m³) from 95°C to 50°C. Cooling water enters the tubes at 15°C and leaves at 38°C. The overall coefficient of heat transfer is 450 W/m² · K. Fouling of the tubes occurs, resulting in one-quarter of the tubes being blocked. The conditions remain the same except for the outlet temperatures. Determine the oil and water exit temperatures under the new conditions.

31. By traveling through a brass tube, 2.5 cm O.D. and 2.2 cm I.D., which is surrounded by steam at 55 kPa and 90% quality, 9.4 kg/s of water is heated from 15°C to 60°C. Assume the inside unit convective coefficient is 2000 W/m² · K and the outside unit convective coefficient is 7950 W/m² · K. Determine (a) the length of tube required; (b) the steam supply in kilograms per hour if no subcooling occurs.

References

Badger, P. H. *Equilibrium Thermodynamics*. Boston: Allyn Bacon, 1967.

Callen, H. B. *Thermodynamics*. New York: Wiley, New York, 1960.

Cohen, H., Roger, G. F. C., and Saravanamuttoo, H. I. H. *Gas Turbine Theory*, 2nd ed. New York: Wiley, 1972.

Dodge, D. F. *Chemical Engineering Thermodynamics*. New York: McGraw-Hill, 1944.

Faires, V. M. *Thermodynamics*. 5th ed. New York: Macmillan, 1970.

Jordan, R. C. and Priester, G. B. *Refrigeration and Air Conditioning*. Englewood Cliffs, N.J.: Prentice-Hall, 1956.

Keenan, J. H. *Thermodynamics*. New York: Wiley, 1941.

Obert, E. F. *Concepts of Thermodynamics*. New York: McGraw-Hill, 1960.

Obert, E. F. *Internal Combustion Engines*, 3rd ed. New York: International Textbook Company, 1968.

Shapiro, A. H. *The Dynamics and Thermodynamics of Compressible Fluid Flow*. New York: Ronald Press, 1953.

Taylor, C. F. *Internal Combustion Engine in Theory and Practice*, 2nd ed. Cambridge, Mass.: M.I.T. Press, 1966.

Van Wylen, G. J. and Sonntag, R. E. *Fundamentals of Classical Thermodynamics*, 2nd ed. New York: Wiley, 1976.

List of Symbols

A	area	$\bar{h}$	enthalpy, specific, mole basis
a	acceleration		
a	acoustic velocity	$\bar{h}_c$	unit convective coefficient
$\mathcal{A}$	availability	J	joules
A.E.	available energy	K	chemical equilibrium constant
C	constant		
c_p	constant pressure specific heat, mass basis	k	ratio of specific heats
		K.E.	kinetic energy
$\bar{c}_p$	constant pressure specific heat, mole basis	L	length
		l	length
c_v	constant volume specific heat, mass basis	ln	natural logarithm
		M	mach number
$\bar{c}_v$	constant volume specific heat, mole basis	M	molecular weight
		m	mass
COP	coefficient of performance	$\dot{m}$	mass flow-rate
d	diameter	n	moles
E	total energy	n	polytropic exponent
e	specific energy	Nu	Nusselt number
$\mathcal{E}$	heat exchanger effectiveness	P.E.	potential energy
		p	pressure
F	force	p_r	reduced pressure
G	Gibbs function, total	Pr	Prandtl number
g	Gibbs function, specific	Q	heat, total
g	local gravitational acceleration	q	heat, specific
		$\dot{Q}$	heat transfer rate
H	enthalpy, total	R	gas constant, specific
h	enthalpy, specific	$\bar{R}$	gas constant, universal

503

Re	Reynolds number	Z	compressibility factor
r	radius	α	thermal diffusivity
$r_{a/f}$	air–fuel ratio	β_s	adiabatic compressibility
r_c	cut-off ratio	Δ	a change in
r_p	pressure ratio	δ	infinitesimal change
r	compression ratio	ε	strain, unit
S	entropy, total	η	efficiency
s	entropy, specific, mass basis	η_{th}	thermal efficiency
		η_c	compressor efficiency
$\bar{s}$	entropy, specific, mole basis	η_t	turbine internal efficiency
		η_m	mechanical efficiency
T	temperature, absolute	η_n	nozzle efficiency
U	internal energy, total	η_d	diffuser efficiency
U.E.	unavailable energy	η_p	propulsion efficiency
u	internal energy, specific, mass basis	η_v	volumetric efficiency
		θ	angle
$\bar{u}$	internal energy, specific, mole basis	λ	thermal conductivity
		μ	Joule-Thompson coefficient, dynamic viscosity
V	volume, total		
v	specific volume	ϕ	entropy function
v	velocity	ϕ	relative humidity
v_r	reduced specific volume	ρ	density
W	work	σ	surface tension
$\dot{W}$	rate of work	τ	time
w	work, specific	ω	specific humidity
x	quality of a wet vapor	ν	kinematic viscosity
z	elevation		

Appendix Tables

Table A.1   Gas Constants and Specific Heats at Low Pressures

Gas	M	c_p, kJ/kg $\cdot$ K	c_v, kJ/kg $\cdot$ K	k	R, kJ/kg $\cdot$ K
Acetylene (C_2H_2)	26.036	1.6947	1.3753	1.232	0.3195
Air	28.97	1.0047	0.7176	1.4	0.287
Ammonia (NH_3)	17.032	2.089	1.5992	1.304	0.4882
Argon (A)	39.95	0.5208	0.3127	1.666	0.2081
Carbon Dioxide (CO_2)	44.01	0.844	0.6552	1.288	0.1889
Carbon Monoxide (CO)	28.01	1.0412	0.7444	1.399	0.2968
Chlorine (Cl_2)	70.914	0.4789	0.3617	1.324	0.1172
Ethane (C_2H_6)	30.068	1.7525	1.4761	1.187	0.2765
Ethylene (C_2H_4)	28.052	1.5297	1.2333	1.24	0.2964
Helium (He)	4.003	5.1954	3.1189	1.666	2.077
Hydrogen (H_2)	2.016	14.3136	10.190	1.4	4.125
Hydrazine (N_2H_4)	32.048	1.6453	1.3815	1.195	0.2594
Methane (CH_4)	16.043	2.1347	1.6164	1.321	0.5183
Neon (Ne)	20.183	1.0298	0.6179	1.666	0.4120
Nitrogen (N_2)	28.016	1.0399	0.7431	1.399	0.2968
Oxygen (O_2)	32	0.9185	0.6585	1.395	0.2598
Propane (C_3H_8)	44.094	1.6683	1.4799	1.127	0.1886
Sulfur Dioxide (SO_2)	64.07	0.6225	0.4927	1.263	0.1298
Water Vapor (H_2O)	18.016	1.8646	1.4033	1.329	0.4615
Xenon (Xe)	131.3	0.1582	0.0950	1.666	0.0633

Table A.2   Properties of Air at Low Pressures

T, K	h, kJ/kg	p_r	u, kJ/kg	v_r	θ, kJ/kg $\cdot$ K
100	99.76	0.029 90	71.06	2230	1.4143
110	109.77	0.041 71	78.20	1758.4	1.5098
120	119.79	0.056 52	85.34	1415.7	1.5971
130	129.81	0.074 74	92.51	1159.8	1.6773
140	139.84	0.096 81	99.67	964.2	1.7515
150	149.86	0.123 18	106.81	812.0	1.8206
160	159.87	0.154 31	113.95	691.4	1.8853
170	169.89	0.190 68	121.11	594.5	1.9461
180	179.92	0.232 79	128.28	515.6	2.0033
190	189.94	0.281 14	135.40	450.6	2.0575
200	199.96	0.3363	142.56	396.6	2.1088
210	209.97	0.3987	149.70	351.2	2.1577
220	219.99	0.4690	156.84	312.8	2.2043
230	230.01	0.5477	163.98	280.0	2.2489
240	240.03	0.6355	171.15	251.8	2.2915
250	250.05	0.7329	178.29	227.45	2.3325
260	260.09	0.8405	185.45	206.26	2.3717
270	270.12	0.9590	192.59	187.74	2.4096
280	280.14	1.0889	199.78	171.45	2.4461
290	290.17	1.2311	206.92	157.07	2.4813
300	300.19	1.3860	214.09	144.32	2.5153
310	310.24	1.5546	221.27	132.96	2.5483
320	320.29	1.7375	228.45	122.81	2.5802
330	330.34	1.9352	235.65	113.70	2.6111
340	340.43	2.149	242.86	105.51	2.6412
350	350.48	2.379	250.05	98.11	2.6704

Table A.2 Properties of Air at Low Pressures (*Continued*)

T, K	h, kJ/kg	p_r	u, kJ/kg	v_r	θ kJ/kg · K
360	360.58	2.626	257.23	91.40	2.6987
370	370.67	2.892	264.47	85.31	2.7264
380	380.77	3.176	271.72	79.77	2.7534
390	390.88	3.481	278.96	74.71	2.7796
400	400.98	3.806	286.19	70.07	2.8052
410	411.12	4.153	293.45	65.83	2.8302
420	421.26	4.522	300.73	61.93	2.8547
430	431.43	4.915	308.03	58.34	2.8786
440	441.61	5.332	315.34	55.02	2.9020
450	451.83	5.775	322.66	51.96	2.9249
460	462.01	6.245	329.99	49.11	2.9473
470	472.25	6.742	337.34	46.48	2.9693
480	482.48	7.268	344.74	44.04	2.9909
490	492.74	7.824	352.11	41.76	3.0120
500	503.02	8.411	359.53	39.64	3.0328
510	513.32	9.031	366.97	37.65	3.0532
520	523.63	9.684	374.39	35.80	3.0733
530	533.98	10.372	381.88	34.07	3.0930
540	544.35	11.097	389.40	32.45	3.1124
550	554.75	11.858	396.89	30.92	3.1314
560	565.17	12.659	404.44	29.50	3.1502
570	575.57	13.500	411.98	28.15	3.1686
580	586.04	14.382	419.56	26.89	3.1868
590	596.53	15.309	427.17	25.70	3.2047
600	607.02	16.278	434.80	24.58	3.2223
610	617.53	17.297	442.43	23.51	3.2397
620	628.07	18.360	450.13	22.52	3.2569
630	638.65	19.475	457.83	21.57	3.2738
640	649.21	20.64	465.55	20.674	3.2905
650	659.84	21.86	473.32	19.828	3.3069
660	670.47	23.13	481.06	19.026	3.3232
670	681.15	24.46	488.88	18.266	3.3392
680	691.82	25.85	496.65	17.543	3.3551
690	702.52	27.29	504.51	16.857	3.3707
700	713.27	28.80	512.37	16.205	3.3861
710	724.01	30.38	520.26	15.585	3.4014
720	734.20	31.92	527.72	15.027	3.4156
730	745.62	33.72	536.12	14.434	3.4314
740	756.44	35.50	544.05	13.900	3.4461
750	767.30	37.35	552.05	13.391	3.4607
760	778.21	39.27	560.08	12.905	3.4751
770	789.10	41.27	568.10	12.440	3.4894
780	800.03	43.35	576.15	11.998	3.5035
790	810.98	45.51	584.22	11.575	3.5174
800	821.94	47.75	592.34	11.172	3.5312
810	832.96	50.08	600.46	10.785	3.5449
820	843.97	52.49	608.62	10.416	3.5584
830	855.01	55.00	616.79	10.062	3.5718
840	866.09	57.60	624.97	9.724	3.5850
850	877.16	60.29	633.21	9.400	3.5981
860	888.28	63.09	641.44	9.090	3.6111
870	899.42	65.98	649.70	8.792	3.6240

Table A.2 Properties of Air at Low Pressures (*Continued*)

T, K	h, kJ/kg	p_r	u, kJ/kg	v_r	θ kJ/kg · K
880	910.56	68.98	658.00	8.507	3.6367
890	921.75	72.08	666.31	8.233	3.6493
900	932.94	75.29	674.63	7.971	3.6619
910	944.15	78.61	682.98	7.718	3.6743
920	955.38	82.05	691.33	7.476	3.6865
930	966.64	85.60	699.73	7.244	3.6987
940	977.92	89.28	708.13	7.020	3.7108
950	989.22	93.08	716.57	6.805	3.7227
960	1000.53	97.00	725.01	6.599	3.7346
970	1011.88	101.06	733.48	6.400	3.7463
980	1023.25	105.24	741.99	6.209	3.7580
990	1034.63	109.57	750.48	6.025	3.7695
1000	1046.03	114.03	759.02	5.847	3.7810
1020	1068.89	123.12	775.67	5.521	3.8030
1040	1091.85	133.34	793.35	5.201	3.8259
1060	1114.85	143.91	810.61	4.911	3.8478
1080	1137.93	155.15	827.94	4.641	3.8694
1100	1161.07	167.07	845.34	4.390	3.8906
1120	1184.28	179.71	862.85	4.156	3.9116
1140	1207.54	193.07	880.37	3.937	3.9322
1160	1230.90	207.24	897.98	3.732	3.9525
1180	1254.34	222.2	915.68	3.541	3.9725
1200	1277.79	238.0	933.40	3.362	3.9922
1220	1301.33	254.7	951.19	3.194	4.0117
1240	1324.89	272.3	969.01	3.037	4.0308
1260	1348.55	290.8	986.92	2.889	4.0497
1280	1372.25	310.4	1004.88	2.750	4.0684
1300	1395.97	330.9	1022.88	2.619	4.0868
1320	1419.77	352.5	1040.93	2.497	4.1049
1340	1443.61	375.3	1059.03	2.381	4.1229
1360	1467.50	399.1	1077.17	2.272	4.1406
1380	1491.43	424.2	1095.36	2.169	4.1580
1400	1515.41	450.5	1113.62	2.072	4.1753
1420	1539.44	478.0	1131.90	1.9808	4.1923
1440	1563.49	506.9	1150.23	1.8942	4.2092
1460	1587.61	537.1	1168.61	1.8124	4.2258
1480	1611.80	568.8	1187.03	1.7350	4.2422
1500	1635.99	601.9	1205.47	1.6617	4.2585
1550	1696.63	691.4	1251.78	1.4948	4.2983
1600	1757.55	791.2	1298.35	1.3485	4.3369
1650	1818.70	902.0	1345.10	1.2197	4.3745
1700	1907.39	1025.1	1392.18	1.1065	4.4112
1750	1941.63	1160.5	1439.38	1.0056	4.4469
1800	2003.36	1310.3	1486.76	0.9164	4.4817
1850	2065.27	1475.0	1534.33	0.8370	4.5156
1900	2127.37	1655.4	1582.09	0.7659	4.5487
1950	2182.72	1830.4	1624.91	0.7084	4.5777
2000	2252.06	2067.9	1678.07	0.6449	4.6127
2500	2883.91	5521	2166.41	0.3019	4.8946
3000	3526.54	12 490	2665.55	0.16015	5.1288
3500	4177.22	25 123	3172.71	0.09289	5.3294

SOURCE: Adapted from J. H. Keenan and J. Kaye, *Gas Tables*, (New York: John Wiley & Sons, 1945).

Table A.3 Products—400 Percent Theoretical Air—at Low Pressures

T, K	h, kJ/kg	p_r	u, kJ/kg	v_r	ϕ (kJ/kg · K)
200	201.2	0.3279	143.8	11782	6.2919
220	221.4	0.4588	158.2	9262.8	6.3884
240	241.7	0.6236	172.8	7433	6.4764
260	262.0	0.8275	187.4	6069	6.5577
280	282.3	1.0756	201.9	5028	6.6330
300	302.6	1.3738	216.5	4218	6.7033
320	323.1	1.7279	231.2	3577	6.7692
340	343.5	2.144	245.9	3063	6.8311
360	364.0	2.629	260.6	2645	6.8897
380	384.6	3.190	275.5	2301	6.9454
400	405.1	3.836	290.3	2014.4	6.9981
420	425.8	4.572	305.2	1774.4	7.0486
440	446.6	5.408	320.2	1571.5	7.0968
460	467.4	6.354	335.3	1398.4	7.1430
480	488.3	7.418	350.4	1249.9	7.1876
500	509.2	8.611	365.7	1121.6	7.2304
520	530.3	9.944	381.0	1010.2	7.2717
540	551.5	11.428	396.4	912.7	7.3116
560	572.8	13.075	412.0	827.4	7.3504
580	594.1	14.897	427.6	752.1	7.3877
600	615.6	16.909	443.3	685.5	7.4241
620	637.1	19.125	459.1	626.3	7.4596
640	658.8	21.56	475.0	573.4	7.4940
660	680.5	24.22	491.0	526.2	7.5275
680	702.4	27.14	507.2	484.0	7.5600
700	724.4	30.32	523.4	445.9	7.5919
720	746.5	33.80	539.7	411.5	7.6230
740	768.6	37.57	556.2	380.5	7.6533
760	790.9	41.66	572.7	352.4	7.6831
780	813.3	46.11	589.4	326.8	7.7122
800	835.8	50.91	606.1	303.6	7.7407
820	858.4	56.10	623.0	282.3	7.7686
840	881.1	61.72	640.0	262.9	7.7959
860	903.9	67.76	657.0	245.2	7.8227
880	926.8	74.26	674.1	228.9	7.8490
900	949.8	81.24	691.4	214.0	7.8749
920	972.8	88.74	708.7	200.26	7.9002
940	996.0	96.78	726.1	187.61	7.9252
960	1019.2	105.39	743.6	175.94	7.9496
980	1042.6	114.60	761.2	165.18	7.9736
1000	1066.0	124.45	778.9	155.21	7.9973
1050	1124.9	152.05	823.5	133.40	8.0549
1100	1184.3	184.33	868.5	115.27	8.1101
1150	1244.2	221.90	914.0	100.11	8.1633
1200	1304.5	265.4	960.0	87.36	8.2148
1250	1365.3	315.3	1006.4	76.56	8.2643
1300	1426.4	372.6	1053.1	67.37	8.3123

Table A.3   Products—400 Percent Theoretical Air—at Low Pressures (*Continued*)

T, K	h, kJ/kg	p_r	u, kJ/kg	v_r	ϕ (kJ/kg · K)
1350	1487.8	438.0	1100.2	59.54	8.3587
1400	1542.8	512.3	1147.6	52.79	8.4037
1450	1611.7	596.2	1195.4	46.98	8.4472
1500	1674.1	690.9	1243.4	41.94	8.4896
1550	1736.7	797.3	1291.7	37.55	8.5306
1600	1799.7	916.4	1340.3	33.73	8.5707
1650	1862.9	1049.4	1389.2	30.37	8.6096
1700	1926.4	1197.4	1438.3	27.42	8.6475
1750	1990.0	1361.7	1487.6	24.83	8.6843
1800	2053.9	1543.5	1537.1	22.53	8.7204
1850	2118.0	1744.5	1586.8	20.49	8.7555
1900	2182.3	1965.7	1636.8	18.67	8.7899

SOURCE: Adapted from J. H. Keenan and J. Kaye, *Gas Tables*, (New York: John Wiley & Sons, 1945).

Table A.4   Products—200 Percent Theoretical Air—at Low Pressures

T, K	h, kJ/kg	p_r	u, kJ/kg	v_r	ϕ, kJ/kg · K
200	202.4	0.3200	145.0	12073	6.2882
220	222.8	0.4491	159.6	9463	6.3854
240	243.3	0.6124	174.4	7571	6.4746
260	263.9	0.8152	189.2	6161	6.5567
280	284.5	1.0630	204.0	5088	6.6331
300	305.1	1.3620	218.9	4255	6.7043
320	325.8	1.7188	233.9	3597	6.7712
340	346.6	2.140	248.9	3070	6.8341
360	367.4	2.632	264.0	2642	6.8936
380	388.3	3.204	279.1	2291	6.9501
400	409.2	3.864	294.3	1999.3	7.0039
420	430.3	4.621	309.6	1755.7	7.0553
440	451.4	5.483	325.0	1550.1	7.1044
460	472.7	6.462	340.5	1375.2	7.1515
480	494.0	7.566	356.1	1225.5	7.1969
500	515.4	8.808	371.7	1096.4	7.2406
520	536.8	10.202	387.5	984.6	7.2829
540	558.5	11.758	403.4	887.2	7.3235
560	580.2	13.490	419.4	801.9	7.3632
580	602.0	15.413	435.4	726.9	7.4014
600	624.0	17.542	451.6	660.8	7.4386
620	646.0	19.894	467.9	602.0	7.4747
640	668.1	22.48	484.3	549.8	7.5099
660	690.4	25.33	500.8	503.3	7.5442
680	712.8	28.46	517.4	461.6	7.5776
700	735.2	31.87	534.2	424.2	7.6101
720	757.8	35.62	551.0	390.5	7.6420
740	780.5	39.68	568.0	360.2	7.6731
760	803.4	44.12	585.1	332.8	7.7034
780	826.3	48.94	602.3	307.9	7.7334
800	849.4	54.17	619.6	285.3	7.7625
820	872.5	59.83	637.0	264.7	7.7911
840	895.8	65.97	654.5	246.0	7.8192
860	919.1	72.60	672.1	228.8	7.8467
880	942.6	79.75	689.8	213.2	7.8736
900	966.2	87.44	707.6	198.81	7.9001
920	989.8	95.73	725.5	185.64	7.9261
940	1013.6	104.63	743.6	173.53	7.9517
960	1037.4	114.19	761.6	162.39	7.9768
980	1061.4	124.44	779.9	152.11	8.0015
1000	1085.4	135.43	798.2	142.63	8.0256
1050	1146.0	166.34	844.4	121.93	8.0848
1100	1207.0	202.71	891.0	104.82	8.1417
1150	1268.6	245.3	938.2	90.57	8.1964
1200	1330.6	294.8	985.9	78.64	8.2492
1250	1393.1	352.0	1034.0	68.57	8.3002
1300	1456.0	418.0	1082.5	60.06	8.3496

Table A.4   Products—200 Percent Theoretical Air—at Lower Pressures (*Continued*)

T, K	h, kJ/kg	p_r	u, kJ/kg	v_r	ϕ, kJ/kg · K
1350	1519.2	493.6	1131.4	52.83	8.3973
1400	1582.9	580.0	1180.7	46.62	8.4436
1450	1646.9	678.1	1230.4	41.30	8.4886
1500	1711.2	789.4	1280.3	36.71	8.5322
1550	1775.8	914.9	1330.6	32.72	8.5746
1600	1840.7	1056.2	1381.1	29.26	8.6159
1650	1906.0	1214.6	1432.0	26.24	8.6560
1700	1971.4	1391.7	1483.1	23.596	8.6951
1750	2037.2	1589.2	1534.5	21.274	8.7333
1800	2103.1	1808.7	1586.0	19.225	8.7704
1850	2169.3	2052.2	1637.8	17.415	8.8068
1900	2235.7	2321.4	1689.9	15.812	8.8422
1950	2302.3	2618	1742.1	14.385	8.8768
2000	2369.0	2946	1794.5	13.114	8.9106
2100	2503.2	3700	1899.9	10.966	8.9760
2200	2638.0	4603	2006.0	9.234	9.0388

SOURCE: Adapted from J. H. Keenan and J. Kaye, *Gas Tables*, (New York: John Wiley & Sons, 1945).

Table A.5 Saturated Steam Temperature Table

Temp. °C T	Press. kPa P	Specific Volume		Internal Energy			Enthalpy			Entropy		
		Sat. Liquid v_f	Sat. Vapor v_g	Sat. Liquid u_f	Evap. u_{fg}	Sat. Vapor u_g	Sat. Liquid h_f	Evap. h_{fg}	Sat. Vapor h_g	Sat. Liquid s_f	Evap. s_{fg}	Sat. Vapor s_g
0.01	0.6113	0.001 000	206.14	.00	2375.3	2375.3	.01	2501.3	2501.4	.0000	9.1562	9.1562
5	0.8721	0.001 000	147.12	20.97	2361.3	2382.3	20.98	2489.6	2510.6	.0761	8.9496	9.0257
10	1.2276	0.001 000	106.38	42.00	2347.2	2389.2	42.01	2477.7	2519.8	.1510	8.7498	8.9008
15	1.7051	0.001 001	77.93	62.99	2333.1	2396.1	62.99	2465.9	2528.9	.2245	8.5569	8.7814
20	2.339	0.001 002	57.79	83.95	2319.0	2402.9	83.96	2454.1	2538.1	.2966	8.3706	8.6672
25	3.169	0.001 003	43.36	104.88	2304.9	2409.8	104.89	2442.3	2547.2	.3674	8.1905	8.5580
30	4.246	0.001 004	32.89	125.78	2290.8	2416.6	125.79	2430.5	2556.3	.4369	8.0164	8.4533
35	5.628	0.001 006	25.22	146.67	2276.7	2423.4	146.68	2418.6	2565.3	.5053	7.8478	8.3531
40	7.384	0.001 008	19.52	167.56	2262.6	2430.1	167.57	2406.7	2574.3	.5725	7.6845	8.2570
45	9.593	0.001 010	15.26	188.44	2248.4	2436.8	188.45	2394.8	2583.2	.6387	7.5261	8.1648
50	12.349	0.001 012	12.03	209.32	2234.2	2443.5	209.33	2382.7	2592.1	.7038	7.3725	8.0763
55	15.758	0.001 015	9.568	230.21	2219.9	2450.1	230.23	2370.7	2600.9	.7679	7.2234	7.9913
60	19.940	0.001 017	7.671	251.11	2205.5	2456.6	251.13	2358.5	2609.6	.8312	7.0784	7.9096
65	25.03	0.001 020	6.197	272.02	2191.1	2463.1	272.06	2346.2	2618.3	.8935	6.9375	7.8310
70	31.19	0.001 023	5.042	292.95	2176.6	2469.6	292.98	2333.8	2626.8	.9549	6.8004	7.7553
75	38.58	0.001 026	4.131	313.90	2162.0	2475.9	313.93	2321.4	2635.3	1.0155	6.6669	7.6824
80	47.39	0.001 029	3.407	334.86	2147.4	2482.2	334.91	2308.8	2643.7	1.0753	6.5369	7.6122
85	57.83	0.001 033	2.828	355.84	2132.6	2488.4	355.90	2296.0	2651.9	1.1343	6.4102	7.5445
90	70.14	0.001 036	2.361	376.85	2117.7	2494.5	376.92	2283.2	2660.1	1.1925	6.2866	7.4791
95	84.55	0.001 040	1.982	397.88	2102.7	2500.6	397.96	2270.2	2668.1	1.2500	6.1659	7.4159

100	0.101 35	0.001 044	1.6729	418.94	2087.6	2506.5	419.04	2257.0	2676.1	1.3069	6.0480	7.3549
105	0.120 82	0.001 048	1.4194	440.02	2072.3	2512.4	440.15	2243.7	2683.8	1.3630	5.9328	7.2958
110	0.143 27	0.001 052	1.2102	461.14	2057.0	2518.1	461.30	2230.2	2691.5	1.4185	5.8202	7.2387
115	0.169 06	0.001 056	1.0366	482.30	2041.4	2523.7	482.48	2216.5	2699.0	1.4734	5.7100	7.1833
120	0.198 53	0.001 060	0.8919	503.50	2025.8	2529.3	503.71	2202.6	2706.3	1.5276	5.6020	7.1296
125	0.2321	0.001 065	0.7706	524.74	2009.9	2534.6	524.99	2188.5	2713.5	1.5813	5.4962	7.0775
130	0.2701	0.001 070	0.6685	546.02	1993.9	2539.9	546.31	2174.2	2720.5	1.6344	5.3925	7.0269
135	0.3130	0.001 075	0.5822	567.35	1977.7	2545.0	567.69	2159.6	2727.3	1.6870	5.2907	6.9777
140	0.3613	0.001 080	0.5089	588.74	1961.3	2550.0	589.13	2144.7	2733.9	1.7391	5.1908	6.9299
145	0.4154	0.001 085	0.4463	610.18	1944.7	2554.9	610.63	2129.6	2740.3	1.7907	5.0926	6.8833
150	0.4758	0.001 091	0.3928	631.68	1927.9	2559.5	632.20	2114.3	2746.5	1.8418	4.9960	6.8379
155	0.5431	0.001 096	0.3468	653.24	1910.8	2564.1	653.84	2098.6	2752.4	1.8925	4.9010	6.7935
160	0.6178	0.001 102	0.3071	674.87	1893.5	2568.4	675.55	2082.6	2758.1	1.9427	4.8075	6.7502
165	0.7005	0.001 108	0.2727	696.56	1876.0	2572.5	697.34	2066.2	2763.5	1.9925	4.7153	6.7078
170	0.7917	0.001 114	0.2428	718.33	1858.1	2576.5	719.21	2049.5	2768.7	2.0419	4.6244	6.6663
175	0.8920	0.001 121	0.2168	740.17	1840.0	2580.2	741.17	2032.4	2773.6	2.0909	4.5347	6.6256
180	1.0021	0.001 127	0.194 05	762.09	1821.6	2583.7	763.22	2015.0	2778.2	2.1396	4.4461	6.5857
185	1.1227	0.001 134	0.174 09	784.10	1802.9	2587.0	785.37	1997.1	2782.4	2.1879	4.3586	6.5465
190	1.2544	0.001 141	0.156 54	806.19	1783.8	2590.0	807.62	1978.8	2786.4	2.2359	4.2720	6.5079
195	1.3978	0.001 149	0.141 05	828.37	1764.4	2592.8	829.98	1960.0	2790.0	2.2835	4.1863	6.4698
200	1.5538	0.001 157	0.127 36	850.65	1744.7	2595.3	852.45	1940.7	2793.2	2.3309	4.1014	6.4323
205	1.7230	0.001 164	0.115 21	873.04	1724.5	2597.5	875.04	1921.0	2796.0	2.3780	4.0172	6.3952
210	1.9062	0.001 173	0.104 41	895.53	1703.9	2599.5	897.76	1900.7	2798.5	2.4248	3.9337	6.3585
215	2.104	0.001 181	0.094 79	918.14	1682.9	2601.1	920.62	1879.9	2800.5	2.4714	3.8507	6.3221
220	2.318	0.001 190	0.086 19	940.87	1661.5	2602.4	943.62	1858.5	2802.1	2.5178	3.7683	6.2861
225	2.548	0.001 199	0.078 49	963.73	1639.6	2603.3	966.78	1836.5	2803.3	2.5639	3.6863	6.2503
230	2.795	0.001 209	0.071 58	986.74	1617.2	2603.9	990.12	1813.8	2804.0	2.6099	3.6047	6.2146
235	3.060	0.001 219	0.065 37	1009.89	1594.2	2604.1	1013.62	1790.5	2804.2	2.6558	3.5233	6.1791
240	3.344	0.001 229	0.059 76	1033.21	1570.8	2604.0	1037.32	1766.5	2803.8	2.7015	3.4422	6.1437
245	3.648	0.001 240	0.054 71	1056.71	1546.7	2603.4	1061.23	1741.7	2803.0	2.7472	3.3612	6.1083

Table A.5 Saturated Steam Temperature Table (Continued)

Temp. °C T	Press. MPa P	Specific Volume Sat. Liquid v_f	Specific Volume Sat. Vapor v_g	Internal Energy Sat. Liquid u_f	Internal Energy Evap. u_{fg}	Internal Energy Sat. Vapor u_g	Enthalpy Sat. Liquid h_f	Enthalpy Evap. h_{fg}	Enthalpy Sat. Vapor h_g	Entropy Sat. Liquid s_f	Entropy Evap. s_{fg}	Entropy Sat. Vapor s_g
250	3.973	0.001 251	0.050 13	1080.39	1522.0	2602.4	1085.36	1716.2	2801.5	2.7927	3.2802	6.0730
255	4.319	0.001 263	0.045 98	1104.28	1496.7	2600.9	1109.73	1689.8	2799.5	2.8383	3.1992	6.0375
260	4.688	0.001 276	0.042 21	1128.39	1470.6	2599.0	1134.37	1662.5	2796.9	2.8838	3.1181	6.0019
265	5.081	0.001 289	0.038 77	1152.74	1443.9	2596.6	1159.28	1634.4	2793.6	2.9294	3.0368	5.9662
270	5.499	0.001 302	0.035 64	1177.36	1416.3	2593.7	1184.51	1605.2	2789.7	2.9751	2.9551	5.9301
275	5.942	0.001 317	0.032 79	1202.25	1387.9	2590.2	1210.07	1574.9	2785.0	3.0208	2.8730	5.8938
280	6.412	0.001 332	0.030 17	1227.46	1358.7	2586.1	1235.99	1543.6	2779.6	3.0668	2.7903	5.8571
285	6.909	0.001 348	0.027 77	1253.00	1328.4	2581.4	1262.31	1511.0	2773.3	3.1130	2.7070	5.8199
290	7.436	0.001 366	0.025 57	1278.92	1297.1	2576.0	1289.07	1477.1	2766.2	3.1594	2.6227	5.7821
295	7.993	0.001 384	0.023 54	1305.2	1264.7	2569.9	1316.3	1441.8	2758.1	3.2062	2.5375	5.7437
300	8.581	0.001 404	0.021 67	1332.0	1231.0	2563.0	1344.0	1404.9	2749.0	3.2534	2.4511	5.7045
305	9.202	0.001 425	0.019 948	1359.3	1195.9	2555.2	1372.4	1366.4	2738.7	3.3010	2.3633	5.6643
310	9.856	0.001 447	0.018 350	1387.1	1159.4	2546.4	1401.3	1326.0	2727.3	3.3493	2.2737	5.6230
315	10.547	0.001 472	0.016 867	1415.5	1121.1	2536.6	1431.0	1283.5	2714.5	3.3982	2.1821	5.5804
320	11.274	0.001 499	0.015 488	1444.6	1080.9	2525.5	1461.5	1238.6	2700.1	3.4480	2.0882	5.5362
330	12.845	0.001 561	0.012 996	1505.3	993.7	2498.9	1525.3	1140.6	2665.9	3.5507	1.8909	5.4417
340	14.586	0.001 638	0.010 797	1570.3	894.3	2464.6	1594.2	1027.9	2622.0	3.6594	1.6763	5.3357
350	16.513	0.001 740	0.008 813	1641.9	776.6	2418.4	1670.6	893.4	2563.9	3.7777	1.4335	5.2112
360	18.651	0.001 893	0.006 945	1725.2	626.3	2351.5	1760.5	720.5	2481.0	3.9147	1.1379	5.0526
370	21.03	0.002 213	0.004 925	1844.0	384.5	2228.5	1890.5	441.6	2332.1	4.1106	.6865	4.7971
374.14	22.09	0.003 155	0.003 155	2029.6	0	2029.6	2099.3	0	2099.3	4.4298	0	4.4298

SOURCE: Reproduced from Joseph H. Keenan, Frederick G. Keyes, Philip G. Hill, and Joan G. Moore, *Steam Tables* (New York: John Wiley & Sons, Inc., 1969) with permission.

Table A.6 Saturated Steam Pressure Table

Press. kPa P	Temp. °C T	Specific Volume		Internal Energy			Enthalpy			Entropy		
		Sat. Liquid v_f	Sat. Vapor v_g	Sat. Liquid u_f	Evap. u_{fg}	Sat. Vapor u_g	Sat. Liquid h_f	Evap. h_{fg}	Sat. Vapor h_g	Sat. Liquid s_f	Evap. s_{fg}	Sat. Vapor s_g
0.6113	0.01	0.001 000	206.14	.00	2375.3	2375.3	.01	2501.3	2501.4	.0000	9.1562	9.1562
1.0	6.98	0.001 000	129.21	29.30	2355.7	2385.0	29.30	2484.9	2514.2	.1059	8.8697	8.9756
1.5	13.03	0.001 001	87.98	54.71	2338.6	2393.3	54.71	2470.6	2525.3	.1957	8.6322	8.8279
2.0	17.50	0.001 001	67.00	73.48	2326.0	2399.5	73.48	2460.0	2533.5	.2607	8.4629	8.7237
2.5	21.08	0.001 002	54.25	88.48	2315.9	2404.4	88.49	2451.6	2540.0	.3120	8.3311	8.6432
3.0	24.08	0.001 003	45.67	101.04	2307.5	2408.5	101.05	2444.5	2545.5	.3545	8.2231	8.5776
4.0	28.96	0.001 004	34.80	121.45	2293.7	2415.2	121.46	2432.9	2554.4	.4226	8.0520	8.4746
5.0	32.88	0.001 005	28.19	137.81	2282.7	2420.5	137.82	2423.7	2561.5	.4764	7.9187	8.3951
7.5	40.29	0.001 008	19.24	168.78	2261.7	2430.5	168.79	2406.0	2574.8	.5764	7.6750	8.2515
10	45.81	0.001 010	14.67	191.82	2246.1	2437.9	191.83	2392.8	2584.7	.6493	7.5009	8.1502
15	53.97	0.001 014	10.02	225.92	2222.8	2448.7	225.94	2373.1	2599.1	.7549	7.2536	8.0085
20	60.06	0.001 017	7.649	251.38	2205.4	2456.7	251.40	2358.3	2609.7	.8320	7.0766	7.9085
25	64.97	0.001 020	6.204	271.90	2191.2	2463.1	271.93	2346.3	2618.2	.8931	6.9383	7.8314
30	69.10	0.001 022	5.229	289.20	2179.2	2468.4	289.23	2336.1	2625.3	.9439	6.8247	7.7686
40	75.87	0.001 027	3.993	317.53	2159.5	2477.0	317.58	2319.2	2636.8	1.0259	6.6441	7.6700
50	81.33	0.001 030	3.240	340.44	2143.4	2483.9	340.49	2305.4	2645.9	1.0910	6.5029	7.5939
75	91.78	0.001 037	2.217	384.31	2112.4	2496.7	384.39	2278.6	2663.0	1.2130	6.2434	7.4564
MPa												
0.100	99.63	0.001 043	1.6940	417.36	2088.7	2506.1	417.46	2258.0	2675.5	1.3026	6.0568	7.3594
0.125	105.99	0.001 048	1.3749	444.19	2069.3	2513.5	444.32	2241.0	2685.4	1.3740	5.9104	7.2844
0.150	111.37	0.001 053	1.1593	466.94	2052.7	2519.7	467.11	2226.5	2693.6	1.4336	5.7897	7.2233
0.175	116.06	0.001 057	1.0036	486.80	2038.1	2524.9	486.99	2213.6	2700.6	1.4849	5.6868	7.1717
0.200	120.23	0.001 061	0.8857	504.49	2025.0	2529.5	504.70	2201.9	2706.7	1.5301	5.5970	7.1271
0.225	124.00	0.001 064	0.7933	520.47	2013.1	2533.6	520.72	2191.3	2712.1	1.5706	5.5173	7.0878

Table A.6 Saturated Steam Pressure Table (*Continued*)

Press. MPa P	Temp. °C T	Specific Volume		Internal Energy			Enthalpy			Entropy		
		Sat. Liquid v_f	Sat. Vapor v_g	Sat. Liquid u_f	Evap. u_{fg}	Sat. Vapor u_g	Sat. Liquid h_f	Evap. h_{fg}	Sat. Vapor h_g	Sat. Liquid s_f	Evap. s_{fg}	Sat. Vapor s_g
0.250	127.44	0.001 067	0.7187	535.10	2002.1	2537.2	535.37	2181.5	2716.9	1.6072	5.4455	7.0527
0.275	130.60	0.001 070	0.6573	548.59	1991.9	2540.5	548.89	2172.4	2721.3	1.6408	5.3801	7.0209
0.300	133.55	0.001 073	0.6058	561.15	1982.4	2543.6	561.47	2163.8	2725.3	1.6718	5.3201	6.9919
0.325	136.30	0.001 076	0.5620	572.90	1973.5	2546.4	573.25	2155.8	2729.0	1.7006	5.2646	6.9652
0.350	138.88	0.001 079	0.5243	583.95	1965.0	2548.9	584.33	2148.1	2732.4	1.7275	5.2130	6.9405
0.375	141.32	0.001 081	0.4914	594.40	1956.9	2551.3	594.81	2140.8	2735.6	1.7528	5.1647	6.9175
0.40	143.63	0.001 084	0.4625	604.31	1949.3	2553.6	604.74	2133.8	2738.6	1.7766	5.1193	6.8959
0.45	147.93	0.001 088	0.4140	622.77	1934.9	2557.6	623.25	2120.7	2743.9	1.8207	5.0359	6.8565
0.50	151.86	0.001 093	0.3749	639.68	1921.6	2561.2	640.23	2108.5	2748.7	1.8607	4.9606	6.8213
0.55	155.48	0.001 097	0.3427	655.32	1909.2	2564.5	655.93	2097.0	2753.0	1.8973	4.8920	6.7893
0.60	158.85	0.001 101	0.3157	669.90	1897.5	2567.4	670.56	2086.3	2756.8	1.9312	4.8288	6.7600
0.65	162.01	0.001 104	0.2927	683.56	1886.5	2570.1	684.28	2076.0	2760.3	1.9627	4.7703	6.7331
0.70	164.97	0.001 108	0.2729	696.44	1876.1	2572.5	697.22	2066.3	2763.5	1.9922	4.7158	6.7080
0.75	167.78	0.001 112	0.2556	708.64	1866.1	2574.7	709.47	2057.0	2766.4	2.0200	4.6647	6.6847
0.80	170.43	0.001 115	0.2404	720.22	1856.6	2576.8	721.11	2048.0	2769.1	2.0462	4.6166	6.6628
0.85	172.96	0.001 118	0.2270	731.27	1847.4	2578.7	732.22	2039.4	2771.6	2.0710	4.5711	6.6421
0.90	175.38	0.001 121	0.2150	741.83	1838.6	2580.5	742.83	2031.1	2773.9	2.0946	4.5280	6.6226
0.95	177.69	0.001 124	0.2042	751.95	1830.2	2582.1	753.02	2023.1	2776.1	2.1172	4.4869	6.6041
1.00	179.91	0.001 127	0.194 44	761.68	1822.0	2583.6	762.81	2015.3	2778.1	2.1387	4.4478	6.5865
1.10	184.09	0.001 133	0.177 53	780.09	1806.3	2586.4	781.34	2000.4	2781.7	2.1792	4.3744	6.5536
1.20	187.99	0.001 139	0.163 33	797.29	1791.5	2588.8	798.65	1986.2	2784.8	2.2166	4.3067	6.5233
1.30	191.64	0.001 144	0.151 25	813.44	1777.5	2591.0	814.93	1972.7	2787.6	2.2515	4.2438	6.4953
1.40	195.07	0.001 149	0.140 84	828.70	1764.1	2592.8	830.30	1959.7	2790.0	2.2842	4.1850	6.4693

1.50	198.32	0.001 154	0.131 77	843.16	1751.3	2594.5	844.89	1947.3	2792.2	2.3150	4.1298	6.4448
1.75	205.76	0.001 166	0.113 49	876.46	1721.4	2597.8	878.50	1917.9	2796.4	2.3851	4.0044	6.3896
2.00	212.42	0.001 177	0.099 63	906.44	1693.8	2600.3	908.79	1890.7	2799.5	2.4474	3.8935	6.3409
2.25	218.45	0.001 187	0.088 75	933.83	1668.2	2602.0	936.49	1865.2	2801.7	2.5035	3.7937	6.2972
2.5	223.99	0.001 197	0.079 98	959.11	1644.0	2603.1	962.11	1841.0	2803.1	2.5547	3.7028	6.2575
3.0	233.90	0.001 217	0.066 68	1004.78	1599.3	2604.1	1008.42	1795.7	2804.2	2.6457	3.5412	6.1869
3.5	242.60	0.001 235	0.057 07	1045.43	1558.3	2603.7	1049.75	1753.7	2803.4	2.7253	3.4000	6.1253
4	250.40	0.001 252	0.049 78	1082.31	1520.0	2602.3	1087.31	1714.1	2801.4	2.7964	3.2737	6.0701
5	263.99	0.001 286	0.039 44	1147.81	1449.3	2597.1	1154.23	1640.1	2794.3	2.9202	3.0532	5.9734
6	275.64	0.001 319	0.032 44	1205.44	1384.3	2589.7	1213.35	1571.0	2784.3	3.0267	2.8625	5.8892
7	285.88	0.001 351	0.027 37	1257.55	1323.0	2580.5	1267.00	1505.1	2772.1	3.1211	2.6922	5.8133
8	295.06	0.001 384	0.023 52	1305.57	1264.2	2569.8	1316.64	1441.3	2758.0	3.2068	2.5364	5.7432
9	303.40	0.001 418	0.020 48	1350.51	1207.3	2557.8	1363.26	1378.9	2742.1	3.2858	2.3915	5.6772
10	311.06	0.001 452	0.018 026	1393.04	1151.4	2544.4	1407.56	1317.1	2724.7	3.3596	2.2544	5.6141
11	318.15	0.001 489	0.015 987	1433.7	1096.0	2529.8	1450.1	1255.5	2705.6	3.4295	2.1233	5.5527
12	324.75	0.001 527	0.014 263	1473.0	1040.7	2513.7	1491.3	1193.6	2684.9	3.4962	1.9962	5.4924
13	330.93	0.001 567	0.012 780	1511.1	985.0	2496.1	1531.5	1130.7	2662.2	3.5606	1.8718	5.4323
14	336.75	0.001 611	0.011 485	1548.6	928.2	2476.8	1571.1	1066.5	2637.6	3.6232	1.7485	5.3717
15	342.24	0.001 658	0.010 337	1585.6	869.8	2455.5	1610.5	1000.0	2610.5	3.6848	1.6249	5.3098
16	347.44	0.001 711	0.009 306	1622.7	809.0	2431.7	1650.1	930.6	2580.6	3.7461	1.4994	5.2455
17	352.37	0.001 770	0.008 364	1660.2	744.8	2405.0	1690.3	856.9	2547.2	3.8079	1.3698	5.1777
18	357.06	0.001 840	0.007 489	1698.9	675.4	2374.3	1732.0	777.1	2509.1	3.8715	1.2329	5.1044
19	361.54	0.001 924	0.006 657	1739.9	598.1	2338.1	1776.5	688.0	2464.5	3.9388	1.0839	5.0228
20	365.81	0.002 036	0.005 834	1785.6	507.5	2293.0	1826.3	583.4	2409.7	4.0139	.9130	4.9269
21	369.89	0.002 207	0.004 952	1842.1	388.5	2230.6	1888.4	446.2	2334.6	4.1075	.6938	4.8013
22	373.80	0.002 742	0.003 568	1961.9	125.2	2087.1	2022.2	143.4	2165.6	4.3110	.2216	4.5327
22.09	374.14	0.003 155	0.003 155	2029.6	0	2029.6	2099.3	0	2099.3	4.4298	0	4.4298

SOURCE: Reproduced from Joseph H. Keenan, Frederick G. Keyes, Philip G. Hill, and Joan G. Moore, *Steam Tables* (New York: John Wiley & Sons, Inc., 1969) with permission.

Table A.7 Superheated Steam Vapor Table

T	$P = .010$ MPa (45.81)				$P = .050$ MPa (81.33)				$P = .10$ MPa (99.63)			
	v	u	h	s	v	u	h	s	v	u	h	s
Sat.	14.674	2437.9	2584.7	8.1502	3.240	2483.9	2645.9	7.5939	1.6940	2506.1	2675.5	7.3594
50	14.869	2443.9	2592.6	8.1749								
100	17.196	2515.5	2687.5	8.4479	3.418	2511.6	2682.5	7.6947	1.6958	2506.7	2676.2	7.3614
150	19.512	2587.9	2783.0	8.6882	3.889	2585.6	2780.1	7.9401	1.9364	2582.8	2776.4	7.6134
200	21.825	2661.3	2879.5	8.9038	4.356	2659.9	2877.7	8.1580	2.172	2658.1	2875.3	7.8343
250	24.136	2736.0	2977.3	9.1002	4.820	2735.0	2976.0	8.3556	2.406	2733.7	2974.3	8.0333
300	26.445	2812.1	3076.5	9.2813	5.284	2811.3	3075.5	8.5373	2.639	2810.4	3074.3	8.2158
400	31.063	2968.9	3279.6	9.6077	6.209	2968.5	3278.9	8.8642	3.103	2967.9	3278.2	8.5435
500	35.679	3132.3	3489.1	9.8978	7.134	3132.0	3488.7	9.1546	3.565	3131.6	3488.1	8.8342
600	40.295	3302.5	3705.4	10.1608	8.057	3302.2	3705.1	9.4178	4.028	3301.9	3704.7	9.0976
700	44.911	3479.6	3928.7	10.4028	8.981	3479.4	3928.5	9.6599	4.490	3479.2	3928.2	9.3398
800	49.526	3663.8	4159.0	10.6281	9.904	3663.6	4158.9	9.8852	4.952	3663.5	4158.6	9.5652
900	54.141	3855.0	4396.4	10.8396	10.828	3854.9	4396.3	10.0967	5.414	3854.8	4396.1	9.7767
1000	58.757	4053.0	4640.6	11.0393	11.751	4052.9	4640.5	10.2964	5.875	4052.8	4640.3	9.9764
1100	63.372	4257.5	4891.2	11.2287	12.674	4257.4	4891.1	10.4859	6.337	4257.3	4891.0	10.1659
1200	67.987	4467.9	5147.8	11.4091	13.597	4467.8	5147.7	10.6662	6.799	4467.7	5147.6	10.3463
1300	72.602	4683.7	5409.7	11.5811	14.521	4683.6	5409.6	10.8382	7.260	4683.5	5409.5	10.5183

T	$P = .20$ MPa (120.23)				$P = .30$ MPa (133.55)				$P = .40$ MPa (143.63)			
	v	u	h	s	v	u	h	s	v	u	h	s
Sat.	.8857	2529.5	2706.7	7.1272	.6058	2543.6	2725.3	6.9919	.4625	2553.6	2738.6	6.8959
150	.9596	2576.9	2768.8	7.2795	.6339	2570.8	2761.0	7.0778	.4708	2564.5	2752.8	6.9299
200	1.0803	2654.4	2870.5	7.5066	.7163	2650.7	2865.6	7.3115	.5342	2646.8	2860.5	7.1706
250	1.1988	2731.2	2971.0	7.7086	.7964	2728.7	2967.6	7.5166	.5951	2726.1	2964.2	7.3789
300	1.3162	2808.6	3071.8	7.8926	.8753	2806.7	3069.3	7.7022	.6548	2804.8	3066.8	7.5662
400	1.5493	2966.7	3276.6	8.2218	1.0315	2965.6	3275.0	8.0330	.7726	2964.4	3273.4	7.8985

P = .20 MPa (120.23)

T	v	u	h	s
500	1.7814	3130.8	3487.1	8.5133
600	2.013	3301.4	3704.0	8.7770
700	2.244	3478.8	3927.6	9.0194
800	2.475	3663.1	4158.2	9.2449
900	2.706	3854.5	4395.8	9.4566
1000	2.937	4052.5	4640.0	9.6563
1100	3.168	4257.0	4890.7	9.8458
1200	3.399	4467.5	5147.3	10.0262
1300	3.630	4683.2	5409.3	10.1982

P = .30 MPa (133.55)

T	v	u	h	s
500	1.1867	3130.0	3486.0	8.3251
600	1.3414	3300.8	3703.2	8.5892
700	1.4957	3478.4	3927.1	8.8319
800	1.6499	3662.9	4157.8	9.0576
900	1.8041	3854.2	4395.4	9.2692
1000	1.9581	4052.3	4639.7	9.4690
1100	2.1121	4256.8	4890.4	9.6585
1200	2.2661	4467.2	5147.1	9.8389
1300	2.4201	4683.0	5409.0	10.0110

P = .40 MPa (143.63)

T	v	u	h	s
500	.8893	3129.2	3484.9	8.1913
600	1.0055	3300.2	3702.4	8.4558
700	1.1215	3477.9	3926.5	8.6987
800	1.2372	3662.4	4157.3	8.9244
900	1.3529	3853.9	4395.1	9.1362
1000	1.4685	4052.0	4639.4	9.3360
1100	1.5840	4256.5	4890.2	9.5256
1200	1.6996	4467.0	5146.8	9.7060
1300	1.8151	4682.8	5408.8	9.8780

P = .50 MPa (151.86)

T	v	u	h	s
Sat.	.3749	2561.2	2748.7	6.8213
200	.4249	2642.9	2855.4	7.0592
250	.4744	2723.5	2960.7	7.2709
300	.5226	2802.9	3064.2	7.4599
350	.5701	2882.6	3167.7	7.6329
400	.6173	2963.2	3271.9	7.7938
500	.7109	3128.4	3483.9	8.0873
600	.8041	3299.6	3701.7	8.3522
700	.8969	3477.5	3925.9	8.5952
800	.9896	3662.1	4156.9	8.8211
900	1.0822	3853.6	4394.7	9.0329
1000	1.1747	4051.8	4639.1	9.2328
1100	1.2672	4256.3	4889.9	9.4224
1200	1.3596	4466.8	5146.6	9.6029
1300	1.4521	4682.5	5408.6	9.7749

P = .60 MPa (158.85)

T	v	u	h	s
Sat.	.3157	2567.4	2756.8	6.7600
200	.3520	2638.9	2850.1	6.9665
250	.3938	2720.9	2957.2	7.1816
300	.4344	2801.0	3061.6	7.3724
350	.4742	2881.2	3165.7	7.5464
400	.5137	2962.1	3270.3	7.7079
500	.5920	3127.6	3482.8	8.0021
600	.6697	3299.1	3700.9	8.2674
700	.7472	3477.0	3925.3	8.5107
800	.8245	3661.8	4156.5	8.7367
900	.9017	3853.4	4394.4	8.9486
1000	.9788	4051.5	4638.8	9.1485
1100	1.0559	4256.1	4889.6	9.3381
1200	1.1330	4466.5	5146.3	9.5185
1300	1.2101	4682.3	5408.3	9.6906

P = .80 MPa (170.43)

T	v	u	h	s
Sat.	.2404	2576.8	2769.1	6.6628
200	.2608	2630.6	2839.3	6.8158
250	.2931	2715.5	2950.0	7.0384
300	.3241	2797.2	3056.5	7.2328
350	.3544	2878.2	3161.7	7.4089
400	.3843	2959.7	3267.1	7.5716
500	.4433	3126.0	3480.6	7.8673
600	.5018	3297.9	3699.4	8.1333
700	.5601	3476.2	3924.2	8.3770
800	.6181	3661.1	4155.6	8.6033
900	.6761	3852.8	4393.7	8.8153
1000	.7340	4051.0	4638.2	9.0153
1100	.7919	4255.6	4889.1	9.2050
1200	.8497	4466.1	5145.9	9.3855
1300	.9076	4681.8	5407.9	9.5575

Table A.7 Superheated Steam Vapor Table (*Continued*)

T	v	u	h	s	v	u	h	s	v	u	h	s
	$P = 1.00$ MPa (179.91)				$P = 1.20$ MPa (187.99)				$P = 1.40$ MPa (195.07)			
Sat.	.194 44	2583.6	2778.1	6.5865	.163 33	2588.8	2784.8	6.5233	.140 84	2592.8	2790.0	6.4693
200	.2060	2621.9	2827.9	6.6940	.169 30	2612.8	2815.9	6.5898	.143 02	2603.1	2803.3	6.4975
250	.2327	2709.9	2942.6	6.9247	.192 34	2704.2	2935.0	6.8294	.163 50	2698.3	2927.2	6.7467
300	.2579	2793.2	3051.2	7.1229	.2138	2789.2	3045.8	7.0317	.182 28	2785.2	3040.4	6.9534
350	.2825	2875.2	3157.7	7.3011	.2345	2872.2	3153.6	7.2121	.2003	2869.2	3149.5	7.1360
400	.3066	2957.3	3263.9	7.4651	.2548	2954.9	3260.7	7.3774	.2178	2952.5	3257.5	7.3026
500	.3541	3124.4	3478.5	7.7622	.2946	3122.8	3476.3	7.6759	.2521	3121.1	3474.1	7.6027
600	.4011	3296.8	3697.9	8.0290	.3339	3295.6	3696.3	7.9435	.2860	3294.4	3694.8	7.8710
700	.4478	3475.3	3923.1	8.2731	.3729	3474.4	3922.0	8.1881	.3195	3473.6	3920.8	8.1160
800	.4943	3660.4	4154.7	8.4996	.4118	3659.7	4153.8	8.4148	.3528	3659.0	4153.0	8.3431
900	.5407	3852.2	4392.9	8.7118	.4505	3851.6	4392.2	8.6272	.3861	3851.1	4391.5	8.5556
1000	.5871	4050.5	4637.6	8.9119	.4892	4050.0	4637.0	8.8274	.4192	4049.5	4636.4	8.7559
1100	.6335	4255.1	4888.6	9.1017	.5278	4254.6	4888.0	9.0172	.4524	4254.1	4887.5	8.9457
1200	.6798	4465.6	5145.4	9.2822	.5665	4465.1	5144.9	9.1977	.4855	4464.7	5144.4	9.1262
1300	.7261	4681.3	5407.4	9.4543	.6051	4680.9	5407.0	9.3698	.5186	4680.4	5406.5	9.2984

T	v	u	h	s	v	u	h	s	v	u	h	s
	$P = 1.60$ MPa (201.41)				$P = 1.80$ MPa (207.15)				$P = 2.00$ MPa (212.42)			
Sat.	.123 80	2596.0	2794.0	6.4218	.110 42	2598.4	2797.1	6.3794	.099 63	2600.3	2799.5	6.3409
225	.132 87	2644.7	2857.3	6.5518	.116 73	2636.6	2846.7	6.4808	.103 77	2628.3	2835.8	6.4147
250	.141 84	2692.3	2919.2	6.6732	.124 97	2686.0	2911.0	6.6066	.111 44	2679.6	2902.5	6.5453
300	.158 62	2781.1	3034.8	6.8844	.140 21	2776.9	3029.2	6.8226	.125 47	2772.6	3023.5	6.7664
350	.174 56	2866.1	3145.4	7.0694	.154 57	2863.0	3141.2	7.0100	.138 57	2859.8	3137.0	6.9563
400	.190 05	2950.1	3254.2	7.2374	.168 47	2947.7	3250.9	7.1794	.151 20	2945.2	3247.6	7.1271
500	.2203	3119.5	3472.0	7.5390	.195 50	3117.9	3469.8	7.4825	.175 68	3116.2	3467.6	7.4317
600	.2500	3293.3	3693.2	7.8080	.2220	3292.1	3691.7	7.7523	.199 60	3290.9	3690.1	7.7024
700	.2794	3472.7	3919.7	8.0535	.2482	3471.8	3918.5	7.9983	.2232	3470.9	3917.4	7.9487

P = 1.60 MPa (201.41) — P = 1.80 MPa (207.15) — P = 2.00 MPa (212.42)

Temp	P = 1.60 MPa (201.41)				P = 1.80 MPa (207.15)				P = 2.00 MPa (212.42)			
	v	u	h	s	v	u	h	s	v	u	h	s
800	.3086	3658.3	4152.1	8.2808	.2742	3657.6	4151.2	8.2258	.2467	3657.0	4150.3	8.1765
900	.3377	3850.5	4390.8	8.4935	.3001	3849.9	4390.1	8.4386	.2700	3849.3	4389.4	8.3895
1000	.3668	4049.0	4635.8	8.6938	.3260	4048.5	4635.2	8.6391	.2933	4048.0	4634.6	8.5901
1100	.3958	4253.7	4887.0	8.8837	.3518	4253.2	4886.4	8.8290	.3166	4252.7	4885.9	8.7800
1200	.4248	4464.2	5143.9	9.0643	.3776	4463.7	5143.4	9.0096	.3398	4463.3	5142.9	8.9607
1300	.4538	4679.9	5406.0	9.2364	.4034	4679.5	5405.6	9.1818	.3631	4679.0	5405.1	9.1329

P = 2.50 MPa (223.99) — P = 3.00 MPa (233.90) — P = 3.50 MPa (242.60)

Temp	P = 2.50 MPa (223.99)				P = 3.00 MPa (233.90)				P = 3.50 MPa (242.60)			
	v	u	h	s	v	u	h	s	v	u	h	s
Sat.	.079 98	2603.1	2803.1	6.2575	.066 68	2604.1	2804.2	6.1869	.057 07	2603.7	2803.4	6.1253
225	.080 27	2605.6	2806.3	6.2639								
250	.087 00	2662.6	2880.1	6.4085	.070 58	2644.0	2855.8	6.2872	.058 72	2623.7	2829.2	6.1749
300	.098 90	2761.6	3008.8	6.6438	.081 14	2750.1	2993.5	6.5390	.068 42	2738.0	2977.5	6.4461
350	.109 76	2851.9	3126.3	6.8403	.090 53	2843.7	3115.3	6.7428	.076 78	2835.3	3104.0	6.6579
400	.120 10	2939.1	3239.3	7.0148	.099 36	2932.8	3230.9	6.9212	.084 53	2926.4	3222.3	6.8405
450	.130 14	3025.5	3350.8	7.1746	.107 87	3020.4	3344.0	7.0834	.091 96	3015.3	3337.2	7.0052
500	.139 98	3112.1	3462.1	7.3234	.116 19	3108.0	3456.5	7.2338	.099 18	3103.0	3450.9	7.1572
600	.159 30	3288.0	3686.3	7.5960	.132 43	3285.0	3682.3	7.5085	.113 24	3282.1	3678.4	7.4339
700	.178 32	3468.7	3914.5	7.8435	.148 38	3466.5	3911.7	7.7571	.126 99	3464.3	3908.8	7.6837
800	.197 16	3655.3	4148.2	8.0720	.164 14	3653.5	4145.9	7.9862	.140 56	3651.8	4143.7	7.9134
900	.215 90	3847.9	4387.6	8.2853	.179 80	3846.5	4385.9	8.1999	.154 02	3845.0	4384.1	8.1276
1000	.2346	4046.7	4633.1	8.4861	.195 41	4045.4	4631.6	8.4009	.167 43	4044.1	4630.1	8.3288
1100	.2532	4251.5	4884.6	8.6762	.210 98	4250.3	4883.3	8.5912	.180 80	4249.2	4881.9	8.5192
1200	.2718	4462.1	5141.7	8.8569	.226 52	4460.9	5140.5	8.7720	.194 15	4459.8	5139.3	8.7000
1300	.2905	4677.8	5404.0	9.0291	.242 06	4676.6	5402.8	8.9442	.207 49	4675.5	5401.7	8.8723

Table A.7 Superheated Steam Vapor Table (*Continued*)

T	v	u	h	s	v	u	h	s	v	u	h	s
	P = 4.0 MPa (250.40)				P = 4.5 MPa (257.49)				P = 5.0 MPa (263.99)			
Sat.	.049 78	2602.3	2801.4	6.0701	.044 06	2600.1	2798.3	6.0198	.039 44	2597.1	2794.3	5.9734
275	.054 57	2667.9	2886.2	6.2285	.047 30	2650.3	2863.2	6.1401	.041 41	2631.3	2838.3	6.0544
300	.058 84	2725.3	2960.7	6.3615	.051 35	2712.0	2943.1	6.2828	.045 32	2698.0	2924.5	6.2084
350	.066 45	2826.7	3092.5	6.5821	.058 40	2817.8	3080.6	6.5131	.051 94	2808.7	3068.4	6.4493
400	.073 41	2919.9	3213.6	6.7690	.064 75	2913.3	3204.7	6.7047	.057 81	2906.6	3195.7	6.6459
450	.080 02	3010.2	3330.3	6.9363	.070 74	3005.0	3323.3	6.8746	.063 30	2999.7	3316.2	6.8186
500	.086 43	3099.5	3445.3	7.0901	.076 51	3095.3	3439.6	7.0301	.068 57	3091.0	3433.8	6.9759
600	.098 85	3279.1	3674.4	7.3688	.087 65	3276.0	3670.5	7.3110	.078 69	3273.0	3666.5	7.2589
700	.110 95	3462.1	3905.9	7.6198	.098 47	3459.9	3903.0	7.5631	.088 49	3457.6	3900.1	7.5122
800	.122 87	3650.0	4141.5	7.8502	.109 11	3648.3	4139.3	7.7942	.098 11	3646.6	4137.1	7.7440
900	.134 69	3843.6	4382.3	8.0647	.119 65	3842.2	4380.6	8.0091	.107 62	3840.7	4378.8	7.9593
1000	.146 45	4042.9	4628.7	8.2662	.130 13	4041.6	4627.2	8.2108	.117 07	4040.4	4625.7	8.1612
1100	.158 17	4248.0	4880.6	8.4567	.140 56	4246.8	4879.3	8.4015	.126 48	4245.6	4878.0	8.3520
1200	.169 87	4458.6	5138.1	8.6376	.150 98	4457.5	5136.9	8.5825	.135 87	4456.3	5135.7	8.5331
1300	.181 56	4674.3	5400.5	8.8100	.161 39	4673.1	5399.4	8.7549	.145 26	4672.0	5398.2	8.7055
	P = 6.0 MPa (275.64)				P = 7.0 MPa (285.88)				P = 8.0 MPa (295.06)			
Sat.	.032 44	2589.7	2784.3	5.8892	.027 37	2580.5	2772.1	5.8133	.023 52	2569.8	2758.0	5.7432
300	.036 16	2667.2	2884.2	6.0674	.029 47	2632.2	2838.4	5.9305	.024 26	2590.9	2785.0	5.7906
350	.042 23	2789.6	3043.0	6.3335	.035 24	2769.4	3016.0	6.2283	.029 95	2747.7	2987.3	6.1301
400	.047 39	2892.9	3177.2	6.5408	.039 93	2878.6	3158.1	6.4478	.034 32	2863.8	3138.3	6.3634
450	.052 14	2988.9	3301.8	6.7193	.044 16	2978.0	3287.1	6.6327	.038 17	2966.7	3272.0	6.5551
500	.056 65	3082.2	3422.2	6.8803	.048 14	3073.4	3410.3	6.7975	.041 75	3064.3	3398.3	6.7240
550	.061 01	3174.6	3540.6	7.0288	.051 95	3167.2	3530.9	6.9486	.045 16	3159.8	3521.0	6.8778
600	.065 25	3266.9	3658.4	7.1677	.055 65	3260.7	3650.3	7.0894	.048 45	3254.4	3642.0	7.0206

524

P = 6.0 MPa (275.64)

T	v	u	h	s
700	.073 52	3453.1	3894.2	7.4234
800	.081 60	3643.1	4132.7	7.6566
900	.089 58	3837.8	4375.3	7.8727
1000	.097 49	4037.8	4622.7	8.0751
1100	.105 36	4243.3	4875.4	8.2661
1200	.113 21	4454.0	5133.3	8.4474
1300	.121 06	4669.6	5396.0	8.6199

P = 7.0 MPa (285.88)

T	v	u	h	s
700	.062 83	3448.5	3888.3	7.3476
800	.069 81	3639.5	4128.2	7.5822
900	.076 69	3835.0	4371.8	7.7991
1000	.083 50	4035.3	4619.8	8.0020
1100	.090 27	4240.9	4872.8	8.1933
1200	.097 03	4451.7	5130.9	8.3747
1300	.103 77	4667.3	5393.7	8.5473

P = 8.0 MPa (295.06)

T	v	u	h	s
700	.054 81	3443.9	3882.4	7.2812
800	.060 97	3636.0	4123.8	7.5173
900	.067 02	3832.1	4368.3	7.7351
1000	.073 01	4032.8	4616.9	7.9384
1100	.078 96	4238.6	4870.3	8.1300
1200	.084 89	4449.5	5128.5	8.3115
1300	.090 80	4665.0	5391.5	8.4842

P = 9.0 MPa (303.40)

T	v	u	h	s
Sat.	.020 48	2557.8	2742.1	5.6772
325	.023 27	2646.6	2856.0	5.8712
350	.025 80	2724.4	2956.6	6.0361
400	.029 93	2848.4	3117.8	6.2854
450	.033 50	2955.2	3256.6	6.4844
500	.036 77	3055.2	3386.1	6.6576
550	.039 87	3152.2	3511.0	6.8142
600	.042 85	3248.1	3633.7	6.9589
650	.045 74	3343.6	3755.3	7.0943
700	.048 57	3439.3	3876.5	7.2221
800	.054 09	3632.5	4119.3	7.4596
900	.059 50	3829.2	4364.8	7.6783
1000	.064 85	4030.3	4614.0	7.8821
1100	.070 10	4236.3	4867.7	8.0740
1200	.075 44	4447.2	5126.2	8.2556
1300	.080 72	4662.7	5389.2	8.4284

P = 10.0 MPa (311.06)

T	v	u	h	s
Sat.	.018 026	2544.4	2724.7	5.6141
325	.019 861	2610.4	2809.1	5.7568
350	.022 42	2699.2	2923.4	5.9443
400	.026 41	2832.4	3096.5	6.2120
450	.029 75	2943.4	3240.9	6.4190
500	.032 79	3045.8	3373.7	6.5966
550	.035 64	3144.6	3500.9	6.7561
600	.038 37	3241.7	3625.3	6.9029
650	.041 01	3338.2	3748.2	7.0398
700	.043 58	3434.7	3870.5	7.1687
800	.048 59	3628.9	4114.8	7.4077
900	.053 49	3826.3	4361.2	7.6272
1000	.058 32	4027.8	4611.0	7.8315
1100	.063 12	4234.0	4865.1	8.0237
1200	.067 89	4444.9	5123.8	8.2055
1300	.072 65	4660.5	5387.0	8.3783

P = 12.5 MPa (327.89)

T	v	u	h	s
Sat.	.013 495	2505.1	2673.8	5.4624
350	.016 126	2624.6	2826.2	5.7118
400	.020 00	2789.3	3039.3	6.0417
450	.022 99	2912.5	3199.8	6.2719
500	.025 60	3021.7	3341.8	6.4618
550	.028 01	3125.0	3475.2	6.6290
600	.030 29	3225.4	3604.0	6.7810
650	.032 48	3324.4	3730.4	6.9218
700	.034 60	3422.9	3855.3	7.0536
800	.038 69	3620.0	4103.6	7.2965
900	.042 67	3819.1	4352.5	7.5182
1000	.046 58	4021.6	4603.8	7.7237
1100	.050 45	4228.2	4858.8	7.9165
1200	.054 30	4439.3	5118.0	8.0987
1300	.058 13	4654.8	5381.4	8.2717

Table A.7 Superheated Steam Vapor Table (*Continued*)

T	v	u	h	s	v	u	h	s	v	u	h	s
	P = 15.0 MPa (342.24)				P = 17.5 MPa (354.75)				P = 20.0 MPa (365.81)			
Sat.	.010 337	2455.5	2610.5	5.3098	.007 920	2390.2	2528.8	5.1419	.005 834	2293.0	2409.7	4.9269
350	.011 470	2520.4	2692.4	5.4421								
400	.015 649	2740.7	2975.5	5.8811	.012 447	2685.0	2902.9	5.7213	.009 942	2619.3	2818.1	5.5540
450	.018 445	2879.5	3156.2	6.1404	.015 174	2844.2	3109.7	6.0184	.012 695	2806.2	3060.1	5.9017
500	.020 80	2996.6	3308.6	6.3443	.017 358	2970.3	3274.1	6.2383	.014 768	2942.9	3238.2	6.1401
550	.022 93	3104.7	3448.6	6.5199	.019 288	3083.9	3421.4	6.4230	.016 555	3062.4	3393.5	6.3348
600	.024 91	3208.6	3582.3	6.6776	.021 06	3191.5	3560.1	6.5866	.018 178	3174.0	3537.6	6.5048
650	.026 80	3310.3	3712.3	6.8224	.022 74	3296.0	3693.9	6.7357	.019 693	3281.4	3675.3	6.6582
700	.028 61	3410.9	3840.1	6.9572	.024 34	3398.7	3824.6	6.8736	.021 13	3386.4	3809.0	6.7993
800	.032 10	3610.9	4092.4	7.2040	.027 38	3601.8	4081.1	7.1244	.023 85	3592.7	4069.7	7.0544
900	.035 46	3811.9	4343.8	7.4279	.030 31	3804.7	4335.1	7.3507	.026 45	3797.5	4326.4	7.2830
1000	.038 75	4015.4	4596.6	7.6348	.033 16	4009.3	4589.5	7.5589	.028 97	4003.1	4582.5	7.4925
1100	.042 00	4222.6	4852.6	7.8283	.035 97	4216.9	4846.4	7.7531	.031 45	4211.3	4840.2	7.6874
1200	.045 23	4433.8	5112.3	8.0108	.038 76	4428.3	5106.6	7.9360	.033 91	4422.8	5101.0	7.8707
1300	.048 45	4649.1	5376.0	8.1840	.041 54	4643.5	5370.5	8.1093	.036 36	4638.0	5365.1	8.0442

T	v	u	h	s	v	u	h	s	v	u	h	s
	P = 25.0 MPa				P = 30.0 MPa				P = 35.0 MPa			
375	.001 973 1	1798.7	1848.0	4.0320	.001 789 2	1737.8	1791.5	3.9305	.001 700 3	1702.9	1762.4	3.8722
400	.006 004	2430.1	2580.2	5.1418	.002 790	2067.4	2151.1	4.4728	.002 100	1914.1	1987.6	4.2126
425	.007 881	2609.2	2806.3	5.4723	.005 303	2455.1	2614.2	5.1504	.003 428	2253.4	2373.4	4.7747
450	.009 162	2720.7	2949.7	5.6744	.006 735	2619.3	2821.4	5.4424	.004 961	2498.7	2672.4	5.1962
500	.011 123	2884.3	3162.4	5.9592	.008 678	2820.7	3081.1	5.7905	.006 927	2751.9	2994.4	5.6282
550	.012 724	3017.5	3335.6	6.1765	.010 168	2970.3	3275.4	6.0342	.008 345	2921.0	3213.0	5.9026
600	.014 137	3137.9	3491.4	6.3602	.011 446	3100.5	3443.9	6.2331	.009 527	3062.0	3395.5	6.1179
650	.015 433	3251.6	3637.4	6.5229	.012 596	3221.0	3598.9	6.4058	.010 575	3189.8	3559.9	6.3010

T	P = 25.0 MPa				P = 30.0 MPa				P = 35.0 MPa			
700	.016 646	3361.3	3777.5	6.6707	.013 661	3335.8	3745.6	6.5606	.011 533	3309.8	3713.5	6.4631
800	.018 912	3574.3	4047.1	6.9345	.015 623	3555.5	4024.2	6.8332	.013 278	3536.7	4001.5	6.7450
900	.021 045	3783.0	4309.1	7.1680	.017 448	3768.5	4291.9	7.0718	.014 883	3754.0	4274.9	6.9886
1000	.023 10	3990.9	4568.5	7.3802	.019 196	3978.8	4554.7	7.2867	.016 410	3966.7	4541.1	7.2064
1100	.025 12	4200.2	4828.2	7.5765	.020 903	4189.2	4816.3	7.4845	.017 895	4178.3	4804.6	7.4057
1200	.027 11	4412.0	5089.9	7.7605	.022 589	4401.3	5079.0	7.6692	.019 360	4390.7	5068.3	7.5910
1300	.029 10	4626.9	5354.4	7.9342	.024 266	4616.0	5344.0	7.8432	.020 815	4605.1	5333.6	7.7653

T	P = 40.0 MPa				P = 50.0 MPa				P = 60.0 MPa			
375	.001 640 7	1677.1	1742.8	3.8290	.001 559 4	1638.6	1716.6	3.7639	.001 502 8	1609.4	1699.5	3.7141
400	.001 907 7	1854.6	1930.9	4.1135	.001 730 9	1788.1	1874.6	4.0031	.001 633 5	1745.4	1843.4	3.9318
425	.002 532	2096.9	2198.1	4.5029	.002 007	1959.7	2060.0	4.2734	.001 816 5	1892.7	2001.7	4.1626
450	.003 693	2365.1	2512.8	4.9459	.002 486	2159.6	2284.0	4.5884	.002 085	2053.9	2179.0	4.4121
500	.005 622	2678.4	2903.3	5.4700	.003 892	2525.5	2720.1	5.1726	.002 956	2390.6	2567.9	4.9321
550	.006 984	2869.7	3149.1	5.7785	.005 118	2763.6	3019.5	5.5485	.003 956	2658.8	2896.2	5.3441
600	.008 094	3022.6	3346.4	6.0114	.006 112	2942.0	3247.6	5.8178	.004 834	2861.1	3151.2	5.6452
650	.009 063	3158.0	3520.6	6.2054	.006 966	3093.5	3441.8	6.0342	.005 595	3028.8	3364.5	5.8829
700	.009 941	3283.6	3681.2	6.3750	.007 727	3230.5	3616.8	6.2189	.006 272	3177.2	3553.5	6.0824
800	.011 523	3517.8	3978.7	6.6662	.009 076	3479.8	3933.6	6.5290	.007 459	3441.5	3889.1	6.4109
900	.012 962	3739.4	4257.9	6.9150	.010 283	3710.3	4224.4	6.7882	.008 508	3681.0	4191.5	6.6805
1000	.014 324	3954.6	4527.6	7.1356	.011 411	3930.5	4501.1	7.0146	.009 480	3906.4	4475.2	6.9127
1100	.015 642	4167.4	4793.1	7.3364	.012 496	4145.7	4770.5	7.2184	.010 409	4124.1	4748.6	7.1195
1200	.016 940	4380.1	5057.7	7.5224	.013 561	4359.1	5037.2	7.4058	.011 317	4338.2	5017.2	7.3083
1300	.018 229	4594.3	5323.5	7.6969	.014 616	4572.8	5303.6	7.5808	.012 215	4551.4	5284.3	7.4837

SOURCE: Reproduced from Joseph H. Keenan, Frederick G. Keyes, Philip G. Hill, and Joan G. Moore, *Steam Tables* (New York: John Wiley & Sons, Inc., 1969) with permission.

Table A.8 Compressed Liquid Table

T	P = 5 MPa (263.99)				P = 10 MPa (311.06)				P = 15 MPa (342.24)			
	v	u	h	s	v	u	h	s	v	u	h	s
Sat.	.001 285 9	1147.8	1154.2	2.9202	.001 452 4	1393.0	1407.6	3.3596	.001 658 1	1585.6	1610.5	3.6848
0	.000 997 7	.04	5.04	.0001	.000 995 2	.09	10.04	.0002	.000 992 8	.15	15.05	.0004
20	.000 999 5	83.65	88.65	.2956	.000 997 2	83.36	93.33	.2945	.000 995 0	83.06	97.99	.2934
40	.001 005 6	166.95	171.97	.5705	.001 003 4	166.35	176.38	.5686	.001 001 3	165.76	180.78	.5666
60	.001 014 9	250.23	255.30	.8285	.001 012 7	249.36	259.49	.8258	.001 010 5	248.51	263.67	.8232
80	.001 026 8	333.72	338.85	1.0720	.001 024 5	332.59	342.83	1.0688	.001 022 2	331.48	346.81	1.0656
100	.001 041 0	417.52	422.72	1.3030	.001 038 5	416.12	426.50	1.2992	.001 036 1	414.74	430.28	1.2955
120	.001 057 6	501.80	507.09	1.5233	.001 054 9	500.08	510.64	1.5189	.001 052 2	498.40	514.19	1.5145
140	.001 076 8	586.76	592.15	1.7343	.001 073 7	584.68	595.42	1.7292	.001 070 7	582.66	598.72	1.7242
160	.001 098 8	672.62	678.12	1.9375	.001 095 3	670.13	681.08	1.9317	.001 091 8	667.71	684.09	1.9260
180	.001 124 0	759.63	765.25	2.1341	.001 119 9	756.65	767.84	2.1275	.001 115 9	753.76	770.50	2.1210
200	.001 153 0	848.1	853.9	2.3255	.001 148 0	844.5	856.0	2.3178	.001 143 3	841.0	858.2	2.3104
220	.001 186 6	938.4	944.4	2.5128	.001 180 5	934.1	945.9	2.5039	.001 174 8	929.9	947.5	2.4953
240	.001 226 4	1031.4	1037.5	2.6979	.001 218 7	1026.0	1038.1	2.6872	.001 211 4	1020.8	1039.0	2.6771
260	.001 274 9	1127.9	1134.3	2.8830	.001 264 5	1121.1	1133.7	2.8699	.001 255 0	1114.6	1133.4	2.8576
280					.001 321 6	1220.9	1234.1	3.0548	.001 308 4	1212.5	1232.1	3.0393
300					.001 397 2	1328.4	1342.3	3.2469	.001 377 0	1316.6	1337.3	3.2260
320									.001 472 4	1431.1	1453.2	3.4247
340									.001 631 1	1567.5	1591.9	3.6546

	P = 20 MPa (365.81)				P = 30 MPa				P = 50 MPa			
T	v	u	h	s	v	u	h	s	v	u	h	s
Sat.	.002 036	1785.6	1826.3	4.0139								
0	.000 990 4	.19	20.01	.0004	.000 985 6	.25	29.82	.0001	.000 976 6	.20	49.03	.0014
20	.000 992 8	82.77	102.62	.2923	.000 988 6	82.17	111.84	.2899	.000 980 4	81.00	130.02	.2848
40	.000 999 2	165.17	185.16	.5646	.000 995 1	164.04	193.89	.5607	.000 987 2	161.86	211.21	.5527
60	.001 008 4	247.68	267.85	.8206	.001 004 2	246.06	276.19	.8154	.000 996 2	242.98	292.79	.8052
80	.001 019 9	330.40	350.80	1.0624	.001 015 6	328.30	358.77	1.0561	.001 007 3	324.34	374.70	1.0440
100	.001 033 7	413.39	434.06	1.2917	.001 029 0	410.78	441.66	1.2844	.001 020 1	405.88	456.89	1.2703
120	.001 049 6	496.76	517.76	1.5102	.001 044 5	493.59	524.93	1.5018	.001 034 8	487.65	539.39	1.4857
140	.001 067 8	580.69	602.04	1.7193	.001 062 1	576.88	608.75	1.7098	.001 051 5	569.77	622.35	1.6915
160	.001 088 5	665.35	687.12	1.9204	.001 082 1	660.82	693.28	1.9096	.001 070 3	652.41	705.92	1.8891
180	.001 112 0	750.95	773.20	2.1147	.001 104 7	745.59	778.73	2.1024	.001 091 2	735.69	790.25	2.0794
200	.001 138 8	837.7	860.5	2.3031	.001 130 2	831.4	865.3	2.2893	.001 114 6	819.7	875.5	2.2634
220	.001 169 3	925.9	949.3	2.4870	.001 159 0	918.3	953.1	2.4711	.001 140 8	904.7	961.7	2.4419
240	.001 204 6	1016.0	1040.0	2.6674	.001 192 0	1006.9	1042.6	2.6490	.001 170 2	990.7	1049.2	2.6158
260	.001 246 2	1108.6	1133.5	2.8459	.001 230 3	1097.4	1134.3	2.8243	.001 203 4	1078.1	1138.2	2.7860
280	.001 296 5	1204.7	1230.6	3.0248	.001 275 5	1190.7	1229.0	2.9986	.001 241 5	1167.2	1229.3	2.9537
300	.001 359 6	1306.1	1333.3	3.2071	.001 330 4	1287.9	1327.8	3.1741	.001 286 0	1258.7	1323.0	3.1200
320	.001 443 7	1415.7	1444.6	3.3979	.001 399 7	1390.7	1432.7	3.3539	.001 338 8	1353.3	1420.2	3.2868
340	.001 568 4	1539.7	1571.0	3.6075	.001 492 0	1501.7	1546.5	3.5426	.001 403 2	1452.0	1522.1	3.4557
360	.001 822 6	1702.8	1739.3	3.8772	.001 626 5	1626.6	1675.4	3.7494	.001 483 8	1556.0	1630.2	3.6291
380					.001 869 1	1781.4	1837.5	4.0012	.001 588 4	1667.2	1746.6	3.8101

SOURCE: Reproduced from Joseph H. Keenan, Frederick G. Keyes, Philip G. Hill, and Joan G. Moore, *Steam Tables* (New York: John Wiley & Sons, Inc., 1969) with permission.

Table A.9 Saturated Ammonia Table

Temp. °C	Abs. Press. kPa P	Specific Volume m³/kg			Enthalpy kJ/kg			Entropy kJ/kg K		
		Sat. Liquid v_f	Evap. v_{fg}	Sat. Vapor v_g	Sat. Liquid h_f	Evap. h_{fg}	Sat. Vapor h_g	Sat. Liquid s_f	Evap. s_{fg}	Sat. Vapor s_g
−50	40.88	0.001 424	2.6239	2.6254	−44.3	1416.7	1372.4	−0.1942	6.3502	6.1561
−48	45.96	0.001 429	2.3518	2.3533	−35.5	1411.3	1375.8	−0.1547	6.2696	6.1149
−46	51.55	0.001 434	2.1126	2.1140	−26.6	1405.8	1379.2	−0.1156	6.1902	6.0746
−44	57.69	0.001 439	1.9018	1.9032	−17.8	1400.3	1382.5	−0.0768	6.1120	6.0352
−42	64.42	0.001 444	1.7155	1.7170	−8.9	1394.7	1385.8	−0.0382	6.0349	5.9967
−40	71.77	0.001 449	1.5506	1.5521	0.0	1389.0	1389.0	0.0000	5.9589	5.9589
−38	79.80	0.001 454	1.4043	1.4058	8.9	1383.3	1392.2	0.0380	5.8840	5.9220
−36	88.54	0.001 460	1.2742	1.2757	17.8	1377.6	1395.4	0.0757	5.8101	5.8858
−34	98.05	0.001 465	1.1582	1.1597	26.8	1371.8	1398.5	0.1132	5.7372	5.8504
−32	108.37	0.001 470	1.0547	1.0562	35.7	1365.9	1401.6	0.1504	5.6652	5.8156
−30	119.55	0.001 476	0.9621	0.9635	44.7	1360.0	1404.6	0.1873	5.5942	5.7815
−28	131.64	0.001 481	0.8790	0.8805	53.6	1354.0	1407.6	0.2240	5.5241	5.7481
−26	144.70	0.001 487	0.8044	0.8059	62.6	1347.9	1410.5	0.2605	5.4548	5.7153
−24	158.78	0.001 492	0.7373	0.7388	71.6	1341.8	1413.4	0.2967	5.3864	5.6831
−22	173.93	0.001 498	0.6768	0.6783	80.7	1335.6	1416.2	0.3327	5.3188	5.6515
−20	190.22	0.001 504	0.6222	0.6237	89.7	1329.3	1419.0	0.3684	5.2520	5.6205
−18	207.71	0.001 510	0.5728	0.5743	98.8	1322.9	1421.7	0.4040	5.1860	5.5900
−16	226.45	0.001 515	0.5280	0.5296	107.8	1316.5	1424.4	0.4393	5.1207	5.5600
−14	246.51	0.001 521	0.4874	0.4889	116.9	1310.0	1427.0	0.4744	5.0561	5.5305
−12	267.95	0.001 528	0.4505	0.4520	126.0	1303.5	1429.5	0.5093	4.9922	5.5015
−10	290.85	0.001 534	0.4169	0.4185	135.2	1296.8	1432.0	0.5440	4.9290	5.4730

Temp (°C)										
−8	315.25	0.001 540	0.3863	0.3878	144.3	1290.1	1434.4	0.5785	4.8664	5.4449
−6	341.25	0.001 546	0.3583	0.3599	153.5	1283.3	1436.8	0.6128	4.8045	5.4173
−4	368.90	0.001 553	0.3328	0.3343	162.7	1276.4	1439.1	0.6469	4.7432	5.3901
−2	398.27	0.001 559	0.3094	0.3109	171.9	1269.4	1441.3	0.6808	4.6825	5.3633
0	429.44	0.001 566	0.2879	0.2895	181.1	1262.4	1443.5	0.7145	4.6223	5.3369
2	462.49	0.001 573	0.2683	0.2698	190.4	1255.2	1445.6	0.7481	4.5627	5.3108
4	497.49	0.001 580	0.2502	0.2517	199.6	1248.0	1447.6	0.7815	4.5037	5.2852
6	534.51	0.001 587	0.2335	0.2351	208.9	1240.6	1449.6	0.8148	4.4451	5.2599
8	573.64	0.001 594	0.2182	0.2198	218.3	1233.2	1451.5	0.8479	4.3871	5.2350
10	614.95	0.001 601	0.2040	0.2056	227.6	1225.7	1453.3	0.8808	4.3295	5.2104
12	658.52	0.001 608	0.1910	0.1926	237.0	1218.1	1455.1	0.9136	4.2725	5.1861
14	704.44	0.001 616	0.1789	0.1805	246.4	1210.4	1456.8	0.9463	4.2159	5.1621
16	752.79	0.001 623	0.1677	0.1693	255.9	1202.6	1458.5	0.9788	4.1597	5.1385
18	803.66	0.001 631	0.1574	0.1590	265.4	1194.7	1460.0	1.0112	4.1039	5.1151
20	857.12	0.001 639	0.1477	0.1494	274.9	1186.7	1461.5	1.0434	4.0486	5.0920
22	913.27	0.001 647	0.1388	0.1405	284.4	1178.5	1462.9	1.0755	3.9937	5.0692
24	972.19	0.001 655	0.1305	0.1322	294.0	1170.3	1464.3	1.1075	3.9392	5.0467
26	1033.97	0.001 663	0.1228	0.1245	303.6	1162.0	1465.6	1.1394	3.8850	5.0244
28	1098.71	0.001 671	0.1156	0.1173	313.2	1153.6	1466.8	1.1711	3.8312	5.0023
30	1166.49	0.001 680	0.1089	0.1106	322.9	1145.0	1467.9	1.2028	3.7777	4.9805
32	1237.41	0.001 689	0.1027	0.1044	332.6	1136.4	1469.0	1.2343	3.7246	4.9589
34	1311.55	0.001 698	0.0969	0.0986	342.3	1127.6	1469.9	1.2656	3.6718	4.9374
36	1389.03	0.001 707	0.0914	0.0931	352.1	1118.7	1470.8	1.2969	3.6192	4.9161
38	1469.92	0.001 716	0.0863	0.0880	361.9	1109.7	1471.5	1.3281	3.5669	4.8950
40	1554.33	0.001 726	0.0815	0.0833	371.7	1100.5	1472.2	1.3591	3.5148	4.8740
42	1642.35	0.001 735	0.0771	0.0788	381.6	1091.2	1472.8	1.3901	3.4630	4.8530
44	1734.09	0.001 745	0.0728	0.0746	391.5	1081.7	1473.2	1.4209	3.4112	4.8322
46	1829.65	0.001 756	0.0689	0.0707	401.5	1072.0	1473.5	1.4518	3.3595	4.8113
48	1929.13	0.001 766	0.0652	0.0669	411.5	1062.2	1473.7	1.4826	3.3079	4.7905
50	2032.62	0.001 777	0.0617	0.0635	421.7	1052.0	1473.7	1.5135	3.2561	4.7696

SOURCE: National Bureau of Standards Circular No. 142, *Tables of Thermodynamic Properties of Ammonia*. Extracted by permission.

Table A.10 Superheated Ammonia Table

Abs. Press. kPa (Sat. Temp.) °C		Temperature, °C											
		−20	−10	0	10	20	30	40	50	60	70	80	100
50 (−46.54)	v	2.4474	2.5481	2.6482	2.7479	2.8473	2.9464	3.0453	3.1441	3.2427	3.3413	3.4397	
	h	1435.8	1457.0	1478.1	1499.2	1520.4	1541.7	1563.0	1584.5	1606.1	1627.8	1649.7	
	s	6.3256	6.4077	6.4865	6.5625	6.6360	6.7073	6.7766	6.8441	6.9099	6.9743	7.0372	
75 (−39.18)	v	1.6233	1.6915	1.7591	1.8263	1.8932	1.9597	2.0261	2.0923	2.1584	2.2244	2.2903	
	h	1433.0	1454.7	1476.1	1497.5	1518.9	1540.3	1561.8	1583.4	1605.1	1626.9	1648.9	
	s	6.1190	6.2028	6.2828	6.3597	6.4339	6.5058	6.5756	6.6434	6.7096	6.7742	6.8373	
100 (−33.61)	v	1.2110	1.2631	1.3145	1.3654	1.4160	1.4664	1.5165	1.5664	1.6163	1.6659	1.7155	1.8145
	h	1430.1	1452.2	1474.1	1495.7	1517.3	1538.9	1560.5	1582.2	1604.1	1626.0	1648.0	1692.6
	s	5.9695	6.0552	6.1366	6.2144	6.2894	6.3618	6.4321	6.5003	6.5668	6.6316	6.6950	6.8177
125 (−29.08)	v	0.9635	1.0059	1.0476	1.0889	1.1297	1.1703	1.2107	1.2509	1.2909	1.3309	1.3707	1.4501
	h	1427.2	1449.8	1472.0	1493.9	1515.7	1537.5	1559.3	1581.1	1603.0	1625.0	1647.2	1691.8
	s	5.8512	5.9389	6.0217	6.1006	6.1763	6.2494	6.3201	6.3887	6.4555	6.5206	6.5842	6.7072
150 (−25.23)	v	0.7984	0.8344	0.8697	0.9045	0.9388	0.9729	1.0068	1.0405	1.0740	1.1074	1.1408	1.2072
	h	1424.1	1447.3	1469.8	1492.1	1514.1	1536.1	1558.0	1580.0	1602.0	1624.1	1646.3	1691.1
	s	5.7526	5.8424	5.9266	6.0066	6.0831	6.1568	6.2280	6.2970	6.3641	6.4295	6.4933	6.6167
200 (−18.86)	v		0.6199	0.6471	0.6738	0.7001	0.7261	0.7519	0.7774	0.8029	0.8282	0.8533	0.9035
	h		1442.0	1465.5	1488.4	1510.9	1533.2	1555.5	1577.7	1599.9	1622.2	1644.6	1689.6
	s		5.6863	5.7737	5.8559	5.9342	6.0091	6.0813	6.1512	6.2189	6.2849	6.3491	6.4732
250 (−13.67)	v		0.4910	0.5135	0.5354	0.5568	0.5780	0.5989	0.6196	0.6401	0.6605	0.6809	0.7212
	h		1436.6	1461.0	1484.5	1507.6	1530.3	1552.9	1575.4	1597.8	1620.3	1642.8	1688.2
	s		5.5609	5.6517	5.7365	5.8165	5.8928	5.9661	6.0368	6.1052	6.1717	6.2365	6.3613
300 (−9.23)	v			0.4243	0.4430	0.4613	0.4792	0.4968	0.5143	0.5316	0.5488	0.5658	0.5997
	h			1456.3	1480.6	1504.2	1527.4	1550.3	1573.0	1595.7	1618.4	1641.1	1686.7
	s			5.5493	5.6366	5.7186	5.7963	5.8707	5.9423	6.0114	6.0785	6.1437	6.2693
	v			0.3605	0.3770	0.3929	0.4086	0.4239	0.4391	0.4541	0.4689	0.4837	0.5129

P (T_sat)		20	30	40	50	60	70	80	100	120	140	160	180
350 (−5.35)	h			1451.5	1476.5	1500.7	1524.4	1547.6	1570.7	1593.6	1616.5	1639.3	1685.2
	s			5.4600	5.5502	5.6342	5.7135	5.7890	5.8615	5.9314	5.9990	6.0647	6.1910
400 (−1.89)	v			0.3125	0.3274	0.3417	0.3556	0.3692	0.3826	0.3959	0.4090	0.4220	0.4478
	h			1446.5	1472.4	1497.2	1521.3	1544.9	1568.3	1591.5	1614.5	1637.6	1683.7
	s			5.3803	5.4735	5.5597	5.6405	5.7173	5.7907	5.8613	5.9296	5.9957	6.1228
450 (1.26)	v			0.2752	0.2887	0.3017	0.3143	0.3266	0.3387	0.3506	0.3624	0.3740	0.3971
	h			1441.3	1468.1	1493.6	1518.2	1542.2	1565.9	1589.3	1612.6	1635.8	1682.2
	s			5.3078	5.4042	5.4926	5.5752	5.6532	5.7275	5.7989	5.8678	5.9345	6.0623
500 (4.14)	v	0.2698	0.2813	0.2926	0.3036	0.3144	0.3251	0.3357	0.3565	0.3771	0.3975		
	h	1489.9	1515.0	1539.5	1563.4	1587.1	1610.6	1634.0	1680.7	1727.5	1774.7		
	s	5.4314	5.5157	5.5950	5.6704	5.7425	5.8120	5.8793	6.0079	6.1301	6.2472		
600 (9.29)	v	0.2217	0.2317	0.2414	0.2508	0.2600	0.2691	0.2781	0.2957	0.3130	0.3302		
	h	1482.4	1508.6	1533.8	1558.5	1582.7	1606.6	1630.4	1677.7	1724.9	1772.4		
	s	5.3222	5.4102	5.4923	5.5697	5.6436	5.7144	5.7826	5.9129	6.0363	6.1541		
700 (13.81)	v	0.1874	0.1963	0.2048	0.2131	0.2212	0.2291	0.2369	0.2522	0.2672	0.2821		
	h	1474.5	1501.9	1528.1	1553.4	1578.2	1602.6	1626.8	1674.6	1722.4	1770.2		
	s	5.2259	5.3179	5.4029	5.4826	5.5582	5.6303	5.6997	5.8316	5.9562	6.0749		
800 (17.86)	v	0.1615	0.1696	0.1773	0.1848	0.1920	0.1991	0.2060	0.2196	0.2329	0.2459	0.2589	
	h	1466.3	1495.0	1522.2	1548.3	1573.7	1598.6	1623.1	1671.6	1719.8	1768.0	1816.4	
	s	5.1387	5.2351	5.3232	5.4053	5.4827	5.5562	5.6268	5.7603	5.8861	6.0057	6.1202	
900 (21.54)	v		0.1488	0.1559	0.1627	0.1693	0.1757	0.1820	0.1942	0.2061	0.2178	0.2294	
	h		1488.0	1516.2	1543.0	1569.1	1594.4	1619.4	1668.5	1717.1	1765.7	1814.4	
	s		5.1593	5.2508	5.3354	5.4147	5.4897	5.5614	5.6968	5.8237	5.9442	6.0594	
1000 (24.91)	v		0.1321	0.1388	0.1450	0.1511	0.1570	0.1627	0.1739	0.1847	0.1954	0.2058	0.2162
	h		1480.6	1510.0	1537.7	1564.4	1590.3	1615.6	1665.4	1714.5	1763.4	1812.4	1861.7
	s		5.0889	5.1840	5.2713	5.3525	5.4292	5.5021	5.6392	5.7674	5.8888	6.0047	6.1159
1200 (30.96)	v			0.1129	0.1185	0.1238	0.1289	0.1338	0.1434	0.1526	0.1616	0.1705	0.1792
	h			1497.1	1526.6	1554.7	1581.7	1608.0	1659.2	1709.2	1758.9	1808.5	1858.2
	s			5.0629	5.1560	5.2416	5.3215	5.3970	5.5379	5.6687	5.7919	5.9091	6.0214

Table A.10 Superheated Ammonia Table (*Continued*)

Abs. Press. kPa (Sat. Temp.) °C		20	30	40	50	60	70	80	100	120	140	160	180
								Temperature, °C					
1400 (36.28)	v			0.0944	0.0995	0.1042	0.1088	0.1132	0.1216	0.1297	0.1376	0.1452	0.1528
	h			1483.4	1515.1	1544.7	1573.0	1600.2	1652.8	1703.9	1754.3	1804.5	1854.7
	s			4.9534	5.0530	5.1434	5.2270	5.3053	5.4501	5.5836	5.7087	5.8273	5.9406
1600 (41.05)	v				0.0851	0.0895	0.0937	0.0977	0.1053	0.1125	0.1195	0.1263	0.1330
	h				1502.9	1534.4	1564.0	1592.3	1646.4	1698.5	1749.7	1800.5	1851.2
	s				4.9584	5.0543	5.1419	5.2232	5.3722	5.5084	5.6355	5.7555	5.8699
1800 (45.39)	v				0.0739	0.0781	0.0820	0.0856	0.0926	0.0992	0.1055	0.1116	0.1177
	h				1490.0	1523.5	1554.6	1584.1	1639.8	1693.1	1745.1	1796.5	1847.7
	s				4.8693	4.9715	5.0635	5.1482	5.3018	5.4409	5.5699	5.6914	5.8069
2000 (49.38)	v				0.0648	0.0688	0.0725	0.0760	0.0824	0.0885	0.0943	0.0999	0.1054
	h				1476.1	1512.0	1544.9	1575.6	1633.2	1687.6	1740.4	1792.4	1844.1
	s				4.7834	4.8930	4.9902	5.0786	5.2371	5.3793	5.5104	5.6333	5.7499

SOURCE: National Bureau of Standards Circular No. 142, *Tables of Thermodynamic Properties of Ammonia.* Extracted by permission.

534

Table A.11 Saturated Freon-12 Table

Temp. °C	Abs. Press. MPa P	Specific Volume m³/kg			Enthalpy kJ/kg			Entropy kJ/kg K		
		Sat. Liquid v_f	Evap. v_{fg}	Sat. Vapor v_g	Sat. Liquid h_f	Evap. h_{fg}	Sat. Vapor h_g	Sat. Liquid s_f	Evap. s_{fg}	Sat. Vapor s_g
−90	0.0028	0.000 608	4.414 937	4.415 545	−43.243	189.618	146.375	−0.2084	1.0352	0.8268
−85	0.0042	0.000 612	3.036 704	3.037 316	−38.968	187.608	148.640	−0.1854	0.9970	0.8116
−80	0.0062	0.000 617	2.137 728	2.138 345	−34.688	185.612	150.924	−0.1630	0.9609	0.7979
−75	0.0088	0.000 622	1.537 030	1.537 651	−30.401	183.625	153.224	−0.1411	0.9266	0.7855
−70	0.0123	0.000 627	1.126 654	1.127 280	−26.103	181.640	155.536	−0.1197	0.8940	0.7744
−65	0.0168	0.000 632	0.840 534	0.841 166	−21.793	179.651	157.857	−0.0987	0.8630	0.7643
−60	0.0226	0.000 637	0.637 274	0.637 910	−17.469	177.653	160.184	−0.0782	0.8334	0.7552
−55	0.0300	0.000 642	0.490 358	0.491 000	−13.129	175.641	162.512	−0.0581	0.8051	0.7470
−50	0.0391	0.000 648	0.382 457	0.383 105	−8.772	173.611	164.840	−0.0384	0.7779	0.7396
−45	0.0504	0.000 654	0.302 029	0.302 682	−4.396	171.558	167.163	−0.0190	0.7519	0.7329
−40	0.0642	0.000 659	0.241 251	0.241 910	−0.000	169.479	169.479	−0.0000	0.7269	0.7269
−35	0.0807	0.000 666	0.194 732	0.195 398	4.416	167.368	171.784	0.0187	0.7027	0.7214
−30	0.1004	0.000 672	0.158 703	0.159 375	8.854	165.222	174.076	0.0371	0.6795	0.7165
−25	0.1237	0.000 679	0.130 487	0.131 166	13.315	163.037	176.352	0.0552	0.6570	0.7121
−20	0.1509	0.000 685	0.108 162	0.108 847	17.800	160.810	178.610	0.0730	0.6352	0.7082
−15	0.1826	0.000 693	0.090 326	0.091 018	22.312	158.534	180.846	0.0906	0.6141	0.7046
−10	0.2191	0.000 700	0.075 946	0.076 646	26.851	156.207	183.058	0.1079	0.5936	0.7014
−5	0.2610	0.000 708	0.064 255	0.064 963	31.420	153.823	185.243	0.1250	0.5736	0.6986
0	0.3086	0.000 716	0.054 673	0.055 389	36.022	151.376	187.397	0.1418	0.5542	0.6960
5	0.3626	0.000 724	0.046 761	0.047 485	40.659	148.859	189.518	0.1585	0.5351	0.6937

Table A.11 Saturated Freon-12 Table (*Continued*)

Temp. °C	Abs. Press. MPa P	Specific Volume m³/kg Sat. Liquid v_f	Evap. v_{fg}	Sat. Vapor v_g	Enthalpy kJ/kg Sat. Liquid h_f	Evap. h_{fg}	Sat. Vapor h_g	Entropy kJ/kg K Sat. Liquid s_f	Evap. s_{fg}	Sat. Vapor s_g
10	0.4233	0.000 733	0.040 180	0.040 914	45.337	146.265	191.602	0.1750	0.5165	0.6916
15	0.4914	0.000 743	0.034 671	0.035 413	50.058	143.586	193.644	0.1914	0.4983	0.6897
20	0.5673	0.000 752	0.030 028	0.030 780	54.828	140.812	195.641	0.2076	0.4803	0.6879
25	0.6516	0.000 763	0.026 091	0.026 854	59.653	137.933	197.586	0.2237	0.4626	0.6863
30	0.7449	0.000 774	0.022 734	0.023 508	64.539	134.936	199.475	0.2397	0.4451	0.6848
35	0.8477	0.000 786	0.019 855	0.020 641	69.494	131.805	201.299	0.2557	0.4277	0.6834
40	0.9607	0.000 798	0.017 373	0.018 171	74.527	128.525	203.051	0.2716	0.4104	0.6820
45	1.0843	0.000 811	0.015 220	0.016 032	79.647	125.074	204.722	0.2875	0.3931	0.6806
50	1.2193	0.000 826	0.013 344	0.014 170	84.868	121.430	206.298	0.3034	0.3758	0.6792
55	1.3663	0.000 841	0.011 701	0.012 542	90.201	117.565	207.766	0.3194	0.3582	0.6777
60	1.5259	0.000 858	0.010 253	0.011 111	95.665	113.443	209.109	0.3355	0.3405	0.6760
65	1.6988	0.000 877	0.008 971	0.009 847	101.279	109.024	210.303	0.3518	0.3224	0.6742
70	1.8858	0.000 897	0.007 828	0.008 725	107.067	104.255	211.321	0.3683	0.3038	0.6721
75	2.0874	0.000 920	0.006 802	0.007 723	113.058	99.068	212.126	0.3851	0.2845	0.6697
80	2.3046	0.000 946	0.005 875	0.006 821	119.291	93.373	212.665	0.4023	0.2644	0.6667
85	2.5380	0.000 976	0.005 029	0.006 005	125.818	87.047	212.865	0.4201	0.2430	0.6631
90	2.7885	0.001 012	0.004 246	0.005 258	132.708	79.907	212.614	0.4385	0.2200	0.6585
95	3.0569	0.001 056	0.003 508	0.004 563	140.068	71.658	211.726	0.4579	0.1946	0.6526
100	3.3440	0.001 113	0.002 790	0.003 903	148.076	61.768	209.843	0.4788	0.1655	0.6444
105	3.6509	0.001 197	0.002 045	0.003 242	157.085	49.014	206.099	0.5023	0.1296	0.6319
110	3.9784	0.001 364	0.001 098	0.002 462	168.059	28.425	196.484	0.5322	0.0742	0.6064
112	4.1155	0.001 792	0.000 005	0.001 797	174.920	0.151	175.071	0.5651	0.0004	0.5655

SOURCE: Copyright 1955 and 1956, E. I. du Pont de Nemours & Company, Inc. Reprinted by permission.

Table A.12 Superheated Freon-12 Table

Temp. °C	0.05 MPa v (m³/kg)	h (kJ/kg)	s (kJ/kg K)	0.10 MPa v (m³/kg)	h (kJ/kg)	s (kJ/kg K)	0.15 MPa v (m³/kg)	h (kJ/kg)	s (kJ/kg K)
−20.0	0.341 857	181.042	0.7912	0.167 701	179.861	0.7401			
−10.0	0.356 227	186.757	0.8133	0.175 222	185.707	0.7628	0.114 716	184.619	0.7318
0.0	0.370 508	192.567	0.8350	0.182 647	191.628	0.7849	0.119 866	190.660	0.7543
10.0	0.384 716	198.471	0.8562	0.189 994	197.628	0.8064	0.124 932	196.762	0.7763
20.0	0.398 863	204.469	0.8770	0.197 277	203.707	0.8275	0.129 930	202.927	0.7977
30.0	0.412 959	210.557	0.8974	0.204 506	209.866	0.8482	0.134 873	209.160	0.8186
40.0	0.427 012	216.733	0.9175	0.211 691	216.104	0.8684	0.139 768	215.463	0.8390
50.0	0.441 030	222.997	0.9372	0.218 839	222.421	0.8883	0.144 625	221.835	0.8591
60.0	0.455 017	229.344	0.9565	0.225 955	228.815	0.9078	0.149 450	228.277	0.8787
70.0	0.468 978	235.774	0.9755	0.233 044	235.285	0.9269	0.154 247	234.789	0.8980
80.0	0.482 917	242.282	0.9942	0.240 111	241.829	0.9457	0.159 020	241.371	0.9169
90.0	0.496 838	248.868	1.0126	0.247 159	248.446	0.9642	0.163 774	248.020	0.9354

Temp. °C	0.20 MPa v (m³/kg)	h (kJ/kg)	s (kJ/kg K)	0.25 MPa v (m³/kg)	h (kJ/kg)	s (kJ/kg K)	0.30 MPa v (m³/kg)	h (kJ/kg)	s (kJ/kg K)
0.0	0.088 608	189.669	0.7320	0.069 752	188.644	0.7139	0.057 150	187.583	0.6984
10.0	0.092 550	195.878	0.7543	0.073 024	194.969	0.7366	0.059 984	194.034	0.7216
20.0	0.096 418	202.135	0.7760	0.076 218	201.322	0.7587	0.062 734	200.490	0.7440
30.0	0.100 228	208.446	0.7972	0.079 350	207.715	0.7801	0.065 418	206.969	0.7658
40.0	0.103 989	214.814	0.8178	0.082 431	214.153	0.8010	0.068 049	213.480	0.7869
50.0	0.107 710	221.243	0.8381	0.085 470	220.642	0.8214	0.070 635	220.030	0.8075
60.0	0.111 397	227.735	0.8578	0.088 474	227.185	0.8413	0.073 185	226.627	0.8276
70.0	0.115 055	234.291	0.8772	0.091 449	233.785	0.8608	0.075 705	233.273	0.8473
80.0	0.118 690	240.910	0.8962	0.094 398	240.443	0.8800	0.078 200	239.971	0.8665
90.0	0.122 304	247.593	0.9149	0.097 327	247.160	0.8987	0.080 673	246.723	0.8853
100.0	0.125 901	254.339	0.9332	0.100 238	253.936	0.9171	0.083 127	253.530	0.9038
110.0	0.129 483	261.147	0.9512	0.103 134	260.770	0.9352	0.085 566	260.391	0.9220

Table A.12 Superheated Freon-12 Table (*Continued*)

Temp. °C	0.40 MPa v m³/kg	0.40 MPa h kJ/kg	0.40 MPa s kJ/kg K	0.50 MPa v m³/kg	0.50 MPa h kJ/kg	0.50 MPa s kJ/kg K	0.60 MPa v m³/kg	0.60 MPa h kJ/kg	0.60 MPa s kJ/kg K
20.0	0.045 836	198.762	0.7199	0.035 646	196.935	0.6999	0.030 422	202.116	0.7063
30.0	0.047 971	205.428	0.7423	0.037 464	203.814	0.7230	0.031 966	209.154	0.7291
40.0	0.050 046	212.095	0.7639	0.039 214	210.656	0.7452	0.033 450	216.141	0.7511
50.0	0.052 072	218.779	0.7849	0.040 911	217.484	0.7667	0.034 887	223.104	0.7723
60.0	0.054 059	225.488	0.8054	0.042 565	224.315	0.7875	0.036 285	230.062	0.7929
70.0	0.056 014	232.230	0.8253	0.044 184	231.161	0.8077	0.037 653	237.027	0.8129
80.0	0.057 941	239.012	0.8448	0.045 774	238.031	0.8275	0.038 995	244.009	0.8324
90.0	0.059 846	245.837	0.8638	0.047 340	244.932	0.8467	0.040 316	251.016	0.8514
100.0	0.061 731	252.707	0.8825	0.048 886	251.869	0.8656	0.041 619	258.053	0.8700
110.0	0.063 600	259.624	0.9008	0.050 415	258.845	0.8840	0.042 907	265.124	0.8882
120.0	0.065 455	266.590	0.9187	0.051 929	265.862	0.9021	0.044 181	272.231	0.9061
130.0	0.067 298	273.605	0.9364	0.053 430	272.923	0.9198			

Temp. °C	0.70 MPa v m³/kg	0.70 MPa h kJ/kg	0.70 MPa s kJ/kg K	0.80 MPa v m³/kg	0.80 MPa h kJ/kg	0.80 MPa s kJ/kg K	0.90 MPa v m³/kg	0.90 MPa h kJ/kg	0.90 MPa s kJ/kg K
40.0	0.026 761	207.580	0.7148	0.022 830	205.924	0.7016	0.019 744	204.170	0.6982
50.0	0.028 100	214.745	0.7373	0.024 068	213.290	0.7248	0.020 912	211.765	0.7131
60.0	0.029 387	221.854	0.7590	0.025 247	220.558	0.7469	0.022 012	219.212	0.7358
70.0	0.030 632	228.931	0.7799	0.026 380	227.766	0.7682	0.023 062	226.564	0.7575
80.0	0.031 843	235.997	0.8002	0.027 477	234.941	0.7888	0.024 072	233.856	0.7785
90.0	0.033 027	243.066	0.8199	0.028 545	242.101	0.8088	0.025 051	241.113	0.7987
100.0	0.034 189	250.146	0.8392	0.029 588	249.260	0.8283	0.026 005	248.355	0.8184
110.0	0.035 332	257.247	0.8579	0.030 612	256.428	0.8472	0.026 937	255.593	0.8376
120.0	0.036 458	264.374	0.8763	0.031 619	263.613	0.8657	0.027 851	262.839	0.8562
130.0	0.037 572	271.531	0.8943	0.032 612	270.820	0.8838	0.028 751	270.100	0.8745
140.0	0.038 673	278.720	0.9119	0.033 592	278.055	0.9016	0.029 639	277.381	0.8923
150.0	0.039 764	285.946	0.9292	0.034 563	285.320	0.9189	0.030 515	284.687	0.9098

1.00 MPa / 1.20 MPa / 1.40 MPa

T (°C)	1.00 MPa			1.20 MPa			1.40 MPa		
50.0	0.018 366	210.162	0.7021	0.014 483	206.661	0.6812			
60.0	0.019 410	217.810	0.7254	0.015 463	214.805	0.7060	0.012 579	211.457	0.6876
70.0	0.020 397	225.319	0.7476	0.016 368	222.687	0.7293	0.013 448	219.822	0.7123
80.0	0.021 341	232.739	0.7689	0.017 221	230.398	0.7514	0.014 247	227.891	0.7355
90.0	0.022 251	240.101	0.7895	0.018 032	237.995	0.7727	0.014 997	235.766	0.7575
100.0	0.023 133	247.430	0.8094	0.018 812	245.518	0.7931	0.015 710	243.512	0.7785
110.0	0.023 993	254.743	0.8287	0.019 567	252.993	0.8129	0.016 393	251.170	0.7988
120.0	0.024 835	262.053	0.8475	0.020 301	260.441	0.8320	0.017 053	258.770	0.8183
130.0	0.025 661	269.369	0.8659	0.021 018	267.875	0.8507	0.017 695	266.334	0.8373
140.0	0.026 474	276.699	0.8839	0.021 721	275.307	0.8689	0.018 321	273.877	0.8558
150.0	0.027 275	284.047	0.9015	0.022 412	282.745	0.8867	0.018 934	281.411	0.8738
160.0	0.028 068	291.419	0.9187	0.023 093	290.195	0.9041	0.019 535	288.946	0.8914

1.60 MPa / 1.80 MPa / 2.00 MPa

T (°C)	1.60 MPa			1.80 MPa			2.00 MPa		
70.0	0.011 208	216.650	0.6959	0.009 406	213.049	0.6794			
80.0	0.011 984	225.177	0.7204	0.010 187	222.198	0.7057	0.008 704	218.859	0.6909
90.0	0.012 698	233.390	0.7433	0.010 884	230.835	0.7298	0.009 406	228.056	0.7166
100.0	0.013 366	241.397	0.7651	0.011 526	239.155	0.7524	0.010 035	236.760	0.7402
110.0	0.014 000	249.264	0.7859	0.012 126	247.264	0.7739	0.010 615	245.154	0.7624
120.0	0.014 608	257.035	0.8059	0.012 697	255.228	0.7944	0.011 159	253.341	0.7835
130.0	0.015 195	264.742	0.8253	0.013 244	263.094	0.8141	0.011 676	261.384	0.8037
140.0	0.015 765	272.406	0.8440	0.013 772	270.891	0.8332	0.012 172	269.327	0.8232
150.0	0.016 320	280.044	0.8623	0.014 284	278.642	0.8518	0.012 651	277.201	0.8420
160.0	0.016 864	287.669	0.8801	0.014 784	286.364	0.8698	0.013 116	285.027	0.8603
170.9	0.017 398	295.290	0.8975	0.015 272	294.069	0.8874	0.013 570	292.822	0.8781
180.0	0.017 923	302.914	0.9145	0.015 752	301.767	0.9046	0.014 013	300.598	0.8955

2.50 MPa / 3.00 MPa / 3.50 MPa

T (°C)	2.50 MPa			3.00 MPa			3.50 MPa		
90.0	0.006 595	219.562	0.6823	0.005 231	220.529	0.6770			
100.0	0.007 264	229.852	0.7103						

Table A.12 Superheated Freon-12 Table (*Continued*)

Temp. °C	2.50 MPa v m³/kg	h kJ/kg	s kJ/kg K	3.00 MPa v m³/kg	h kJ/kg	s kJ/kg K	3.50 MPa v m³/kg	h kJ/kg	s kJ/kg K
110.0	0.007 837	239.271	0.7352	0.005 886	232.068	0.7075	0.004 324	222.121	0.6750
120.0	0.008 351	248.192	0.7582	0.006 419	242.208	0.7336	0.004 959	234.875	0.7078
130.0	0.008 827	256.794	0.7798	0.006 887	251.632	0.7573	0.005 456	245.661	0.7349
140.0	0.009 273	265.180	0.8003	0.007 313	260.620	0.7793	0.005 884	255.524	0.7591
150.0	0.009 697	273.414	0.8200	0.007 709	269.319	0.8001	0.006 270	264.846	0.7814
160.0	0.010 104	281.540	0.8390	0.008 083	277.817	0.8200	0.006 626	273.817	0.8023
170.0	0.010 497	289.589	0.8574	0.008 439	286.171	0.8391	0.006 961	282.545	0.8222
180.0	0.010 879	297.583	0.8752	0.008 782	294.422	0.8575	0.007 279	291.100	0.8413
190.0	0.011 250	305.540	0.8926	0.009 114	302.597	0.8753	0.007 584	299.528	0.8597
200.0	0.011 614	313.472	0.9095	0.009 436	310.718	0.8927	0.007 878	307.864	0.8775

Temp. °C	4.00 MPa v m³/kg	h kJ/kg	s kJ/kg K
120.0	0.003 736	224.863	0.6771
130.0	0.004 325	238.443	0.7111
140.0	0.004 781	249.703	0.7386
150.0	0.005 172	259.904	0.7630
160.0	0.005 522	269.492	0.7854
170.0	0.005 845	278.684	0.8063
180.0	0.006 147	287.602	0.8262
190.0	0.006 434	296.326	0.8453
200.0	0.006 708	304.906	0.8636
210.0	0.006 972	313.380	0.8813
220.0	0.007 228	321.774	0.8985
230.0	0.007 477	330.108	0.9152

SOURCE: Copyright 1955 and 1956, E. I. du Pont de Nemours & Company, Inc. Reprinted by permission.

Table A.13 Properties of Selected Materials at 20°C

Material	λ, W/m · K	c, kJ/kg · K	ρ kg/m³	ε
Metals				
Aluminum	236	0.896	2702	0.037 (polished)
Copper	399	0.383	8933	0.035 (slightly tarnished)
Gold	316	0.129	19 300	0.446 (not polished)
Iron	81	0.452	7870	0.741 (oxidized smooth)
Lead	35	0.129	11 340	0.226 (gray oxidized)
Magnesium	156	1.017	1740	
Nickel	91	0.446	8900	0.389 (oxidized)
Silver	427	0.234	10 500	0.03 (polished)
Zinc	121	0.385	7140	0.237 (tarnished)
Alloys				
Brass (70% Cu, 30% Zn)	111	0.385	8522	0.209 (tarnished)
Cast Iron ($\approx$ 4% C)	52	0.420	7272	0.64 (oxidized)
Stainless steel	14	0.461	7817	0.14
Steel, 1% C	43	0.473	7801	0.921 (oxidized rough)
Insulating				
Asbestos	0.113	0.816	383	
Brick, common	0.42	0.840	1800	0.911 (red rough)
Cork, boards	0.042	1.880	150	
Glass fiber	0.035		220	
85% magnesia	0.022		270	
Fireclay brick	0.347	0.963	2322	0.75
Chrome brick	0.774	0.837	3011	
Wood (oak)	0.069	2.386	528	0.840 (planed)

Table A.14 Physical Properties of Selected Fluids

Dry Air at Atmospheric Pressure

Temperature T, K	Density ρ, kg/m³	Specific Heat c_p, J/kg·K	Thermal Conductivity λ, W/m·K	Thermal Diffusivity $\alpha \times 10^6$, m²/s	Absolute Viscosity $\mu \times 10^6$, N·s/m²	Kinematic Viscosity $\nu \times 10^6$, m²/s	Prandtl Number Pr
273	1.252	1011	0.0237	19.2	17.456	13.9	0.71
293	1.164	1012	0.0251	22.0	18.240	15.7	0.71
313	1.092	1014	0.0265	24.8	19.123	17.6	0.71
333	1.025	1017	0.0279	27.6	19.907	19.4	0.71
353	0.968	1019	0.0293	30.6	20.790	21.5	0.71
373	0.916	1022	0.0307	33.6	21.673	23.6	0.71
473	0.723	1035	0.0370	49.7	25.693	35.5	0.71
573	0.596	1047	0.0429	68.9	39.322	49.2	0.71
673	0.508	1059	0.0485	89.4	32.754	64.6	0.72
773	0.442	1076	0.0540	113.2	35.794	81.0	0.72
1273	0.268	1139	0.0762	240	48.445	181	0.74

Water at Saturation Pressure

Temperature T, K	Density ρ, kg/m³	Specific Heat c_p, J/kg·K	Thermal Conductivity λ, W/m·K	Thermal Diffusivity $\alpha \times 10^6$, m²/s	Absolute Viscosity $\mu \times 10^6$, N·s/m²	Kinematic Viscosity $\nu \times 10^6$, m²/s	Prandtl Number Pr
273	999.3	4226	0.558	0.131	1794	1.789	13.7
293	998.2	4182	0.597	0.143	993	1.006	7.0
313	992.2	4175	0.633	0.151	658	0.658	4.3
333	983.2	4181	0.658	0.159	472	0.478	3.00
353	971.8	4194	0.673	0.165	352	0.364	2.25
373	958.4	4211	0.682	0.169	278	0.294	1.75
473	862.8	4501	0.665	0.170	139	0.160	0.95
573	712.5	5694	0.564	0.132	92.2	0.128	0.98

SOURCE: From K. Raznjevic, *Handbook of Thermodynamic Tables and Charts* (New York: McGraw-Hill Book Company, 1976).

Refrigerant R 12 (CCl_2F_2), Saturated Liquid

Temperature T, K	Density ρ, kg/m^3	Specific Heat c_p, J/kg·K	Thermal Conductivity λ, W/m·K	Thermal Diffusivity $\alpha \times 10^8$, m^2/s	Absolute Viscosity $\mu \times 10^4$, N·s/m^2	Kinematic Viscosity $\nu \times 10^6$, m^2/s	Prandtl Number Pr
223	1547	875.0	0.067	5.01	4.796	0.310	6.2
233	1519	884.7	0.069	5.14	4.238	0.279	5.4
243	1490	895.6	0.069	5.26	3.770	0.253	4.8
253	1461	907.3	0.071	5.39	3.433	0.235	4.4
263	1429	920.3	0.073	5.50	3.158	0.221	4.0
273	1397	934.5	0.073	5.57	2.990	0.214	3.8
283	1364	949.6	0.073	5.60	2.769	0.203	3.6
293	1330	965.9	0.073	5.60	2.633	0.198	3.5
303	1295	983.5	0.071	5.60	2.512	0.194	3.5
313	1257	1001.9	0.069	5.55	2.401	0.191	3.5
323	1216	1021.6	0.067	5.45	2.310	0.190	3.5

Unused Engine Oil, Saturated Liquid

Temperature T, K	Density ρ, kg/m^3	Specific Heat c_p, J/kg·K	Thermal Conductivity λ, W/m·K	Thermal Diffusivity $\alpha \times 10^{10}$, m^2/s	Absolute Viscosity $\mu \times 10^3$, N·s/m^2	Kinematic Viscosity $\nu \times 10^6$, m^2/s	Prandtl Number Pr $\times 10^{-2}$
273	899.1	1796	0.147	911	3848	4280	471
293	888.2	1880	0.145	872	799	900	104
313	876.1	1964	0.144	834	210.	240.	28.7
333	864.0	2047	0.140	800	72.5	83.9	10.5
353	852.0	2131	0.138	769	32.0	37.5	4.90
373	840.0	2219	0.137	738	17.1	20.3	2.76
393	829.0	2307	0.135	710	10.3	12.4	1.75
413	816.9	2395	0.133	686	6.54	8.0	1.16
433	805.9	2483	0.132	663	4.51	5.6	0.84

SOURCE: From E. R. G. Eckert and R. M. Drake, *Analysis of Heat and Mass Transfer* (New York: McGraw-Hill Book Company, 1972).

Table B.1   Mollier (Enthalpy-Entropy) Diagram for Steam

Entropy, s, kJ/kgK

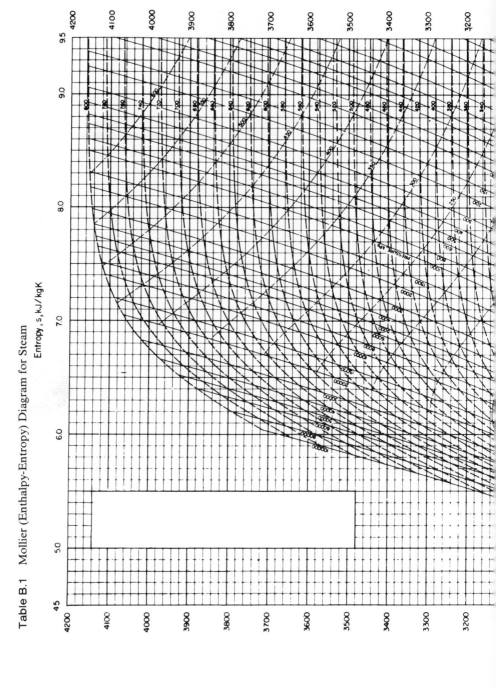

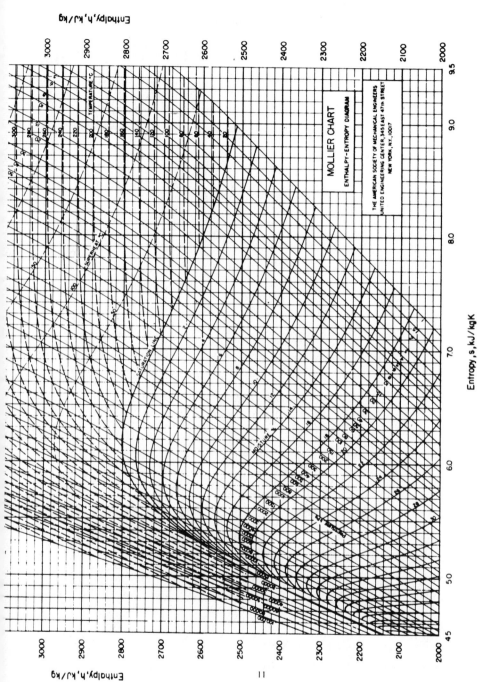

MOLLIER CHART

ENTHALPY–ENTROPY DIAGRAM

THE AMERICAN SOCIETY OF MECHANICAL ENGINEERS
UNITED ENGINEERING CENTER, 345 EAST 47th STREET
NEW YORK, N.Y., 10017

Entropy, s, kJ/kgK

Enthalpy, h, kJ/kg

SOURCE: Reprinted by permission from ASME Steam Tables, American Society of Mechanical Engineers, N. Y. 1967.

Table B.2   Temperature-Entropy Diagram for Steam

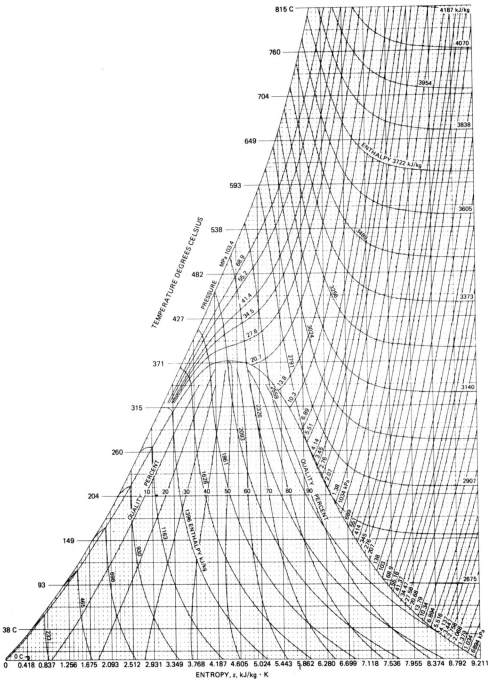

SOURCE: Reprinted by permission from ASME Steam Tables, American Society of Mechanical Engineers, N. Y. 1967.

Table B.3 Ammonia-Water Equilibrium Chart SOURCE: Reprinted, by permission, from *Refrigerating Engineering*, Vol. 58, No. 10, Oct. 1950.

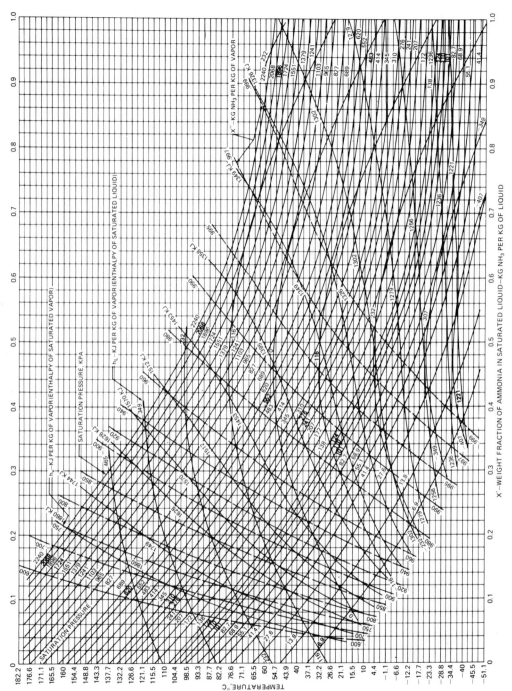

Table B.4 Psychrometric Chart

SOURCE: Reproduced by permission of Carrier Corporation.

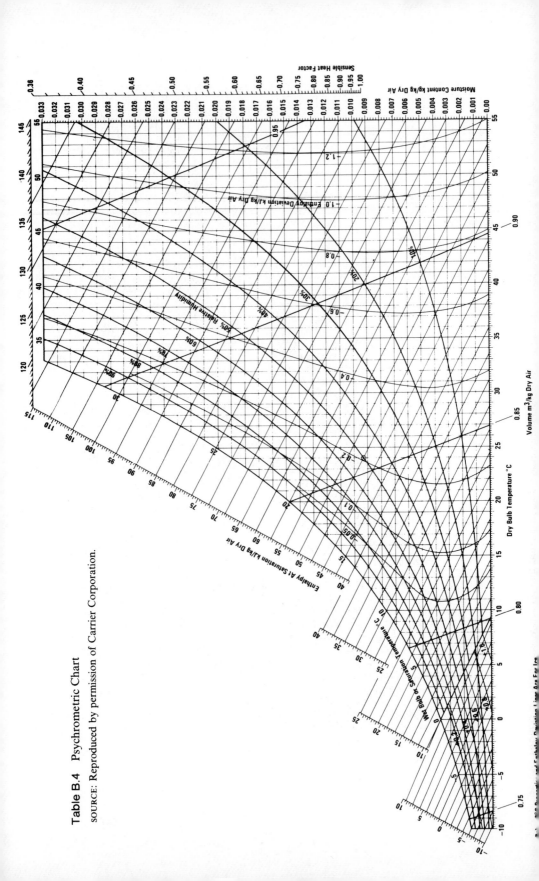

Table C.1 Enthalpies of Formation, Gibbs Function of Formation, and Absolute Entropy at 25°C and 1 atm Pressure

Substance	M	$\bar{h}_f^\circ$, kJ/kgmol	$\bar{g}_f^\circ$, kJ/kgmol	$\bar{s}^\circ$ kJ/kgmol · K
Acetylene $C_2H_2(g)$	26.038	226 866	209 290	200.98
Ammonia $NH_3(g)$	17.032	−45 926	−16 390	192.72
Benzene $C_6H_6(g)$	78.108	82 976	129 732	269.38
Butane $C_4H_{10}(g)$	58.124	−126 223	−17 164	310.32
Carbon C graphite	12.011	0	0	5.69
Carbon Dioxide $CO_2(g)$	44.01	−393 757	−344 631	213.83
Carbon Monoxide $CO(g)$	28.01	−110 596	−137 242	197.68
Ethane $C_2H_6(g)$	30.07	−84 718	−32 905	229.64
Ethene $C_2H_4(g)$	28.054	52 315	68 159	219.58
Hydrazine $N_2H_4(g)$	32.048	95 410	159 260	238.77
Methane $CH_4(g)$	16.043	−74 917	−50 844	186.27
Octane $C_8H_{18}(g)$	114.23	−208 581	16 536	467.04
Octane $C_8H_{18}(l)$	114.23	−250 102	6 614	361.03
Propane $C_3H_8(g)$	44.097	−103 909	−23 502	270.09
Water $H_2O(g)$	18.016	−241 971	−228 729	188.85
Water $H_2O(l)$	18.016	−286 010	−237 327	69.98
Dodecane $C_{12}H_{26}(g)$	170.328	−290 971		
Dodecane $C_{12}H_{26}(l)$	170.328	−394 199		

SOURCE: *JANAF Thermochemical Tables*, Document PB 168–370, Clearinghouse for Federal Scientific and Technical Information, 1965.

Table C.2 Enthalpy of Formation at 25°C, Ideal-Gas Enthalpy, and Absolute Entropy at 0.1 MPa Pressure

	Nitrogen, Diatomic (N₂)		Nitrogen, Monatomic (N)	
	$(\bar{h}_f^\circ)_{298} = 0$ kJ/kmol $M = 28.013$		$(\bar{h}_f^\circ)_{298} = 472\ 646$ kJ/kmol $M = 14.007$	
Temp. K	$(\bar{h}^\circ - \bar{h}^\circ_{298})$ kJ/kmol	$\bar{s}^\circ$ kJ/kmol K	$(\bar{h}^\circ - \bar{h}^\circ_{298})$ kJ/kmol	$\bar{s}^\circ$ kJ/kmol K
0	−8 669	0	−6 197	0
100	−5 770	159.813	−4 117	130.596
200	−2 858	179.988	−2 042	145.006
298	0	191.611	0	153.302
300	54	191.791	38	153.432
400	2 971	200.180	2 117	159.411
500	5 912	206.740	4 197	164.051
600	8 891	212.175	6 276	167.842
700	11 937	216.866	8 351	171.047
800	15 046	221.016	10 431	173.821
900	18 221	224.757	12 510	176.268
1000	21 460	228.167	14 590	178.461
1100	24 757	231.309	16 669	180.440
1200	28 108	234.225	18 749	182.247
1300	31 501	236.941	20 824	183.803
1400	34 936	239.484	22 903	185.452
1500	38 405	241.878	24 983	186.887
1600	41 903	244.137	27 062	188.230
1700	45 430	246.275	29 142	189.490
1800	48 982	248.304	31 217	190.678
1900	52 551	250.237	33 296	191.799
2000	56 141	252.078	35 376	192.866

Table C.2 Enthalpy of Formation at 25°C, Ideal-Gas Enthalpy, and Absolute Entropy at 0.1 MPa Pressure (*Continued*)

Temp. K	Nitrogen, Diatomic (N_2) $(\bar{h}_f^\circ)_{298} = 0$ kJ/kmol $M = 28.013$		Nitrogen, Monatomic (N) $(\bar{h}_f^\circ)_{298} = 472\ 646$ kJ/kmol $M = 14.007$	
	$(\bar{h}^\circ - \bar{h}^\circ_{298})$ kJ/kmol	$\bar{s}^\circ$ kJ/kmol K	$(\bar{h}^\circ - \bar{h}^\circ_{298})$ kJ/kmol	$\bar{s}^\circ$ kJ/kmol K
2100	59 748	253.836	37 455	193.883
2200	63 371	255.522	39 535	194.850
2300	67 007	257.137	41 614	195.774
2400	70 651	258.689	43 698	196.661
2500	74 312	260.183	45 777	197.511
2600	77 973	261.622	47 861	198.326
2700	81 659	263.011	49 949	199.113
2800	85 345	264.350	52 036	199.875
2900	89 036	265.647	54 124	200.607
3000	92 738	266.902	56 220	201.318
3200	100 161	269.295	60 421	202.674
3400	107 608	271.555	64 647	203.954
3600	115 081	273.689	68 906	205.171
3800	122 570	275.714	73 199	206.330
4000	130 076	277.638	77 534	207.443
4200	137 603	279.475	81 923	208.514
4400	145 143	281.228	86 370	209.548
4600	152 699	282.910	90 881	210.552
4800	160 272	284.521	95 462	211.527
5000	167 858	286.069	100 115	212.477
5200	175 456	287.559	104 847	213.401
5400	183 071	288.994	109 663	214.309
5600	190 703	290.383	114 558	215.201
5800	198 347	291.726	119 537	216.075
6000	206 008	293.023	124 600	216.933

Table C.2 Enthalpy of Formation at 25°C, Ideal-Gas Enthalpy, and Absolute Entropy at 0.1 MPa Pressure (*Continued*)

Temp. K	Oxygen, Diatomic (O_2) $(\overline{h}_f^\circ)_{298} = 0$ kJ/kmol $M = 31.999$		Oxygen, Monatomic (O) $(\overline{h}_f^\circ)_{298} = 249\ 195$ kJ/kmol $M = 16.00$	
	$(\overline{h}^\circ - \overline{h}^\circ_{298})$ kJ/kmol	$\overline{s}^\circ$ kJ/kmol K	$(\overline{h}^\circ - \overline{h}^\circ_{298})$ kJ/kmol	$\overline{s}^\circ$ kJ/kmol K
0	−8 682	0	−6 728	0
100	−5 778	173.306	−4 519	135.947
200	−2 866	193.486	−2 188	152.156
298	0	205.142	0	161.060
300	54	205.322	42	161.198
400	3 029	213.874	2 209	167.432
500	6 088	220.698	4 343	172.202
600	9 247	226.455	6 460	176.063
700	12 502	231.272	8 569	179.314
800	15 841	235.924	10 669	182.118
900	19 246	239.936	12 770	184.590
1000	22 707	243.585	14 862	186.795
1100	26 217	246.928	16 949	188.787
1200	29 765	250.016	19 041	190.603
1300	33 351	252.886	21 125	192.272
1400	36 966	255.564	23 213	193.820
1500	40 610	258.078	25 296	195.260
1600	44 279	260.446	27 380	196.603
1700	47 970	262.685	29 464	197.866
1800	51 689	264.810	31 547	199.059
1900	55 434	266.835	33 631	200.184
2000	59 199	268.764	35 715	201.251
2100	62 986	270.613	37 798	202.268
2200	66 802	272.387	39 882	203.238
2300	70 634	274.090	41 961	204.163
2400	74 492	275.735	44 045	205.050
2500	78 375	277.316	46 133	205.899
2600	82 274	278.848	48 216	206.720
2700	86 199	280.329	50 304	207.506
2800	90 144	281.764	52 392	208.268
2900	94 111	283.157	54 484	209.000
3000	98 098	284.508	56 576	209.711
3200	106 127	287.098	60 768	211.063
3400	114 232	289.554	64 973	212.339
3600	122 399	291.889	69 191	213.544
3800	130 629	294.115	73 425	214.686

Table C.2 Enthalpy of Formation at 25°C, Ideal-Gas Enthalpy, and Absolute Entropy at 0.1 MPa Pressure (*Continued*)

	Oxygen, Diatomic (O_2)		Oxygen, Monatomic (O)	
	$(\bar{h}_f^\circ)_{298} = 0$ kJ/kmol $M = 31.999$		$(\bar{h}_f^\circ)_{298} = 249\ 195$ kJ/kmol $M = 16.00$	
Temp. K	$(\bar{h}^\circ - \bar{h}_{298}^\circ)$ kJ/kmol	$\bar{s}^\circ$ kJ/kmol K	$(\bar{h}^\circ - \bar{h}_{298}^\circ)$ kJ/kmol	$\bar{s}^\circ$ kJ/kmol K
4000	138 913	296.236	77 676	215.778
4200	147 248	298.270	81 948	216.820
4400	155 628	300.219	86 236	217.816
4600	164 046	302.094	90 546	218.774
4800	172 502	303.893	94 876	219.694
5000	180 987	305.621	99 224	220.585
5200	189 502	307.290	103 596	221.439
5400	198 037	308.901	107 985	222.267
5600	206 593	310.458	112 395	223.071
5800	215 166	311.964	116 821	223.849
6000	223 756	313.420	121 269	224.602

	Carbon Dioxide (CO_2)		Carbon Monoxide (CO)	
	$(\bar{h}_f^\circ)_{298} = -393\ 757$ kJ/kmol $M = 44.01$		$(\bar{h}_f^\circ)_{298} = -110\ 596$ kJ/kmol $M = 28.01$	
Temp. K	$(\bar{h}^\circ - \bar{h}_{298}^\circ)$ kJ/kmol	$\bar{s}^\circ$ kJ/kmol K	$(\bar{h}^\circ - \bar{h}_{298}^\circ)$ kJ/kmol	$\bar{s}^\circ$ kJ/kmol K
0	−9 364	0	−8 669	0
100	−6 456	179.109	−5 770	165.850
200	−3 414	199.975	−2 858	186.025
298	0	213.795	0	197.653
300	67	214.025	54	197.833
400	4 008	225.334	2 975	206.234
500	8 314	234.924	5 929	212.828
600	12 916	243.309	8 941	218.313
700	17 761	250.773	12 021	223.062
800	22 815	257.517	15 175	227.271
900	28 041	263.668	18 397	231.066
1000	33 405	269.325	21 686	234.531
1100	38 894	274.555	25 033	237.719
1200	44 484	279.417	28 426	240.673

Table C.2 Enthalpy of Formation at 25°C, Ideal-Gas Enthalpy, and Absolute Entropy at 0.1 MPa Pressure (*Continued*)

Temp. K	Carbon Dioxide (CO_2) $(\overline{h}_f^\circ)_{298} = -393\,757$ kJ/kmol $M = 44.01$		Carbon Monoxide (CO) $(\overline{h}_f^\circ)_{298} = -110\,596$ kJ/kmol $M = 28.01$	
	$(\overline{h}^\circ - \overline{h}^\circ_{298})$ kJ/kmol	$\overline{s}^\circ$ kJ/kmol K	$(\overline{h}^\circ - \overline{h}^\circ_{298})$ kJ/kmol	$\overline{s}^\circ$ kJ/kmol K
1300	50 158	283.956	31 865	243.426
1400	55 907	288.216	35 338	245.999
1500	61 714	292.224	38 848	248.421
1600	67 580	296.010	42 384	250.702
1700	73 492	299.592	45 940	252.861
1800	79 442	302.993	49 522	254.907
1900	85 429	306.232	53 124	256.852
2000	91 450	309.320	56 739	258.710
2100	97 500	312.269	60 375	260.480
2200	103 575	315.098	64 019	262.174
2300	109 671	317.805	67 676	263.802
2400	115 788	320.411	71 346	265.362
2500	121 926	322.918	75 023	266.865
2600	128 085	325.332	78 714	268.312
2700	134 256	327.658	82 408	269.705
2800	140 444	329.909	86 115	271.053
2900	146 645	332.085	89 826	272.358
3000	152 862	334.193	93 542	273.618
3200	165 331	338.218	100 998	276.023
3400	177 849	342.013	108 479	278.291
3600	190 405	345.599	115 976	280.433
3800	202 999	349.005	123 495	282.467
4000	215 635	352.243	131 026	284.396
4200	228 304	355.335	138 578	286.241
4400	241 003	358.289	146 147	287.998
4600	253 734	361.122	153 724	289.684
4800	266 500	363.837	161 322	291.299
5000	279 295	366.448	168 929	292.851
5200	292 123	368.963	176 548	294.349
5400	304 984	371.389	184 184	295.789
5600	317 884	373.736	191 832	297.178
5800	330 821	376.004	199 489	298.521
6000	343 791	378.205	207 162	299.822

Table C.2 Enthalpy of Formation at 25°C, Ideal-Gas Enthalpy, and Absolute Entropy at 0.1 MPa Pressure (*Continued*)

Temp. K	Water (H_2O) $(\bar{h}^\circ_f)_{298} = -241\,971$ kJ/kmol, $M = 18.015$		Hydroxyl (OH) $(\bar{h}^\circ_f)_{298} = 39\,463$ kJ/kmol, $M = 17.007$	
	$(\bar{h}^\circ - \bar{h}^\circ_{298})$ kJ/kmol	$\bar{s}^\circ$ kJ/kmol K	$(\bar{h}^\circ - \bar{h}^\circ_{298})$ kJ/kmol	$\bar{s}^\circ$ kJ/kmol K
0	−9 904	0	−9 171	0
100	−6 615	152.390	−6 138	149.587
200	−3 280	175.486	−2 975	171.591
298	0	188.833	0	183.703
300	63	189.038	54	183.892
400	3 452	198.783	3 033	192.465
500	6 920	206.523	5 991	199.063
600	10 498	213.037	8 941	204.443
700	14 184	218.719	11 903	209.004
800	17 991	223.803	14 878	212.979
900	21 924	228.430	17 887	216.523
1000	25 978	232.706	20 933	219.732
1100	30 167	236.694	24 025	222.677
1200	34 476	240.443	27 158	225.405
1300	38 903	243.986	30 342	227.949
1400	43 447	247.350	33 568	230.342
1500	48 095	250.560	36 840	232.598
1600	52 844	253.622	40 150	234.736
1700	57 685	256.559	43 501	236.769
1800	62 609	259.371	46 890	238.702
1900	67 613	262.078	50 308	240.551
2000	72 689	264.681	53 760	242.325
2100	77 831	267.191	57 241	244.020
2200	83 036	269.609	60 752	245.652
2300	88 295	271.948	64 283	247.225
2400	93 604	274.207	67 839	248.739
2500	98 964	276.396	71 417	250.200
2600	104 370	278.517	75 015	251.610
2700	109 813	280.571	78 634	252.974
2800	115 294	282.563	82 266	254.296
2900	120 813	284.500	85 918	255.576
3000	126 361	286.383	89 584	256.819
3200	137 553	289.994	96 960	259.199
3400	148 854	293.416	104 387	261.450
3600	160 247	296.676	111 859	263.588

Table C.2 Enthalpy of Formation at 25°C, Ideal-Gas Enthalpy, and Absolute
Entropy at 0.1 MPa Pressure (*Continued*)

Temp. K	Water (H$_2$O) $(\bar{h}_f^\circ)_{298}) = -241\,971$ kJ/kmol $M = 18.015$		Hydroxyl (OH) $(\bar{h}_f^\circ)_{298} = 39\,463$ kJ/kmol $M = 17.007$	
	$(\bar{h}^\circ - \bar{h}^\circ_{298})$ kJ/kmol	$\bar{s}^\circ$ kJ/kmol K	$(\bar{h}^\circ - \bar{h}^\circ_{298})$ kJ/kmol	$\bar{s}^\circ$ kJ/kmol K
3800	171 724	299.776	119 378	265.618
4000	183 280	302.742	126 934	267.559
4200	194 903	305.575	134 528	269.408
4400	206 585	308.295	142 156	271.182
4600	218 325	310.901	149 816	272.885
4800	230 120	313.412	157 502	274.521
5000	241 957	315.830	165 222	276.099
5200	253 839	318.160	172 967	277.617
5400	265 768	320.407	180 736	279.082
5600	277 738	322.587	188 531	280.500
5800	289 746	324.692	196 351	281.873
6000	301 796	326.733	204 192	283.203

Temp. K	Hydrogen, Diatomic (H$_2$) $(\bar{h}_f^\circ)_{298} = 0$ kJ/kmol $M = 2.016$		Hydrogen, Monatomic (H) $(\bar{h}_f^\circ)_{298} = 217\,986$ kJ/kmol $M = 1.008$	
	$(\bar{h}^\circ - \bar{h}^\circ_{298})$ kJ/kmol	$\bar{s}^\circ$ kJ/kmol K	$(\bar{h}^\circ - \bar{h}^\circ_{298})$ kJ/kmol	$\bar{s}^\circ$ kJ/kmol K
0	−8 468	0	−6 197	0
100	−5 293	102.145	−4 117	92.011
200	−2 770	119.437	−2 042	106.417
298	0	130.684	0	114.718
300	54	130.864	38	114.847
400	2 958	139.215	2 117	120.826
500	5 883	145.738	4 197	125.466
600	8 812	151.077	6 276	129.257
700	11 749	155.608	8 351	132.458
800	14 703	159.549	10 431	135.236
900	17 682	163.060	12 510	137.684
1000	20 686	166.223	14 590	139.872

Temp. K	Hydrogen, Diatomic (H$_2$) $(\bar{h}_f^\circ)_{298} = 0$ kJ/kmol $M = 2.016$		Hydrogen, Monatomic (H) $(\bar{h}_f^\circ)_{298} = 217\,986$ kJ/kmol $M = 1.008$	
	$(\bar{h}^\circ - \bar{h}_{298}^\circ)$ kJ/kmol	$\bar{s}^\circ$ kJ/kmol K	$(\bar{h}^\circ - \bar{h}_{298}^\circ)$ kJ/kmol	$\bar{s}^\circ$ kJ/kmol K
1100	23 723	169.118	16 669	141.855
1200	26 794	171.792	18 749	143.662
1300	29 907	174.281	20 824	145.328
1400	33 062	176.620	22 903	146.867
1500	36 267	178.833	24 983	148.303
1600	39 522	180.929	27 062	149.641
1700	42 815	182.929	29 142	150.905
1800	46 150	184.833	31 217	152.093
1900	49 522	186.657	33 296	153.215
2000	52 932	188.406	35 376	154.281
2100	56 379	190.088	37 455	155.294
2200	59 860	191.707	39 535	156.265
2300	63 371	193.268	41 610	157.185
2400	66 915	194.778	43 689	158.072
2500	70 492	196.234	45 769	158.922
2600	74 090	197.649	47 848	159.737
2700	77 718	199.017	49 928	160.520
2800	81 370	200.343	52 007	161.277
2900	85 044	201.636	54 082	162.005
3000	88 743	202.887	56 162	162.708
3200	96 199	205.293	60 321	164.051
3400	103 738	207.577	64 475	165.311
3600	111 361	209.757	68 634	166.499
3800	119 064	211.841	72 793	167.624
4000	126 846	213.837	76 948	168.691
4200	134 700	215.753	81 107	169.704
4400	142 624	217.594	85 266	170.670
4600	150 620	219.372	89 420	171.595
4800	158 682	221.087	93 579	172.482
5000	166 808	222.744	97 734	173.327
5200	174 996	224.351	101 893	174.143
5400	183 247	225.907	106 052	174.930
5600	191 556	227.418	110 207	175.683
5800	199 924	228.886	114 365	176.415
6000	208 346	230.313	118 524	177.118

Table C.2   Enthalpy of Formation at 25°C, Ideal-Gas Enthalpy, and Absolute Entropy at 0.1 MPa Pressure (*Continued*)

Temp. K	Nitric Oxide (NO) $(\bar{h}_f^\circ)_{298} = 90\ 592$ kJ/kmol $M = 30.006$		Nitrogen Dioxide (NO$_2$) $(\bar{h}_f^\circ)_{298} = 33\ 723$ kJ/kmol $M = 46.005$	
	$(\bar{h}^\circ - \bar{h}^\circ_{298})$ kJ/kmol	$\bar{s}^\circ$ kJ/kmol K	$(\bar{h}^\circ - \bar{h}^\circ_{298})$ kJ/kmol	$\bar{s}^\circ$ kJ/kmol K
0	−9 192	0	−10 196	0
100	−6 071	177.034	−6 870	202.431
200	−2 950	198.753	−3 502	225.732
298	0	210.761	0	239.953
300	54	210.950	67	240.183
400	3 042	219.535	3 950	251.321
500	6 058	226.267	8 150	260.685
600	9 146	231.890	12 640	268.865
700	12 309	236.765	17 368	276.149
800	15 548	241.091	22 288	282.714
900	18 857	244.991	27 359	288.684
1000	22 230	248.543	32 552	294.153
1100	25 652	251.806	37 836.	299.190
1200	29 121	254.823	43 196	303.855
1300	32 627	257.626	48 618	308.194
1400	36 166	260.250	54 095	312.253
1500	39 731	262.710	59 609	316.056
1600	43 321	265.028	65 157	319.637
1700	46 932	267.216	70 739	323.022
1800	50 559	269.287	76 345	326.223
1900	54 204	271.258	81 969	329.265
2000	57 861	273.136	87 613	332.160
2100	61 530	274.927	93 274	334.921
2200	65 216	276.638	98 947	337.562
2300	68 906	278.279	104 633	340.089
2400	72 609	279.856	110 332	342.515
2500	76 320	281.370	116 039	344.846
2600	80 036	282.827	121 754	347.089
2700	83 764	284.232	127 478	349.248
2800	87 492	285.592	133 206	351.331
2900	91 232	286.902	138 942	353.344
3000	94 977	288.174	144 683	355.289
3200	102 479	290.592	156 180	359.000
3400	110 002	292.876	167 695	362.490
3600	117 545	295.031	179 222	365.783
3800	125 102	297.073	190 761	368.904

Table C.2 Enthalpy of Formation at 25°C, Ideal-Gas Enthalpy, and Absolute Entropy at 0.1 MPa Pressure (*Continued*)

	Nitric Oxide (NO)		Nitrogen Dioxide (NO₂)	
	$(\bar{h}_f^\circ)_{298} = 90\ 592$ kJ/kmol $M = 30.006$		$(\bar{h}_f^\circ)_{298} = 33\ 723$ kJ/kmol $M = 46.005$	
Temp. K	$(\bar{h}^\circ - \bar{h}^\circ_{298})$ kJ/kmol	$\bar{s}^\circ$ kJ/kmol K	$(\bar{h}^\circ - \bar{h}^\circ_{298})$ kJ/kmol	$\bar{s}^\circ$ kJ/kmol K
4000	132 675	299.014	202 309	371.866
4200	140 260	300.864	213 865	374.682
4400	147 863	302.634	225 430	377.372
4600	155 473	304.324	237 003	379.946
4800	163 101	305.947	248 580	382.410
5000	170 736	307.508	260 161	384.774
5200	178 381	309.006	271 751	387.046
5400	186 042	310.449	283 340	389.234
5600	193 707	311.847	294 934	391.343
5800	201 388	313.194	306 532	393.376
6000	209 074	314.495	318 130	395.343

SOURCE: *JANAF Thermochemical Tables*, Document PB 168–370, Clearinghouse for Federal Scientific and Technical Information, 1965.

Table C.3 Enthalpy of Combustion (Heating Value) of Various Compounds

	Liquid Hydrocarbon		Gaseous Hydrocarbon	
Fuel	Negative of Higher Heating Value $H_2O(l)$, kJ/kg	Negative of Lower Heating Value $H_2O(v)$, kJ/kg	Negative of Higher Heating Value $H_2O(l)$ kJ/kg	Negative of Lower Heating Value $H_2O(v)$ kJ/kg
Methane CH_4			−55 491	−50 005
Ethane C_2H_6			−51 870	−47 479
Propane C_3H_8	−49 977	−45 984	−50 346	−46 346
Butane C_4H_{10}	−49 149	−45 363	−49 519	−45 732
Pentane C_5H_{12}	−48 637	−44 977	−49 005	−45 346
Hexane C_6H_{14}	−48 307	−44 728	−48 674	−45 095
Heptane C_7H_{16}	−48 065	−44 551	−48 430	−44 916
Octane C_8H_{18}	−47 886	−44 419	−48 249	−44 781
Decane $C_{10}H_{22}$	−47 637	−44 232	−47 995	−44 593
Dodecane $C_{12}H_{26}$	−47 465	−44 102	−47 823	−44 460
Acetylene C_2H_2			−49 907	−48 219
Benzene C_6H_6	−41 825	−40 137	−42 260	−40 572
Hydrogen H_2			−141 856	−120 012

SOURCE: *JANAF Thermochemical Tables*, Document PB 168–370, Clearinghouse for Federal Scientific and Technical Information, 1965.

Table C.4 Natural Logarithm of Equilibrium Constant K

For the reaction $\nu_A A + \nu_B B \rightleftharpoons \nu_C C + \nu_D D$ the equilibrium constant K is defined as

Temperature °K	$H_2 \rightleftharpoons 2H$	$O_2 \rightleftharpoons 2O$	$N_2 \rightleftharpoons 2N$	$H_2O \rightleftharpoons H_2 + \frac{1}{2}O_2$	$H_2O \rightleftharpoons \frac{1}{2}H_2 + OH$	$CO_2 \rightleftharpoons CO + \frac{1}{2}O_2$	$\frac{1}{2}N_2 + \frac{1}{2}O_2 \rightleftharpoons NO$
298	−164.018	−186.988	−367.493	−92.214	−106.214	−103.768	−35.052
500	−92.840	−105.643	−213.385	−52.697	−60.287	−57.622	−20.295
1000	−39.816	−45.163	−99.140	−23.169	−26.040	−23.535	−9.388
1200	−30.887	−35.018	−80.024	−18.188	−20.289	−17.877	−7.569
1400	−24.476	−27.755	−66.342	−14.615	−16.105	−13.848	−6.270
1600	−19.650	−22.298	−56.068	−11.927	−13.072	−10.836	−5.294
1800	−15.879	−18.043	−48.064	−9.832	−10.663	−8.503	−4.536
2000	−12.853	−14.635	−41.658	−8.151	−8.734	−6.641	−3.931
2200	−10.366	−11.840	−36.404	−6.774	−7.154	−5.126	−3.433
2400	−8.289	−9.510	−32.024	−5.625	−5.838	−3.866	−3.019
2600	−6.530	−7.534	−28.317	−4.654	−4.725	−2.807	−2.671
2800	−5.015	−5.839	−25.130	−3.818	−3.769	−1.900	−2.372
3000	−3.698	−4.370	−22.372	−3.092	−2.943	−1.117	−2.114
3200	−2.547	−3.085	−19.950	−2.457	−2.218	−0.435	−1.888
3400	−1.529	−1.948	−17.813	−1.897	−1.582	0.163	−1.690
3600	−0.622	−0.939	−15.911	−1.398	−1.014	0.695	−1.513
3800	0.189	−0.032	−14.212	−0.951	−0.507	1.170	−1.356
4000	0.921	0.783	−12.673	−0.548	−0.050	1.593	−1.216
4500	2.473	2.500	−9.427	0.306	0.914	2.484	−0.921
5000	3.712	3.882	−6.820	0.990	1.683	3.191	−0.686
5500	4.730	5.010	−4.679	1.554	2.312	3.765	−0.497
6000	5.577	5.950	−2.878	2.026	2.837	4.239	−0.341

SOURCE: *JANAF Thermochemical Tables*, Document PB 168–370, Clearinghouse for Federal Scientific and Technical Information, 1965.

Answers to Selected Problems

Chapter 2

1. 520 kPa, 310 kPa   **3.** 4704 N   **5.** 5 kg/s, 90 min   **7.** 10.34 m   **9.** 8400 N
11. 1687.6 mm Hg abs   **13.** -2731.5 X   **15.** 24.5 kg, 24.5 kg   **17.** 456.9 kPa
19. 80.1 kg

Chapter 3

1. 68.44 kJ   **3.** 69.99 kg   **5.** 7.67 m/s   **7.** 562.5 kJ, 350 kJ   **9.** 7.7 kJ   **9.** 7.7
kJ   **11.** -481.64 kJ/kg   **13.** -10.91 kW   **15.** 68.6 kJ   **17.** 1432.33 kJ
19. 2450 kJ, 2450 kJ, 2450 kJ   **21.** -121.1 kW   **23.** 41 kg/day   **25.** -22817.9
kW, 493.9 kg/s   **27.** 772.2 m/s

Chapter 4

1. (a) $p = 7.384$ kPa, $m = 0.102$ kg; (b) $p = 1554.33$ kPa, $m = 24.0$ kg; (c) $p = 960.7$ kPa,
$m = 110.1$ kg   **7.** 2599.1 kJ/kg   **9.** 122 589 kJ   **11.** -441.5 kJ/kg, 441.5 kJ/kg,
-336.4 kJ/kg   **13.** (a) 178.1 kg, 11.54 kg; (b) 0.0304 m³, 0.4606 m³   **15.** 1553.9
kJ/kg, 0.2258 m³/kg   **17.** 16.51 MPa, -308.9 kJ/kg   **19.** 3.128 kg/s, 0.672 kg/s
21. 524.93°C   **23.** 2.6–2.65 MPa, 260–270°C   **25.** 0.9435   **27.** 0.9956
29. 11.08 kW

Chapter 5

1. 1.111 kJ/kg-K   **3.** 0.01921 kg, 0.7124 kg, 1.882 kPa   **5.** 45.66 kg, 337.1 kPa
7. 371.3 kPa, 371.0 kPa   **9.** 308.9 kPa, 308.7 kPa   **11.** 0.0259 m³/kg, 0.0072 m³/kg,
1.383 m³/kg   **13.** 1.026 kJ/kg · K, 0.7467 kJ/kg · K   **15.** 32.75 kg/kgmol, 0.2539
kJ/kg · K, 0.8461 kJ/kg · K   **17.** 862.3 kJ/kg, 654.1 kJ/kg   **19.** 3.861 km
21. 15.036 kg, 16.674 kg   **23.** 354.3 K   **25.** 0.426 m³, 8.88 kJ   **27.** 4875 N
29. 13.387 m

Chapter 6

1. 52.35 K   **3.** 2256 kPa   **5.** 1032.5 kJ, 757.68 kPa   **7.** 917.1 kJ/kg   **9.** 284 m/s
11. 21.2 kg/s   **13.** 40.08 kJ, 27.96 kJ   **15.** 576 K, 310.0 kJ, 310.0 kJ   **17.** 900°C
19. 221 887 kW, 2655.4 kg/s   **21.** 1948 cycles   **23.** 268.9 m/s, 262.3 kPa
25. -29.73 kJ/kg, 248.8 K, -53.33 kJ/kg   **27.** 275.55 kJ, 78 kJ, 275.55 kJ, 197.55 kJ
29. 2218.1 kJ, 2218.1 kJ, 0, 0   **31.** 83.45 kJ, 89.46 kJ, 14 kJ, 7 kJ, 250 kPa
33. 482.1 K, -374.1 kW   **35.** 450 kJ, 270 kJ   **37.** 531.3 K, 1.662×10^4 kW, 34.24
kg/s   **39.** 676.1 K, 703.1 kJ/kg, 83.94 kg/s   **41.** 255.42 kJ   **43.** (b) 1916.4 kJ/kg;
(c) 1560 kJ/kg   **45.** (b) 0, -4984.8 kJ, 2005.7 kJ; (c) 9491.1 kJ   **47.** 465.4 K
49. 51.97 kJ

Chapter 7

1. 516.5 K, 0.375, 429.4 kPa, -20 kJ per cycle   **3.** 0.67, -17.49 kJ, 35.51 kJ   **5.** 0.70, 285.7 kW, 0.02 m^3 **7.** 0.84   **9.** 95.5 kJ, -31.83 kJ, 530.6 kW, 0.184 m^3, 22.94 kPa, 0.667   **11.** 632.45 K, 183.75 kJ, 116.25 kJ, 200 kJ   **13.** 8.4%   **15.** 1.162 kW, 0.094 kW, heating   **17.** 197.7 K, 1.2727, 1400 kJ   **19.** 7.657, 1.021 kg/s   **21.** 714 K

Chapter 8

1. -0.987 kJ/K, -0.8291 kJ/K, -5.1038 kJ/K   **3.** 1.66, 1.22, 2.05   **5.** -0.2815 kJ/kg-K, -84.45 kJ/kg   **7.** -20.57 kW/K, -6275 kW   **9.** -0.2908 kJ/K **11.** 300 K   **15.** 313.65 K, 0.0682 kJ/K, 26.57 kJ, 21.1 kJ   **17.** 7.2212 kJ/kg · K **19.** 935.6 kJ, 2.2 MPa, 524.7 kJ, 1.1736 kJ/K   **21.** 620 kPa, 17.94 kJ, 42.2 liters **23.** 475.7 K, 0.0847 kg/s, -31.76 kW   **25.** 333.3 kPa, 311 K, -6.005 kJ/K, 0, 0 **27.** 0.0086 kJ/kg · K   **29.** 994.72 kJ, -370.88 kJ, 2.103 kJ/K

Chapter 9

1. -518.1 kJ/kg   **3.** -268.7 kJ, 3.316 kJ/K   **5.** 7.84%   **7.** 590.76 kJ, 1.3928 kJ/K, 394.16 kJ, 196.6 kJ   **9.** -0.0111 kJ/K, -1.44 kJ   **11.** -97.2 kJ **13.** 22 096 kW, -15 605.4 kW, 9501.0 kW, 20.76 kW/K   **15.** -52.1 kJ/kg steam **17.** 1286.2 kW, 0   **19.** -60.6 kJ/kg, 193.3 kJ/kg   **21.** 4802.2 kW, -7079.8 kW, 7.73 kW/K   **23.** 4.11 kg/s, -748.9 kW, 76.8%   **25.** 0.965

Chapter 10

7. 1889.5 kJ/kg, 1100.2 kJ/kg, 143.9 kJ/kg   **13.** $a = 119$, $b = 0.0363$ **15.** 0.00328 K^{-1}, 0.00339 K^{-1} **17.** -123.6 kJ/kg, -123.4 kJ/kg, 0.327 kJ/kg · K   **19.** 18.39 kg   **23.** 0.00969 K · m^2/kN

Chapter 11

1. 0.442, 968.4 kJ/kg, -303.8 kJ/kg   **3.** 0.396, 0.42   **5.** 26.97 kJ/kg, 25.89 kJ/kg, 25.53 kJ/kg   **7.** 0.438, 33.28 kg/s, 0.1434, 0.809   **9.** 908.9 kg/s, 2342 MW, 66.91 kg/s, 0.1066, 0.427   **11.** 1392.8 kJ/kg, 13.0%, 38.2%   **13.** 74.2%, 0.448   **15.** 0.159, 0.457, 46.58 kg/s   **17.** 0.512, 610.4 kg/s, 18 124.1 kg/s, 20.02 MW   **19.** 0.37, 1018.4 kJ/kg, 49.38 kg/s, 20 338 kW   **21.** 12.6%   **23.** 105.5 kJ/kg Hg, 911.9 kJ/kg stm, 299.9 kg/s, 0.557   **25.** 996.0 kJ/kg   **27.** 2.6 kg/s   **29.** 1.49 × 10^4 kW, 293.0 kg/s, 0.943, 0.229   **31.** 129.8 kW   **33.** 0.7997 kg/s, 444.5 kW, 37.88 kg/s, 14.3% **35.** 550 kPa, 5.878 kJ/kg water, 9.37%

Chapter 12

1. 0.777, 9.95 tons, 4.48   **3.** T_h **5.** 42.97 tons   **7.** 43.05 kJ/kg   **9.** 4.01, 0.244 kg/s   **11.** 0.218 kg/s, 5.26 kW, 29.87 kW, 4.68   **13.** 1.499, 0.702 kg/s, 35.19 kW, 505 kg/day   **15.** 0.163 kg/s, 2.69, 1076 kJ/kg, 4.9 C, 65.3 kW   **17.** 0.2156, 0.09976 kg/s, -8.37 kW, 0.0804 m^3/s, 17.2 cm   **19.** 409 kPa, 0.365 kg/s, 0.286 kg/s, 3.19, 110.1, 0.096   **21.** 2.76, 0.227, 127.2 kW   **23.** 315.7 kPa, 26.7 tons, 2.61, 35.95 kW **25.** yes   **27.** 5.44, 4.44, 18.68 kW, 22.82 kW   **29.** 0.8637 kg/s, 0.164, 4273.6 kW, 0.1276 m^3/s   **31.** 9.75 liter/s, 6.43 kW, 4.66, 1.57.7 C   **33.** 0.0817 kg/s, 2.36, 37.28 kW

Chapter 13

1. (a) 0.27, 0.494, 0.236; (b) 40 kPa, 30 kPa; (c) 35.61 kg/kgmol   **3.** (a) 0.1139 kJ/kg · K, 73 kg/kgmol; (b) 62.6%, 37.4%   **5.** (a) 0.94, 0.06; (b) 141 kPa, 9 kPa
7. (a) 0.346, 0.425, 0.229; (b) 0.552, 0.431, 0.017   **9.** 290.65 kPa, 0.5472 kJ/K, 0
11. 9432 kPa, 0.1334 kJ/K   **13.** 0.01754 kg, 4.3245 kPa, 136.2 kPa, −76 kJ   **15.** 60.06 C, 0.0691 kg water/kg air, 64%   **17.** 61.5 kg/h   **19.** 1.13 m³/s, 61.6 kW   **21.** 7.05 m³/s, 0.0342 kg/s, 7.66 m³/s   **23.** 0.0408 kg/s, 234.3 kW   **25.** 23%, 0.89 m³/kg, 228.6 kW   **27.** 74 m³/s, 2.94 kg/s   **29.** 0.0505 kg/h   **31.** −309.3 kW

Chapter 14

1. (a) 859.5 kg; (b) 10.5%, 11.9%, 74.3%, 3.3%   **3.** 3586 kg   **5.** 2.14 kPa, 36 kg
7. 15.0 kg air/kg fuel, 88.06 kgmol air/kgmol fuel, 0.726 kg fuel/kg water, 30.42 kg/kgmol, 28.65 kg/kgmol, 0.942   **11.** 26 476 kJ/kg   **13.** 30.82 kg air/kg fuel, −2 460 457 kJ/kgmol fuel, 41°C   **15.** 22%   **17.** 3110 K, 980 kPa   **19.** (a) 0.08, 0.151, 0.039, 0.73; (b) 54°C; (c) −313.1 kW; (d) 1.61 m/s   **23.** 9770.7 kg/h
25. (a) −45 741 kJ/kg, −49 529 kJ/kg; (b) −45 805 kJ/kg, −49 593 kJ/kg
27. −394 650 kJ/kgmol   **29.** 48 855 kJ/kg   **31.** 59.5%   **33.** 2214 K   **35.** 2.09 volts, 1.025   **37.** 0.00099 kg/s

Chapter 15

1. −519 kW, 990 kPa, 575.8 K, 1105 K, 216.4 kW   **5.** 0.464   **7.** 0.825, 0.796, 5.5%
9. 0.831, 0.0641 m³ per stroke, −93.5 kW, 0.434 m   **13.** −20 919 kW, 460 K, −16 155 kW   **15.** 36 kW   **17.** 43.86 m/s, 80.7 m/s, 0.872, −2310.6 kW   **19.** 0.59 m³/s, 405.1 K, 354.7 kPa, 1.411 kg/s, −184.9 kW, 0.6915 m, 0.484 m, 278.2 kW   **21.** −18.3 kW
23. 612.5 W   **25.** 0.267 m³/s, 0.192 m³/s, 582.1 kPa

Chapter 16

1. 0.565, 1057 kPa   **3.** 6.31   **5.** 623.6 kJ/kg, 0.491   **7.** 2000 K, 0.475, 290.05 kJ, −152.36 kJ, −71.12 kJ   **9.** 0.223   **11.** 1994.6 kJ/kg, 1209.2 kJ/kg   **13.** 18.47, 1894 K, 1.94   **15.** 827 kJ/kg, 0.577, 1650.5 K, 786.4 K   **17.** 223 kJ/kg, −44.76 kJ/kg, 0.80   **19.** −202.3 kJ/kg   **21.** 0.0314 kg fuel/kg air, 0.00522 kg/s, 130.2 kW, 0.1409 kg/s   **23.** 106.2 km/h, 7.16 km/liter, 22.5%   **25.** 330 kJ/kg, −251 kJ/kg, −388.8 kJ/kg, 0.8531 kJ/kg · K   **27.** 12.12, 0.00435 kg/s, 0.417, 80.14 kW, 0.228 kg/s
29. 169.1 kW, 0.02424 m³/s, 902 kW, 541 kPa   **31.** 0.424 m, 0.509 m, 0.00048 m³, 0.0375 kg/s, 0.336   **33.** 307 kPa, 12.69 kW, 5.12 kW, 0.59   **35.** 0.566, 0.419, $6.58

Chapter 17

1. 0.448, 375 kJ/kg, 837.1 kJ/kg   **3.** 255 kJ/kg, 637.6 K, 4.6638 m³/kg   **5.** 671.6 K, 0.606   **7.** 3   **9.** 794 K, 0.386, 0.631 kg/s   **11.** 0.366, 296.7 kJ/kg, 810.2 kJ/kg
13. 0.296, 0.961, 0.921, 1062.9 K, 2147.8 kW, 1072.3 kW   **15.** 0.1148 kg/s, 0.358, 824 K, −1339.7 kW   **17.** 0.0229 kg fuel/kg air   **19.** 0.019 25 kg fuel/kg air, 729 K, 0.519, 655 K, 0.407   **21.** 0.548, 0.019 40 kg fuel/kg air, 688.14 kJ/kg, −230.5 kJ/kg, 247.13 kJ/kg, −139.1 kJ/kg, 0.534   **23.** 0.877, 0.967, 3226.1 kJ/kg, 0.435, −1823.4 kJ/kg, 71.29 kg/s   **25.** −280 kJ/kg, 468.4 kJ/kg, 564.2 kJ/kg, 0.334, 8478 kW, −73.41 kJ/kg
27. 16 349 kW, 0.3308, 0.484, 29 261 kW, 0.7176   **29.** −93.05 kJ/kg, 0.027 43 kg fuel/kg air, 78.8 kPa, 657 N · s/kg, 0.149, 21 024 kN   **31.** 205.1 K, 0.821, 5308 kW

Chapter 18

1. 338.5 K, 162.3 kPa   **3.** 75.3 kPa, 264.2 K, 153.3 cm^2 **5.** 314.7 m/s, 0.78
7. 870.1 m/s, 1.51, 45.2 kg/s   **9.** 448.6 K, 23 cm^2, 534.3 K, 16.79 cm^2, 1.58
11. 3107.3 kJ/kg, 567.8 m/s, 14.195 kN, 76 nozzles   **13.** 2.49 kg/s, 0.90, 6.15 cm^2,
0.0879 m^3/kg   **15.** 2623.5 kJ/kg, 1.1995 m^3/kg, 808.6 m/s, 0.753   **17.** 0.1925 m^3/kg,
253 K, 0.01406 m, 5.195 kg/s   **19.** 0.564, 1039.8 m/s, 1160.2 K, 2522.9 m/s, 11.88
21. 152°C, 176.1°C   **23.** 2573.7 kPa, 930.6 K, 20.9 cm^2 **25.** 0.0797 m^3/s
27. 4732.7 kW, 0.422   **29.** 2644.9 kWh

Chapter 19

1. 3.53 W/m · K   **3.** 468.7 W   **5.** 0.535 m   **7.** 64.5 cm, 642.7 W/m^2 **9.** 14.57
kW/m^2 **11.** 3029 K   **13.** 0.03 kg/s   **15.** 2.43 m   **17.** 4.7 cm   **19.** 11.7
W/m^2 · K, 181.69°C, 4252.5 kW   **21.** 207.3°C, 27.49 m^2, 22.7 m^2 **23.** 7 tubes
25. 754.4 m^2, 0.78, 178.7°C   **27.** (a)   **29.** 165.6°C, 0.704, 1417.8 kW, yes
31. 345.1 m, 0.854 kg/s

Index

Absolute entropy, 151
  table, 549
Absolute pressure, 20
Absolute temperature, 24
Absorption refrigeration cycle,
  259
  performance factor, 260
Acceleration, 14
  gravitational, 16
Acoustic velocity, 442
Actual gases, 80, 93
Additive pressures, 278
Additive volumes, 277
Adiabatic, 39
  reversible process, 113
Adiabatic compressibility, 444
Adiabatic compression efficiency,
  357
Adiabatic flame temperature, 318
  with dissociation, 333
Adiabatic saturation process, 287
Air, composition, 305
  table, 507
Air compressors, 344
Air-fuel ratio, 307
Air standard cycles, 366
  Brayton, 404
  Diesel, 373
  Dual, 381
  Ericsson, 378
  Otto, 367
  Stirling, 378

Amagat's law, 277
Ammonia, 71
  table, 530
Ammonia-water equilibrium
  chart, 547
Atmospheric pressure, 21
Availability, 173, 176
Available energy, 166, 168
  closed system, 166
Axial flow compressor, 358

Beattie-Bridgeman, 80
Bernoulli's equation, 447
Berthelot, 207
Binary vapor cycle, 232
Boundaries, 11
Bourdon tube pressure gage, 18
Boyle's law, 84
Brake engine efficiency, 392
Brayton cycle, 404
  maximum work, 406
  with regeneration, 417

Calorimeters, for steam quality,
  69
Carbon dioxide, table, 553
Carbon monoxide, table, 553
Carnot, Nicholas, L.S., 7
  cycle, 131
  efficiency, 134
  reversed, 137, 247
Cascade refrigeration system, 254

Celsius temperature scale, 24
Centrifugal compressor, 358
Charles' law, 84
Charts
  ammonia-water equilibrium,
    547
  compressibility factor, 82, 83
  Mollier, steam, 157, 544
  psychrometric, 548
  temperature-entropy, 546
Chemical equilibrium, 327
Chemical potential, 328
Choked flow, 450
Clapeyron equation, 199
Clausius, Rudolf J., 7
Clausius inequality, 146
Clearance volume, 347
Closed system, 11
  first law, 43
Coal, combustion, 308
Coefficient of performance, 138,
    262
Coefficient of thermal expansion,
    201
Cold air standard, 369
Combined cycle, 423
Combustion, 304
  adiabatic flame temperature,
    318
  analysis of products, 310
  constant volume, 317
  efficiency, 413
  excess and deficient air, 306
Combustor, 411
Compressed liquid, water, 61, 69,
    528
Compressibility chart, 81
Compressibility factor, 81, 83
Compression efficiency, 357
Compression ratio, 368
Compressor, 344
  axial, 358
  capacity, 352
  centrifugal, 358
  efficiency, 357
  multistage, 353
  performance factors, 356
  reciprocating, 347
  rotary, 357

Condenser, 214
Condition curve, 236
Conduction, 470
Conservation of energy, 43, 46
Conservation of mass, 28, 438
Conservation of momentum, 439
Constant pressure process, 99,
    108
  specific heat, 88
Constant specific volume process,
    110
Constant temperature process,
    105, 109
Constant volume process, 103
  specific heat, 87
Continuity equation, 31
Continuum, 18
Control volume, 29
Convection, 473
Convective coefficient, 474
Convertability of energy, 40
Cooling tower, 296
Critical point, 62
Critical pressure, 62
  ratio, 450
Critical properties, 62
Critical radius, 484
Critical temperature, 62
Cryogenic, 269
Cut-off ratio, diesel cycle, 377
Cycle, definition, 13
  three process, 116

Dalton's law, 278
Dead state, 176
Degradation of energy level, 128
Density, 18
Detonation, 390
Dew point, 283, 308
Diesel cycle, air standard, 373
  open system, 385
Diesel engine
  comparison with Otto, 389
  four-stroke, 386
  two-stroke, 388
Dieterici, 207
Differential
  exact, 189
  inexact, 34, 189

Diffuser, 458
  efficiency, 459
Displacement volume, 347
Dissociation, 331
Dry-bulb temperature, 289
Dual combustion cycle, 381

Effectiveness, heat exchanger, 493
  regenerator, gas turbine, 413
Efficiency
  combustion, 413
  compression, 356
  compressor, 356
  diffuser, 459
  engine, 392
  nozzle, 452
  propulsive, 430
  pump, 219
  steam generator, 334
  thermal, 133
  turbine, 219
  volumetric, 351
Energy forms, 31
  internal, 42
  kinetic, 41
  potential, 41
Energy level, 128
Engine efficiencies, 392
Enthalpy, 49, 196
  of combustion, 320, 559
  of formation, 312, 549
  of gases, 89
  stagnation, 444
  of wet mixture, 287, 290
Entropy, 150
  absolute, 151
  change, open system, 160
  change for ideal gas, 153
  general remarks, 152
  of isolated system, 151
  of mixing, 279
  production, net increase, 169
Equation of state, 77
  Beattie-Bridgeman, 80
  Berthelot, 207
  Dieterici, 207
  generalized, 82
  ideal gas, 77
  Van der Waals, 80

Equilibrium, 60, 97
  constant, 331, 560
  chemical, 327
  mechanical, 13
  phase, 63
  processes, 97
  state, 151
  thermal, 13
  thermodynamic, 13
Equivalent diameter, 476
Ericsson cycle, 378
Euler's equation, 447
Excess air, 306
Expansion valve, 250
Extensive properties, 12
Extraction steam, 220

Feedwater heater, 220
First law of thermodynamics, 31
  closed system, 43
  open system, 46
Flow measurement, 460
Flow work, 48
Fluid property table, 542
Force, 14
400% theoretical air table, 510
Fourier's law of heat conduction,
  470
Free air, 352
Free energy, 178
Friction, 53
Fuel-air ratio, 307
Fuel cell, 335
  efficiency, 336
  voltage, 337
Fuels, 304

Gage pressure, 20
Gas constant, 78
  table, 507
  in terms of specific heats, 90
  universal, 78
Gas-liquid equilibrium, 262
Gas turbine, 403
  aircraft, 426
  combined cycle, 423
  efficiency, 405, 407
  ideal regenerative heating, 413
  intercooling, 419

Gas turbine (*Continued*)
open cycle, 410
reheat, 419
Gas-vapor mixtures, 282
Generalized compressibility chart, 83
factors, 82
Geothermal energy, 238
flashed steam system, 240
Gibbs, J. Willard, 8
Gibbs function, 178, 323
of formation, 323

Heat, 39
Heat engine, 117
Heat exchangers, 484
correction factor, 489
counterflow, 487
effectiveness, 493
log mean temperature difference, 486
multiple pass, 489
number of transfer units, 493
parallel flow, 485
shell and tube, 491
Heating value, 320
Heat pump, 267
coefficient of performance, 268
Heat rate, 334
Heat transfer, 469
combined modes, 478
conduction, 470
convection, 473
cylindrical coordinates, 480
overall coefficient, 479
radiation, 472
Helmholtz function, 193
Hot air standard, 369
Humidity ratio, 283
Hydrocarbon fuels, 303
Hydrogen, tables, 556
Hydroxyl, table, 555

Ideal gas, 77
enthalpy, 89
entropy change, 153
equation of state, 77
internal energy, 88

processes, 97
Ignition delay, 390
Imperfect gases, 80
compressibility factor, 81
equations of state, 80
generalized equation of state, 82
reduced coordinates, 83
Impulse turbine, 237
Indicated engine efficiency, 392
Individual gas constant, 78
Intensive property, 12
Intercooler, compressor, 354
Internal combustion engines, 366
four-stroke cycle, 386
two-stroke cycle, 388
Internal energy, 42, 196
of combustion, 320
of ideal gas, 88
Irreversibility, 173, 325
Irreversible process, 36, 102
Isentropic process, 155
Isolated system, 152
Isothermal compressibility, 201
Isothermal process, 105, 109

Jet propulsion, 427
diffuser, 427
impulse-momentum principle, 428
maximum work, 430
propulsive efficiency, 430
ram effects, 427
specific thrust, 430
Joule, James Prescott, 8
Joule's experiment, 40, 87
Joule-Thomson coefficient, 202

Kelvin, Lord, 8
Kelvin temperature scale, 25
Kinetic energy, 41

Latent heat, 63
Lavoisier, Antoine Laurent, 7
Linde liquefier, 270
Liquefying gases, 269
Log mean temperature difference, 486

Mach number, 446
Macroscopic, 9
Manometer, 21
Mass, conservation of, 28
    fraction, 276
Material property table, 541
Maximum work, 167, 173, 326
Maxwell relations, 194
Mean effective pressure, 135, 395
Mechanical efficiency, 357, 393
Melting, 13, 62
Microscopic, 9
Minimum work of compression,
    354
Mixtures
    of gases, 276
    of gases and vapors, 282
Mole, 78
Mole fraction, 276
Mollier diagram, 157, 544
Moment of momentum, 441
Momentum equation, 441
Multistage compression, 353

Natural gas, 304
Newton's equation, 14
    unit, 15
Nitric oxide, table, 558
Nitrogen, table, 550
Nitrogen dioxide, table, 558
Nozzle, 447
    choked flow, 450
    critical pressure ratio, 450
    efficiency, 452
    shape, 448
    supersaturated flow, 454
Nusselt number, 476

Open system, 11
Orsat analysis, 310
Otto cycle, 367
    analysis of processes, 367
    compression ratio, 368
    efficiency, 368
    open system, 382
Overall coefficient of heat
    transfer, 479
Oxygen, table, 552

Partial derivatives, 188
Partial pressure, 278
Path, 13
Percentage clearance, 347
Perpetual motion, 139
Phase diagram, 62, 65
Phases of a substance, 13, 62
Platen-Munters refrigeration
    system, 267
Point function, 189
Polytropic process, 111
    effect of varying n, 111
Potential energy, 41
Power, 37
    measurement, 394
Prandtl number, 476
Pressure, 19
    absolute, 20
    critical, 62
    gage, 20
    ratio, 450
    reduced, 206
    relative, 155
    stagnation, 445
    vacuum, 21
Problem answers, 561
Process, 13
    irreversible, 35
    quasi-equilibrium, 36
    reversible, 36
Products, 304
Products of combustion, 309
    analysis, 310
    dew point, 308
Property, definition, 12
Psychrometer, 289
Psychrometric chart, 548
    use of, 290
Pump, 215
Pure substance, 10, 59

Quality, 64
Quasi-equilibrium process, 36

Radiation, heat 472
    emittance, 473
Ram effect, 427
Rankine cycle, 193

Rankine cycle (*Continued*)
  components, 212
  efficiencies, 218
  improving efficiency, 220
Reactants, 304
Reduced properties, 206
Refrigerants, 71, 248
  tables, 530, 535
Refrigeration, 247
  absorption, 259
  Carnot, 31
  cascade, 254
  coefficient of performance,
    138, 247, 260
  Platen-Munters, 267
  two-stage vapor compression,
    254
  vapor compression, 250
Regenerative gas turbine, 413
Regenerative steam cycle, 220
  number of stages, 224
  optimal heater location, 222
Regenerator effectiveness, 413
Reheat cycle, steam, 228
Reheat factor, 236
Reheat-regenerative cycle, 229
Relative humidity, 283
Relative pressure, 155
Relative specific volume, 155
Reversible process, 97
Reversible work, 34, 172
Reynolds number, 475
Rotative compressors, 357

Saturated air, 283
Saturated liquid, 60
Saturated mixture, 60
Saturated steam, 66
Second law efficiency, 179
  cycle, 182
  process, 179
Second law of thermodynamics,
  128
  closed system, 153
  for a cycle, 129
  first corollary, 139
  open system, 160
  second corollary, 140

Shock waves, 457
Sign convention
  heat, 39
  work, 32
SI units, 16
Specific fuel consumption, 334
Specific heat, 87, 195
  constant pressure, 88
  constant volume, 87
  curves, 92
  difference, 90, 198
  table, 507
  variation with temperature, 91
Specific humidity, 283
Specific volume, 17
Stage efficiency, 236
Stagnation properties, 444
State of substance, 12
Steady flow, 30, 48
Steady state, 48
Steam
  Mollier diagram, 157, 544
  properties of, 514
  temperature-entropy diagram,
    157, 546
Steam generators, 217
  efficiency, 334
Steam power plant, 1, 209
Steam turbine, 212
Stirling cycle, 378
Subcooled liquid, 61
  water, 69
Sublimation, 62
Substance, 10
Supercritical vapor power cycle,
  232
Superheated steam, 68
Supersaturation, 454
Supersonic velocity, 449, 457
Surface tension, 37
Surroundings, 11
System, 10
  boundaries, 11
  closed, 11
  open, 11

Tank
  charging, 121

discharging, 119
Temperature, 23
  absolute, 24
  critical, 62
  dry-bulb, 289
  reduced, 206
  scales, 24
  stagnation, 445
  thermodynamic, 140
  wet-bulb, 289
Theoretical combustion, 305
Thermal conductivity, 470
  table, 541
Thermal efficiency, 133
Thermal equilibrium, 13
Thermal expansion coefficient,
  201
Thermodynamic surface, 65, 98
Thermodynamics, definition, 9
Third law of thermodynamics,
  151
Three-dimensional surface, 64
Three-process cycle, 116
Throttling process, 69
  in calorimeter, 69
  for gases, 202
Tons of refrigeration, 251
Torque, 37, 395
Transient flow, 118
Triple point, 63
Turbine, 174, 212
  condition curve, 236
  efficiency, 219
Turbojet engine, 427
200% theoretical air, table,
  512

Unavailable energy, 166, 168
Units, 17
Universal gas constant, 77

Vacuum pressure, 22
Van der Waals' equation of state,
  80, 205
Van't Hoff equilibrium box, 329
Vapor compression refrigeration
  system, 250
Vaporization, 63
Vapor pressure curve, 61
Velocity
  of light, 442
  of sound, 442
Venturi, 461
Volume
  reduced, 206
  specific, 17
Volumetric analysis, 277, 308
Volumetric efficiency, 351

Wankel engine, 391
Wet-bulb temperature, 289
Windmill, 52, 462
  maximum efficiency, 462
Windpower, 461
Work, 31
  flow, 48
  irreversible, 36, 102
  net, 101
  reversible, 56, 173
  sign convention, 32

Zeroth law of thermodynamics,
  23

82 83 84 85 9 8 7 6 5 4 3 2